# Textbook of Aquatic Ecology

# Textbook of Aquatic Ecology

Edited by **Simon Oakenfold**

**S**YRAWOOD
PUBLISHING HOUSE

New York

Published by Syrawood Publishing House,
750 Third Avenue, 9th Floor,
New York, NY 10017, USA
www.syrawoodpublishinghouse.com

**Textbook of Aquatic Ecology**
Edited by Simon Oakenfold

International Standard Book Number: 978-1-68286-115-8 (Hardback)

Printed in the United States of America.

# Contents

**Permissions**

**List of Contributors**

# Preface

The purpose of the book is to provide a glimpse into the dynamics and to present opinions and studies of some of the scientists engaged in the development of new ideas in the field from very different standpoints. This book will prove useful to students and researchers owing to its high content quality.

Aquatic ecology refers to the ecosystem that is present within a particular water body. This book explores various aquatic ecologies and their relation to Earth's ecosystem in the larger context. It strives to provide a fair idea about this discipline and help develop a better understanding of the emerging concepts and issues within this field with topics such as climate change, marine biology and biodiversity, biogeochemistry, etc. This book is an essential guide for both academicians and those who wish to pursue this discipline further.

At the end, I would like to appreciate all the efforts made by the authors in completing their chapters professionally. I express my deepest gratitude to all of them for contributing to this book by sharing their valuable works. A special thanks to my family and friends for their constant support in this journey.

**Editor**

# Solar energy capture and transformation in the sea

David M. Karl[1]*

[1]Center for Microbial Oceanography: Research and Education, University of Hawaii, Honolulu, Hawaii, United States
*dkarl@hawaii.edu

*"Everything is based on energy. Energy is the source and control of all things, all value, and all the actions of human beings and nature."*

*H. T. Odum and E. C. Odum (1976)*

Domain Editor-in-Chief
Jody W. Deming, University of Washington

Knowledge Domain
Ocean Science

## Ecological energy flow

Solar energy ultimately drives all biogeochemical cycles and sustains planetary habitability. All life forms and processes on Earth, including human economic and social systems, exist within a complex network of energy flow. In the sea, microorganisms comprise most of the genetic and metabolic diversity, and are responsible for a majority of the system energy flow including solar energy capture, transformation, and dissipation. All of these processes involve conversion of low quality forms of energy into a smaller fraction of higher quality energy plus degraded heat, in accordance with the basic laws of thermodynamics. Energy flow is at the core of ecosystem analysis (Odum 1968).

Sunlight is the most abundant form of energy for marine microorganisms, and biophysical/biochemical mechanisms for solar energy capture have evolved by natural selection during eons of Earth's history (Brown and Ulgiati 2004; Nealson and Rye 2003). Marine ecosystems, especially the expansive subtropical gyres, have an enormous capacity for solar energy capture and transformation. Ecologists often use the term "carbon and energy flow" to describe solar energy capture, organic matter transformation, and heat dissipation through the food web via the coupled processes of photosynthesis and respiration. A number of different methods have been used to track the flow of carbon and associated bioelements (e.g., nitrogen, phosphorus, oxygen, and sulfur), but energy flow is rarely if ever measured in field studies. An untested assumption is that matter and energy flow are inextricably and quantitatively linked in space and time in the open sea.

Howard T. Odum, largely in collaboration with his brother Eugene P. Odum, pioneered the discipline of systems ecology. He observed and studied a variety of aquatic ecosystems and was the first to characterize them as networks of energy circuits (e.g., Silver Springs, Florida; Odum 1956). Odum later developed an explicit energy circuit language and set of symbols that could be used to represent interactive energy capture, transformation, and dissipation in both natural and manmade systems (Odum 1983a). While some scientists have criticized "Odum's conjectures" and his energy-centric approach to the study of ecosystems (e.g., Månsson and McGlade 1993), the debate centers on the formidable obstacles to a comprehensive, quantitative analysis and understanding of ecological energy flow rather than a challenge to its fundamental importance in ecosystem analysis.

In a pioneering essay on the relationship of energy flow to evolution, Alfred Lotka concluded that natural selection will operate to preserve and expand those species "possessing superior energy-capturing and directing devices" (Lotka 1922). Consequently, he reasoned, as long as there is a residue of untapped available energy, "the total organic mass of the system, the rate of circulation of mass through the system, and the total energy flux" will be maximized. This reasoning has since become known as the maximum power principle (Odum and Pinkerton 1955; Odum 1983b), and has led to vigorous debate over the validity and implications of what some have termed the fourth law of thermodynamics (see Sciubba 2011 for a recent assessment).

The Earth is an energetically open system where solar energy input is balanced by radiative heat loss. There are numerous connections among the hydrosphere, lithosphere, and atmosphere such that materials and energy can be easily exchanged. In a thought-provoking commentary, *On certain unifying principles in ecology*, Ramon Margalef concluded that the energy required to maintain an ecosystem is inversely proportional to complexity, with a trend for a decreasing flow of energy per unit biomass as succession occurs. He suggested that climax ecosystems with complex structures and high information content could be maintained with a relatively low expenditure of energy (Margalef 1963). Despite great scientific progress in the intervening half-century, we are still unable to examine this unifying principle in ocean ecosystems due, in large part, to inadequate methodology for energy flux estimation and an incomplete understanding of the pathways of solar energy capture and transformation in marine microbial assemblages. Ironically, two pioneering methods developed to analyze key aspects of planktonic microbial communities, the adenosine triphosphate (ATP) assay (Holm-Hansen and Booth 1966) and the electron transport system (ETS) assay (Packard 1971), were laboratory calibrated to yield estimates of living carbon and oxygen consumption, respectively, rather than the equally appropriate parameters of total energy content and flux.

## Phototrophy in the deep blue sea

In marine planktonic systems, green plant-like photosynthesis, termed oxygenic phototrophy (OP; Table 1 and Figure 1), is generally assumed to be the primary, if not exclusive, process of solar energy capture. Light-driven oxidation of water provides electrons and hydrogen ions that partially conserve the solar energy absorbed by chlorophyll $a$ to catalyze the production of a variety of reduced inorganic and organic molecules. Estimates of the magnitude of chlorophyll $a$ -based solar energy capture can be made by measuring the fundamental biophysical properties of photosynthetic energy conversion (e.g., by fast repetition rate fluorescence; Kolber et al. 1998) or gross oxygen production during in situ incubations using $^{18}$O-labeled $H_2O$ (Grande et al. 1989), and applying assumptions regarding energy-to-oxygen stoichiometry. Based upon the energy requirements for OP and cell maintenance, it has been estimated that approximately $5 \times 10^{14}$ W of solar energy are captured per year to support oceanic primary production (Kolber 2007). This power expenditure of marine phytoplankton is nearly two orders of magnitude greater than that expended by the human-based global economies, but as Kolber (2007) laments it is still less than 1% of the full potential of solar radiation incident on the ocean's surface. Consequently, marine systems appear to be relatively inefficient at using the solar energy that is available to them. Ultimately, solar energy capture in the sea by OP is controlled by chlorophyll $a$ concentration (e.g., phytoplankton biomass) and phytoplankton growth rate, which in turn are controlled primarily by nutrient availability. Global distributions of OP and, hence, patterns of solar energy capture coincide with maximum nutrient rather than maximum solar fluxes (e.g., coastal upwelling regions; Kolber 2007). However, even in nutrient-poor environments like the oligotrophic North Pacific Subtropical Gyre (NPSG), the solar energy that is absorbed appears to be very efficiently utilized in photosynthesis (Karl et al. 2002).

Table 1. Solar energy capture in marine microbial assemblages via complementary energy and carbon flow pathways.

| Sample organisms | Metabolic type | Primary (secondary) energy source(s) | Primary (secondary) electron source(s) | Primary (secondary) carbon source(s) | Comments |
|---|---|---|---|---|---|
| Diatoms | Oxygenic phototroph (OP) | Light | $H_2O$ | $CO_2$ | Obligate photolithoautotrophy may not exist in nature |
| *Prochlorococcus* | Oxygenic phototroph (OP) | Light (Org-C[a]) | $H_2O$ | $CO_2$ (Org-C[a]) | Facultative mixotrophy; can also grow photolithoautotrophically |
| *Roseobacter, Erythrobacter* | Aerobic anoxygenic phototroph (AAP) | Light and Org-C[a] | Org-C[a] | Org-C[a] | Facultative photoorganoheterotrophy (mixotrophy); can also grow chemoorganoheterotrophically, but not photolithoautotrophically |
| *Flavobacter, Pelagibacter, Vibrio* | Proteorhodopsin-based phototroph (PR) | Light and Org-C[a] | Org-C[a] | Org-C[a] | Facultative photoorganoheterotrophy (mixotrophy); can also grow chemoorganoheterotrophically, but not photolithoautotrophically |

[a]Organic carbon

During the past decade, there has been a "quiet revolution" in our conceptualization of energy flow in marine systems. This change has resulted in large part from two independent discoveries of unexpected pathways of phototrophy that supplement the better understood OP pathway (Karl 2002; Figure 1 and Table 1). These two novel pathways differ significantly in the mechanism of solar energy capture and in the quantitative and mechanistic role that light energy plays in cellular metabolism. For example, both aerobic anoxygenic phototrophy (AAP) and proteorhodopsin (PR) phototrophy appear to be facultative solar energy capture processes that supplement an otherwise chemoorganoheterotrophic metabolism (Figure 1). Quantitative analysis of energy flow through these alternate pathways will require the development of new instrumentation and methodology, and will likely lead to a new paradigm of energy flow in the sea.

Bacteriochlorophyll $a$-containing marine bacteria were first reported by Shiba et al. (1979) from coastal marine habitats. They were later rediscovered in the oligotrophic waters of the North Pacific Ocean using a

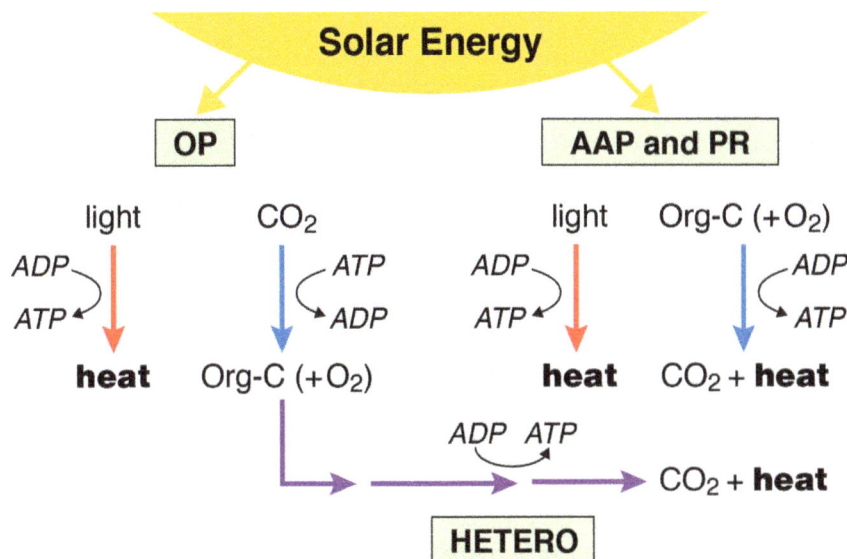

Figure 1

Schematic view of the flow of energy (red arrows), carbon (blue arrows), or energy plus carbon (purple arrows) through a hypothetical marine system.

Solar energy capture processes of OP (oxygenic phototrophy), AAP (aerobic anoxygenic phototrophy), and PR (proteorhodopsin-based phototrophy) convert solar energy into chemical bond energy as ATP plus heat, and in the case of OP a portion of the energy gain is used to reduce carbon dioxide ($CO_2$) to organic carbon (Org-C). The light-independent heterotrophic (HETERO) flow of carbon and energy ultimately dissipates the potential energy in Org-C to heat.

purpose-built infrared fast repetition rate (IRFRR) fluorometer that detected variable fluorescence transients as evidence of bacterial photosynthetic electron transport (Kolber et al. 2000). AAPs differ from oxygenic phototrophs in that their photosynthetic activity neither reduces carbon dioxide nor evolves oxygen (Koblížek et al. 2010). Subsequent laboratory and field investigations of AAP bacteria determined that they are facultative photoheterotrophs, using light energy when organic matter is limiting for growth (Kolber et al. 2001; Koblížek et al. 2003). Laboratory study of an isolated member of the *Roseobacter* clade (strain COL2P) determined that respiration decreased by ~ 30% in the light for an equivalent production rate (Koblížek et al. 2010), suggesting a much more efficient metabolism when the microorganism was living as a mixotroph (e.g., part phototroph, part chemotroph). AAPs are numerically abundant, globally distributed, and metabolically active with respect to solar energy capture where they appear to coexist with OPs (Kolber et al. 2001). However, measurement of total energy flow through these bacteriochlorophyll-containing microbial assemblages (solar plus organic matter) will require a novel experimental approach.

The second novel pathway of solar energy capture in the sea is via PR-containing microorganisms (Béjà et al. 2000; 2001). These novel microbes contain transmembrane light-driven proton pumps that produce ATP. During PR phototrophy, photon capture leads to a conformational change in the retinal molecule resulting in proton export from the cell. This process generates a proton motive force across the cell membrane, setting up the possibility for ATP production. Gómez-Consarnau et al. (2007) were the first to demonstrate enhanced growth efficiency in the light for a PR-containing marine bacterium. Martinez et al. (2007) documented PR-based photophosphorylation (ATP production) in a genetically transformed heterologous host, and PR-dependent proton pumping activity sufficient to generate ATP has also been demonstrated in laboratory-reared marine bacteria (Wang et al. 2012; Yoshizawa et al. 2012). PR-phototrophy has since been shown to increase the long-term survival of diverse bacteria that favor either carbon-rich (e.g., *Vibrio*; Gómez-Consarnau et al. 2010) or carbon-poor (e.g., *Pelagibacter*; Steindler et al. 2011) marine habitats. PR genes have also been found in archaea (Frigaard et al. 2006) and eukaryotes, including both photosynthetic (Marchetti et al. 2012) and predatory (Slamovits et al. 2011) protists. Recently, Kimura et al. (2011) surveyed the transcriptional and growth responses of a PR-containing marine flavobacterium during carbon-limited growth under light and dark conditions. Their results demonstrated a direct role for retinal-based, PR-phototrophy and the previously observed light-enhanced growth response. Our current, but still evolving, understanding is that the light-driven, PR-based proton pumps enhance survival, and therefore may be a significant pathway for the flow of energy in oligotrophic habitats (DeLong and Béjà 2010). This phototrophic process, which is now believed to be ubiquitous in the well-lit portions of the euphotic zone, enhances solar energy capture in the ecosystem without direct impacts on either carbon dioxide or oxygen reservoirs. The indirect effects, however, will scale on the total energy budgets of these unique phototrophic microbes, specifically on the role of PRs in the efficiency of heterotrophic metabolism of dissolved organic matter.

Recently, Kirchman and Hanson (2013) have reviewed the current state-of-knowledge concerning the bioenergetics of photoheterotrophic bacteria in the sea, including both bacteriochlorophyll-(AAP) and PR-based pathways. In the absence of any in situ experimental data, they provided a theoretical cost vs. benefit analysis of phototrophy. This scholarly assessment included estimates of the number of photosynthetic units per cell, the absorption cross-sectional area and wavelength dependent absorption of light, the quantum efficiency, and the number of protons pumped per photon of absorbed light. Most of the data used in these calculations were derived from laboratory studies of model organisms, and in some cases were "best guesses"— but it is a great start. The theoretical gross energy yield via these two types of photoheterotrophy was compared to a model marine cyanobacterium, *Synechococcus elongatus* (MacKenzie et al. 2004), grown as an obligate photolithoautotroph (i.e., OP; Table 1). By their analysis, a "typical" PR-containing phototrophic bacterium would gain ~ 10% of energy captured by a "typical" AAP bacterium, and ~ 0.2–0.3% of the cyano-bacterium (Kirchman and Hanson 2013). These estimates may be a bit misleading because they are expressed on a per cell basis. Given the fact that many PR-containing marine bacteria have biovolumes that are only a few percent of the model cyanobacterium used for comparison, the potential solar energy gain relative to their total energy demand for the PR-containing microbes may be greater than reported.

## Quantitative assessments of energy flow

This brief commentary has focused on new pathways for solar energy capture and transformation in the deep blue sea. However, there are also other important aspects to marine system energy flow that need to be considered. These include the cycling of: (1) dissolved organic matter (DOM), especially labile products of photosynthesis; (2) reduced biogenic gases, especially methane and hydrogen; and (3) reduced inorganic derivatives of nitrogen, phosphorus, and sulfur. These reservoirs store and shunt potential energy, and enhance the overall magnitude and efficiency of solar energy capture and transformation in the sea. In the case of DOM, the energy content of the fairly large reservoir (~ 75–100 mmol C per cubic meter in the euphotic zone) greatly exceeds the daily capture of solar energy and may regulate energy flow in stable oceanic com-munities. Even the more refractory portion (~ 30–40%) of the total surface DOM pool, with a mean age of a few thousand years, may represent a longer term potential energy reservoir for the growth of "low-energy" specialists. Any quantitative assessment of ocean system energy flow must be able to measure all possible pathways of energy capture, transformation, and dissipation and needs to integrate over both space and time. Two possible experimental approaches have been employed to estimate total energy flow through marine planktonic assemblages: (1) heat flow via microcalorimetry, and (2) total ATP pool turnover rate, but additional methods need to be devised, calibrated, and field-tested. Future developments in the emergent field of metabolomics, especially energy transduction and energy storage molecules, will likely provide new opportunities for field application.

Microcalorimetry has only rarely been used to estimate heat flow in marine ecosystems (Pamatmat et al. 1981; Pamatmat 1982), primarily in coastal benthic habitats where metabolic activities are relatively high. Comparison of heat flow estimated from dark rates of oxygen uptake using assumptions regarding organic substrate composition and utilization efficiencies can lead to large discrepancies with direct heat flow estima-tion (Pamatmat 2003). In one of the few published studies of heat flow in marine plankton (Friday Harbor, Washington), the direct calorimetric-based value of 200–300 µW per liter was 4–6 times larger than that derived from extrapolation based on oxygen utilization (Pamatmat 2003). However, calorimetry currently suffers from several limitations. First, the nature of the differential microcalorimeters used for ecological studies cannot be used to assess solar energy for capture directly or to resolve biotic versus abiotic reactions. Second, the specialized nature of differential microcalorimeters limits sample throughput and replication. Finally, the relatively insensitive limits of heat detection preclude measurements in most open ocean plank-tonic ecosystems and, even for those systems that can be measured, calorimetry requires fairly long incubation periods which may bias estimates of in situ energy flow.

Recently, Djamali et al. (2012) have employed a purpose-built, differential microcalorimeter to measure the heat output of the marine microbial food web with an emphasis on the role of viral lysis. They experimented with aquarium-reared, size-fractionated model systems that were diluted to provide treatments with or without viruses. Their results indicated that approximately 25% of the total heat flow in their artificial planktonic communities could be attributed to viral activities. While the claim is made that their novel instrument is capable of measuring the heat produced from open ocean assemblages of ~ $10^5$ bacterial cells ml$^{-1}$ without pre-concentration (Djamali et al. 2012), no such data are presented, or to my knowledge published elsewhere. Nevertheless, recent improvements in technology are very encouraging for possible use in future field studies (see review by Braissant et al. 2010). The three major limitations of calorimetry, however, remain: (1) inability to resolve biotic from abiotic processes, (2) difficulty measuring light versus dark energy heat fluxes, and (3) low sample throughput and lack of sample and reference replication for most commercial microcalorimeters.

An alternative to direct estimation of heat flow is the measurement of the turnover rate of the ATP pool in the microbial community (Karl and Bossard 1985). The central role of ATP in the stoichiometric coupling of all energy transforming metabolic reactions (phototrophic as well as chemotrophic) has been known since the pioneering work of Lipmann (1941). While intracellular ATP concentrations (i.e., the so-called "ATP pool") are fairly well buffered at 1–3 mM, the turnover rate of the pool tracks metabolic energy flow. ATP pool turnover results from the hydrolysis of one or both "high energy" phosphate bonds, followed by regeneration of ATP via substrate level, oxidative, or photophosphorylation. Because ATP is the common energy currency in all organisms and because the free energy of ATP hydrolysis is well constrained (46 ± 4 kJ per mol; Bridger and Henderson 1983), direct measurements of ATP pool turnover coupled with ATP concentration should provide a quantitative estimation of biological energy flux (Karl 1993). Both heat flow and ATP pool turnover might be viewed as the epitome of reductionism because neither approach provides explicit information on which organisms or which pathways are most important in natural systems. Clearly in order to be useful, energy flow measurements need to be part of the holistic study of ecosystems and used as a tool in experimental perturbation studies to learn more about the controls on energy capture, transformation, and dissipation in marine systems.

## Future research prospectus

As we move further into the anthropocene and continue to alter the sea around us, we need to have the capacity to monitor changes in the most fundamental property of the system, namely energy flow. The future ocean will be warmer, more stratified and nutrient starved, more acidic, and less oxygenated as a consequence of anthropogenic forcing by greenhouse gas emissions (Gruber 2011). These habitat changes will impact solar energy capture and transformation by microbial assemblages, so there is an urgent need to improve our conceptual understanding and quantitative assessments of energy flow in the open sea. I consider this to be one of the greatest contemporary challenges in microbial oceanography and marine ecology. The Center for Microbial Oceanography: Research and Education (C-MORE) is poised to begin a systematic two-year study (2014–2015) of planktonic community energy flow in the NPSG with an emphasis on pathways and controls. Once a comprehensive energy budget is available for the NPSG microbial assemblage, other fundamental properties including the maximum empower selection principle (Odum 1983b; Sciubba 2011), net metabolic balance (Ducklow and Doney 2013), the concept of energy equivalents and transformity (Odum 1983a), and the enigma of microbial production of recalcitrant organic matter (Jiao et al. 2010) can be systematically investigated. The development of a new theoretical framework for solar energy capture and energy flow via microorganisms in the sea may also be of practical value for policy makers and society as a whole (Prosser et al. 2007). As our demands for renewable energy continue to increase, a better understanding of the unique evolutionary adaptations of our magnificent marine microbes might improve our standard of living and extend our survival as a species.

## References

Béjà O, Aravind L, Koonin EV, Suzuki MT, Hadd A, et al. 2000. Bacterial rhodopsin: Evidence for a new type of phototrophy in the sea. *Science* **289**(5486): 1902–1906.

Béjà O, Spudich EN, Spudich JL, Leclerc M, DeLong EF. 2001. Proteorhodopsin phototrophy in the ocean. *Nature* **411**(6839): 786–789.

Braissant O, Wirz D, Göpfert B, Daniels AU. 2010. Use of isothermal microcalorimetery to monitor microbial activities. *FEMS Microbiol. Lett* **303**: 1–8.

Bridger WA, Henderson JF. 1983. *Cell ATP*. New York: John Wiley & Sons.

Brown MT, Ulgiati S. 2004. Energy quality, emergy, and transformity: H. T. Odum's contributions to quantifying and understanding systems. *Ecol. Model* **178**(1–2): 201–213.

DeLong EF, Béjà O. 2010. The light-driven proton pump Proteorhodopsin enhances bacterial survival during tough times. *PLoS Biol* **8**(4): e1000359. doi: 10.1371/journal.pbio.1000359

Djamali E, Nulton JD, Turner PJ, Rohwer F, Salamon P. 2012. Heat output by marine microbial and viral communities. *J. Non-Equilib. Thermodyn* **37**(3): 291–313.

Ducklow HW, Doney SC. 2013. What is the metabolic state of the oligotrophic ocean? A debate. *Ann. Rev. Mar. Sci* **5**: 525–53.

Frigaard NU, Martinez A, Mincer TJ, DeLong EF. 2006. Proteorhodopsin lateral gene transfer between marine planktonic Bacteria and Archaea. *Nature* **439**(7078): 847–850.

Gómez-Consarnau L, González JM, Coll-Lladó M, Gourdon P, Pascher T, et al. 2007. Light stimulates growth of proteorhodopsin-containing marine Flavobacteria. *Nature* **445**(7124): 210–213.

Gómez-Consarnau L, Akram N, Lindell K, Pedersen A, Neutze R, et al. 2010. Proteorhodopsin phototrophy promotes survival of marine bacteria during starvation. *PLoS Biol* **8**(4): e1000358. doi: 10.1371/journal.pbio.1000358

Grande KD, Williams PJLeB, Marra J, Purdie DA, Heinemann K, et al. 1989. Primary production in the North Pacific gyre: a comparison of rates determined by the $^{14}$C, $O_2$ concentration and $^{18}$O methods. *Deep-Sea Res. Part A* **36**(11): 1621–1634.

Gruber N. 2011. Warming up, turning sour, losing breath: ocean biogeochemistry under global change. *Phil. Trans. Royal Soc. A — Math. Phys. Eng. Sci* **369**(1943): 1980–1996.

Holm-Hansen O, Booth CR. 1966. The measurement of adenosine triphosphate in the ocean and its ecological significance. *Limnol. Oceanogr* **11**(4): 510–519.

Jiao N, Herndl GJ, Hansell DA, Benner R, Kattner G, et al. 2010. Microbial production of recalcitrant dissolved organic matter: long-term carbon storage in the global ocean. *Nature Rev. Microbiol* **8**(8): 593–598.

Karl DM. 1993. Adenosine triphosphate (ATP) and total adenine nucleotide (TAN) pool turnover rates as measures of energy flux and specific growth rate in natural populations of microorganisms, in Kemp PF, Sherr BF, Sherr EB, Cole JJ, eds., *Current Methods in Aquatic Microbial Ecology*. Boca Raton, Florida: Lewis Publishers: p. 483–494.

Karl DM. 2002. Hidden in a sea of microbes. *Nature* **415**(6872): 590–591.

Karl DM, Bossard P. 1985. Measurement and significance of ATP and adenine nucleotide pool turnover in microbial cells and environmental samples. *J. Microbiol. Meth* **3**(3–4): 125–139.

Karl DM, Bidigare RR, Letelier RM. 2002. Sustained and aperiodic variability in organic matter production and phototrophic microbial community structure in the North Pacific Subtropical Gyre, in Williams PJleB, Thomas DR, Reynolds CS, eds., *Phytoplankton Productivity and Carbon Assimilation in Marine and Freshwater Ecosystems*. London: Blackwell Publishers: p. 222–264.

Kimura H, Young CR, Martinez A, DeLong EF. 2011. Light-induced transcriptional responses associated with proteorhodopsin-enhanced growth in a marine flavobacterium. *ISME J* 5(10): 1641–1651.

Kirchman DL, Hanson TE. 2013. Bioenergetics of photoheterotrophic bacteria in the oceans. *Environ. Microbiol. Reports* 5(2): 188–199.

Koblížek M, Béjà O, Bidigare RR, Christensen S, Benitez-Nelson B, et al. 2003. Isolation and characterization of *Erythrobacter* sp. Strains from the upper ocean. *Arch. Microbiol* 180(5): 327–338.

Koblížek M, Mlčoušková J, Kolber Z, Kopecký J. 2010. On the photosynthetic properties of marine bacterium COL2P belonging to *Roseobacter* clade. *Arch. Microbiol* 192(1): 41–49.

Kolber ZA, Van Dover CL, Niederman RA, Falkowski PG. 2000. Bacterial photosynthesis in surface waters of the open ocean. *Nature* 407(6801): 177–179.

Kolber Z. 2007. Energy cycle in the ocean: Powering the microbial world. *Oceanogr* 20(2): 79–88.

Kolber ZS, Prášil O, Falkowski PG. 1998. Measurements of variable chlorophyll fluorescence using fast repetition rate techniques: defining methodology and experimental protocols. *Biochim. Biophys. Acta* 1367(1–3): 88–106.

Kolber ZA, Plumley FG, Lang AS, Beatty JT, Blankenship RE, et al. 2001. Contribution of aerobic photoheterotrophic bacteria to the carbon cycle of the ocean. *Science* 292(5526): 2492–2495.

Lipmann F. 1941. Metabolic generation and utilization of phosphate bond energy. *Adv. Enzymol* 1: 99–162.

Lotka AJ. 1922. Natural selection as a physical principle. *Proc. Natl. Acad. Sci. USA* 8(6): 151–154.

Margalef R. 1963. On certain unifying principles in ecology. *Amer. Naturalist* 97(897): 357–374.

MacKenzie TDB, Burns RA, Campbell DA. 2004. Carbon status constrains light acclimation in the cyanobacterium *Synechococcus elongatus*. *Plant Physiol* 136(2): 3301–3312.

Månsson BA, McGlade JM. 1993. Ecology, thermodynamics and H. T. Odum's conjectures. *Oecologia* 93(4): 582–596.

Marchetti A, Schruth DM, Durkin CA, Parker MS, Kodner RB, et al. 2012. Comparative metatranscriptomics identifies molecular bases for the physiological responses of phytoplankton to varying iron availability. *Proc. Natl. Acad. Sci. USA* 109(6): E317–E325.

Martinez A, Bradley AS, Waldbauer JR, Summons RE, DeLong EF. 2007. Proteorhodopsin photosystem gene expression enables photophosphorylation in a heterologous host. *Proc. Natl. Acad. Sci. USA* 104(13): 5590–5595.

Nealson KH, Rye R. 2003. Evolution of metabolism, in Schlesinger WH, ed., *Biogeochemistry*. Oxford, UK: Elsevier: p. 41–61.

Odum HT. 1956. Primary production in flowing waters. *Limnol. Oceanogr* 1(2): 102–117.

Odum EP. 1968. Energy flow in ecosystems: A historical review. *Am. Zoologist* 8(1): 11–18.

Odum HT. 1983a. *Systems Ecology: An Introduction*. New York: John Wiley & Sons.

Odum HT. 1983b. Maximum power and efficiency: A rebuttal. *Ecol. Model* 20(1): 71–82.

Odum HT, Odum EC. 1976. *Energy Basis for Man and Nature*. New York: McGraw-Hill Book Co.

Odum HT, Pinkerton RC. 1955. Time's speed regulator: The optimum efficiency for maximum power output in physical and biological systems. *Amer. Scientist* 43(2), 331–343.

Packard TT. 1971. The measurement of respiratory electron transport activity in marine plankton. *J. Mar. Res* 29: 235–244.

Pamatmat MM. 1982. Heat production by sediment: ecological significance. *Science* 215(4531): 395–397.

Pamatmat MM. 2003. Heat-flow measurements in aquatic ecosystems. *J. Plankton Res* 25(4): 461–464.

Pamatmat MM, Graf G, Bengtsson W, Novak CS. 1981. Heat production, ATP concentration and electron transport activity of marine sediments. *Mar. Ecol. Prog. Ser* 4: 135–143.

Prosser JI, Bohannan BJM, Curtis TP, Ellis RJ, Firestone MK, et al. 2007. The role of ecological theory in microbial ecology. *Nature Rev. Micro* 5(5): 384–392.

Sciubba E. 2011. What did Lotka really say? A critical reassessment of the "maximum power principle". *Ecol. Model* 222(8): 1347–1353.

Shiba T, Simidu U, Taga N. 1979. Distribution of aerobic bacteria which contain Bacteriochlorophyll *a*. *Appl. Environ. Microbiol* 38(1): 43–45.

Slamovits CH, Okamoto N, Burri L, James ER, Keeling PJ. 2011. A bacterial proteorhodopsin proton pump in marine eukaryotes. *Nature Comm* 2(Article 183): doi:10.1038/ncomms1188

Steindler L, Schwalbach MS, Smith DP, Chan F, Giovannoni SJ. 2011. Energy starved *Candidatus* Pelagibacter ubique substitutes light-mediated ATP production for endogenous carbon respiration. *PLoS One* 6(5): e19725. doi: 10.1371/journal.pone.0019725

Wang Z, O'Shaughnessy TJ, Soto CM, Rahbar AM, Robertson KL, et al. 2012. Function and regulation of *Vibrio campbellii* proteorhodopsin: Acquired phototrophy in a classical organoheterotroph. *PLoS ONE* 7(6): e38749.

Yoshizawa S, Kawanabe A, Ito H, Kandori H, Kogure K. 2012. Diversity and functional analysis of proteorhodopsin in marine *Flavobacteria*. *Environ. Microbiol*. doi: 10.1111/j.1462-2920.2012.02702.x

## Acknowledgments

I thank Jody Deming, Ocean Science domain editor, for the cordial invitation to contribute a commentary to *Elementa: Science of the Anthropocene*, Karin Bjorkman and Sam Wilson for comments on an earlier draft of the manuscript, and the National Science Foundation (EF04-24599) and the Gordon and Betty Moore Foundation for their generous support of my research.

# 2

# Sea ice algal biomass and physiology in the Amundsen Sea, Antarctica

Kevin R. Arrigo[1]* • Zachary W. Brown[1] • Matthew M. Mills[1]

[1]Stanford University, Stanford, California, United States
*arrigo@stanford.edu

Domain Editor-in-Chief
Jody W. Deming, University of Washington

Associate Editor
Jean-Éric Tremblay, Université Laval

Knowledge Domain
Ocean Science

## Abstract

Sea ice covers approximately 5% of the ocean surface and is one of the most extensive ecosystems on the planet. The microbial communities that live in sea ice represent an important food source for numerous organisms at a time of year when phytoplankton in the water column are scarce. Here we describe the distributions and physiology of sea ice microalgae in the poorly studied Amundsen Sea sector of the Southern Ocean. Microalgal biomass was relatively high in sea ice in the Amundsen Sea, due primarily to well developed surface communities that would have been replenished with nutrients during seawater flooding of the surface as a result of heavy snow accumulation. Elevated biomass was also occasionally observed in slush, interior, and bottom ice microhabitats throughout the region. Sea ice microalgal photophysiology appeared to be controlled by the availability of both light and nutrients. Surface communities used an active xanthophyll cycle and effective pigment sunscreens to protect themselves from harmful ultraviolet and visible radiation. Acclimation to low light microhabitats in sea ice was facilitated by enhanced pigment content per cell, greater photosynthetic accessory pigments, and increased photosynthetic efficiency. Photoacclimation was especially effective in the bottom ice community, where ready access to nutrients would have allowed ice microalgae to synthesize a more efficient photosynthetic apparatus. Surprisingly, the pigment-detected prymnesiophyte *Phaeocystis antarctica* was an important component of surface communities (slush and surface ponds) where its acclimation to high light may precondition it to seed phytoplankton blooms after the sea ice melts in spring.

## Introduction

Over the course of an annual cycle, the sea ice that forms on the surface of polar oceans extends over an area of $15$–$22 \times 10^6$ km$^2$. This enormous surface area ranks sea ice as one of the most expansive ecosystems on Earth, covering approximately 4.1–6.1% of the surface area of the global ocean (Arrigo, 2014). Much of this ice is found in the Southern Hemisphere, expanding in size around the continent of Antarctica from a minimum extent of $3 \times 10^6$ km$^2$ in February to a maximum area of $19 \times 10^6$ km$^2$ in September. Sea ice ecosystems are home to a diverse community of bacteria, archaea, microalgae, protists, and metazoan grazers within the numerous microhabitats that are formed during its lifetime.

Sea ice microbial communities grow best in microhabitats that are in close proximity to seawater nutrients and receive enough light for net microalgal photosynthesis. Microhabitats within Antarctic pack ice inhabited by ice microalgae include surface melt ponds, slush (melted or flooded snow at the surface of the ice), gap layers, internal ice, and bottom ice (Legendre et al., 1992). Surface ponds form either when snowmelt collects in discrete ponds on the surface of relatively flat ice (melt ponds) or when the ice surface is forced below the freeboard level due to ice rafting or snow loading and becomes flooded with seawater (deformation ponds). While melt ponds usually contain relatively little biomass owing to their low nutrient concentrations, deformation ponds can support high algal biomass (Garrison et al., 2003). Internal layers of relatively solid undeformed ice are generally the most inhospitable habitats for microbial life in sea ice. While these layers often receive ample light, they can be very cold with brine salinities too high for microalgal growth (Arrigo and Sullivan, 1992) and brine volumes too low for adequate nutrient exchange (Golden et al., 1998, 2007; Garrison et al., 2003). When the skeletal layer (the actively growing region at the base of growing sea ice) is present, bottom ice is often the most biologically productive sea ice habitat owing to its ubiquity, proximity to seawater nutrients, and mild temperature and salinity gradients (Grossi et al., 1987).

Sea ice communities are responsible for a small but important fraction of total primary production in Southern Ocean waters (Arrigo et al., 1997, 2008; Lizotte, 2001). They provide food for protists, ctenophores, annelids, and a variety of crustaceans, including copepods and euphausiids (Garrison and Buck, 1989; Daly, 1990; Gowing and Garrison, 1992; Guglielmo et al., 2007; Kiko et al., 2008; Caron and Gast, 2010). In particular, high krill densities have been observed beneath the ice throughout the year as they feed on microalgae within and at the base of the sea ice (Flores et al., 2011, 2012).

As the sea ice melts, the microbial community is rapidly released as a large pulse into surface waters (Grossi et al., 1987; Suzuki et al., 2001; Juul-Pedersen et al., 2008). Some of the algal cells can provide seed stock for phytoplankton blooms at the receding ice edge (Haecky et al., 1998; Mangoni et al., 2009). Much of the remaining biomass is eaten by pelagic grazers as it sinks through the water column (Brown and Belt, 2012). Ice algal food is rich in polyunsaturated and other essential fatty acids (McMahon et al., 2006; Søreide et al., 2010) that are necessary for zooplankton growth and reproduction. Uneaten ice microalgae can settle on the seafloor and are consumed by benthic invertebrates (Ratkova and Wassmann, 2005; Boetius et al., 2013). Ice microalgae are often enriched in $^{13}C$ relative to pelagic phytoplankton (Rau et al., 1991) and their $^{13}C$ signatures have been used to assess the proportion of sea ice microalgae in the diet of benthic invertebrates (Wing et al., 2012).

Although the number of observations of sea ice microbial communities has increased in recent years, there are still large areas around the Antarctic for which few samples are available. These include the south Pacific and south Indian oceans and the Amundsen Sea (Meiners et al., 2012). Here we present results of a study of sea ice algal biomass and physiology along a 17-station transect from the Amundsen Sea to the Ross Sea (Figure 1). The focus of the study was to assess spatial variability in ice algal biomass and determine what factors control their distributions. We were also interested in investigating how ice algal physiology varied in different microhabitats within the ice, in part to better understand how this fragile ecosystem might respond to future changes in sea ice conditions.

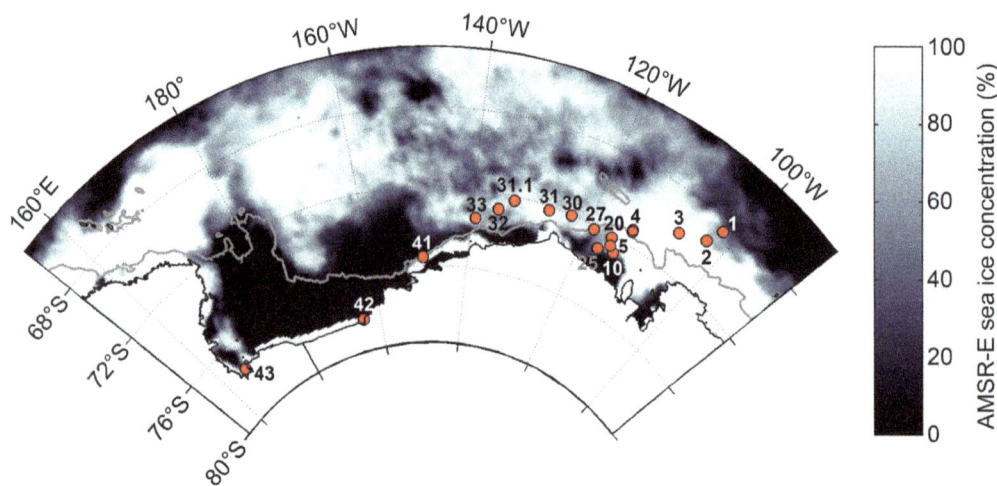

**Figure 1**

**Map of the Ross and Amundsen Seas.**

Locations of sea ice stations and the mean sea ice concentration at the time of the cruise are shown. Gray line indicates the 1000-m isobath.

## Methods

Samples were collected from the Swedish icebreaker *Oden* along a roughly zonal transect extending from approximately 100°W to 166°E (Figure 1). Sea ice was sampled between 16 December 2010 and 10 January 2011 by deployment of personnel directly onto the ice pack. Sea ice samples were obtained using a SIPRE corer (0.076 m interior diameter). Ice cores longer than 0.2 m were sectioned at 0.1 to 0.2 m intervals and each segment was placed in individual labeled polyethylene bags. Slush and surface ponds were sampled by scooping a known volume into 4 L dark polyethelene containers and processed immediately upon return to the ship.

All sea ice samples were stored in a thermally insulated cooler until they could be processed. Once onboard (within 1 h of collection) a sufficient quantity of 0.2 μm-filtered seawater was added to each ice core section to maintain salinity > 28 (to minimize osmotic shock to the microbial community), and the samples were allowed to melt in the dark (< 24 hr) prior to further analysis. No seawater was added to surface pond or slush samples. The exact quantity of seawater added to the ice sections was recorded so that concentrations of solutes could be corrected to their undiluted values. Comparison of stored samples with samples that were processed immediately after collection showed that there was little change in physiology or pigment composition resulting from this treatment. Water column samples were collected from 1–2 m below the bottom of the ice using a bilge pump.

## Snow and ice thickness

Before drilling ice cores, snow thickness was measured by inserting a ruler through the snow to the snow/ice interface. At least four holes (usually spaced 1 m apart) were drilled at each sea ice station for determination of sea ice thickness. A tape measure attached to the center of a brass rod was inserted into each hole, and the tape was pulled tight until the brass rod held securely to the bottom sea ice surface. The thickness was then read off the tape measure at the snow/ice interface. The mean ice and snow thickness for each station was calculated by averaging the thickness at all core locations at that station.

## Sea ice salinity and temperature

The temperature of each ice core section was measured by inserting the tip of a digital temperature probe (Corning Science Products) approximately 0.01 m into the ice. The reading was taken after the temperature had stabilized (ca. 10 s). Sea ice salinity for each core section was measured using a refractometer (accuracy 1 ppt) after ice core sections had melted and were corrected for seawater dilution.

## Pigments, POC, PON, $\delta^{13}C$, and $\delta^{15}N$ analyses

### Pigments

Samples for fluorometric analysis of chlorophyll $a$ (Chl $a$) were filtered onto 25 mm Whatman GF/F filters (nominal pore size 0.7 $\mu$m), placed in 5 mL of 90% acetone, and extracted in the dark at 3°C for 24 hrs. Chl $a$ was measured fluorometrically (Holm-Hansen et al., 1965) using a Turner 10-AU fluorometer (Turner Designs, Inc.). The fluorometer was calibrated using a pure Chl $a$ standard (Sigma). For samples of relatively high biomass, high performance liquid chromatography (HPLC) analysis of pigment composition, including chlorophylls, phaeopigments, and carotenoids, was performed using the method of Wright et al. (1991) as described in DiTullio and Smith (1996).

Fluorometric Chl $a$ concentrations generally exceed those measured by HPLC by ~ 50%. Because we only conducted HPLC analyses on a subset of the total samples, all Chl $a$ concentrations reported here were measured fluorometrically. Pigment ratios, however, were determined from HPLC data.

### Particulate organic carbon and nitrogen

Particulate organic carbon (POC) and nitrogen (PON) samples were collected by filtering water onto pre-combusted (450°C for 4 hrs) 25 mm Whatman GF/F filters. Filter blanks were produced by passing ~50 ml of 0.2 $\mu$m filtered seawater through a GF/F. All filters were then immediately dried at 60°C and stored dry until analysis. Prior to analysis, samples and blanks were fumed with concentrated HCl, dried at 60°C and packed into tin capsules (Costech Analytical Technologies, Inc.) for elemental analysis on a Elementar Vario EL Cube (Elementar Analysensysteme GmbH, Hanau, Germany) interfaced to a PDZ Europa 20–20 isotope ratio mass spectrometer (Sercon Ltd., Cheshire, UK). Standards included peach leaves and glutamic acid. The stable carbon ($\delta^{13}C$) and nitrogen ($\delta^{15}N$) isotopes of POC and PON in high biomass samples were simultaneously measured using the elemental analyzer and mass spectrometer system. Isotopic compositions were calibrated against the NBS-21 and IAEA-N1 standards that were run before and after each set of analyses. Isotopic reproducibility was on the order of 0.11‰.

## Sea ice algal photophysiology

Photosynthesis versus irradiance (P-E) relationships for microalgae released from their sea ice matrix were determined using a modification of the $^{14}C$-bicarbonate technique of Lewis and Smith (1983) as described by Arrigo et al. (2010a). Microalgae were inoculated with 0.925 MBq $^{14}C$-bicarbonate and each 2 ml aliquot was exposed to one of 20 irradiances ranging from <1 to >500 $\mu$mol photons m$^2$ s$^{-1}$ for one hour at $0.0 \pm 0.5$°C. The DIC concentration used to calculate C-uptake rates was measured as described in Tortell et al. (2012). The photosynthetic parameters $P^*_m$ (maximum photosynthetic rate, mg C mg$^{-1}$ Chl $a$ hr$^{-1}$) and $\alpha^*$ (photosynthetic efficiency, mg C mg$^{-1}$ Chl $a$ hr$^{-1}$ ($\mu$mol photons m$^{-2}$ s$^{-1}$)$^{-1}$) were calculated by normalizing uptake rates to fluorometric Chl $a$ concentration; values were estimated from a fit of P-E data to the equation of Platt et al. (1980). The photoacclimation index ($E_k$, $\mu$mol photons m$^{-2}$ s$^{-1}$) was calculated as $P^*_m/\alpha^*$.

The particle absorption coefficient ($a_p$) from 300 to 800 nm was determined spectrophotometrically (Perkin Elmer Lambda 18 with a RSA-PE-18 integrating sphere) on fresh samples by collecting particles onto a 25 mm filter (Whatman GF/F) and measuring its optical density relative to a blank reference filter. Spectral absorption coefficients were calculated as described in Mitchell (1990). Following measurement of $a_p$, sample filters were extracted in 90% methanol and re-measured to yield detrital absorption ($a_d$). Microalgal absorption ($a_{ph}$) was determined by difference as $a_{ph} = a_p - a_d$. Chl $a$-specific microalgal absorption ($a_{ph}^*$) was calculated as $a_{ph}$/Chl $a$, where Chl $a$ was determined fluorometrically.

Although light levels within and beneath the ice were not measured during this project, the potential light environment for each microhabitat was characterized by calculating a light index based on the amount of overlying sea ice and snow. Because snow attenuates light approximately 10-fold higher than sea ice, a simple metric was produced wherein the light index was set to $1/(10 \cdot$ snow thickness + ice thickness). For interior communities, only the amount of ice above the community (rather than total ice thickness) was used in the calculation. Based on this simple metric, microhabitats were classified from high to low light as surface pond, slush, high light interior ice, low light interior ice, bottom ice, under-ice high light, and under-ice low light.

# Results

## Snow

Snow thickness varied considerably throughout the study region, averaging $0.30 \pm 0.24$ m. Snow cover along our transect ranged from virtually snow free conditions in a few locations to snow cover as thick as 0.82 m (Table 1). In general, snow was thickest along the eastern section of the transect, averaging $0.42 \pm 0.22$ m, and much thinner to the west of 130°W ($0.08 \pm 0.07$ m). There was no apparent relationship between ice thickness and snow thickness for the stations we sampled. Six of the 17 stations sampled (2, 3, 27, 30, 31, and 41) had slush layers between the snow and ice. These stations had the thickest snow cover, suggesting that slush formation was primarily due to surface flooding as the thick snow cover forced the surface of the ice below freeboard.

**Table 1.** Physical and biological characteristics of sea ice stations

| Station | Date | Latitude | Longitude | Snow depth | Ice depth | Depth of maximum biomass | Temp. | Salinity | Chl a | POC | PON |
|---|---|---|---|---|---|---|---|---|---|---|---|
| | | (°S) | (°E or °W) | (m) | (m) | (m) | (°C) | | (mg m⁻²) | (mg m⁻²) | (mg m⁻²) |
| 1 | 16 Dec 2010 | 68.588 | 102.142 W | 0.28 | 0.66 | 0.05 | | 4.3 (1.5) | 3.12 | 821.6 | 97.1 |
| 2 | 17 Dec 2010 | 69.461 | 103.072 W | 0.33 | 0.90 | 0.05 | − 0.91 (0.27) | 4.2 (2.5) | 8.44 | 1465.3 | 228.8 |
| 3 | 18 Dec 2010 | 70.025 | 106.944 W | 0.00 | 1.31 | 0.75 | − 1.35 (0.55) | 4.3 (1.7) | 9.96 | 2408.7 | 279.2 |
| 4 | 19 Dec 2010 | 71.064 | 112.983 W | 0.68 | 1.98 | 0.65 | − 1.27 (0.24) | 7.7 (2.8) | 72.2 | 3211.2 | 598.8 |
| 5 | 20 Dec 2010 | 72.448 | 115.357 W | 0.37 | 1.05 | | − 1.92 (0.25) | 8.4 (1.4) | | | |
| 10 | 21 Dec 2010 | 72.773 | 114.171 W | 0.35 | 1.26 | 1.15 | − 1.23 (0.23) | 8.0 (1.3) | 7.14 | 1189.9 | 153.6 |
| 20 | 24 Dec 2010 | 72.121 | 115.612 W | 0.48 | 1.16 | 0.25 | − 1.30 (0.18) | 4.7 (2.7) | 38.7 | 3141.5 | 507.0 |
| 25 | 26 Dec 2010 | 72.958 | 116.964 W | 0.52 | 1.81 | 0.55 | − 1.58 (0.21) | 4.5 (1.9) | 25.2 | 2683.8 | 454.8 |
| 27 | 27 Dec 2010 | 72.185 | 118.958 W | 0.82 | 2.26 | 0.65 | − 1.43 (0.07) | 5.2 (1.8) | 30.4 | 1968.8 | 388.9 |
| 30 | 29 Dec 2010 | 72.042 | 123.173 W | 0.31 | 0.65 | 0.58 | − 1.29 (0.06) | 4.7 (0.9) | 13.6 | 1615.8 | 161.3 |
| 31 | 30 Dec 2010 | 72.163 | 127.081 W | 0.40 | 1.54 | 1.41 | − 1.60 (0.13) | 4.9 (1.0) | 11.5 | 1893.6 | 171.1 |
| 31.1 | 2 Jan 2011 | 72.221 | 133.200 W | 0.12 | 0.79 | 0.75 | − 0.23 (0.13) | 3.6 (1.5) | 16.6 | 2277.3 | 246.8 |
| 32 | 3 Jan 2011 | 72.805 | 135.578 W | 0.15 | 1.25 | 1.17 | − 1.10 (0.16) | 3.3 (1.6) | 4.63 | 1743.6 | 120.0 |
| 33 | 4 Jan 2011 | 73.427 | 139.305 W | 0.06 | 1.65 | 1.55 | − 0.87 (0.40) | 3.7 (1.0) | 7.97 | 2287.3 | 170.3 |
| 41 | 6 Jan 2011 | 75.544 | 149.300 W | 0.15 | 1.35 | 1.25 | − 1.64 (0.37) | 4.9 (1.2) | 9.41 | 1627.9 | 149.4 |
| 42 | 8 Jan 2011 | 78.635 | 164.295 W | 0.00 | 3.85 | 2.50 | − 4.50 (0.61) | 3.9 (2.1) | 15.9 | 2791.2 | 355.2 |
| 43 | 10 Jan 2011 | 77.590 | 165.708 E | 0.01 | 1.52 | 1.35 | − 1.58 (0.39) | 3.8 (1.7) | 1.80 | 1532.0 | 175.7 |

Temperature and salinity (mean ± standard deviation) are vertical averages within the ice core; Chl a, POC, and PON are vertical integrals from the top to the bottom of the core.

## Sea ice

Most of the sea ice sampled during this study was first year ice that ranged in thickness from 0.65 m to 2.26 m (Table 1) and averaged $1.47 \pm 0.76$ m. The lone exception was at station 42, located along the eastern side of the Ross Ice Shelf, where multiyear ice nearly 4 m thick was observed. Ice thickness increased from 0.66 m along the northern ice edge (station 1) to 1.31 m at the interior of the pack (station 4), in conjunction with the increase in sea ice concentration (Figure 1). However, within the interior of the ice pack, there was no apparent spatial pattern in sea ice thickness distribution.

Due to our sampling in early summer, the ice pack was largely isothermal and near the freezing point of seawater at the time of sampling, with the vertically-averaged ice core temperature at all but one station ranging from –0.23°C to –1.92°C (Table 1). The single exception was at the multiyear ice station 42 where the vertically-averaged temperature of this extremely thick ice was substantially colder at –4.50°C. The skeletal layer was generally but not always present.

The bulk salinity of sea ice sampled during our study averaged 5.07 ± 1.61, with significantly higher salinity on the eastern portion of the transect. East of 130°W, bulk salinity averaged 5.70 ± 1.66, while to the west, bulk salinity averaged only 3.90 ± 0.54. There was no significant relationship between mean bulk sea ice salinity and either temperature, snow thickness, or sea ice thickness at a given station.

The salinity of surface ponds, which were generally less than 0.2 m deep, averaged 20.0 ± 2.8. Slush layers were approximately 0.05 m thick with a mean bulk salinity of 26.3 ± 7.3. These relatively high salinities compared to bulk sea ice salinity suggest that seawater was infiltrating the surface of the ice, due either to surface flooding or percolation through the ice. Temperatures of both melt ponds and slush layers were just above the freezing point.

## Depth-integrated microalgal biomass

### Chlorophyll a

Depth-integrated Chl $a$ biomass in sea ice ranged from 1.80 to 72.2 mg m$^{-2}$ during our study, averaging 17.3 ± 17.8 mg m$^{-2}$ (Table 1). The highest values were concentrated in the region of the Amundsen Sea between 113°W and 118°W (Figure 1, Table 1). Interestingly, depth-integrated Chl $a$ was significantly positively correlated with snow depth (Figure 2a) but not with the ratio of snow depth:ice thickness (not shown), with the four stations with the highest algal biomass also having the thickest snow cover (Table 1). Thicker snow was also associated with the depth of the microalgal biomass maximum that was nearer the sea ice surface (Table 1). There was no apparent relationship between sea ice thickness and Chl $a$ biomass (Figure 2b).

### Particulate organic carbon

The mean depth-integrated POC within the sea ice during our study was 2041 ± 687 mg m$^{-2}$. Total POC varied from 822 mg m$^{-2}$ in the thin ice at the northern ice edge to 3211 mg m$^{-2}$ in the interior of the pack at station 4 (Table 1), which also had the highest depth-integrated Chl $a$ concentration. POC was positively, but non-linearly, correlated with Chl $a$ (Figure 3a), rising rapidly at Chl $a$ values below 20 mg m$^{-2}$ and then rising more slowly at higher levels of Chl $a$. Unlike Chl $a$, depth-integrated POC exhibited no statistically significant relationship with either snow depth (Figure 2c) or ice thickness (Figure 2d).

The depth-integrated POC/Chl $a$ ratio averaged 214.1 ± 191 (g:g) throughout the study region. Values ranged from 44.5 at station 4, where the ice algal bloom was most intense, to 852 at station 43, which had very low Chl $a$ accumulation despite a moderate amount of POC. The POC/Chl $a$ ratio was significantly negatively correlated with snow depth (Figure 2e) but exhibited no apparent relationship with sea ice thickness (Figure 2f).

### Particulate organic nitrogen

Depth-integrated PON averaged 266 ± 151 mg m$^{-2}$ in our study area, ranging from 97.1 mg m$^{-2}$ at the northern ice edge to 599 mg m$^{-2}$ at station 4 within the interior of the ice pack (Table 1). Depth-integrated PON was highly correlated with POC (Figure 3b), with the POC/PON ratio averaging 8.72 ± 2.76 within our study region. Like Chl $a$, PON was positively correlated with snow depth but exhibited no statistically significant relationship with ice thickness. Because of the positive correlation between snow depth and PON, the POC/PON ratio was significantly negatively correlated with snow depth (Figure 2g) and exhibited no relationship with ice thickness (Figure 2h).

## Microalgal physiology

Many assays used to characterize algal physiology require an ample supply of algal biomass to produce a measurable biological signal. Because algal biomass is often heterogeneously distributed within an individual sea ice core, we could best measure algal pigments and physiological properties on those sections of each ice core that had a sufficient amount of algal biomass. Therefore, the values reported below do not reflect averages over the entire core, but are indicative of values in microhabitats where ice microalgae were most prevalent. In descending order of incident light intensity (estimated as described in the methods), the microhabitats sampled in this study include surface pond, slush, high light interior ice, low light interior ice, bottom ice, under-ice high light, and under-ice low light. The stations used to quantify algal pigment ratios and physiology in each microhabitat are given in Table 2.

### Pigment concentrations

The concentration of Chl $a$ associated with microalgal blooms in sea ice microhabitats was greatest in deformation ponds (105 ± 116 mg m$^{-3}$) and within the low light interior ice (111 ± 127 mg m$^{-3}$) (Table 2). Surprisingly, Chl $a$ concentrations were substantially lower in the bottom ice and in slush associated with

Figure 2

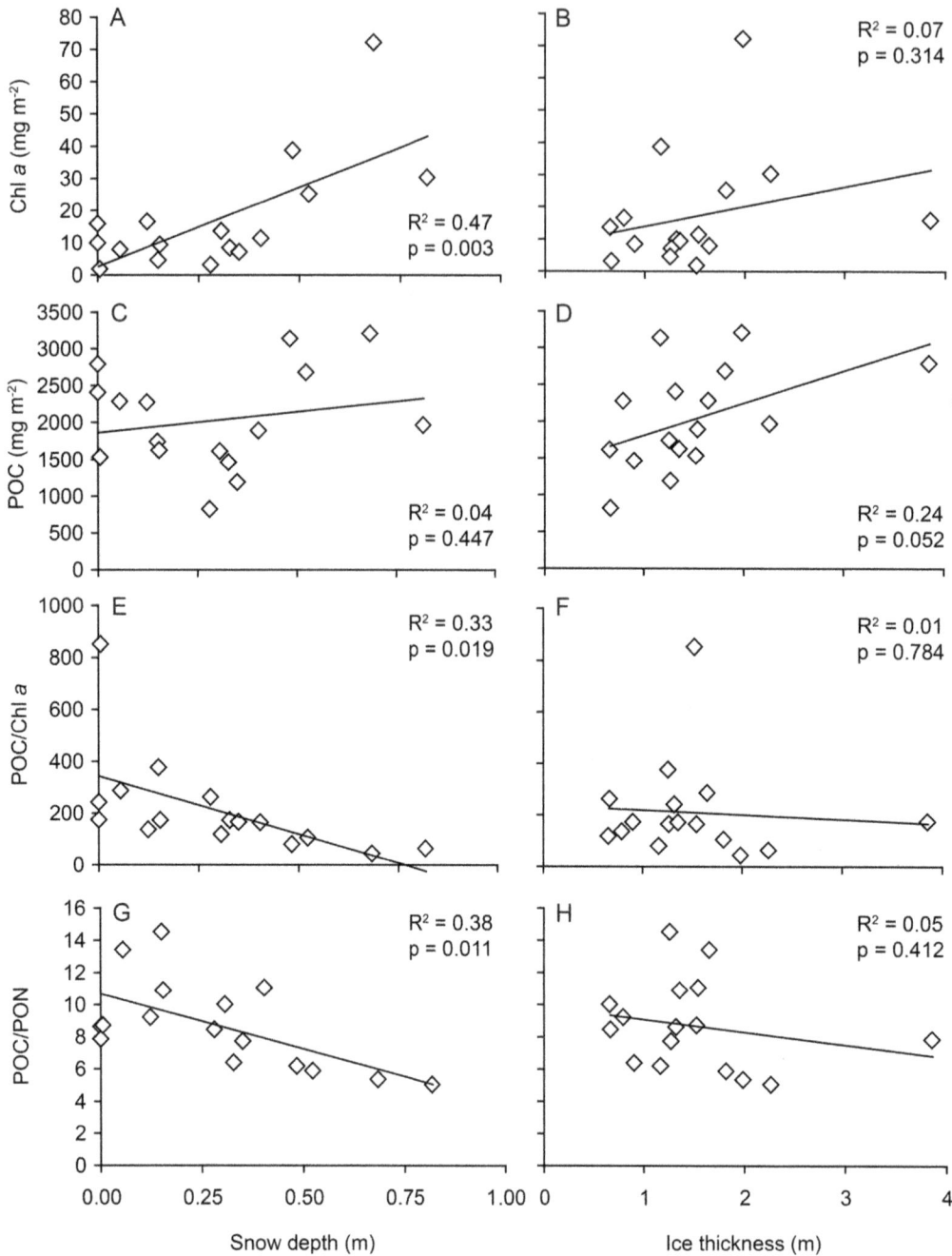

**Factors controlling algal biomass in sea ice.**

Regressions of depth-integrated chlorophyll *a* (Chl *a*) versus (a) snow depth and (b) sea ice thickness, depth-integrated particulate organic carbon (POC) versus (c) snow depth and (d) sea ice thickness, the POC/Chl *a* ratio (g:g) versus (e) snow depth and (f) sea ice thickness, and the ratio (g:g) of POC to particulate organic nitrogen (PON) versus (g) snow depth and (h) sea ice thickness.

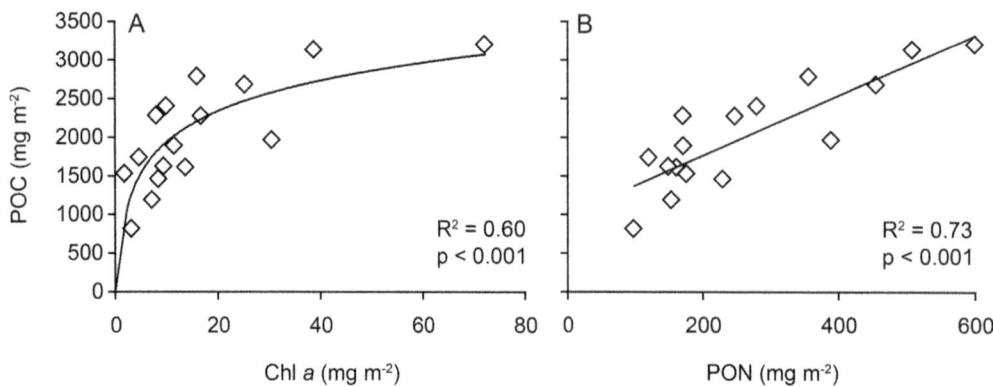

Figure 3

**Relationship between POC and other biomass metrics.**

Particulate organic carbon (POC) concentration versus (a) chlorophyll *a* (Chl *a*) concentration (POC = 570.09 ln(Chl *a*) + 645.18, n = 16) and (b) particulate organic nitrogen (PON) concentration in sea ice.

Table 2. Mean (± standard deviation) pigment concentrations for algae living in different sea ice microhabitats

| Microhabitat | Stations | Chl $a$ | Phaeo | Fuco:Chl $a$ | (DD + DT):Chl $a$ | 19-Hex:Chl $a$ |
|---|---|---|---|---|---|---|
| | | (mg m$^{-3}$) | (mg $^{-3}$) | (g:g) | (g:g) | (g:g) |
| Surface pond | 31.1, 32, 33 | 105.0 (115.9) | 6.37 (8.15) | 0.191 (0.155) | 0.301 (0.094) | 0.198 (0.187) |
| Surface slush | 1, 2, 27, 30, 31, 41 | 19.5 (24.5) | 1.33 (3.79) | 0.404 (0.247) | 0.150 (0.042) | 0.277 (0.088) |
| High light interior ice | 1, 3, 5, 10, 31, 31.1, 33, 41 | 34.1 (45.7) | 0.83 (1.16) | 0.405 (0.402) | 0.039 (0) | 0.189 (0.218) |
| Low light interior ice | 4, 20, 25, 27 | 111.2 (127.2) | 4.35 (6.79) | 0.670 (0.142) | 0.074 (0.051) | 0.092 (0.060) |
| Bottom ice | 2, 30, 32 | 31.5 (22.6) | 1.74 (1.91) | 0.836 (0.211) | 0.081 (0.047) | 0.014 (0.019) |
| Under-ice water, high light | 10, 30, 32, 33 | 2.5 (4.1) | 0.21 (0.09) | 0.344 (0.090) | 0.069 (0.049) | 0 |
| Under-ice water, low light | 25, 27, 31 | 2.2 (2.6) | 0.36 (0.28) | 0.453 (0.122) | 0.076 (0.018) | 0.053 (0.036) |

Chl $a$ = chlorophyll $a$, Phaeo = phaeopigments, Fuco = fucoxanthin, DD = diadinoxanthin, DT = diatoxanthin, 19-Hex = 19'-hexanoyloxyfucoxanthin
Microhabitats are ordered from highest light to lowest, based on light index described in the Methods

surface flooding (20–30 mg Chl $a$ m$^{-3}$), two habitats that typically support high microalgal biomass. Maximum Chl $a$ concentrations in sea ice were more than 40-fold higher than Chl $a$ concentrations measured in the under-ice water (2.2–2.5 mg m$^{-3}$). The concentrations of both phaeopigments (fluorometry) and the sum of Chl $a$ allomer and epimer, chlorophyllide $a$, monovinyl-chlorophyllide $a$, pheophorbide $a$ and pheophytin $a$ (HPLC) in sea ice habitats were generally 2–8% of Chl $a$ concentrations, indicating that little degradation of Chl $a$ had taken place by the time of sampling.

Fucoxanthin (Fuco), a photosynthetic accessory pigment associated primarily with diatoms in sea ice, varied markedly within the various microhabitats when normalized by Chl $a$ concentration (Table 2). Consistent with the role of fucoxanthin in light harvesting, the Fuco/Chl $a$ ratio was highest in low light environments such as the bottom ice and low light interior ice (Table 2). The lowest Fuco/Chl $a$ ratios were measured in deformation ponds and in high light under-ice environments.

The concentration of the xanthophyll cycle pigments diatoxanthin (DT) and diadinoxanthin (DD), which are used by microalgae for photoprotection against excessive irradiance, were also measured in the sea ice (Table 2). In contrast to the photosynthetic pigment fucoxanthin, the ratio (g:g) of DD + DT to Chl $a$ was greatly elevated in the environments with the highest light levels, including deformation ponds (0.30) and slush layers (0.15). Interior ice and bottom ice, as well as at the under-ice water column, all exhibited relatively low (DD + DT)/Chl $a$ ratios ranging from 0.04 to 0.08 (Table 2).

Finally, the photosynthetic accessory pigment 19'-hexanoyloxyfucoxanthin (19-Hex), which has been confirmed microscopically to be an effective marker pigment in the Southern Ocean to identify *Phaeocystis antarctica* in both the water column and the sea ice (Arrigo et al., 1999, 2002, 2003), also varied considerably with sea ice habitat (Table 2). The 19-Hex/Chl $a$ ratio (g:g) was highest in microhabitats within the upper ice, such as deformation ponds, slush layers, and high light interior ice, and lowest in bottom ice. This vertical pattern suggests that *Phaeocystis antarctica* grows best in regions of the ice that receive the most light.

### POC, PON, and isotopic ratios
POC concentrations ranged from an average of 1354 ± 869 mg m$^{-3}$ in ice algal blooms associated with bottom ice to 9304 ± 5811 mg m$^{-3}$ in ice algal blooms associated with deformation ponds (Table 3). These values are 13–90-fold higher than POC concentrations measured in the under-ice water column. PON showed a similar vertical distribution, with bloom concentrations being greatest in both low light interior ice (1179 mg m$^{-3}$) and deformation ponds (1032 mg m$^{-3}$) and lowest in bottom ice (210 mg m$^{-3}$). Like both Chl $a$ and POC, bloom values for PON in sea ice were substantially higher (6 to 33-fold) than in the under-ice water column.

POC/PON ratios in sea ice algal blooms ranged from 5.5 to 9.2 (Table 3) and tended to be higher in high light environments located closer to the sea ice surface (e.g., surface pond, slush, and high light interior). Values in sea ice were much greater than those measured in the under-ice water column (POC/PON = 2.4–3.2).

$\delta^{13}$C of particulate matter in sea ice ranged from −21.8 to −14.3 ‰. $^{13}$C enrichment of POC was significantly positively correlated with POC concentration (Figure 4), indicating higher levels of enrichment

Table 3.  Mean (± standard deviation) particulate organic carbon (POC) and nitrogen (PON) content and stable isotopic ratios of C and N for different sea ice microhabitats

| Microhabitat | POC | $\delta^{13}$C | PON | $\delta^{15}$N | POC/PON |
|---|---|---|---|---|---|
| | (mg m$^{-3}$) | (‰) | (mg m$^{-3}$) | (‰) | (g:g) |
| Surface pond | 9304 (5811) | −14.3 (1.6) | 1032 (528.0) | 3.44 (1.11) | 9.0 |
| Surface slush | 1921 (1501) | −19.2 (2.6) | 280.7 (224.0) | 2.57 (1.41) | 6.8 |
| High light interior ice | 2674 (2085) | −21.8 (2.3) | 290.7 (178.3) | 1.66 (2.02) | 9.2 |
| Low light interior ice | 6442 (7641) | −19.3 (1.3) | 1179 (1169) | 1.16 (1.45) | 5.5 |
| Bottom ice | 1354 (869) | −20.9 (1.8) | 210.1 (111.2) | 2.13 (0.59) | 6.4 |
| Under-ice, high light | 64.6 (7.8) | −22.4 (2.3) | 27.2 (22.0) | −1.86 (1.28) | 2.4 |
| Under-ice, low light | 139.0 (120.6) | −20.3 (4.8) | 43.9 (24.0) | 0.54 (1.58) | 3.2 |

Microhabitats are ordered from highest light to lowest, based on light index described in the Methods.

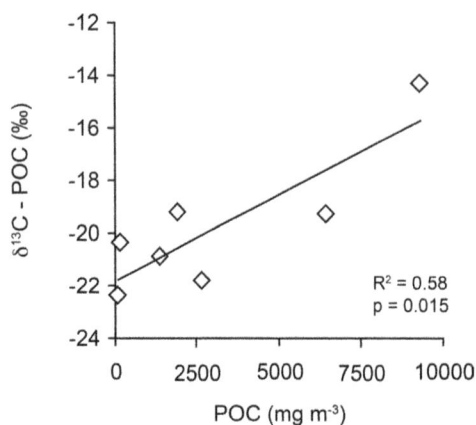

Figure 4
$\delta^{13}$C-POC.

$\delta^{13}$C of particulate organic carbon ($\delta^{13}$C-POC) versus POC concentration in sea ice.

at higher levels of ice algal biomass. There was no such correlation between $\delta^{15}$N of particulate matter and algal biomass, although $^{15}$N was significantly enriched in sea ice environments compared to the under-ice water column (Table 3).

### Photosynthetic parameters

Maximum light-saturated photosynthetic rates (P$^{*}_{m}$) among the sea ice microhabitats were greatest in the slush layers, bottom ice, and deformation ponds and lowest in interior ice (Table 4). Maximum values for P$^{*}_{m}$ in sea ice were not as high as those measured in the under-ice water column, the latter of which were also more highly variable. Photosynthetic efficiency ($\alpha^{*}$) was highest in the bottom ice, exceeding values in other sea ice habitats by a factor of 3–10. Values for $\alpha^{*}$ in both the high light and low light interior ice were by far the lowest of any habitat sampled. The photoacclimation parameter E$_k$ varied from 20 to 96 µmol photons m$^{-2}$ s$^{-1}$ and was highest in habitats nearer the sea ice surface and declined with depth within the ice (Table 4). The highest values for E$_k$ in the sea ice were greater than the highest values measured in the under-ice water column.

## Absorption parameters

The ratio of the magnitude of the algal absorption peaks at blue (e.g., 443 nm) and red (e.g., 676 nm) wavelengths provides an indication of the level of pigment packaging, which increases with larger microalgal cell size and higher intracellular Chl $a$ concentration (Morel and Bricaud, 1981). Lower blue/red absorption ratios indicate a greater level of pigment packaging (Figure 5a, Table 4). Within the sea ice, blue/red absorption ratios were lowest in the interior ice and highest in the surface communities (slush and deformation ponds), with values very similar to those measured in the under-ice water column (Table 4). Interestingly, the blue/red ratio in the bottom ice fell between these two extremes.

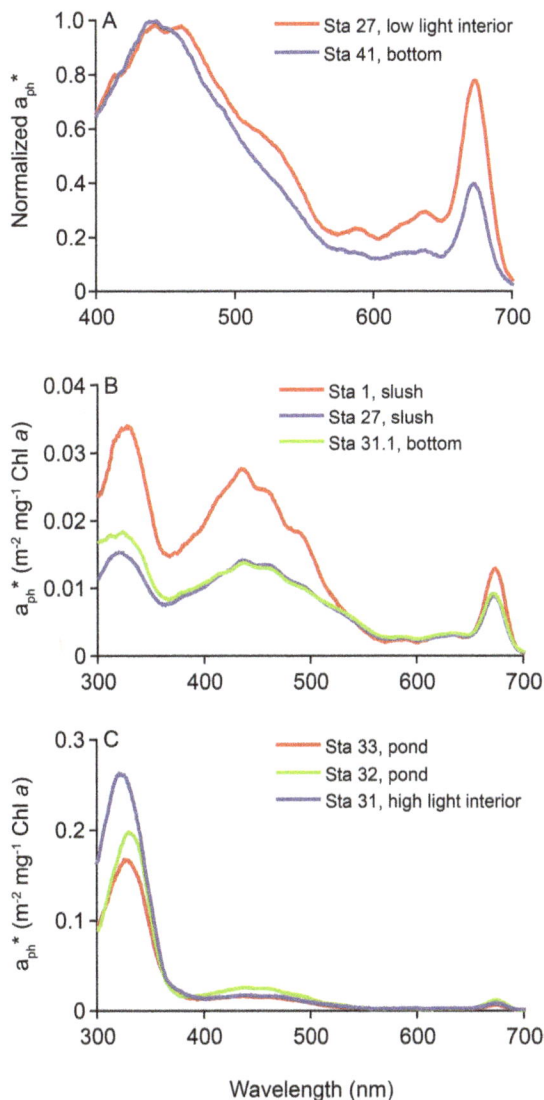

**Figure 5**

**Microalgal absorption.**

Chlorophyll $a$-specific algal absorption ($a_{ph}^*$) spectra, (a) normalized to the maximum absorption within the visible, (b) for high biomass stations with moderate mycosporine-like amino acid (MAA) absorption peaks (~ 330 nm), and (C) for high biomass stations with high MAA absorption peaks.

Absorption peaks at ultraviolet (UV) wavelengths varied markedly within the sea ice (Table 4), suggesting differential production of mycosporine-like amino acids (MAAs) by the microbial community. MAA absorption was generally low in bottom ice and in both low light and high light interior ice communities, ranging from approximately 0.02–0.03 m² mg⁻¹ Chl $a$ (Figure 5b, Table 4). In surface slush layers, MAA absorption was 3–4-fold higher than elsewhere in the ice, averaging $0.070 \pm 0.053$ m² mg⁻¹ Chl $a$. The highest MAA absorption values were measured in deformation ponds (Figure 5c), which exceeded values in the slush layer by a factor of two and values elsewhere in the ice by nearly a factor of five (Table 4).

Table 4. Mean (± standard deviation) photosynthetic and absorption parameters for algae living in different sea ice microhabitats

| Microhabitat | $P'_m$ | $\alpha^*$ | $E_k$ | Blue:Red peak | MAA peak |
|---|---|---|---|---|---|
| Surface pond | 1.02 (0.17) | 0.016 (0.005) | 66.9 (23.9) | 2.39 (0.19) | 0.145 (0.073) |
| Surface slush | 1.68 (1.66) | 0.017 (0.014) | 95.8 (58.4) | 2.06 (0.27) | 0.070 (0.053) |
| High light interior ice | 0.07 (0.04) | 0.003 (0.003) | 27.7 (21.2) | 1.90 (0.40) | 0.032 (0.013) |
| Low light interior ice | 0.23 (0.14) | 0.007 (0.006) | 36.7 (8.7) | 1.58 (0.02) | 0.029 (0.014) |
| Bottom ice | 1.14 | 0.059 | 19.5 | 2.02 (0.46) | 0.019 (0.020) |
| Under-ice, high light | 3.19 (2.93) | 0.039 (0.032) | 78.7 (28.5) | 2.43 (0.46) | 0.032 (0.013) |
| Under-ice, low light | 0.60 (0.35) | 0.013 (0.007) | 44.7 (20.6) | 2.19 (0.27) | 0.029 (0.014) |

$P'_m$ = mg C mg$^{-1}$ Chl $a$ hr$^{-1}$, $\alpha^*$ = mg C mg$^{-1}$ Chl $a$ hr$^{-1}$ ($\mu$mol photons m$^{-2}$ s$^{-1}$)$^{-1}$, $E_k$ = $\mu$mol photons m$^{-2}$ s$^{-1}$, MAA peak = m$^2$ mg$^{-1}$ Chl $a$
Microhabitats are ordered from highest light to lowest, based on light index described in the Methods.

# Discussion

## Microalgal biomass

The mean microalgal biomass in sea ice within our study region (17.3 ± 17.8 mg Chl $a$ m$^{-2}$), which consisted primarily of stations sampled in the Amundsen Sea during summer, was comparable to the circumpolar summer mean (12.9 ± 16.5 mg Chl $a$ m$^{-2}$) calculated from historical data in the Antarctic Sea Ice Processes and Climate (ASPeCt)-Bio database (Meiners et al., 2012). Like the ASPeCt-Bio database, peak Chl $a$ biomass during our study was distributed in microhabitats throughout the ice column. However, in the AntarcASPeCt – Biology (ASPeCt-Bio) database, surface, internal, and bottom microhabitats contained approximately equal fractions of total depth-integrated Chl $a$. This pattern contrasts with our study in which bottom communities were well developed in only 18% of the stations sampled, while surface and interior ice communities were well developed at 53% and 71% of stations, respectively (some stations had multiple communities so the percentages sum to > 100%). In this respect, vertical distributions of Chl $a$ observed in our study share a greater similarity with those measured in the Ross Sea, where surface and interior communities also were relatively more abundant than bottom communities (Meiners et al., 2012). It is possible that, because the Amundsen Sea is aggregated with the Bellingshausen Sea in the ASPeCt-Bio database, the vertical distributions of Chl $a$ are being skewed by the large number of samples near the Antarctic Peninsula, which may differ from those in the Amundsen and Ross Seas. In addition, the ASPeCt-Bio database includes data from all months of the year, while our study was conducted in summer, so some temporal bias may have been introduced in the comparison.

Snow depth appears critical in determining both the depth-integrated Chl $a$ (higher with deeper snow) and distribution (dominance of slush communities in deeper snow) of ice microalgae. This observation is almost certainly a reflection of the impact of snow loading on nutrient availability via surface flooding rather than its impact on light availability, since more snow leads to reduced light levels (Perovich, 1990; Arrigo et al., 1997; Saenz and Arrigo, 2012). Low snow and a high freeboard means that the greatest nutrients can be found down at the ice/water interface, while heavy snow depresses the freeboard and causes flooding/nutrient renewal near the ice surface. Such surface flooding occurs over 15–30% of the ice pack in Antarctica (Wadhams et al., 1987).

These results are consistent with model results that demonstrate the importance of snow loading and surface flooding to the development and maintenance of high algal biomass in surface communities (Saenz and Arrigo, 2012; Saenz and Arrigo, in press). However, our results for the Amundsen Sea contrast with previous observations from the Ross Sea, which showed higher microalgal biomass at lower snow thicknesses (Arrigo et al., 2003). The difference between the two studies is likely explained by the lower snow accumulation in the Ross Sea, which rarely exceeded 0.1 m (Arrigo et al., 2003) and was unlikely to depress the surface of the ice below freeboard and cause surface flooding. In contrast, flooded sea ice in the Amundsen

Sea during our study had snow accumulations of 0.3–0.8 m (Table 1) and well developed slush and surface pond communities associated with relatively high salinities, indicative of flooding. Thus, snow accumulation appears to be detrimental to algal growth due to light attenuation (as demonstrated in the Ross Sea) until it becomes sufficiently heavy to cause flooding, at which point the detrimental effect of reduced light would be more than compensated for by the advantage of nutrient renewal. Concurrent measurements of nutrient resupply and microalgal parameters would confirm this scenario.

Our results support previous observations showing that microalgal biomass in the ice is much greater and more highly concentrated than in the under-ice water column below (Garrison et al., 1990; Legendre et al., 1992; Arrigo et al., 2003). As a result, microhabitats within sea ice represent an important food source during those times of year when pelagic production is still low. Given the seasonal importance of sea ice biology, factors that control the amount of biomass within the ice can have a profound impact on other parts of the marine ecosystem (Arrigo et al., 2010b). For example, models predict that in the future there will be increased precipitation in the Southern Ocean resulting in greater snow cover on sea ice (e.g., Sarmiento et al., 1998). Our results suggest that such a change in forcing could increase the amount of algal biomass within the ice by enhancing nutrient supplies via surface flooding. This effect might be especially important in places like the Ross Sea where snow depths are currently too low to promote surface flooding.

## Microalgal physiology

The photosynthetic rates of sea ice microalgae are controlled by a combination of sea ice temperature, brine salinity, and light and nutrient availability (Arrigo and Sullivan, 1992). At the time of our study, the sea ice was almost isothermal, so differences in sea ice temperature and brine salinity were likely having little impact on the spatial variability of sea ice microalgal physiological state. Therefore, the differences in elemental composition, pigment content, and photosynthetic parameters that we observed for sea ice microalgae were likely driven almost exclusively by vertical and horizontal gradients in light or nutrient availability.

For example, the POC/Chl a ratio in sea ice was lower at stations with a thicker snow cover (Figure 2e), indicative of microalgae increasing their cellular pigment concentrations in an effort to harvest more of the available light (Moisan and Mitchell, 1999; Kropuenske et al., 2009). Although the magnitude of these ratios suggests that the samples contained some detritus, complicating a possible physiological interpretation, the non-linear relationship observed between POC and Chl a is consistent with a photoacclimation response to reduced light levels (Figure 3a). As microalgal biomass (e.g., POC) increased, light availability was reduced due to self-shading, and the microalgae increased their cellular Chl a concentration to compensate. Consequently, the amount of additional Chl a necessary to sustain net photosynthesis at high biomass was higher than would have been necessary in the absence of microalgal self-shading.

While the observed POC/Chl a ratio in our study likely reflects acclimation to light availability, the variation in the POC/PON ratio with snow depth (Figure 2g) is also consistent with a microalgal response to nutrient availability. While it is true that the decreasing POC/PON ratio with increasing snow depth is consistent with an enhancement of N-rich photosystems by microalgae growing at reduced light levels (Klausmeier et al., 2004), the magnitude of the POC/PON ratio at low snow thicknesses (> 10) is very high relative to the ratio expected for N-replete cells (5–7). This finding suggests that the microalgal community growing beneath a thin cover of snow also may have been nutrient-stressed due to a lack of surface flooding.

Our other pigment data also provide strong evidence of acclimation by microalgae to the different light regimes associated with various microhabitats within the sea ice. Enhanced pigment packaging is manifested as a flattening of the algal absorption spectrum (i.e., lower blue/red peak ratios) as more light is absorbed by a microalgal cell, either due to a larger cell size or a greater amount of pigment per cell (Morel and Bricaud, 1981). The degree of pigment packaging was low in surface pond and slush communities and greatest in interior ice (Table 4), suggesting that microalgae growing in these low light environments either were larger or had higher pigment content than algae growing nearer the sea ice surface. A higher pigment content per cell is supported by the vertical distribution of the ratio of the photosynthetic accessory pigment fucoxanthin to Chl a, which showed a clear increase with depth within the ice (Table 2). Interestingly, pigment packaging was also low in the under-ice water column beneath the ice. Because this environment would be expected to experience relatively low light, the small amount of packaging could be the result of this pelagic community being dominated by smaller sized cells than are usually found in sea ice (Gradinger and Ikävalko, 1998; Riaux-Gobin et al., 2011).

Like pigment concentration, the types of pigments present in various sea ice microhabitats also reflect acclimation to different light environments. The xanthophyll cycle is used by many microalgal species to protect themselves from excessive irradiance. For sea ice microalgae, the xanthophyll cycle consists of enzymatic de-epoxidation of the carotenoid pigment DD to DT, the latter of which thermally dissipates excess energy (Olaizola and Yamamoto, 1994; Demmig-Adams and Adams, 2006; Goss et al. 2006). This photoprotective mechanism is particularly important for microalgae that get mixed periodically into surface waters (Alderkamp et al., 2013) or that live in the upper reaches of the sea ice. In our study, by far the highest (DD + DT)/Chl a ratios were measured in deformation ponds and surface slush layers, habitats that can be subjected to very high light levels throughout the 24-hour day. Xanthophyll cycle pigments were also found in microalgae within interior and bottom microhabitats within the ice and in the under-ice water column, but at much reduced concentrations relative to near-surface communities. This vertical pattern demonstrates clearly that microalgae living near the sea ice surface were using xanthophyll cycling as a mechanism to withstand excessive irradiance. However, it should be noted that even the lowest (DD + DT)/Chl $a$ ratios measured in the sea ice during our study were 3-fold higher than values measured previously in ice-free waters of the Amundsen Sea (Alderkamp et al., 2013).

Like xanthophyll pigment concentrations, absorption of UV radiation by MAAs was very high in surface ice communities (pond and slush) and declined with depth within the ice (Table 2). Production of MAAs by microalgae can be photoinduced by elevated levels of UVB (280–320 nm) radiation, and to a lesser extent by UVA (320–400 nm) and visible (400–700 nm) radiation (Hannach and Sigleo, 1998; Klisch and Häder, 2001; Sinha et al., 2001), although the relative response to UVB, UVA, and visible radiation varies by algal taxa (Riegger and Robinson, 1997; Moisan and Mitchell, 2001). Algae produce at least nine different MAAs whose spectral absorption peak ranges from 310 nm to 386 nm (Riegger and Robinson, 1997). The vertical pattern of MAA absorption observed in ice during our study, which was only high in near surface samples, suggests that UV radiation was attenuated rapidly within the snow and ice and was not an important factor influencing rates of microalgal growth within the ice interior and sea ice bottom. The heterogeneous vertical distributions of both MAAs and xanthophyll cycle pigments demonstrate that, while sea ice is generally considered to be a low light habitat, microalgae living there must be able to acclimate to a wide range of light intensities.

The variations in photosynthetic parameters in the different sea ice microhabitats also demonstrate microalgal acclimation to both light and nutrient levels. Although both $P^*_m$ and $\alpha^*$ can vary as a function of either light or nutrient availability, in our study nutrient availability seemed to be the dominant controlling variable. Both $P^*_m$ and $\alpha^*$ were higher in the under-ice water column than in the ice, a likely reflection of greater nutrient availability (although we cannot rule out differences in phytoplankton species composition). This conclusion also is consistent with the vertical patterns observed within the ice, with both $P^*_m$ and $\alpha^*$ being higher in flooded surface communities and at the bottom of the ice (where nutrient concentrations would be high) than within the sea ice interior (Table 4). If light had been the controlling variable, $P^*_m$ would have declined and $\alpha^*$ would have increased with depth within the ice, a pattern contrary to our observations.

Interestingly, while both $P^*_m$ and $\alpha^*$ measured in sea ice indicate a system in which photosynthetic activity is most likely controlled by nutrient availability, the photoacclimation parameter ($E_k$) shows that light levels are also important. Because $E_k$ is calculated as $P^*_m/\alpha^*$, it reflects a balance between the dark and light reactions of photosynthesis (Cullen, 1990). Because it is advantageous to maintain an optimal balance between light harvesting machinery and carbon fixation machinery, $E_k$ often reflects the mean light history experienced by microalgal cells. The relationship between light history and $E_k$ is apparent in our dataset as a progressive decrease in $E_k$ with depth within the ice. $E_k$ was highest in deformation ponds where light levels were highest and, unlike both $P^*_m$ and $\alpha^*$, was lowest at the bottom of the ice where light levels were lowest. Ironically, the maintenance of this balance between $E_k$ and light history in the bottom ice community would be made possible by abundant nutrients. Although $P^*_m$ was relatively high in this microhabitat (Table 4), $\alpha^*$ was even higher (relative to other microhabitats). Maintaining such a high photosynthetic efficiency (e.g., $\alpha^*$) when light is low requires a large pigment bed capable of harvesting additional photons and transferring them to the reaction centers, the synthesis of which requires a sufficiently large nutrient supply. Because nutrients were likely to be relatively high in the bottom community due to its proximity to seawater, microalgae growing there could elevate their photosynthetic efficiency far above that of internal communities (where light was also low). These high values for $\alpha^*$ are what allowed bottom communities to reduce their $E_k$ values to match their low light environment. Thus, we conclude that the balance between light and nutrients was critical in shaping sea ice microalgal metabolism in the Amundsen Sea during our study.

Although greater nutrient availability likely explains the higher biomass and more active microalgal physiology we observed in some microhabitats, there was also evidence that these high biomass microenvironments may have depleted their resources. We observed a statistically significant relationship between POC and the $\delta^{13}C$ of POC in sea ice (Figure 4), which is likely the result of depletion of dissolved inorganic carbon by extremely high accumulations of ice algal biomass and consequent enrichment of the inorganic carbon pool (Rau et al., 1991; Kennedy et al., 2002; Arrigo et al., 2003). This relationship is similar to that observed in bottom ice in the Beaufort Sea region of the Arctic Ocean (Pineault et al., 2013), although both our POC and $\delta^{13}C$ values were higher, indicating greater depletion of inorganic carbon by a larger algal community. Although there was no significant relationship between PON concentration and the $\delta^{15}N$ of PON in sea ice, $\delta^{15}N$ of particulate matter was significantly enriched in sea ice environments compared to the under-ice water column. Enrichment could be due to the presence of heterotrophs in sea ice (DeNiro and Epstein, 1981; Teranes and Bernasconi, 2000), but it is most likely due to $NO_3$ being nearly depleted in ice cores such that both light and heavy isotopes of N were assimilated (while in the under-ice water column, mostly the light $NO_3$ was utilized, Berg et al. 2011). This level of nutrient depletion was possible because, unlike the water column where iron (Fe) concentrations are insufficient to support complete $NO_3$ drawdown by microalgae, microbial communities in the sea ice have enough Fe to more efficiently utilize the available macronutrients (Edwards and Sedwick, 2001). Thus, the biological pump associated with sea ice in the Southern Ocean is likely to be more efficient (but not stronger) than that of the water column, even on productive continental shelves. Given the more complete utilization of $NO_3$ in ice relative to the water column, it might be worth exploring whether $\delta^{15}N$ of PON in sediments can be used as a proxy for the presence of sea ice in a given area.

Finally, we were surprised to find a relatively large proportion of the pigment 19-Hex in sea ice microhabitats exposed to high light. 19-Hex is a marker pigment for *Phaeocystis antarctica*, a colonial prymnesiophyte that forms dense blooms in open waters on Antarctic continental shelves (Arrigo et al., 1999). Although this species has been observed previously in newly formed sea ice (Arrigo et al., 2003), and can tolerate prolonged darkness and freezing (Tang et al., 2009), the results of this study are the first to imply that it was physiologically active in the upper layers of sea ice in late spring/early summer. Its presence in recently flooded sea ice may indicate that it was introduced to the ice relatively recently and was able to grow there because of its ability to withstand high light levels. *P. antarctica* has an active xanthophyll cycle and is capable of efficiently repairing photodamage incurred during periods of high light stress (Kropuenske et al., 2009, 2010), allowing it to outcompete other species under highly variable light conditions (Arrigo et al., 2010a). It also is able to regulate its MAA content to protect itself from excessive radiation, including UV (Riegger and Robinson, 1997). Because the release from the sea ice into surface waters can expose microalgae to high radiation levels, high MAA content and enhanced xanthophyll cycle pigment concentrations may pre-acclimate *P. antarctica* to survive the transition from the sea ice to the water column as the sea ice melts in spring. In this way, we suggest that high xanthophyll cycle pigment and MAA content may facilitate the role that sea ice algae play in seeding ice-edge phytoplankton blooms and thereby structuring phytoplankton communities.

# References

Alderkamp A-C, Mills MM, van Dijken GL, Arrigo KR. 2013. Photoacclimation and non-photochemical quenching under in situ irradiance in natural phytoplankton assemblages from the Amundsen Sea, Antarctica. *Mar Ecol Prog Ser* **475**: 15–34.

Arrigo KR, Sullivan CW. 1992. The influence of salinity and temperature covariation on the photophysiological characteristics of Antarctic sea ice microalgae. *J Phycol* **28**: 746–56.

Arrigo KR, Lizotte MP, Worthen DL, Dixon P, Dieckmann G. 1997. Primary production in Antarctic sea ice. *Science* **276**: 394–97.

Arrigo KR, Robinson DH, Worthen DL, Dunbar RB, DiTullio GR, et al. 1999. Phytoplankton community structure and the drawdown of nutrients and $CO_2$ in the Southern Ocean. *Science* **283**: 365–367.

Arrigo KR, Dunbar RB, Lizotte MP, Robinson DH. 2002. Taxon-specific differences in C/P and N/P drawdown for phytoplankton in the Ross Sea, Antarctica. *Geophys Res Lett* **29**(20).

Arrigo KR, Robinson DH, Dunbar RB, Leventer AR, Lizotte MP. 2003. Physical control of chlorophyll *a*, POC, and PON distributions in the pack ice of the Ross Sea, Antarctica. *J Geophys Res* **108**(C10): 3316. doi: 10.1029/2001JC001138

Arrigo KR, van Dijken GL, Bushinsky S. 2008. Primary Production in the Southern Ocean, 1997–2006. *J Geophys Res* **113**, C08004. doi:10.1029/2007JC004551

Arrigo KR, Mills MM, Kropuenske LR, van Dijken GL, Alderkamp A-C, Robinson DH. 2010a. Photophysiology in two major Southern Ocean phytoplankton taxa: Photosynthesis and growth of *Phaeocystis antarctica* and *Fragilariopsis cylindrus* under different irradiance levels. *Integr Comp Biol* **50**: 950–966.

Arrigo KR, Lizotte MP, Mock T. 2010b. Primary producers and sea ice., in Thomas DN, Dieckmann GS eds., *Sea Ice*, 2nd Edition. Oxford, UK: Blackwell Science, Ltd.: pp. 283–326.

Arrigo KR. 2014. Sea ice ecosystems. *Ann Rev Mar Sci* **6**: 13.1–13.29. doi:10.1146/annurev-marine-010213-135103

Berg GM, Arrigo KR, Mills MM, Long MC, Bellerby R, et al. 2011. Variation in particulate C and N isotope composition following iron fertilization in two successive phytoplankton communities in the Southern Ocean. *Glob Biogeochem Cycles* **25**: GB3013. doi:10.1029/2010GB003824

Boetius A, Albrecht S, Bakker K, Bienhold C, Felden J, et al. 2013. Export of algal biomass from the melting Arctic sea ice. *Science* **339**: 1430–1432.

Brown TA, Belt ST. 2012. Closely linked sea ice-pelagic coupling in the Amundsen Gulf revealed by the sea ice diatom biomarker IP$_{25}$. *J Plankton Res* **34**: 647–54.

Caron DA, Gast RJ. 2010. Heterotrophic protists associated with sea ice, in Thomas DN, Dieckmann GS eds., *Sea Ice*, 2nd Edition. Oxford, UK: Blackwell Science, Ltd.: pp. 327–56.

Cullen JJ. 1990. On models of growth and photosynthesis in phytoplankton. *Deep-Sea Res* **37**: 667–683.

Daly KL. 1990. Overwintering development, growth, and feeding of larval *Euphausia superba* in the Antarctic marginal ice zone. *Limnol Oceanogr* **35**(7): 1564–1576.

Demmig-Adams B, Adams W. 2006. Photoprotection in an ecological context: the remarkable complexity of thermal energy dissipation. *New Phytol* **172**: 11–21.

DeNiro MJ, Epstein S. 1981. Influence of diet on the distribution of nitrogen isotopes in animals. *Geochim Cosmochim Acta* **45**: 341–351.

DiTullio GR. Smith WO. 1996. Spatial patterns in phytoplankton biomass and pigment distributions in the Ross Sea. *J Geophys Res* **101**: 18467–18477.

Edwards R, Sedwick P. 2001. Iron in East Antarctic snow: Implications for atmospheric iron deposition and algal production in Antarctic waters. *Geophys Res Lett* **28**(20): 3907–3910.

Flores H, van Franeker JA, Cisewski B, Leach H, Van de Putte AP, et al. 2011. Macrofauna under sea ice and in the open surface layer of the Lazarev Sea, Southern Ocean. *Deep-Sea Res Part II* **58**: 1948–1961.

Flores H, van Franeker JA, Siegel V, Haraldsson M, Strass V, et al. 2012. The association of Antarctic krill *Euphausia superba* with the under-ice habitat. *PLoS ONE* **7**: 1–11.

Garrison DL, Buck KR. 1989. The biota of Antarctic pack ice in the Weddell Sea and Antarctic Peninsula region. *Polar Biol* **10**: 237–239.

Garrison DL, Close AR, Reimnitz E. 1990. Algae concentrated by frazil ice: evidence from laboratory and field measurements. *Antarct Sci* **1**: 313–316.

Garrison DL, Jeffries MO, Gibson A, Coale SL, Neenan D, et al. 2003. Development of sea ice microbial communities during autumn ice formation in the Ross Sea. *Mar Ecol Prog Ser* **259**: 1–15.

Golden KM, Ackley SF, Lytle VI. 1998. The percolation phase transition in sea ice. *Science* **282**: 2238–2241.

Golden KM, Eicken H, Heaton AL, Miner J, Pringle DJ, et al. 2007. Thermal evolution of permeability and microstructure in sea ice. *Geophys Res Lett* **34**: L16501.

Goss R, Pinto AE, Wilhem C, Richter M. 2006. The importance of a highly active and ΔpH-regulated diatoxanthin epoxidase for the regulation of the PS II antenna function in diadinoxanthin cycle containing algae. *J Plant Physiol* **163**: 1008–1021.

Gowing MM, Garrison DL. 1992. Abundance and feeding ecology of larger protozooplankton in the ice edge zone of the Weddell and Scotia seas during the austral winter. *Deep-Sea Res Part A* **39**: 893–919.

Gradinger R, Ikävalko J. 1998. Organism incorporation into newly forming Arctic sea ice in the Greenland Sea. *J Plank Res* **20**(5): 871–886.

Grossi SM, Kottmeier ST, Moe RL, Taylor GT, Sullivan CW. 1987. Sea ice microbial communities. 6. Growth and primary production in bottom ice under graded snow cover. *Mar Ecol Prog Ser* **35**: 153–164.

Guglielmo L, Zagami G, Saggiomo V, Catalano G, Granata A. 2007. Copepods in spring annual sea ice at Terra Nova Bay (Ross Sea, Antarctica). *Polar Biol* **30**: 747–58.

Haecky P, Jonsson S, Andersson A. 1998. Influence of sea ice on the composition of the spring phytoplankton bloom in the northern Baltic Sea. *Polar Biol* **20**: 1–8.

Hannach G, Sigleo AC. 1998. Photoinduction of UV-absorbing compounds in six species of marine phytoplankton. *Mar Ecol Prog Ser* **174**: 207–222.

Holm-Hansen O, Lorenzen CJ, Holmes RW, Strickland JDH. 1965. Fluorometric determination of chlorophyll. *ICES J Mar Sci* **30**: 3–15.

Juul-Pedersen T, Michel C, Gosselin M, Seuthe L. 2008. Seasonal changes in the sinking export of particulate material under first-year sea ice on the Mackenzie Shelf (western Canadian Arctic). *Mar Ecol Prog Ser* **353**: 13–25.

Kennedy H, Thomas DN, Kattner G, Haas C, Dieckmann GS. 2002. Particulate organic matter in Antarctic summer sea ice: concentration and stable isotopic composition. *Mar Ecol Prog Ser* **238**: 1–13.

Kiko R, Michels J, Mizdalski E, Schnack-Schiel SB, Werner I. 2008. Living conditions, abundance and composition of the metazoan fauna in surface and sub-ice layers in pack ice of the western Weddell Sea during late spring. *Deep-Sea Res Part II* **55**: 1000–1014.

Klausmeier CA, Litchman E, Levin SA. 2004. Phytoplankton growth and stoichiometry under multiple nutrient limitation. *Limnol Oceanogr* **49**: 1463–1470.

Klisch M and Häder D-P. 2001. Mycosporine-like amino acids in the marine dinoflagellate *Gyrodinium dorsum*: induction by ultraviolet irradiation. *J Plant Physiol* **158**: 1449–1454.

Kropuenske LR, Mills MM, van Dijken GL, Bailey S, Robinson DH, Welschmeyer NA, Arrigo KR. 2009. Photophysiology in two major Southern Ocean phytoplankton taxa: Photoprotection in *Phaeocystis antarctica* and *Fragilariopsis cylindrus*. *Limnol Oceanogr* **54**(4): 1176–1196.

Kropuenske LR, Mills MM, van Dijken GL, Alderkamp A-C, Berg GM, Robinson DH, Welschmeyer NA, Arrigo KR. 2010. Strategies and rates of photoacclimation in two major Southern Ocean phytoplankton taxa: *Phaeocystis antarctica* (Haptophyta) and *Fragilariopsis cylindrus* (Bacillariophyceae). *J Phycol* **46**: 1138–1151. doi:10.1111/j.1529-8817.2010.00922.x

Legendre L, Ackley SF, Dieckmann GS, Gulliksen B, Horner R, Hoshiai T, Melnikov IA, Reeburgh WS, Spindler M, Sullivan CW. 1992. Ecology of sea ice biota. *Polar Biol* **12**(3–4): 429–444. doi:10.1007/BF00243114

Lewis MR, Smith JC. 1983. A small volume, short incubation-time method for measurement of photosynthesis as a function of incident irradiance. *Mar Ecol Prog Ser* **13**: 99–102.

Lizotte MP. 2001. The contribution of sea ice algae to Antarctic marine primary production. *Am Zool* **41**: 57–73.

Mangoni O, Saggiomo M, Modigh M, Catalano G, Zingone A, Saggiomo V. 2009. The role of platelet ice microalgae in seeding phytoplankton blooms in Terra Nova Bay (Ross Sea, Antarctica): a mesocosm experiment. *Polar Biol* **32**: 311–323.

McMahon KW, Ambrose WG Jr, Johnson BJ, Sun MY, Lopez GR, et al. 2006. Benthic community response to ice algae and phytoplankton in Ny Ålesund, Svalbard. *Mar Ecol Prog Ser* **310**: 1–14.

Meiners KM, Vancoppenolle M, Thanassekos S, Dieckmann GS, Thomas DN, et al. 2012. Chlorophyll *a* in Antarctic sea ice from historical ice core data. *Geophys Res Lett* **39**: L21602, doi:10.1029/2012GL053478

Mitchell BG. 1990. Algorithms for determining the absorption coefficient of aquatic particulates using the quantitative filter technique (QFT), *Proc International Soc Optical Engineers* **10**: 137–148.

Moisan TA, Mitchell BG 1999. Photophysiological acclimation of *Phaeocystis antarctica* Karsten under light limitation. *Limnol Oceanogr* **44**: 247–258.

Moisan TA, Mitchell BG. 2001. UV absorption by mycosporine-like amino acids in *Phaeocystis antarctica* Karsten induced by photosynthetically available radiation. *Mar Biol* **13**: 217–227.

Morel A, Bricaud A. 1981. Theoretical results concerning light absorption in a discrete medium, and application to specific absorption of phytoplankton. *Deep-Sea Res* **28**: 1375–1393.

Olaizola M, Yamamoto HY. 1994. Short-term response of the diadinoxanthin cycle and fluorescence yield to high irradiance in *Chaetoceros muelleri* (Bacillariophyceae). *J Phycol* **30**: 606–612.

Perovich DK. 1990. Theoretical estimates of light reflection and transmission by spatially complex and temporally varying sea ice covers. *J Geophys Res* **95**: 9557–9567.

Pineault S, Tremblay J-E, Gosselin M, Thomas H, Shadwick E. 2013. The isotopic signature of particulate organic C and N in bottom ice: Key influencing factors and applications for tracing the fate of ice-algae in the Arctic Ocean. *J Geophys Res* **118**: 287–300.

Platt T, Gallegos CL, Harrison WG. 1980. Photoinhibition of photosynthesis in natural assemblages of marine phytoplankton. *J Mar Res* **38**: 687–701.

Ratkova TN, Wassmann P. 2005. Sea ice algae in the White and Barents seas: composition and origin. *Polar Res* **24**: 95–110.

Rau GH, Sullivan CW, Gordon LI. 1991. $\delta^{13}C$ and $\delta^{15}N$ variations in Weddell Sea particulate organic matter. *Mar Chem* **35**: 355–369.

Riaux-Gobin C, Poulin M, Dieckmann G, Labrune C, Vétion G. 2011. Spring phytoplankton onset after the ice break-up and sea-ice signature (Adélie Land, East Antarctica). *Polar Res* **30**: 5910. doi:10.3402/polar.v30i0.5910

Riegger Land Robinson D. 1997. Photoinduction of UV-absorbing compounds in Antarctic diatoms and *Phaeocystis antarctica*. *Mar Ecol Prog Ser* **160**: 13–25.

Saenz B, Arrigo KR. 2012. Simulation of a sea ice ecosystem using a hybrid model for slush layer desalination. *J Geophys Res* **117**: C05007. doi:10.1029/2011JC007544

Saenz BT, Arrigo KR. Primary production in Antarctic sea ice from a sea ice state estimate. *J Geophys Res* (in press).

Sarmiento JL, Hughes TMC, Stouffer RJ, Manabe S. 1998. Simulated response of the ocean carbon cycle to anthropogenic climate warming. *Nature* **393**: 245–249, doi:10.1038/30455

Sinha RP, Klisch M, Helbling EW, Häder D-P. 2001. Induction of mycosporine-like amino acids (MAAs) in cyanobacteria by solar ultraviolet-B radiation. *J Photochem Photobiol B-Biol* **60**: 129–135.

Søreide JE, Leu E, Berge J, Graeve M, Falk-Petersen S. 2010. Timing of blooms, algal food quality and *Calanus glacialis* reproduction and growth in a changing Arctic. *Glob. Change Biol* **16**: 3154–3163.

Suzuki H, Sasaki H, Fukuchi M. 2001. Short-term variability in the flux of rapidly sinking particles in the Antarctic marginal ice zone. *Polar Biol* **24**: 697–705.

Tang KW, Smith WO, Shields AR, Elliott DT. 2009. Survival and recovery of *Phaeocystis antarctica* (Prymnesiophyceae) from prolonged darkness and freezing. *Proc Royal Soc B-Biol Sci* **276**(1654): 81–90. doi:10.1098/rspb.2008.0598

Teranes J, Bernasconi SM. 2000. The record of nitrate utilization and productivity limitation provided by $\delta^{15}N$ values in lake organic matter—A study of sediment trap and core sediments from Baldeggersee, Switzerland. *Limnol Oceanogr* **45**(4): 801–813.

Tortell PD, Mills MM, Payne CD, Maldonado MT, Chierici M, et al. 2013. Inorganic C utilization and C isotope fractionation by pelagic and sea ice algal assemblages along the Antarctic continental shelf. *Mar Ecol Prog Ser* **483**: 47–66.

Wadhams P, Lange MA, Ackley SF. 1987. The ice thickness distribution across the Atlantic sector of the Antarctic Ocean in midwinter. *J Geophys Res* **92**(C13): 14535–14552. doi:10.1029/JC092iC13p14535

Wing SR, McLeod RJ, Leichter JJ, Frew RD, Lamare MD. 2012. Sea ice microbial production supports Ross Sea benthic communities: influence of a small but stable subsidy. *Ecology* **93**: 314–323.

Wright SW, Jeffrey SW, Mantoura RFC, Llewellyn CA, Bjornland T, Repeta D, Welschmeyer N. 1991. Improved HPLC method for the analysis of chlorophylls and carotenoids from marine-phytoplankton. *Mar Ecol Prog Ser* **77**(2–3): 183–196. doi:10.3354/meps077183

## Contributions

- Contributed to conception and design: KRA
- Contributed to acquisition of data: ZWB, MMM
- Contributed to analysis and interpretation of data: KRA, ZWB, MMM
- Drafted and revised the article: KRA
- Approved and submitted version for publication: KRA

## Acknowledgments

We thank the captain and crew of the Swedish icebreaker *Oden* as well as S. Ackley and Katrina Abrahamsson (and their research teams) for their assistance during the cruise.

## Funding information

The research was supported by a National Science Foundation Office of Polar Programs grant to KR Arrigo (ANT-0838872).

## Competing interests

The authors have no competing interests, as defined by *Elementa*, that might be perceived to influence the research presented in this manuscript.

## Data accessibility statement

All of the data used in this study are included in Tables 1–4 of this manuscript.

# 3

# Methods for biogeochemical studies of sea ice: The state of the art, caveats, and recommendations

Lisa A. Miller[1]* • Francois Fripiat[2,3] • Brent G.T. Else[4,25] • Jeff S. Bowman[5] • Kristina A. Brown[6] • R. Eric Collins[7] • Marcela Ewert[5] • Agneta Fransson[8] • Michel Gosselin[9] • Delphine Lannuzel[10] • Klaus M. Meiners[11,12] • Christine Michel[13] • Jun Nishioka[14] • Daiki Nomura[14] • Stathys Papadimitriou[15] • Lynn M. Russell[16] • Lise Lotte Sørensen [17,18] • David N. Thomas [15,18,19] • Jean-Louis Tison[2] • Maria A. van Leeuwe[20] • Martin Vancoppenolle[21] • Eric W. Wolff [22] • Jiayun Zhou[23,24]

[1]Institute of Ocean Sciences, Fisheries and Oceans Canada, Sidney, British Columbia, Canada
[2]Laboratoire de Glaciologie, Université Libre de Bruxelles, Brussels, Belgium
[3]Analytical, Environmental and Geo-Chemistry, Earth Sciences Research Group, Vrije Universiteit Brussel, Brussels, Belgium
[4]Department of Geography, University of Calgary, Calgary, Alberta, Canada
[5]School of Oceanography, University of Washington, Seattle, Washington, United States
[6]Department of Earth, Ocean and Atmospheric Sciences, University of British Columbia, Vancouver, British Columbia, Canada
[7]School of Fisheries and Ocean Sciences, University of Alaska Fairbanks, Fairbanks, Alaska, United States
[8]Norwegian Polar Institute, Fram Centre, Tromsø, Norway
[9]Institut des sciences de la mer, Université du Québec à Rimouski, Rimouski, Quebec, Canada
[10]Institute for Marine and Antarctic Studies, University of Tasmania, IMAS–Sandy Bay, Hobart, Tasmania, Australia
[11]Australian Antarctic Division, Dept. of the Environment, Kingston, Tasmania, Australia
[12]Antarctic Climate and Ecosystems Cooperative Research Centre, University of Tasmania, Hobart, Tasmania, Australia
[13]Freshwater Institute, Fisheries and Oceans Canada, Winnipeg, Manitoba, Canada
[14]Institute of Low Temperature Science, Hokkaido University, Sapporo, Japan
[15]School of Ocean Sciences, Bangor University, Menai Bridge, Anglesey, United Kingdom
[16]Scripps Institution of Oceanography, La Jolla, California, United States
[17]Department of Environmental Science, Aarhus University, Roskilde, Denmark
[18]Arctic Research Centre, Aarhus University, Aarhus, Denmark
[19]Finnish Environment Institute (SYKE), Helsinki, Finland
[20]Laboratory of Plant Physiology, University of Groningen, Groningen, The Netherlands
[21]Laboratoire d'Océanographie et du Climat (LOCEAN-IPSL), Sorbonne Universités (UPMC Paris 6, CNRS, IRD, MNHN), Paris, France
[22]Department of Earth Sciences, University of Cambridge, Cambridge, United Kingdom
[23]Laboratoire de Glaciologie, Université Libre de Bruxelles, Brussels, Belgium
[24]Unité d'océanographie chimique, Université de Liège, Liège, Belgium
[25]Centre for Earth Observation Science, University of Manitoba, Winnipeg, MB, Canada

*lisa.miller@dfo-mpo.gc.ca

**Domain Editor-in-Chief**
Jody W. Deming, University of Washington

**Associate Editor**
Stephen F. Ackley, University of Texas at San Antonio

**Knowledge Domain**
Ocean Science

## Abstract

Over the past two decades, with recognition that the ocean's sea-ice cover is neither insensitive to climate change nor a barrier to light and matter, research in sea-ice biogeochemistry has accelerated significantly, bringing together a multi-disciplinary community from a variety of fields. This disciplinary diversity has contributed a wide range of methodological techniques and approaches to sea-ice studies, complicating comparisons of the results and the development of conceptual and numerical models to describe the important biogeochemical processes occurring in sea ice. Almost all chemical elements, compounds, and biogeochemical processes relevant to Earth system science are measured in sea ice, with published methods available for determining biomass, pigments, net community production, primary production, bacterial activity, macronutrients, numerous natural and anthropogenic organic compounds, trace elements, reactive and inert gases, sulfur species, the carbon dioxide system parameters, stable isotopes, and water-ice-atmosphere fluxes of gases, liquids, and solids. For most of these measurements, multiple sampling and processing techniques are available, but to date there has been little intercomparison or intercalibration between methods. In addition, researchers collect different types of ancillary data and document their samples differently, further confounding comparisons between studies. These problems are compounded by the heterogeneity of sea ice, in which even adjacent cores can have dramatically different biogeochemical compositions. We recommend that, in future investigations, researchers design their programs based on nested sampling patterns, collect a core suite of ancillary measurements, and employ a standard approach for sample identification and documentation. In addition, intercalibration exercises are most critically needed for measurements of biomass, primary production, nutrients, dissolved and particulate organic matter (including exopolymers), the $CO_2$ system, air-ice gas fluxes, and aerosol production. We also encourage the development of *in situ* probes robust enough for long-term deployment in sea ice, particularly for biological parameters, the $CO_2$ system, and other gases.

# 1. The rise of sea-ice biogeochemical studies

Sea ice covers up to 8% of the Earth's ocean surface (Steele et al., 2001), and despite global warming trends, both polar oceans are still mainly covered by sea ice in winter (Comiso, 2010) and likely will continue to be for the foreseeable future. The changes in sea-ice extent and physical structure associated with a warming climate (Perovich and Richter-Menge, 2009; Massom and Stammerjohn, 2010) are causing dramatic shifts in sea-ice ecosystems and the interactions between sea ice and both the atmosphere and the underlying waters. Long assumed to be a passive barrier to both light and matter, sea ice was relatively neglected in biogeochemical studies until the early 1990s. Since then, intensive research in the Arctic and Southern Oceans, as well as in subpolar seas, has shown that, in reality, sea ice is an active player in biogeochemical processes, making significant contributions to regional and possibly global cycles of many elements (*e.g.*, Arrigo et al., 2010; Deming, 2010; Thomas et al., 2010; Loose et al., 2011; Rysgaard et al., 2011; Vancoppenolle et al., 2013). Future changes in the sea-ice environment will be accompanied by changes to these biogeochemical cycles, generating an urgent need to better understand the chemical-physical-biological function of the ocean-ice-atmosphere system.

Although this review is focused on recent methodological developments in sea-ice biogeochemistry, the first formal biological studies of sea ice date back to the mid-19th century (Horner, 1985), and chemical studies extend back to the early 20th century (*e.g.*, Ringer, 1928; Wiese, 1930). Post-war studies of sea-ice chemistry were mainly motivated by efforts to understand sea-ice structural properties in support of cold-war military operations and potential industrial development in the polar regions (*e.g.*, Nelson and Thompson, 1954; Assur, 1958; Bennington, 1963; Tsurikov, 1965), but that work also provided extremely useful information on the geochemistry of ice brines. Incremental work continued at a slow pace for several decades, but with recent increases in access to polar regions and technological developments, as well as with the growing urgency in climate-change research (*e.g.*, Post et al., 2013) and a need to improve representation of sea-ice processes in numerical models at all spatial and temporal scales, the study of sea-ice biogeochemistry has expanded rapidly. Scientists have come to this field from a variety of disciplines, including glaciology, oceanography, sedimentology, and even tundra ecology; as we attempt to understand the complex biogeochemical processes occurring in sea ice, many creative modifications have been applied to methods not originally designed for sea-ice applications.

Sea ice presents a particularly challenging environment for biogeochemical studies (see Petrich and Eicken, 2010, for a comprehensive description of sea-ice types, characteristics, and life-cycles). Perhaps most significantly, the sea-ice environment is cold. Standard seawater with a salinity of 35 g kg$^{-1}$ freezes at −1.9 °C (Petrich and Eicken, 2010). Sea-ice temperatures below -30 °C have been measured (*e.g.*, Miller et al., 2011b), while the air temperatures above the ice, where people and instruments must operate, can drop to −60 °C or even lower. Standard oceanographic equipment, however, has been built to operate at temperatures only as low as 0 °C, while most chemical analyses and instrumentation are designed to operate between 20 and 25 °C. If it is not cold, *i.e.*, if the temperatures are around 0 °C, as occurs during the spring and autumn transition seasons, the sea ice is often very thin or deteriorating, making sampling dangerous. Additional difficulties arise from the heterogeneity of sea ice, which is a complex mixture of pure ice, solid salts and particulate organic matter, liquid brines, and gas bubbles. Even the boundary conditions are variable, with snow, frost flowers, and melt ponds at the top and skeletal ice, platelet ice, and algal mats at the bottom. In many ways, soils or sediments are a better conceptual model than seawater for describing sea-ice spatial variability, for multi-phase theories are required to describe sea-ice physical-chemical properties (*i.e.*, a "mushy layer"; Vancoppenolle et al., 2010; Hunke et al., 2011). Finally, the high and variable salinities of sea-ice brines compromise chemical analyses by complicating calibrations and corroding delicate instrument components.

Comprehensive guides for biological and physical methods in sea-ice research have already been published (Horner et al., 1992; Eicken et al., 2009; Michel and Niemi, 2009); we will not attempt to reiterate them here. Rather, we are accepting the challenge issued by Eicken et al. (2009) in their preface, wherein they hoped their book would "spark broader collaboration among sea-ice researchers to document and refine the best-practice approaches to sea-ice field studies." We discuss not only some remaining ambiguities in determining biomass, nutrient concentrations, and the rates of biological processes in sea ice, but also the challenges of quantifying gas concentrations and fluxes, aerosol emissions, trace metals and their chemical speciation, the complex inorganic carbon and organic sulfur systems, and the sticky problem of organic matter in sea ice, including the difficulties in defining and distinguishing between the dissolved, colloidal, and particulate fractions. We also make concrete recommendations for what ancillary physical data should be collected in conjunction with biogeochemical measurements, to allow effective interpretation of the resulting data sets, as well as for the most critical and potentially useful directions for future methodological developments. We do not intend this paper to be a stand-alone methods manual. Rather, we direct readers to the sources in which individual methods are described.

## 2. General considerations

Each element, compound, or process has a unique set of sampling and analytical requirements for accurate, precise, and useful results. The bulk of this paper addresses the specialized procedures for sampling and analyzing a wide range of parameters in sea ice. However, there are several generalities in sampling ice and its components that are worth discussing before tackling the specifics.

### 2.1. Patchiness and scaling

Sea-ice physical, chemical, and biological properties are highly variable, both temporally and spatially. During most seasons, sea ice is characterized by strong vertical gradients in temperature, brine salinity, habitable pore space, and permeability (Eicken, 1992; Petrich and Eicken, 2010; Vancoppenolle et al., 2013). Particularly during spring and, possibly, autumn, those gradients can change rapidly on daily or even hourly timescales (e.g., Mundy et al., 2005; Nomura et al., 2010a). Horizontal variability is also extremely high; for example, biomass can vary by an order of magnitude on the sub-meter scale (Spindler and Dieckmann, 1986; Steffens et al., 2006). Horizontal patchiness of ice algae has been mainly attributed to the spatial variability in physical sea-ice properties (Eicken et al., 1991) and light exposure (Raymond et al., 2009), which are affected by ice formation processes, parent seawater salinity, and meteorological events, among other factors (Gosselin et al., 1986; Rysgaard et al., 2001; Granskog et al., 2005a; Fritsen et al., 2011). This sea-ice heterogeneity has consequences for sampling design: not only do the goals of the project dictate the type of ice that should be sampled (i.e., first-year, multi-year, smooth, ridged, young, melting, etc.), but the most appropriate sampling scheme and the minimum number of samples required will depend on the representativeness of any single sample.

The scales of horizontal spatial variability in biological parameters in different sea-ice regimes have been investigated using transects and nested equilateral triangle sampling patterns combined with parametric and non-parametric statistical analysis techniques (e.g., Gosselin et al., 1986; Swadling et al., 1997; Granskog et al., 2005a; Steffens et al., 2006; Søgaard et al., 2010, 2013). We recommend that, to the extent possible and relevant to the specific study, researchers design their sampling using a nested approach that facilitates extrapolation of detailed information to larger scales by distinguishing hierarchical layers of detail (Figure 1). In a nested sampling regime, the primary scale defines the study area, in all its variation, and the secondary scale serves to determine the representativeness of each site within the study area. The tertiary scale (i.e., the number of individual sampling sites), along with the number of replicates (the quaternary scale), defines the accuracy of a parameter.

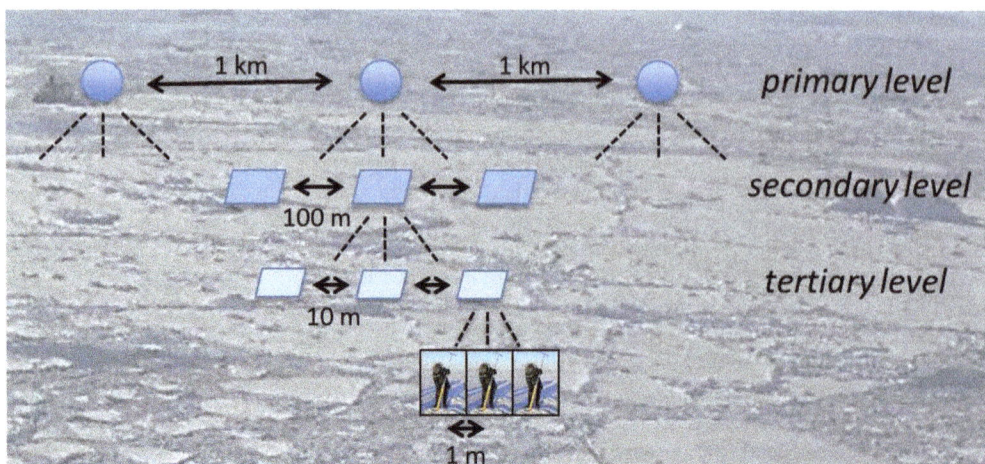

Figure 1

Hierarchical sea-ice sampling design.

Photos: J. Stefels; D. Leitch.

Designing a nested sampling program begins with a visual survey of the sampling area to establish the various spatial scales of the variability and determine how many sites and samples are required at each hierarchical level. There is no simple, universal algorithm that can be used; every sampling site must be assessed in relation to the goals and resources of the project. For example, Sturm (2009) provides recommendations for sampling densities required to minimize errors in mean snow depth (see his Figure 3.1.13), but the ease of snow-depth measurements allows for a large number of data points that are not always practical for studies of biogeochemical properties.

In addition, the practical and safety requirements associated with ice coring unavoidably produce a bias towards stable ice floes with low levels of deformation. The true scale of the horizontal variability can be determined more accurately by combining methods that provide information on varying scales (*e.g.*, ice coring surveys combined with optical investigations from underwater platforms; Williams et al., 2013). State-of-the-art methods to determine floe-scale sea-ice physical properties, such as sea-ice surface elevation, snow thickness, freeboard, and sea-ice draft, need to be linked with new methods in sea-ice ecology (*e.g.*, determination of ice algal biomass from transmitted under-ice irradiance spectra; Mundy et al., 2007) to evaluate coupled physical-biological sea-ice processes on relevant scales. Developing ice buoy networks and autonomous underwater vehicle technology, combined with improved physical (*e.g.*, CTDs, upward looking sonars), chemical (*e.g.*, oxygen and pH sensors), and bio-optical (*e.g.*, hyperspectral radiometers, fluorometers) sensors, is an active field of research and promises a step-change in our understanding of horizontal patchiness and physical-chemical controls over biological properties in sea ice.

## 2.2 Sampling techniques and considerations

In general, because sea ice is highly heterogeneous (section 2.1), collecting different samples for analyses of different parameters makes it difficult to confidently link the biogeochemistry of those parameters. Therefore, to the extent possible, each individual sample should be analyzed for as many parameters as is feasible. However, this ideal goal is severely constrained by realities of required sample volumes for analyses, incompatible sample processing requirements, and vulnerabilities to different contamination sources. Here we summarize the basic techniques to collect samples from sea-ice environments, noting that the best approach may vary depending on the project goals and the sea-ice conditions. With this in mind, we have also included what we feel are important considerations to take into account when sampling each particular medium.

### 2.2.1 Bulk ice

A standard approach to processing sea-ice samples is to collect a core and divide it into sections using a clean stainless steel saw, depending on the scientific question and the demands of analytical sample volume and of collaborative cooperation between researchers (Figure 2). The core sections are then melted and analyzed as any other aquatic samples. The thickness of the core sections may vary, according to the needs of the project, but parameters to be compared should be analyzed on sections of comparable thickness. Cores are usually sectioned from the bottom, to limit brine loss from more permeable parts of the ice (the vertical extent of

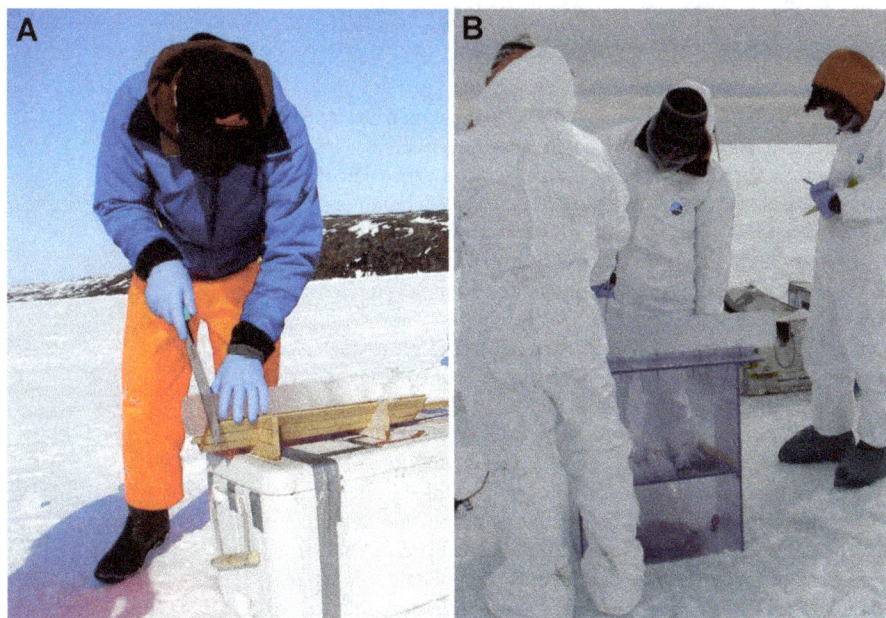

**Figure 2**

**Sectioning ice cores in the field.**

(a) Organic-clean methods off Nuuk, Greenland, April 2013. Photo: N.-X. Geilfus. (b) Trace metal-clean methods on McMurdo Sound, November 2012. Photo: T. Goossens.

Figure 3
Sampling young ice.

Using a ship's basket (a) and a flat-bottomed b oat ( b) i n t he Beaufort Sea, October 2003. Photos: M. Poulin, J. Ehn.

which depends on ice temperature, salinity, and texture), although the individual sections should be identified according to their depth from the top of the ice (see section 2.3). Core extraction invariably results in at least some brine loss (*e.g.*, Notz et al., 2005), contributing to losses of both dissolved and particulate matter, and therefore, cores must be sampled and processed quickly, preferably sectioned into melt containers in the field. Divers have collected sea ice from below in attempts to more quantitatively recover the components of bottom ice (*e.g.*, Welch et al., 1988; Horner et al., 1992; McMinn and Hegseth, 2007). In addition, although analyses of melted ice cores are most common, many interesting analytes and processes (such as microorganism abundances and species compositions, metabolic rates, gas partitioning, salt precipitation and dissolution, *etc.*) are strongly affected by the drastic changes in temperature and salinity that result when sea ice melts (see section 3.2, below); more complex sampling and analysis methods are often required to avoid melting samples.

The early stages of ice formation play a fundamental role in partitioning material and organisms between the atmosphere, ice, and underlying water (*e.g.*, Giannelli et al., 2001; Notz and Worster, 2009; Müller et al., 2013), but sampling young ice that cannot bear a load (*i.e.*, frazil, grease, nilas, or pancake ice) requires special safety considerations. Thin, broken ice (brash ice and small pancakes) can be sampled from a dinghy (*e.g.*, Grossmann and Dieckmann, 1994) or with a bucket or basket lowered from the side of the ship (*e.g.*, Gradinger and Ikävalko, 1998). Intact, young ice sheets can be cored or cut directly by researchers from a ship's basket (Figure 3a) or using a flat-bottomed boat (Figure 3b). Young, poorly consolidated ice samples contain high quantities of interstitial water and brines that are easily lost, which makes it important to record

**Figure 4**

**Sampling sea-ice brines.**

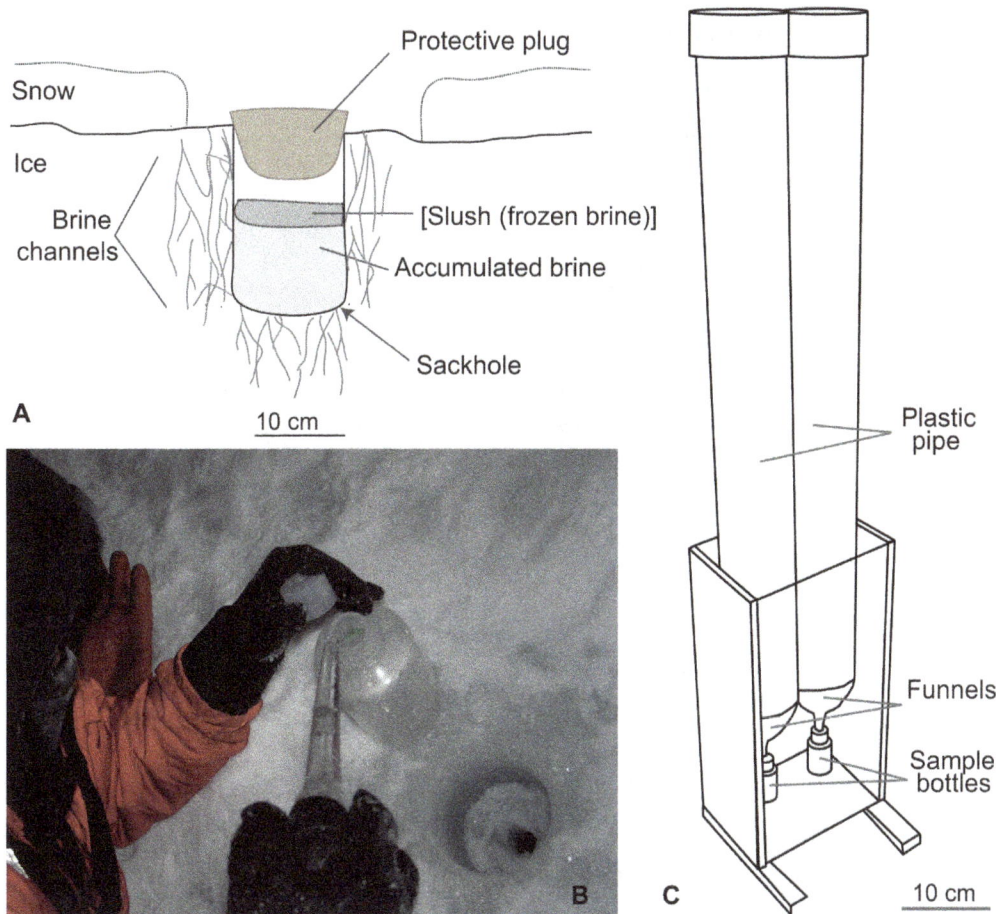

(a) Schematic diagram of a sackhole for sampling sea-ice brines (slush on top of collected brine not always present). (b) Sampling sea-ice brines accumulated overnight in a sackhole using a clean baster to reach the bottom of the hole. Photo: M. Ewert. (c) Apparatus for collecting brines from whole cores by gravity drainage.

whether young ice samples were drained to separate the pore water or if special care was taken to retain the brines (Grossmann and Dieckmann, 1994). Cottier et al. (1999) and Smedsrud and Skogseth (2006) described specialized tools for collecting frazil and young sea ice. High-salinity brine waters surrounding frazil and brash ice as it is forming can also be collected using open-mouth jars covered with mesh (Miller et al., 2011a). Similarly, Kristiansen et al. (1998) sampled the infiltration community at the snow-ice interface of flooded ice floes by immediately filtering the slush, at low temperature, through a 200 μm net to remove ice crystals.

Thick, ridged sea ice is even more severely undersampled than young, thin ice, due to a number of significant difficulties with travelling over and through ridged ice, accessing ridge keels and sails, and collecting representative samples from such a varied environment (see section 2.1). However, ridged ice constitutes a large fraction of the total area and volume of sea ice and, according to the limited data available, represents an important biomass pool (Gradinger et al., 2010; Meiners et al., 2012). The methods used for sampling thick, deformed ice have been largely *ad hoc*; standardized approaches are needed.

### 2.2.2 Brines

The fractionation between the solid (ice and particulates) and liquid (brine) phases of the ice is often important to understanding sea-ice biogeochemical cycles, but effectively collecting and accurately analyzing representative brine samples from mature sea ice has been an exceptional challenge. Traditionally, sea-ice brines have been collected by drilling sackholes (partial core holes) to a desired depth within the ice and then allowing brines from the surrounding ice to drain into the hole (Figure 4a). The most obvious problem with this approach is that, because of the three-dimensional structure and variable connectivity of the brine network, the sackhole brine integrates the geochemical properties of numerous individual brine channels from an undefined volume of the sea ice surrounding the hole. Therefore, sackhole brines provide data on the vertical and horizontal macro-scale (over several tens of centimeters, depending on the depth of the sackhole); our ability to obtain data on the brines at the micro-scale level (a few centimeters or less) in sea ice is still limited for the majority of solutes of biogeochemical interest. Also, if the ice is warm and highly permeable, the collected brines can be contaminated with upward seeping seawater or draining snowmelt and meltpond water; the significance of this problem can be assessed by comparing the measured brine salinity to that estimated from the *in situ* temperature profile. Conversely, the colder the ice, the less brine it contains, the less connected the brine network is, and the longer it takes for a sufficient volume of brine to accumulate in the sackhole. At very

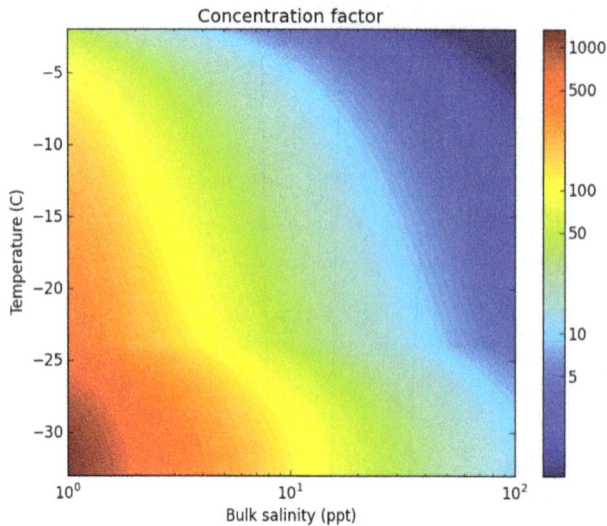

Figure 5

Theoretical concentration factors
in sea-ice brines.

Concentration factors (equivalent
to $S_B$:$S_I$; Equation 1) modelled as
the inverse of the brine volume
fraction for temperatures between
−2 and −35 °C and brine salinities
between 1 and 200, encompassing
the range observed in sea-ice
environments, including frost
flowers. Brine volume fraction was
calculated according to Equation
2.6 of Petrich and Eicken (2010).
This calculation represents the ideal
case, where the solute is entirely
partitioned into the brine phase and
does not precipitate. Experimental
measurements of brine salinity
between -2 and -23 °C show a
similar concentration factor for
salts, although measured brine
salinities deviate from the ideal case
at the lowest temperatures, due to
salt precipitation.

cold temperatures, it may not be possible to collect enough brine within a suitable timeframe. Furthermore, the longer it takes for the brines to accumulate, the greater the risk that the sample will be compromised by interaction with the air, which is likely at a different temperature (often much lower) than the ice interior, causing the brines to freeze further within the sackhole. Potential brine-air exchange is a particular problem for analyses of insoluble gases (section 4.4). To minimize such air-brine interactions, the sackhole can be capped with plugs made from thick insulated material. When sampling sackholes, any snow needs to be removed from the sampling site before coring, and care is required to prevent ice core shavings from entering the hole and contaminating the brines. The accumulated brine can be extracted from the sackhole with a large pipette (a "turkey baster"; Figure 4b), with tubing attached to a syringe or a peristaltic pump, or simply by dipping a bottle into the brine.

Other methods used to collect sea-ice brines include gravity-draining full cores into containers (Figure 4c; Nomura et al., 2009) or crushing and/or centrifuging ice samples (e.g., Grossmann and Dieckmann, 1994; McMinn et al., 2009; Munro et al., 2010). However, both of those approaches generally deliver relatively small volumes that are suitable for only a limited number of analyses and may not be statistically representative (see section 2.1). Also, the pressure generated by centrifugation can melt some of the ice, diluting the extracted brines (Papadimitriou et al., 2004), apparently by as much as 15%.

Particulates (including organisms) and organic matter appear to be under-represented in brines collected in sackholes (e.g., Weissenberger, 1992; Sime-Ngando et al., 1997; Lannuzel et al., 2008) by up to 98% (Becquevort et al., 2009). Likely explanations include preferential adsorption onto the ice walls, "filtration" by the brine channel network, and impeded transport by sticky, gelatinous exopolymeric substances (EPS) or aggregates of ice algae (Meiners et al., 2004; Krembs et al., 2011).

Concentrations of dissolved materials in brines can also be estimated from calculated brine salinity (based on in situ temperature; Cox and Weeks, 1983; Petrich and Eicken, 2010) and measured bulk ice concentrations (Figure 5), using the equation

$$C_B = C_I \left( \frac{S_B}{S_I} \right), \tag{1}$$

where C indicates analyte concentration, S is salinity, and the subscripts B and I represent brine and bulk ice (e.g., Dieckmann et al., 1991a; Norman et al., 2011). This calculation assumes that none of the analyte is in solid or gaseous form and that, therefore, the concentration of the analyte in the brine is directly proportional to salinity, an assumption that is probably only valid for highly soluble substances. Not only do supersaturated brines precipitate salts (even the highly soluble ions $Cl^-$ and $Na^+$ precipitate from sea-ice brine solutions to a significant extent at temperatures below −24 °C; Assur, 1958) and release gas bubbles, but both dissolved and particulate organic matter can adsorb onto the surface of the brine channels. In addition, high concentrations of organic matter, particularly EPS (section 4.2.3), can impact sea-ice microstructure (Krembs et al., 2011), with hypothesized (Ewert and Deming, 2013) but still unknown implications for the sea-ice equations of state (Cox and Weeks, 1983).

### 2.2.3 Gases
The third phase in sea ice, gas bubbles, are particularly important in carbon, oxygen, and sulfur cycling, but gas inclusions are even more challenging to recover and analyze than brines. The methods that have been developed to tackle this problem are summarized in section 4.4.

Figure 6
Sampling frost-flowers with a clean spatula.

Beaufort Sea, January 2008. Note that this sampling method does not separate frost flowers from the underlying brine skim layer. Photo: M. Ewert.

### 2.2.4 Snow and frost flowers

On top of sea ice, the snow cover thermally insulates the ice, limits the penetration of visible and UV radiation, and exchanges brines and gases with the underlying ice (*e.g.*, Kelley and Gosink, 1985; Massom et al., 2001; Zemmelink et al., 2008; Ewert et al., 2013). The snow cover is generally sampled in layers by excavating snow pits, as described by Sturm (2009). Established sampling methods for other environments can usually be adapted to snow over sea ice. For example, clean methods used for sampling trace elements in snow over land have been successfully implemented for sampling of halogens and trace elements in snow over sea ice (Simpson et al., 2005, Poulain et al., 2007).

Frost flowers are usually collected by simply scraping or scooping them into sample containers (Figure 6; Obbard et al., 2009; Bowman and Deming, 2010; Miller et al., 2011a; Aslam et al., 2012; Douglas et al., 2012; Bowman et al., 2013; Fransson et al., 2013; Granfors et al., 2013a), although Alvarez-Aviles et al. (2008) used tweezers. Brine skims on the top of the ice (often associated with frost flowers) can be collected with a scooped spatula (Bowman and Deming, 2010; Roscoe et al., 2011) or an eyedropper (Alvarez-Aviles et al., 2008). However, exclusive sampling of frost flowers separately from brine skims remains a challenge, nor do current approaches likely adequately capture volatile components. Effective study of these sea-ice micro-environments can benefit from application of non-invasive techniques, such as infrared imaging (Barber et al., 2014), and from development of sensitive but robust microsensors for *in situ* analyses.

### *2.3. Record-keeping*

Sea-ice scientists still use non-standardized, *ad hoc* systems for identifying samples, but if we are to establish comprehensive, accessible, and useful sea-ice biogeochemical databases, then instituting basic standards to classify samples will be necessary. These metadata requirements apply not only to observations at the time of sampling, but also to sample processing and preparing data sets for archiving. In an effort to begin to meet the need for standardized record-keeping, we encourage researchers to always identify their samples with the following information:

- latitude and longitude;
- date and time (preferably UTC, but if local time is used, clearly identify the time zone and whether it is under Summer, or Daylight Savings, Time);
- weather conditions, including air temperature, cloud fraction, wind speed and direction, and contact information for complete meteorological data from a nearby ship or station;
- water depth, particularly in coastal waters;
- ice description, including approximate age (*i.e.*, whether it is multi-year or first-year ice, and for young ice, estimates of time since initial ice formation), thickness, freeboard, and texture (according to Eicken et al., 2009);
- if applicable, depth in the ice core, measured from the top, in cm; and
- if applicable, estimates of snow or melt-pond coverage and measurements of their depths on the surface of the ice.

For each core, a standard data sheet should be prepared, giving the important metadata, information on how the core was processed, and what subsamples were taken for what analyses (Figure 7); eventually these sheets should be also populated with the analysis results for archiving. As it is usually best to collect several ice samples or cores, the distance between samples or the approximate total area from which the samples originated must also be specified in the metadata. If possible, replicates should be collected according to the guidelines for nested sampling design in section 2.1.

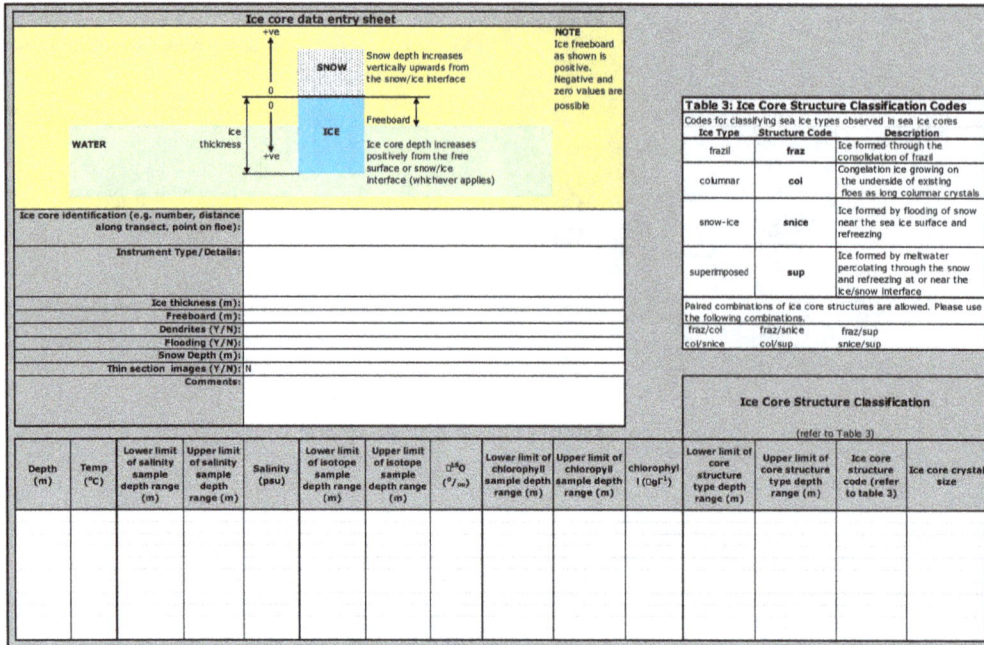

**Figure 7**

**Example ice core data sheet.**

Developed by the Antarctic Sea Ice Processes & Climate (ASPeCt) program. Additional data columns should be added for all project-specific parameters. This and other standardized sea-ice data entry templates are available for download via the sea-ice physical data portal of the Australian Antarctic Data Centre.

To facilitate data archiving and retrieval, we suggest that each sample be identified (at least in data files, if not also on the samples themselves) by an expedition code, followed by the date in YYYYMMDD format (with a lower-case letter for each sampling 'event' on a given day, associated with a unique time and location), followed by an identifier of the sample type (*i.e.*, "ice" for sea-ice cores, "br" for sackhole brines, "ff" for frost flowers, "sn" for snow, "gap" for gap layers, and "pond" for melt ponds), with each replicate core or sample receiving a different sequential number. Individual core sections should then be identified by the depth from the air-ice or snow-ice interface. For example, a specific core section from an Antarctic sea-ice camp in the year 2015 might be identified as BS2015-20150409b-ice-03-20-30, where BS2015 would be the expedition code, followed by the date (April 9th) and "b" indicating that it was the second location sampled that day, the 3rd replicate core from that location, and the 20–30 cm section down in that core.

## 2.4. Ancillary measurements

Any biogeochemical study of sea ice should include a number of ancillary measurements (Table 1). Beyond the needs for working up and interpreting our own results, when archived data are used by later researchers investigating questions we have not yet conceived, the ancillary data may prove critical.

**Table 1. Ancillary measurements for sea-ice biogeochemical studies**

| Importance | Parameter |
|---|---|
| Required | Temperature[a] |
| | Bulk salinity[a] |
| Strongly recommended | Brine salinity[a] |
| | Ice texture[a] |
| | $\delta^{18}O$[b] |
| | Snow thickness[a] |
| Recommended | Macronutrients[b] |
| | Chlorophyll $a$[b] |
| | Brine volume[a] |
| | Snow biogeochemistry[b] |
| | Radiative forcing fluxes (light and heat)[a] |

[a]Methods given by Eicken et al., 2009.

[b]Methods reviewed in this paper.

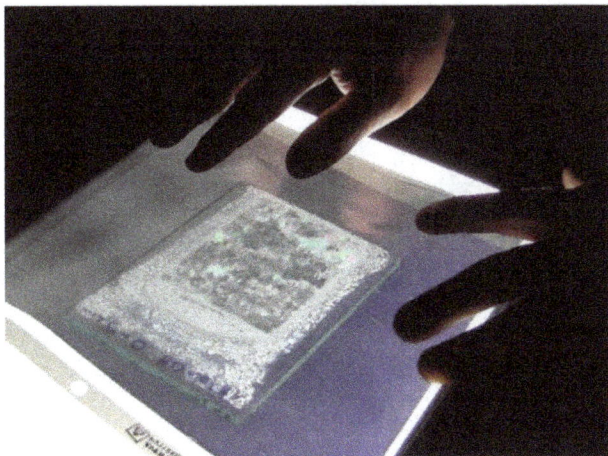

Figure 8

Sea-ice texture observed under a polarized filter in a thin slice of ice.

Cut lengthwise from an ice core section in a -20 °C cold room. Photo: M. Ewert.

Most importantly, the physical properties of the ice should be described in as much detail as possible (see Eicken et al., 2009, for standard methods). At a minimum, the *in situ* ice temperature and bulk salinity should be determined and reported, preferably in 5-cm vertical increments throughout the depth of the ice. Indeed, bulk salinity is sufficiently important (and simple enough to measure) that it should be determined on all sections analyzed for any biogeochemical parameter.

We also encourage sea-ice investigators to report brine salinity, which can be calculated from temperature (assuming thermodynamic equilibrium; Petrich and Eicken, 2010), and the brine volume as a fraction of the bulk sea-ice volume, which provides insight into the connectivity of the brine channel network (Golden et al., 1998). Estimating the brine volume fraction (Cox and Weeks, 1983, for $T < -2°C$; Leppäranta and Manninen, 1988, for $T > -2°C$) requires that in addition to temperature and bulk salinity, the ice density be measured (Eicken, 2009) or estimated based on an assumed air volume content. It is also helpful to describe the ice texture throughout the full profile (Eicken, 2009), at least based on a direct visual inspection of the thicknesses and sequence of granular and columnar layers (which can be recorded with photographs). A polarized light source, if available, can provide more detailed information (Figure 8). The $\delta^{18}O$ ratio of the sea ice (section 4.8) helps identify its meteoric versus marine origins (*e.g.*, Macdonald et al., 1999; Tison et al., 2008). Finally, macronutrient (nitrate, nitrite, phosphate, silicic acid, ammonium; section 4.1) and chlorophyll *a* (Chl *a*; section 3.3.2) measurements, in combination, allow us to assess the overall status of the biological community in the ice (*i.e.*, bloom versus non-bloom states; degree of nutrient limitation, *etc.*).

At a minimum, the presence of a snow cover and its thickness need to be reported. If resources allow, the snow cover should be fully characterized, as described by Sturm (2009), including basic stratigraphic analysis, with measurements of thickness, temperature, density, hardness, and grain properties (shape and size) for each layer. Snow samples from different strata should be also collected and analyzed for $\delta^{18}O$ (for insight into brine exchange with the underlying sea ice), as well as for the same analyses as those planned for the ice samples. More specialized information on atmospheric forcing, both light and heat, is also valuable, and standard methods are given by Perovich (2009).

# 3. Biological factors

Sea ice contains an abundant and taxonomically diverse community, including Bacteria and Archaea (hereafter referred to jointly as bacteria), autotrophic algae, hetero- and mixo-trophic protists, fungi, and metazoans. Algal cell concentrations in bulk sea ice range over six orders of magnitude (from $10^1$ to $10^6$ cells/mL; Arrigo et al., 2010), whereas those of bacteria range over four orders of magnitude (from $10^3$ to $10^7$ cells/mL; Deming, 2010) and those of viruses over three orders of magnitude (from $10^5$ to $10^8$ viral particles/mL; Deming, 2010). In order to understand and budget the exchanges of matter and energy within this rich sea-ice microbial community and between it and the surrounding environment, including the atmosphere, the water column, and sediments (*i.e.*, the paleorecord), it is essential to have good measurements of biomass, its chemical composition (C, N, P, Si, *etc.*), taxonomic composition, and metabolic functions and transformation rates. Such measurements are challenging in any environment, and particularly so in sea ice.

## 3.1. Sampling

Sampling considerations for determining the biological properties of sea ice depend on the parameter, organism of interest, and type of ice sampled. When sampling for microorganisms, especially bacteria, application of sterile aseptic technique, to the extent possible in the field, is needed; *e.g.*, the use of ethanol-sterilized tools for coring, cutting or otherwise collecting the samples, and sterile receptacles (*e.g.*, Bowman et al., 2012). Tools must be ethanol-cleaned between samples to prevent cross-contamination. Samples should be kept as close to the *in situ* temperature as possible, to prevent thermal or osmotic shock to the microorganisms, and light exposure needs to be minimized during transport and storage to limit growth, light shock, and photochemical degradation.

When the planned analysis requires full, intact ice cores, the cores should be placed into sterilized, black plastic sleeves, to retain brines and protect organisms from light stress, immediately after extraction from the core barrel (*e.g.*, Maas et al., 2012). Typically, however, the core is divided into sections to determine detailed vertical profiles (section 2.2.1). When cutting ice cores into sections, contamination may be prevented by placing them on autoclaved foil (Bowman et al., 2012) or, as when sampling for organic compounds (section 4.2.1), removing the outer layers of the cores (Song et al., 2011; Fripiat et al., 2014a). If the bottom ice community is of interest, the core must be handled gently to preserve the lower skeletal layer. Dieckmann et al. (1992) developed an apparatus for sampling unconsolidated platelet layers that can be deployed through a hole in the ice only 5 cm in diameter and was successfully deployed by Arrigo et al. (1995) and Robinson et al. (1998) to study the biogeochemistry and photophysiology of the platelet ice community. When large sample volumes are required, sections from different cores can be pooled during melting, although information on the horizontal heterogeneity (section 2.1) is lost.

## 3.2. Sample processing

Ideally, we would study the sea-ice biological community *in situ* in order to fully understand the relationship between the organisms and their environment (Junge et al., 2001; Krembs et al., 2002). However, most standard methods for evaluating biomass and activity are unable to accommodate the bulk ice matrix; investigators usually melt ice samples before examining the biological community. Unfortunately, large temperature changes or osmotic stress, as the salinity drops during ice melt, can cause sympagic cells to burst, a significant concern in biological sea-ice studies. In addition, even if cells remain intact, photosynthetic stress has been observed in sea-ice algae subjected to dramatic salinity decreases (Ralph et al., 2007). Therefore, samples should generally be processed in the cold (*i.e.*, at or only slightly above the freezing point of the final melt solution) to limit temperature changes (Deming, 2010; Mikkelsen and Witkowski, 2010), but limiting osmotic shock is more difficult. A variety of methods are used to melt sea ice, with a range of potential impacts on the sea-ice community. Available protocols include simply melting the ice (direct melt), melting in filtered or artificial seawater (seawater melt), and melting in concentrated brine to give a final salinity similar to the *in situ* brine salinity (brine, or isohaline, melt). Use of direct and seawater melts in samples collected for analyses of nutrients, organics, extracellular polysaccharides, and sulfur species is discussed in sections 4.1, 4.2.1, 4.2.3, and 4.5, respectively.

Although some studies have observed no difference between direct and seawater melts for Chl *a* measurements (Dieckmann et al., 1998; Kaartokallio, 2004) or diatom counts (Mikkelsen and Witkowski, 2010), or for the culturable fraction of sea-ice bacteria (Helmke and Weyland, 1995), direct melting can cause loss of anywhere from 13 to 97% of eukaryotic cells, with ciliates and flagellates most susceptible to bursting (Garrison and Buck, 1986; Mikkelsen and Witkowski, 2010). However, Mikkelsen and Witkowski (2010) found that slow, direct melting under refrigerated conditions appeared to be suitable for most of the eukaryotic cells, with no significant differences from seawater-buffered melts, except for one flagellate group. Winter ice, which can contain sharp vertical gradients in brine salinity, varying from salinities similar to seawater (approximately 35 g kg$^{-1}$) at the bottom to over 200 g kg$^{-1}$ near the upper surface, requires particular care in melting. In a comparison of direct and brine melts for different sections of the winter ice column, Ewert et al. (2013) found that direct melts resulted in a significant loss of up to 55% of bacterial cells in the upper ice column but no difference in the lower sections, indicating that *in situ* brine salinity is a key factor in selecting the appropriate melting method. An important caveat to melting under refrigerated or buffered conditions is that it can take several days (Mikkelsen and Witkowski, 2010), during which ongoing biological processes can modify the sample and its community structure in ways that are still difficult to assess.

As a general recommendation, the choice of melt protocol should be based on the target measurement and the expected change in salinity resulting from the melting process. Although isohaline melts often appear to be the best approach to melting sea-ice samples for analyses of biological parameters, isohaline melting at low temperature is time consuming, which may introduce artifacts into the analyses (even while protecting against others). In addition, adding seawater or brines (either natural or artificial) can dilute or contaminate the sample, both in terms of organisms and analytes. Therefore, usually the concentrations of relevant analytes (such as macronutrients) need to be quantified in those seawater or brine solutions, which in turn must be pretreated to remove or inactivate contaminants (*e.g.*, by filtration or UV oxidation). Finally, whereas the loss of a specific group of organisms can unacceptably bias the community composition, it may not necessarily impact total biomass determinations (POC, PON, bSiO$_2$, Chl *a*) to a significant extent, particularly if the community is dominated by diatoms.

### 3.3. Biomass and community structure

#### 3.3.1. Particulate organic matter and biominerals in sea ice

Particulate organic matter, which includes both living and non-living material, has been measured in sea-ice melts at concentrations up to $10^3$ µmol C $L^{-1}$ (*e.g.*, Gradinger, 1999; Kennedy et al., 2002). Such maximum concentrations are up to two orders of magnitude higher than in the surface open ocean (Martiny et al., 2013), confirming that sea ice is an important pool of biogenic organic matter in the polar oceans.

The differentiation between the particulate and dissolved phases in sea-ice melts and brines is arbitrary and based on the filters used, a problem in all aquatic biogeochemical studies (see Hilmer and Bate, 1989; Knefelkamp et al., 2007, and Wang et al., 2007, for reviews of the best practices for filtration in aquatic science). At the molecular level, the distinction between dissolved, colloidal, and particulate material is physically ambiguous and variable, particularly at high concentrations, such as in sea-ice brines. Also, the filtration process, itself, can cause particulate organic matter to break apart, while dissolved macromolecules can adsorb onto the filters (*e.g.*, Wangersky, 1993). This latter effect can be a particular problem for sea-ice samples with high concentrations of EPS that might clog the filters and reduce their effective pore size. Most but not all bacteria are captured by 0.2 µm pore-size filters; capturing viruses requires filters of even smaller pore size (0.02 µm; *e.g.*, Wells and Deming, 2006). In practice, many researchers rely on the convenience of glass and quartz fiber filters (*e.g.*, GF/F, 0.7 µm nominal pore-size, pre-combusted at high temperatures to remove organic contamination), although they, of course, are not suitable for analyses of biogenic silica (see below).

After melting (section 3.2) and filtration, the C and N contents of the particulate organic matter (particulate organic carbon, POC, and nitrogen, PON) in sea-ice samples are generally analyzed by combustion to $CO_2$ and $N_2$, the standard method used in seawater (*e.g.*, Ehrhardt and Koeve, 1999). Although studies of Chl *a* in sea ice have found no difference between cores melted with or without filtered seawater (section 3.2), to our knowledge, similar confirmation that melt procedures do not impact POC and PON analyses has not yet been published.

Measurements of cellular abundance in sea ice can be reported in different units for different purposes (Horner et al., 1992). For bulk macroscale analyses important in biogeochemical modelling, depth-integrated abundances (*i.e.*, cells $m^{-2}$) can be useful, as can further converting cellular abundance to biomass (typically in mmol C $m^{-2}$), either by direct measurement of elemental composition or by using a published conversion factor (*e.g.*, Miller et al., 2011b). For insight into the ecology of the organisms, results are sometimes scaled to the *in situ* brine volume (*e.g.*, cells $mL^{-1}$ brine; Junge et al., 2004; Wells and Deming, 2006; Collins et al., 2008).

Diatoms are the only ecologically significant group in sea ice producing biogenic silica ($bSiO_2$), which is filtered from ice melts using polycarbonate membranes (Fripiat et al., 2007). Analysis generally follows a double/single wet-alkaline digestion method (Ragueneau et al., 2005) to also assess lithogenic contamination, which could be an issue in landfast sea ice. Biogenic calcium carbonate ($CaCO_3$), in the form of foraminifera, has been observed and quantified by visual counting in Southern Ocean sea ice (Spindler and Dieckmann, 1986; Dieckmann et al., 1991b; Eicken et al., 1991; Thomas et al., 1998). The analysis of abiotic $CaCO_3$ minerals is discussed in section 4.6.5.

#### 3.3.2. Ice algal pigments and absorption spectra

Algal pigments provide both quantitative and qualitative information on the composition of the sea-ice community over a variety of temporal and spatial scales; Chl *a*, a ubiquitous pigment in algae and phytoplankton, is the most commonly used proxy of viable algal biomass in sea ice (*e.g.*, Dieckmann et al., 1998; Meiners et al., 2012). Pigments are collected by filtering melted ice core sections in the dark using either GF/F or polycarbonate membrane filters; in general, GF/F filters capture nearly all of the algae and are compatible with standard pigment extraction methods (Mantoura et al., 1997; Roy et al., 2011). Investigators usually melt the ice in filtered seawater to prevent cell rupture (*e.g.*, Garrison and Buck, 1986; Becquevort et al., 2009), although some have found no significant difference in Chl *a* concentrations between samples melted with and without filtered seawater (section 3.2). Exposure to light can affect the algal pigment composition; therefore, melting should always take place in the dark.

Chlorophyll *a* is usually determined by fluorometric detection following an acetone/methanol extraction (*e.g.*, Arar and Collins, 1997; Gosselin et al., 1997). Many investigators apply a constant ratio to convert Chl *a* to carbon biomass, but in sea ice (as well as in other environments), this ratio is highly variable among photosynthetic organisms, as well as with changes in light, temperature, and nutrient concentrations (*e.g.*, Arrigo et al., 2010). Therefore, although POC is a more expensive analysis than Chl *a*, sea-ice investigations benefit from parallel measurements of POC and Chl *a*.

High Performance Liquid Chromatography (HPLC) allows simultaneous analyses of a number of other pigments, in addition to Chl *a*. Standard methods are presented by Bidigare et al. (2005) and Roy et al. (2011). The large variations in light conditions throughout the sea-ice column and horizontal heterogeneity (section 2.1) strongly affect cellular pigment contents, so that special care is required in implementing the

Table 2. Extraction of nucleic acids (DNA or RNA) from sea ice for sequencing

| DNA or RNA | Extraction method[a] | Season | Sea ice types[b] | References |
|---|---|---|---|---|
| DNA | PC | Spring/summer | FYI and MYI | Brown and Bowman, 2001 |
| DNA | DNeasy | Summer/autumn/winter | FYI and MYI | Brinkmeyer et al., 2003 |
| DNA | DNeasy | Summer | MYI | Gerdes et al., 2005 |
| DNA | PC | Winter/spring | FYI | Brakstad et al., 2008 |
| DNA | PC | Winter/spring | FYI | Kaartokallio et al., 2008 |
| DNA | PC | Winter | FYI | Collins et al., 2010 |
| Both | PC, RNeasy | Summer | MYI | Koh et al., 2010 |
| Both | DN-, RNeasy | Summer | MYI | Cowie, 2011 |
| DNA | PC | Summer | MYI | Martin et al., 2011 |
| DNA | PC | Summer | MYI | Bowman et al., 2012 |
| DNA | PC | Spring | FYI | Maas et al., 2012 |
| DNA | PC | Spring | YI and FF | Bowman et al., 2013 |
| DNA | PC | Winter | FYI, YI, FF | Barber et al., 2014 |

[a]PC = phenol chloroform; DNeasy and RNeasy available from Qiagen.

[b]FYI = first year sea ice, MYI = multiyear sea ice, YI = young sea ice, FF = frost flowers.

complex algorithms used to process HPLC data (Wright and Jeffrey, 2006; Latasa, 2007; Alou-Font et al., 2013). In particular, different layers of an ice core may have to be treated as separate community groups.

### 3.3.3. Bacteria and viruses
Bacteria and viruses contribute directly to the particulate nutrient and carbon pools in sea ice, participate in the still poorly understood sympagic food web, and mediate changes in the chemical composition of the ice brines. Knowing the abundance of bacteria and viruses is the first step in assimilating these microorganisms into conceptual and mathematical models of biogeochemical cycling in sea ice. Additional information on microbial metabolic rates (section 3.4.3) and diversity (sections 3.3.2 and 3.3.4) further facilitate the conceptual integration of biological and chemical perspectives on sea-ice processes.

Epifluorescence microscopy of melted ice samples (subject to the limitations and ambiguities discussed in section 3.2) is the principal technique by which bacterial and viral abundances are determined in sea ice. The methods have not changed substantially since they were first introduced in the late 1970s (*e.g.*, Hobbie et al., 1977; Porter and Feig, 1980; Noble and Fuhrman, 1998; Kaartokallio et al., 2008). Challenges faced by users of this method in sea-ice samples include the difficulty of microscopic observations aboard moving vessels, high background fluorescence (attributed to non-specific staining or staining of nucleic acids in EPS), and extremely low or high cellular or viral abundances in some melted ice samples. Flow cytometry, with its high sample throughput, is commonly used for measurements of bacterial and viral abundance in marine systems and can also be useful in sea-ice biology (Riedel et al., 2007a; Kaartokallio et al., 2013; Piwosz et al., 2013), although high EPS concentrations could interfere with the analysis in some sea-ice samples.

### 3.3.4. Genetic community assessments
Since 2001, investigators have extracted environmental DNA from sea ice for various sequencing or prokaryotic community fingerprinting applications, enabling determination of prokaryotic community composition, structure, and metabolic potential within sea-ice samples (Table 2). In a few cases, RNA has been extracted with DNA to determine the structure and function of the active prokaryotic community.

Samples are often size-fractionated with filters of different pore sizes to separate major components of the community before nucleic acid extraction. To facilitate intercomparison with the GOS (Global Ocean Sampling Expedition) dataset, we recommend standard pore sizes of 0.1, 0.8, and 3.0 μm (Rusch et al., 2007), although there is little evidence that these cutoffs correspond to natural ecological boundaries in sea ice. For accurate DNA extraction from cells, free from background contamination, it is essential that cells not be lysed until after capture on the filter, making the sample melting conditions critical (section 3.2). Low biomass samples, such as winter sea ice and some multi-year sea ice, require large melt volumes (greater than 1 L), while higher biomass samples can have an overabundance of eukaryotic DNA, along with interfering compounds such as EPS and non-specific humics. Because these compounds are chemically similar to nucleic acids, they often co-extract and interfere with downstream applications, such as the polymerase chain reaction (*e.g.*, Tebbe and Vahen, 1993). An ideal nucleic acid extraction protocol for sea ice would produce a high yield of the target molecule, reducing the bias toward or against any member of the microbial community and minimizing the co-extraction of EPS, humics, and other interfering compounds.

The available extraction methods fall into three broad categories: phenol chloroform (PC), kit-based, and electrophoretic methods. Although electrophoretic extraction is a promising new technology that may overcome some of the challenges regarding biomass and interfering compounds (*e.g.*, So et al., 2010), to date, only PC and kit-based methods have been applied to sea-ice samples. Cowie (2011) evaluated PC and kit-based methods in Antarctic sea ice and found that PC (using the methods of Moeseneder et al., 2001) is suitable for sea-ice samples. The same study also found that the RNA extraction kit RNeasy (Qiagen) in combination with bead beating was the most effective RNA extraction method, when samples were preserved with RNAlater (Qiagen). Although this and other studies show that nucleic acids can be extracted from sea ice, we encourage more comprehensive intercomparisons. In the meantime, investigators should carefully test their methods on the specific ice types they are studying.

## 3.4 Metabolic processes

Numerous studies have attempted to adapt methods for determining community metabolic rates in aquatic systems to the sea-ice environment (Table 3). None of these methods has been entirely satisfactory. Inter-calibration experiments and further method development are high priorities in sea-ice biogeochemistry.

### 3.4.1. Primary production and elemental uptake rates

The term "primary production" largely refers to organic matter synthesis by photosynthetic organisms, harvesting light to convert inorganic to organic carbon. The conversion of inorganic to organic carbon by chemosynthetic microorganisms in sea ice, in particular by nitrifying bacteria (which are also chemosynthetic), has been implicated in studies using stable isotopes (*e.g.*, Fripiat et al., 2014a) and DNA sequencing (Barber et al., 2014), but chemosynthesis is generally considered a minor contribution to overall primary production in sea ice. Several methods exist to directly estimate photosynthesis-based primary production (gross and net) in aquatic systems (*e.g.*, Falkowski and Raven, 2007); each method has its own assumptions, ambiguities, and biases, which have been extensively discussed in the oceanographic literature (*e.g.*, Bender et al., 1987; Laws et al., 2002). However, the complexity of the sea-ice/brine matrix presents particular problems in quantifying metabolic rates.

Incubations for determining primary production in sea ice, usually based on the incorporation of a tracer into particulate organic matter over a known amount of time, can be conducted either *in vitro*, in refrigerated incubators with spectral filters to mimic natural light conditions, or *in situ*, by embedding inoculated samples back into the sea-ice environment (*e.g.*, Horner and Schrader, 1982; Mock and Gradinger, 1999; McMinn and Hegseth, 2007; Gradinger, 2009). *In vitro* incubations remove the *in situ* variability, allowing

Table 3. Methods used for estimating metabolic rates in sea ice

| Approach | Method | Targeted processes | Timeframe | Spatial scale (m²) | Location in ice cover | Comments[a] | Example references |
|---|---|---|---|---|---|---|---|
| Incubations | $^{14}$C, $^{13}$C | Gross-net primary production[b] | days | 0.01 | Interior | Invasive | Arrigo et al., 2003; Gradinger, 2009 |
| | $^{15}$N | Nutrient uptake, remineralization | days | 0.01 | Interior | Invasive | Kristiansen et al., 1992; 1998 |
| | $^{3}$H-leucine, $^{3}$H-thymidine | Bacterial production | days | 0.01 | Interior | Invasive; requires tracer-to-carbon conversion factors | Kaartokallio, 2004 |
| | Dissolved $O_2$ | Gross primary production + respiration | days | 0.01 | Interior[c] | Invasive | Satoh and Watanabe, 1988 |
| Oxygen fluxes | $O_2$:Ar ratio | Net community production | vegetative season | 0.01 | Interior | Non-invasive; physical biases | Zhou et al., 2014b |
| | Optodes | Gross primary production | days | 0.01 | Interior | Non-invasive; physical biases; placement unknown | Mock et al., 2003 |
| | Microelectrodes | Gross primary production | days | 1 | Bottom | Non-invasive; physical biases | McMinn et al., 2000; 2007 |
| | Under-ice eddy covariance | Net community production | days | 100 | Bottom | Non-invasive; spatial integration; physical biases | Long et al., 2012 |

[a]Physical biases include bubble formation, sea ice-atmosphere exchange, and solubility changes.

[b]Rate depends on the incubation time, with shorter incubation times more closely approximating gross primary production (*e.g.*, Laws et al., 2002).

[c]Satoh and Watanabe (1988) incubated algae scraped off the bottom of ice cores, but the method should be applicable to any depth in the ice core.

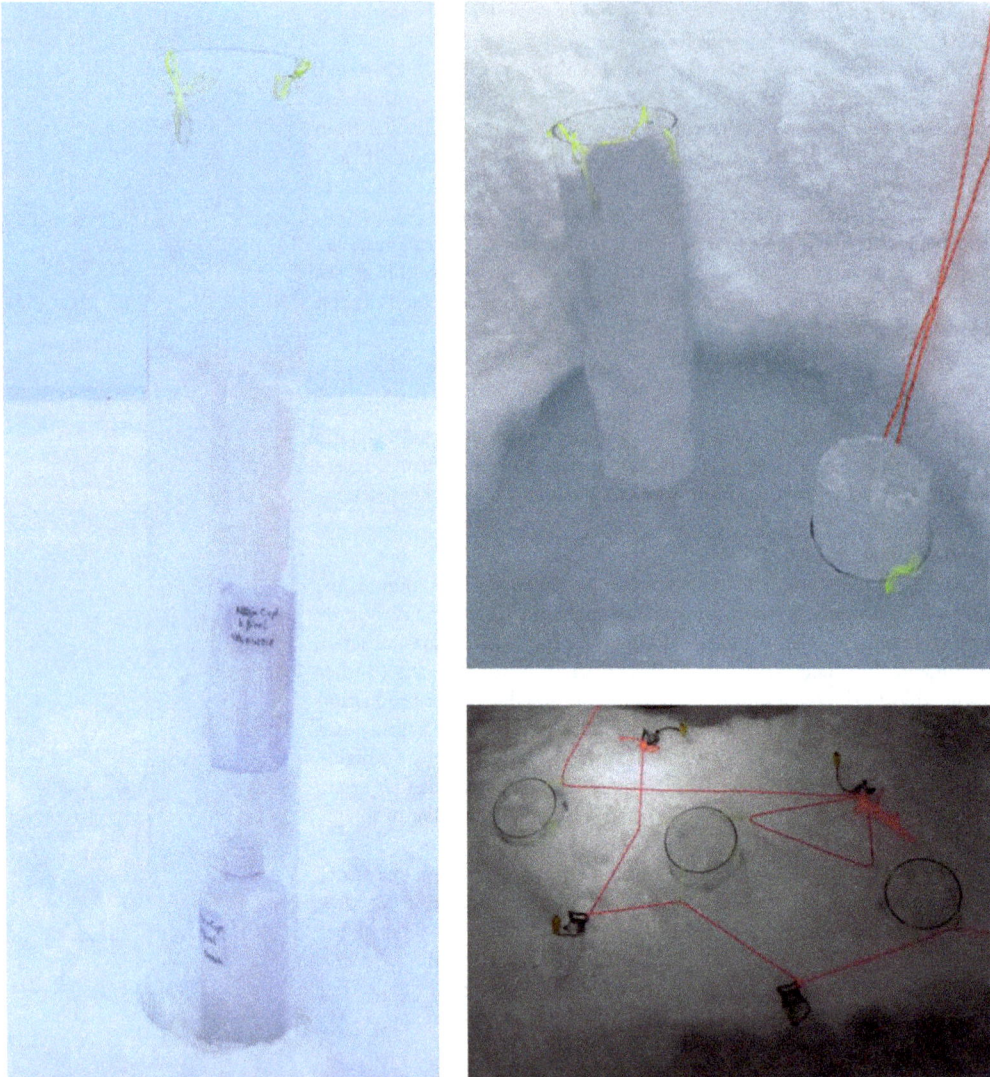

Figure 9

*In situ* incubations for determining metabolic rates in sea-ice communities.

Ice sections were crushed, placed in polycarbonate bottles, and spiked with enriched isotopes ($^{15}$N and $^{13}$C). Bottles were then incubated *in situ* using Plexiglas tubes (filled with untreated ice sections and incubation bottles) re-inserted into the core holes. Antarctic pack ice, the Sea Ice Physics and Ecosystem eXperiment (SIPEX2). Photos: A. Roukaerts.

easier comparison between different experiments, and are still the best method to determine maximum photosynthetic rates and efficiencies as functions of light intensity (*e.g.*, Burkholder and Mandelli, 1965; Arrigo and Sullivan, 1992). By combining these parameters with measurements of Chl *a* and light in the field, primary production can be derived (Arrigo et al., 2010). On the other hand, *in situ* incubations more directly assess primary production under the specific conditions to which the natural community is exposed.

For an incubation to be informative, the sample needs to represent the *in situ* community and environment as well as possible. Although brines collected from sackholes or by centrifugation are convenient to handle and provide a solution that represents the environment experienced by sea-ice algae, the biomass collected with such brines is not representative of the sea-ice community (section 2.2.2). On the other hand, while bulk sea-ice melts seem to provide representative biomass of some taxonomic groups, the dramatic changes in temperature and salinity associated with melting, even if the ice samples are melted in seawater, destroy the natural habitat, as well as often rupturing cells (particularly the flagellated taxa; section 3.2). In addition, the melting process at low temperature can take some time (often more than a day), further alienating the community from *in situ* conditions. The cost-benefit balance of the length of time required for melting (longest for isohaline melts if also isothermal; Junge et al., 2004) versus the modification of the samples (greatest for direct melts) is unknown.

Some investigators have incubated whole ice sections in closed containers filled with seawater bearing the relevant isotopic tracer either *in vitro* (*e.g.*, Grossi et al., 1987) or *in situ* by replacing the inoculated ice samples, sealed in transparent containers, into the core holes from which the ice had been extracted (Figure 9; Mock and Gradinger, 1999; Mock, 2002). These methods using unmelted ice sections have the advantage of maintaining a representative sample, but questions remain as to whether the tracer is adequately distributed within the

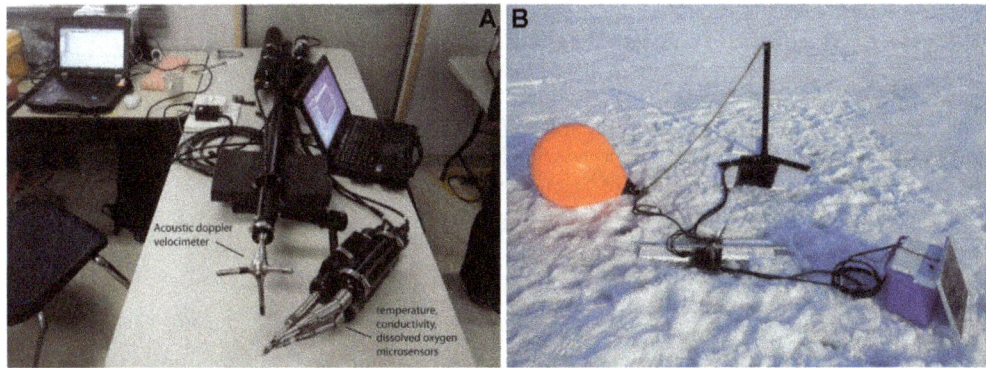

Figure 10

An underwater eddy covariance system for measuring fluxes of dissolved oxygen, salt, heat, and momentum.

In the laboratory prior to deployment (a) and deployed through 60-cm thick sea ice in southeast Greenland (b), March 2013. Photos: B. Else.

brine network. If the tracer preferentially remains in the surrounding seawater or brines, primary production could be severely underestimated. Crushing the ice (Rysgaard et al., 2007) should improve homogenization.

Changes in $O_2$ concentration can be used to estimate net community production (NCP), defined as primary production minus respiration. This method has been applied to sea-ice communities using *in situ* oxygen optodes (Mock et al., 2003) and microelectrodes (McMinn and Ashworth, 1998; McMinn et al., 2000; 2007; McMinn and Hegseth, 2007). Determining NCP solely from $O_2$ dynamics in sea ice is complicated by exchange with the atmosphere arising from solubility changes, as temperature and salinity vary with ice formation and melt (Glud et al., 2002). In seawater, Ar measurements are used to correct for such physical contributions to the $O_2$ concentration changes (*e.g.*, Craig and Hayward, 1987); Zhou et al. (2014b) have used $O_2/Ar$ ratio measurements to estimate NCP in winter and spring sea ice.

With recent technological advances, underwater eddy covariance (EC) has potential for investigating sea-ice primary production (Long et al., 2012). The technique (which is analogous to the atmospheric technique described in section 5.1.3) requires the under-ice deployment of a three-dimensional current velocimeter in conjunction with a fast-response oxygen electrode (Figure 10). The instruments can be deployed through a hole in the ice and are usually positioned within 1 m of the ice-seawater interface. By correlating oxygen concentrations with vertical current velocity, oxygen fluxes representative of conditions in an area upstream of the instruments (typically on the order of 100 $m^2$) can be calculated. The technique is thus a true *in situ* measurement, it does not disturb the ice under investigation, and it avoids some of the small-scale patchiness issues associated with other techniques. However, at best, EC only provides a measurement of net productivity (as respiration cannot be separated from photosynthesis). At worst, EC may be strongly influenced by abiotic processes, such as gas rejection (during ice growth) and dilution (during melt), which need to be accounted for in certain cases. The technique is also only useful for measuring production in the bottom-most layers of the ice (*i.e.*, those that are freely exchanging with the underlying seawater).

Incubations with other isotopic tracers, such as $^{15}N$ (Harrison et al., 1990; Kristiansen et al., 1992; 1998), can be used to assess the nature of primary production (*e.g.*, new versus regenerated; Dugdale and Goering, 1967) and concomitant biogeochemical dynamics (such as N-uptake and nitrification). Because diatoms are important members of sea-ice communities, the information on Si uptake and $bSiO_2$ dissolution that could be generated by adapting methods for incubations with $^{30}Si$ and $^{32}Si$ (Fripiat et al., 2009) to sea-ice samples would be particularly valuable. Such incubations depend on many of the same assumptions as carbon-based primary production incubations and, therefore, suffer from the same challenges and limitations. In addition, because these macronutrients occur in sea-ice brines at much smaller concentrations than inorganic carbon, labelled N and Si substrates are added to the incubations in 'trace' quantities (generally, 10% of the ambient level is recommended; Dudgale and Goering, 1967).

### 3.4.2. Variable fluorescence methods to determine ice algal photosynthetic parameters

Over the last 20 years, chlorophyll *a* variable fluorescence methods have proven to be useful tools for understanding photophysiological properties of marine algae. The application of pulse amplitude modulated (PAM) fluorometry to study sea-ice algae was pioneered by Kühl et al. (2001) in the Arctic and McMinn et al. (2003) in Antarctica. This approach measures the energy conversion efficiency of photosystem II (the quantum efficiency) to derive maximum relative electron transport rates and estimate photosynthetic efficiency and the photoadaptive index of the algae (Ralph and Gademann, 2005). In general, because PAM fluorometers do not provide a direct measurement of carbon fixation, the overall value in PAM fluorescence techniques lies more in their ability to measure ice algal photophysiological responses to varying physicochemical conditions, particularly on small scales relevant to the sea-ice skeletal layer and brine channels (*e.g.*, Hawes et al., 2012).

The most widely used PAM methods in sea-ice research involve ice shavings, brines, and melted ice samples that are analyzed *in vitro* (McMinn et al., 2007; Ralph et al., 2007; Manes and Gradinger, 2009; Meiners et al., 2009; Hawes et al., 2012; Granfors et al., 2013a); these methods are thus biased by sea-ice sampling procedures (sections 2.2 and 3.2). To bypass sample extraction artifacts, divers have successfully deployed non-invasive PAM fluorometers under sea ice to measure the spatial variability of bottom ice algal biomass and algal photosynthetic parameters in Greenland fast ice (Kühl et al., 2001). Detailed studies utilizing PAM techniques have also confirmed vertical variability in ice algal distributions and photosynthetic properties (Manes and Gradinger, 2009; Hawes et al., 2012).

A second type of variable fluorescence instruments, so-called Fast Repetition Rate (FRR) fluorometers, are increasingly used in phytoplankton research but have not yet been employed widely in sea-ice research. In contrast to PAM fluorometers, FRR fluorometers provide measurements of quantum efficiency, absorption cross section, and turnover times of photosystem II and can, therefore, be used to estimate primary production (Robinson et al., 1998; Suggett et al., 2003).

### 3.4.3. Bacterial production

Marine bacterial biomass production is usually estimated from incubations with radioisotope-labelled precursors of DNA or proteins (*e.g.*, Ducklow, 2000). As with primary production measurements (section 3.4.1), incubations of melted ice samples are of limited utility, and methods that attempt to maintain the *in situ* conditions during the incubations are preferred. Tritiated-thymidine incorporation (TTI) into DNA (*e.g.*, Smith and Clement, 1990; Deming, 2010) and $^3$H-leucine (LEU) incorporation into protein (*e.g.*, Kaartokallio, 2004; Paterson and Laybourn-Parry, 2012) are the most common techniques applied to sea-ice samples. While TTI and LEU incorporation often co-vary in the marine environment, suggesting that both methods address bacterial production-related processes, the two methods measure distinctly separate physiological processes; a range of ratios of LEU incorporation to TTI has been reported, with very high ratios in some sea-ice habitats (Mock et al., 1997; Kaartokallio et al., 2008; 2013). High ratios have been attributed to unbalanced growth (*e.g.*, investment in cell growth, measured by LEU incorporation, versus cell division, measured by TTI), but also to incorporation of the radiolabeled LEU tracer into other, non-protein components of sea-ice organisms. Bacterial activity in sea ice has also been estimated from transformations of $^{15}$N-labelled nitrogen substrates (Rysgaard and Glud, 2004) and from uptake and reduction of the dye 5-cyano-2,3-ditolyl tetrazolium chloride (CTC; Junge et al., 2004; Meiners et al., 2008).

Estimating bacterial biomass production from the uptake of labelled substrates requires tracer-to-carbon conversion factors, but such conversion factors are not easy to measure routinely; conversion factor determination usually involves incubating natural samples for several days. Therefore, literature values from open-water studies are often applied to sea-ice data, although the accuracy of the conversion factor is strongly affected by community composition and physiology. In order to accurately estimate sea-ice bacterial carbon production, conversion factors for both the LEU incorporation and TTI methods need to be determined for a variety of sea-ice communities in different habitats and regions. In addition, bacterial processes can be particularly sensitive to temperature (*e.g.*, Rivkin and Legendre, 2001); the impact of incubation temperature on measured metabolic rates specifically for sympagic organisms warrants further investigation.

# 4. Chemical components

Methods for analyzing the major seawater ions (Cl⁻, Na⁺, SO₄²⁻, Mg²⁺, Ca²⁺, K⁺) in sea ice, its overlying snow, and frost flowers are well established and uncontroversial (*e.g.*, Nelson and Thompson, 1954; Domine et al., 2004; Granskog et al., 2004; Douglas et al., 2012). Modern methods almost universally utilize ion chromatography. However, the analyses of nearly all other chemical parameters in sea ice are still subject to debate. Our focus here is on those methods in which we have less confidence.

Our lack of confidence in most chemical analyses in sea ice stems from difficulties in sampling, handling, and processing the samples, and rarely from the actual analyses of the aqueous samples. An important exception is analyses of undiluted brine samples, in which the high salinities often reduce precision and complicate calibrations.

## *4.1. Inorganic macronutrients*

Fundamental information on the nutritional state of the sympagic biological community can be derived from the distributions of the inorganic macronutrients (phosphate, PO₄³⁻; silicic acid, Si(OH)₄; and the nitrogen species: nitrate, NO₃⁻; nitrite, NO₂⁻; and ammonium, NH₄⁺) in sea ice (*e.g.*, Dieckmann et al., 1992; Kaartokallio, 2001). Variations in their concentrations throughout the ice column and with time indicate biological activity and exchange with the underlying water. Most sea-ice nutrient analyses are based on spectrophotometric analysis, using a continuous flow analyzer (*e.g.*, Hydes et al., 2010). A large range of nutrient concentrations occurs in sea ice (Thomas et al., 2010), from depleted to replete conditions, and sampling requires the usual precautions against contamination. To the extent possible, equipment preparation and analyses should follow standard repeat hydrography protocols (*e.g.*, Granskog et al., 2005b; Hydes et al., 2010).

Nutrients are often measured in melted bulk sea-ice samples, but brines for nutrient analyses have also been collected both by centrifuging ice samples (*e.g.*, McMinn et al., 2009; Munro et al., 2010) and from sackholes (*e.g.*, Gleitz et al., 1995; Papadimitriou et al., 2007). The costs versus benefits of filtering ice melts before nutrient analysis are unclear. Particulate (intracellular) nutrient concentrations can be quite high in sea ice, because of high biomass accumulation, particularly near the ice-water interface (*e.g.*, Arrigo et al., 2010). On the other hand, particle-bound nutrients can be released to solution during melting (section 3.2), although Thomas et al. (1998) found no evidence during their spring-time study that cell lysis significantly impacted dissolved nutrient measurements of bulk ice samples melted without adding seawater (direct melt). Large cells generally would have higher intracellular nutrient contents to release upon lysis than small cells, but large diatoms, which often dominate sea-ice communities, appear to have reduced susceptibility to lysis during melting (section 3.2; Mikkelsen and Witkowski, 2010), minimizing the effect of cell lysis on nutrient concentration estimates for diatom-rich ice. Additional intercalibration exercises are needed to more thoroughly assess the effects of core melting protocols and filtration on measured nutrient concentrations with different microbial communities and in different seasons.

Ideally, nutrient samples, particularly those for ammonium, should be analyzed immediately after sampling to avoid artifacts associated with biological growth or decay during sample storage (Holmes et al., 1999; Hydes et al., 2010). When analysis cannot be completed within hours, the samples generally should be stored frozen (< -20°C) in the dark. However, silicic acid in seawater polymerizes when it freezes and only very slowly redissolves when the sample is thawed for analysis, possibly resulting in underestimation of the original dissolved silicic acid concentration (*e.g.*, Hydes et al., 2010). The impacts of this phenomenon on estimates of *in situ* concentrations of biologically available silicic acid in sea-ice brines is unknown. Therefore, filtration to remove Si-bearing diatoms and storage at room temperature in the dark are more suitable procedures for silicic acid analyses of sea-ice samples (Fripiat et al., 2014b). Alternatively, filtered samples for nutrient analyses can be poisoned with agents such as mercuric chloride and stored refrigerated but unfrozen until analysis (*e.g.*, Kattner, 1999).

## 4.2. Organic compounds

Sea ice acquires organic matter from seawater during ice formation and through *in situ* biological production (Thomas et al., 1998; 2001a; Giannelli et al., 2001; Granskog et al., 2004; Riedel et al., 2007a; Stedmon et al., 2007; 2011; Müller et al., 2013). Extremely high concentrations of dissolved organic carbon, several thousand $\mu$mol L$^{-1}$ (up to two orders of magnitude higher than seawater values; Hansell et al., 2009), have been reported in sea-ice samples (*e.g.*, Thomas et al., 2001a; Junge et al., 2004; Riedel et al., 2008). The resulting sea-ice organic pool is a complex mixture of living and non-living particulate material and dissolved compounds. Operational definitions of dissolved, colloidal, and particulate carbon are particularly tenuous in sea ice, as the temperature and salinity changes associated with melting can cause phase changes in the organic matter. In addition, we do not know how well sampled brines represent the organic content of brine pockets and channels within the undisturbed ice; even if not rigorously particulate, the colloidal and dissolved organic matter may still be "sticky" and adhere to brine channel walls.

In most cases, concentrations should be measured and reported in units of moles of carbon (or nitrogen or phosphorous) and not grams of bulk organic matter. This approach allows the data to be used in biogeochemical cycling studies, including incorporation into numerical models, without unverified assumptions about carbon content and the C:N:P ratios of the organic matter.

Most studies of organic biogeochemistry in sea ice to date have focused on the bulk parameters of total, dissolved, and particulate organic matter. Methods for determining the particulate fractions are detailed in section 3.3.1. Here we address sample collection for the total (unfiltered) and dissolved (filtered) aqueous fractions. Photochemically active colored dissolved organic matter (CDOM) is discussed in section 4.7; halocarbons are discussed in section 5.2.

### 4.2.1. Sample handling

Organic matter concentrations in sea ice vary over an order of magnitude, from low values similar to those observed in the deep ocean to high values in spring brines and gap layers (*e.g.*, Thomas et al., 1995; Song et al., 2011). When dissolved organic matter (DOM) is present in high concentrations in sea-ice samples, some of the stringent protocols required to successfully sample seawater for DOM are eased. In low-DOM sea-ice samples, however, care is required to avoid not only environmental contamination during sampling and processing, but also cross-contamination between samples during both processing and analyses. Therefore, the exact requirements for preventing contamination of sea-ice DOM samples have not yet been established and probably depend on the samples; that is, upper levels of multiyear ice probably need to be handled more rigorously than gap-layer slush samples. To eliminate potential contamination during core extraction and manipulation, the outer layers of cores can be removed before subsampling for organic matter (Granskog et al., 2004; Song et al., 2011; Fripiat et al., 2014a). In addition, although glass equipment that has been combusted at high temperatures is preferred, plastics are being used more often and may be acceptable under some conditions (*e.g.*, Thomas et al., 1998; Miller et al., 2011b), particularly if contact times are kept short. Rigorous experiments are still required to confirm the conditions under which plastics can be used.

The question of whether or not to filter samples for organic analysis can only be answered within the context of each specific study. Particularly in concentrated brines, the physical and behavioral distinction between "dissolved" and "particulate" organic matter may have little relationship with the operational definition based on the type of filter used. Filtration also introduces artifacts, as dissolved organic matter can stick to filters and turbulence or pressure gradients associated with the filtration process can break apart particles (section 3.3.1).

Furthermore, the salinity changes associated with melting ice can cause coagulation and disaggregation of organic matter, as well as cell lysis (section 3.2). On the other hand, while melting the ice in artificial or filtered seawater will minimize artifacts associated with salinity changes, it is almost impossible to generate truly "organic-free" seawater and thus avoid contributing to the analysis in ways that are difficult to quantify.

### 4.2.2. Total organic carbon and nitrogen (TOC and TON)

The generic, undifferentiated pool of organic matter has been analyzed in melted bulk ice (e.g., Thomas et al., 2001a; Cozzi, 2008; Dumont et al., 2009) and in brines collected from sackholes (e.g., Papadimitriou et al., 2007; Meiners et al., 2009). In parallel to seawater methods, the samples are often passed through precombusted glass fiber filters, in which case the resulting analytical results are termed "dissolved" organic carbon or nitrogen (DOC, DON), although this pool also includes some bacteria and viruses (Deming, 2010).

High temperature catalytic oxidation is generally used for TOC and DOC analyses (Qian and Mopper, 1996; Spyres et al., 2000), while DON is often inferred from the difference between total dissolved nitrogen (TDN; Bronk et al., 2000) and the inorganic nitrogen species ($NO_3^-$, $NO_2^-$, $NH_4^+$; section 4.1). Sea-ice studies have generally measured TDN using UV oxidation (Thomas et al., 2001a; Papadimitriou et al., 2007; Cozzi, 2008), although other methods may also be suitable (e.g., chemical oxidation with persulfate; Bronk et al., 2000; Fripiat et al., 2014a). Less common are studies of urea in sea ice (Harrison et al., 1990; Kristiansen et al., 1998; Conover et al., 1999; Garrison et al., 2003; Papadimitriou et al., 2009), generally measured using the urease or diacetyl monoxime methods (e.g., Price and Harrison, 1987).

### 4.2.3. Exopolymeric substances (EPS)

Over the past decade, extracellular or exopolymeric substances (EPS) have been recognized as extremely important components of sea ice. This high-C, low-N material, similar to the transparent exopolymeric particles (TEP) found in seawater, may act as a cryoprotectant for sympagic biota and has a measurable effect on the microstructure and salinity of sea ice (Krembs et al., 2011). Also present in surface sea-ice environments, such as frost flowers and snow, EPS has additional implications for air-ice interactions (Bowman and Deming, 2010; Ewert et al., 2013).

Ice core samples for EPS analysis have been melted (section 3.2) directly (e.g., Krembs et al., 2002; Juhl et al., 2011), in seawater (e.g., Meiners et al., 2003; Riedel et al., 2006), and in concentrated brines (e.g., Collins et al., 2008; Ewert et al., 2013). Direct melts are convenient, because lower salinities simplify further chemical analyses, but additional studies are required to confirm whether the melting protocol has an effect on measured EPS content or on the physical and chemical properties of this complex material.

After melting, EPS is separated into dissolved (dEPS) and particulate (pEPS) fractions by filtration through different types of filters (Table 4). Additional size fractionation can be achieved with sequential precipitation of dEPS fractions of varying solubility across an ethanol gradient (Underwood *et al.* 2010; 2013; Aslam et al., 2012). Three methods are commonly used to quantify particulate and dissolved EPS in sea-ice research (Table 4): the standard colorimetric Alcian blue method developed for TEP analysis of seawater (Passow, 2002), the colorimetric TPTZ (2,4,6-tripyridyl-*s*-triazine) method of Myklestad et al. (1997), and the phenol-sulfuric acid assay (PSA) of Dubois et al. (1956), modified for small sample volumes. In a direct comparison, van der Merwe et al. (2009) found that results from analyses of Antarctic sea ice using the Alcian blue and PSA methods agreed at high but not at low EPS concentrations near the detection limits. More such intercomparisons are needed to confirm the validity and comparability of results from the Alcian blue, TPTZ, and PSA methods under varying conditions. Some investigators have quantified specific components of EPS using a carbazole assay for uronic acids (Bitter and Muir, 1962) and hydrolysis followed by gas chromatography for the neutral monosaccharide composition (Table 4). Methods need to be developed to more thoroughly characterize EPS in sea ice, including: polymer composition, structure and molecular size in different EPS types; interactions between EPS and other components of the sea-ice biogeochemical system (salts, trace metals, other forms of organic matter); and the potential for EPS to interfere with other chemical analyses.

Filters stained with Alcian blue can also be analyzed microscopically to determine particle abundance, size distribution, and the number of EPS-associated bacteria (Meiners et al., 2003; 2004; 2008). Krembs et al. (2002; 2011) also used *in situ* visualization of EPS distribution within brine pockets and channels to directly examine the association between EPS and ice biota.

Table 4. EPS analyses in sea ice

| Approach | Method | Comments | References |
|---|---|---|---|
| Size fractionation | GF/F filters (nominally 0.7 μm) | Consistent with POC/DOC methods | Dumont et al., 2009 |
| | Polycarbonate membranes (0.4 μm) | Compatible with microscopic observation and PSA assay | Krembs et al., 2002; 2011 |
| | | | Meiners et al., 2003 |
| | | | Riedel et al., 2006; 2007a;b; 2008 |
| | | | Collins et al., 2008 |
| | | | van der Merwe et al., 2009 |
| | | | Bowman and Deming, 2010 |
| | | | Juhl et al., 2011 |
| | | | Ewert et al., 2013 |
| | | | Barber et al., 2014 |
| | Polycarbonate membranes (0.2 μm) | Compatible with bacterial capture | van der Merwe et al., 2009 |
| | Consecutive filtration | Multiple size fractions | Ewert et al., 2013 |
| | Ethanol gradient precipitation | Recovery of higher quantities of EPS | Underwood et al., 2010; 2013 |
| | | | Krembs et al., 2011 |
| | | | Aslam et al., 2012 |
| Chemical analysis | TPTZ | High sensitivity | Herborg et al., 2001 |
| | Alcian blue | Consistent with seawater TEP methods and microscopic analysis | Krembs et al., 2002; 2011 |
| | | | Riedel et al., 2006; 2007a;b; 2008 |
| | | | Collins et al., 2008 |
| | | | Dumont et al., 2009 |
| | | | van der Merwe et al., 2009 |
| | Phenol sulfuric acid (PSA) assay | Commonly used | van der Merwe et al., 2009 |
| | | | Bowman and Deming, 2010 |
| | | | Underwood et al., 2010; 2013 |
| | | | Juhl et al., 2011 |
| | | | Krembs et al., 2011 |
| | | | Aslam et al., 2012 |
| | | | Ewert et al., 2013 |
| | Carbazole assay w/gas chromatography | Acidic component, neutral monosaccharides | Underwood et al., 2010 |
| | | | Aslam et al., 2012 |
| Microscopic/visual observation | Alcian blue on filters | Size distribution | Meiners et al., 2003; 2004; 2008 |
| | Alcian blue in sea ice | Observations of unmelted/melting ice | Krembs et al., 2002; 2011 |
| | | | Juhl et al., 2011 |

### 4.2.4. Specific organic compounds

Very few studies have attempted to measure specific organic classes or compounds, either natural or anthropogenic, in sea ice. The steps required to limit contamination, gas exchange, or brine loss strongly depend on the nature of the compounds of interest: their concentration ranges, sources, volatility, and particle affinity.

Herborg et al. (2001) and Dumont et al. (2009) distinguished between the mono- and polycarbohydrate fractions of the DOC pool in sea ice. Belt et al. (2013) measured the lipid paleo-biomarker IP25, sterols, and fatty acids in filtered bulk sea-ice melts. Mycosporine-like amino acids (MAAs), which may serve as photoprotectants under some circumstances, have been determined in both Arctic and Antarctic sea ice (Ryan et al., 2002; Uusikivi et al., 2010; Mundy et al., 2011). Stedmon et al. (2007; 2011) and Granskog et al. (2015) used fluorescence to quantify humics and "amino acid-like" organic matter in sea-ice melts and brines. A number of organic anthropogenic contaminants have been measured in sea-ice melts (Rahm et al., 1995; Pućko et al., 2010a;b), in brines (Pućko et al., 2010b), and in snow (Garbarino et al., 2002), frost flowers, and brine skims (Douglas et al., 2012) over sea ice.

Table 5. Trace metal analyses in sea ice

| Element | Samples | Fractions | Reference |
|---|---|---|---|
| Al | Bulk ice | Particulate | Hölemann et al., 1999 |
| | Bulk ice | Total | Granskog and Virkanen, 2001 |
| | Bulk ice | Total | Granskog et al., 2004 |
| | Snow | Dissolved | Garbarino et al., 2002 |
| | Brine | Dissolved, particulate | Hendry et al., 2010a |
| | Snow, bulk ice | Dissolved, particulate | Lannuzel et al., 2011 |
| | Snow, bulk ice | Particulate | de Jong et al., 2013 |
| | Bulk ice | Dissolved, particulate, colloidal | Lannuzel et al., 2014 |
| Ti | Bulk ice | Particulate | Hölemann et al., 1999 |
| V | Bulk ice | Particulate | Hölemann et al., 1999 |
| | Bulk ice | Total | Tovar-Sánchez et al., 2010 |
| Cr | Bulk ice | Particulate | Hölemann et al., 1999 |
| | Snow | Dissolved | Garbarino et al., 2002 |
| | Snow, bulk ice | Dissolved, particulate | Lannuzel et al., 2011 |
| Mn | Bulk ice | Total, dissolved | Campbell and Yeats, 1982 |
| | Bulk ice | Particulate | Hölemann et al., 1999 |
| | Snow | Dissolved | Garbarino et al., 2002 |
| | Bulk ice | Dissolved, particulate | Grotti et al., 2005 |
| | Snow, bulk ice | Dissolved, particulate | Lannuzel et al., 2011 |
| | Bulk ice | Dissolved, particulate, colloidal | Lannuzel et al., 2014 |
| Fe | Bulk ice | Total, dissolved | Campbell and Yeats, 1982 |
| | Snow | Total | Westerlund and Öhman, 1991 |
| | Snow, bulk ice, brine | Total dissolvable | Löscher et al., 1997 |
| | Bulk ice | Particulate | Hölemann et al., 1999 |
| | Bulk ice | Organic complexes | Boye et al., 2001 |
| | Snow | Total dissolvable | Edwards and Sedwick, 2001 |
| | Bulk ice | Total, dissolved | Granskog and Virkanen, 2001 |
| | Snow | Dissolved | Garbarino et al., 2002 |
| | Bulk ice | Total | Granskog et al., 2004 |
| | Bulk ice | Dissolved, particulate | Grotti et al., 2005 |
| | Snow, bulk ice, brine | Total dissolvable, dissolved | Lannuzel et al., 2006; 2007 |
| | Bulk ice | Dissolved | Aguilar-Islas et al., 2008 |
| | Snow, bulk ice, brine | Total dissolvable, dissolved, particulate | Lannuzel et al., 2008 |
| | Snow, bulk ice, brine | Dissolved | van der Merwe et al., 2009 |
| | Bulk ice | Dissolved | Lannuzel et al., 2010 |
| | Bulk ice | Total | Tovar-Sánchez et al., 2010 |
| | Snow, bulk ice, brine | Total dissolvable, dissolved, particulate | van der Merwe et al., 2011a;b |
| | Snow, bulk ice | Dissolved, particulate | de Jong et al., 2013 |
| | Bulk ice | Dissolved, particulate, colloidal | Lannuzel et al., 2014 |
| | Snow | Dissolved | Winton et al., 2014 |
| Co | Bulk ice | Particulate | Hölemann et al., 1999 |
| | Snow | Dissolved | Garbarino et al., 2002 |
| | Bulk ice | Total | Tovar-Sánchez et al., 2010 |
| Ni | Bulk ice | Total, dissolved | Campbell and Yeats, 1982 |
| | Bulk ice | Particulate | Hölemann et al., 1999 |
| | Bulk ice | Total | Granskog and Virkanen, 2001 |
| | Snow | Dissolved | Garbarino et al., 2002 |
| | Bulk ice | Total | Tovar-Sánchez et al., 2010 |

| Element | Samples | Fractions | Reference |
|---|---|---|---|
| Cu | Bulk ice | Total, dissolved | Campbell and Yeats, 1982 |
| | Bulk ice | Particulate | Hölemann et al., 1999 |
| | Bulk ice | Particulate | Frache et al., 2001 |
| | Bulk ice | Total, dissolved | Granskog and Virkanen, 2001 |
| | Snow | Dissolved | Garbarino et al., 2002 |
| | Bulk ice | Total | Granskog et al., 2004 |
| | Bulk ice | Dissolved, particulate | Grotti et al., 2005 |
| | Bulk ice | Total | Tovar-Sánchez et al., 2010 |
| | Snow, bulk ice | Dissolved, particulate | Lannuzel et al., 2011 |
| Zn | Bulk ice | Particulate | Hölemann et al., 1999 |
| | Snow | Dissolved | Garbarino et al., 2002 |
| | Bulk ice | Total | Granskog et al., 2004 |
| | Bulk ice | Total | Tovar-Sánchez et al., 2010 |
| | Snow, bulk ice | Dissolved, particulate | Lannuzel et al., 2011 |
| As | Bulk ice | Particulate | Hölemann et al., 1999 |
| Rb | Bulk ice | Particulate | Tütken et al., 2002 |
| Sr | Bulk ice | Particulate | Hölemann et al., 1999 |
| | Snow | Dissolved | Garbarino et al., 2002 |
| | Bulk ice | Particulate | Tütken et al., 2002 |
| Mo | Bulk ice | Particulate | Hölemann et al., 1999 |
| | Snow | Dissolved | Garbarino et al., 2002 |
| | Bulk ice | Total | Tovar-Sánchez et al., 2010 |
| | Snow, bulk ice | Dissolved, particulate | Lannuzel et al., 2011 |
| Cd | Bulk ice | Total, dissolved | Campbell and Yeats, 1982 |
| | Bulk ice | Particulate | Hölemann et al., 1999 |
| | Bulk ice | Particulate | Frache et al., 2001 |
| | Snow | Dissolved | Garbarino et al., 2002 |
| | Snow, bulk ice | Dissolved | Nedashkovskii, 2002 |
| | Snow, bulk ice | Total | Granskog and Kaartokallio, 2004 |
| | Bulk ice | Total | Granskog et al., 2004 |
| | Bulk ice | Dissolved, particulate | Grotti et al., 2005 |
| | Brine | Dissolved | Hendry et al., 2010b |
| | Snow, bulk ice | Dissolved, particulate | Lannuzel et al., 2011 |
| Sn | Bulk ice | Particulate | Hölemann et al., 1999 |
| Sb | Bulk ice | Particulate | Hölemann et al., 1999 |
| Cs | Bulk ice | Particulate | Hölemann et al., 1999 |
| Ba | Bulk ice | Particulate | Hölemann et al., 1999 |
| | Snow | Dissolved | Garbarino et al., 2002 |
| | Snow, bulk ice | Dissolved, particulate | Lannuzel et al., 2011 |
| Nd | Bulk ice | Particulate | Tütken et al., 2002 |
| Hg | Snow | Total, dissolved | Garbarino et al., 2002 |
| | Snow, frost flowers | Total[a] | Douglas et al., 2005 |
| | Snow | Total, particulate | Poulain et al., 2007 |
| | Snow | Total | Douglas et al., 2008 |
| | Snow, bulk ice, brine | Total | Chaulk et al., 2011 |
| | Snow, bulk ice, brine | Dissolved | Cossa et al., 2011 |
| | Snow, frost flowers | Total, stable isotopes | Sherman et al., 2012 |
| | Bulk ice | Dissolved, particulate | Burt et al., 2013 |

| Element | Samples | Fractions | Reference |
|---|---|---|---|
| Tl | Snow | Dissolved | Garbarino et al., 2002 |
| Pb | Bulk ice | Particulate | Hölemann et al., 1999 |
|  | Bulk ice | Particulate | Frache et al., 2001 |
|  | Bulk ice | Total | Granskog and Virkanen, 2001 |
|  | Snow | Dissolved | Garbarino et al., 2002 |
|  | Snow, bulk ice | Total, particulate | Nedashkovskii, 2002 |
|  | Snow, bulk ice | Total | Granskog and Kaartokallio, 2004 |
|  | Bulk ice | Total | Granskog et al., 2004 |
|  | Bulk ice | Dissolved, particulate | Grotti et al., 2005 |
| Th | Bulk ice | Particulate | Hölemann et al., 1999 |
| U | Bulk ice | Particulate | Hölemann et al., 1999 |
|  | Snow | Dissolved | Garbarino et al., 2002 |
|  | Bulk ice, brine | Total | Not et al., 2012 |

ªMethods not specified

## 4.3. Trace metals

The seasonal ice cover represents a key reservoir, storing and transporting potentially bio-active trace metals and thus likely playing major roles in not only trace metal cycles but also the carbon cycle in polar and subpolar oceans (*e.g.*, Lannuzel et al., 2010). However, trace metal research in sea ice is subject to the same draconian restrictions required in seawater to avoid contamination (Figure 11; *e.g.*, Bruland and Rue, 2001; Cutter et al., 2010). In general, all equipment, preparations, and analyses should follow standard GEOTRACES protocols (Cutter et al., 2010). Several studies have successfully developed sampling and measurement techniques for trace metals in the cryospheric environment and have reported data for numerous elements in sea ice, snow, and brines in the Arctic and Southern Oceans (Table 5).

Samples for trace metal analyses need to be recovered from a dedicated sampling site, upwind from all other operations, and specific sampling procedures must be performed against the wind. Clean-room garments and plastic gloves should be worn over cold-weather clothing. While trace metal-clean sampling is relatively straightforward for snow above sea ice (*e.g.*, Edwards and Sedwick, 2001; Lannuzel et al., 2006), ice core sampling carries a greater risk of contamination. Ideally, a titanium or an electropolished stainless steel corer (Lannuzel et al., 2006) should be used, although standard corers can also be used, if the outer layer of the core can be removed without contamination or substantial brine loss (Figure 11b; Hölemann et al., 1999; Granskog and Kaartokallio, 2004; Granskog et al., 2004; Grotti et al., 2005; Aguilar-Islas et al., 2008). The ice should be cored by hand, although electric auger motors have also been used with the generators downwind (*e.g.*, de Jong et al., 2013). A number of investigators have collected brine samples from sackholes for trace metal analyses (Lannuzel et al., 2006; 2007; van der Merwe et al., 2009; 2011a; Chaulk et al., 2011; Cossa et al., 2011).

Changes to the *in situ* chemical speciation and fractionation between oxidation states and particulate/colloidal/soluble phases during sample collection and processing are exceptionally problematic in sea-ice studies of trace metals. In particular, iron speciation is very poorly understood in sea ice, and the definitions of what is actually measured are highly operational (*e.g.*, Bruland and Rue, 2001). To date, trace metal speciation

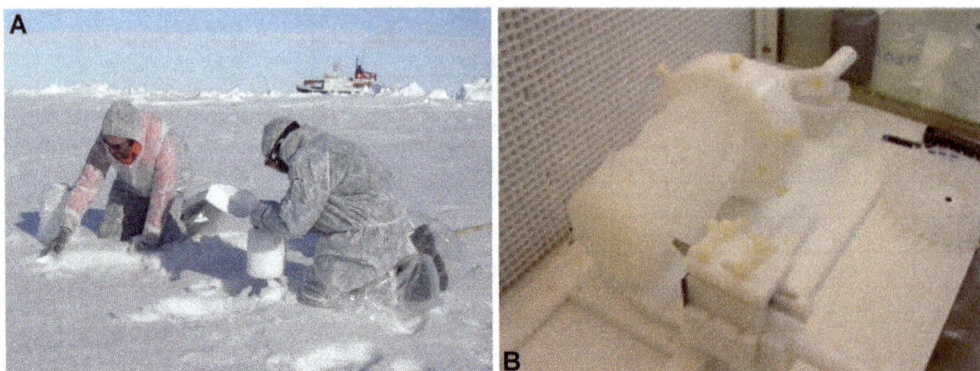

**Figure 11**
**Trace metal-clean sample handling.**

(a) Collecting snow on top of sea ice. Weddell Sea, January, 2005. Photo: J-L. Tison. (b) A trace metal-clean lathe (polypropylene, with titanium blade and ceramic handle) for removing contaminated outer layers of cores (with a core section), mounted in a laminar flow bench in a cold lab. Photo: D. Lannuzel

measurements in sea ice and brines have been limited mainly to separations between operationally-defined particulate and dissolved fractions, separated by filtration and varying dissolution procedures (*e.g.*, Grotti et al., 2005; Lannuzel et al., 2006; 2007; 2011; van der Merwe et al., 2011a; Hendry et al., 2010a). Boye et al. (2001) analyzed iron-organic complexation in one sea-ice sample and confirmed that a large fraction of the iron in sea ice can be complexed by organic matter, indicating that organic complexation may be as important in sea ice as in seawater.

Mercury is another special case, involving not only a highly contamination-prone metal in solution, but also gaseous and organic phases. Most studies of Hg in the marine cryosphere have been focused above the ice, analyzing snow, frost flower, or surface brine skim samples (Garbarino et al., 2002; Douglas et al., 2005; 2008; Poulain et al., 2007; Sherman et al., 2012). The analyses utilize either standard cold vapor atomic fluorescence spectroscopy (EPA, 2002) or atomic absorption spectroscopy methods. Chaulk et al. (2011) and Cossa et al. (2011) measured total Hg and dissolved Hg chemical speciation, respectively, within sea ice. Sherman et al. (2012) also used stable mercury isotopes ($\Delta^{199}$Hg, analyzed by inductively coupled plasma mass spectrometry) to investigate air-ice mercury fluxes.

## 4.4. Gases

With the realization that sea ice is porous comes an understanding that it could serve as a source or sink of climatically active gases. Most of the gases measured in sea ice, to date, have been found at relatively high concentrations; in general, the precision of the analyses has been a greater challenge than detection limits. The

Table 6. Gas analyses in sea ice

| Gas | Samples | Extraction method | Analysis | References |
|---|---|---|---|---|
| Total gas content | Bulk ice | Thaw/freeze cycling | Toepler pump | Tison et al., 2002 |
| | | Melting in artificial seawater | Tygon burette | Rysgaard and Glud, 2004 |
| $O_2$ | Bulk ice | Thaw/freeze cycling | GC[a] | Matsuo and Miyake, 1966 |
| | | *In situ* | Optodes | Mock et al., 2002; 2003 |
| | | | | Rysgaard et al., 2008 |
| | | Dry crushing | GC[a] | Tison et al., 2002 |
| | | Melting in artificial seawater | Winkler titration, GC[a] | Rysgaard and Glud, 2004 |
| | | Direct melting | Winkler titration | Søgaard et al., 2010 |
| | Brine | Sackholes | Winkler titration | Gleitz et al., 1995 |
| | | | | Delille et al., 2007 |
| | | | | Papadimitriou et al., 2007 |
| | | Gravity drainage | Winkler titration | Nomura et al., 2009 |
| | Bubbles | Melting in artificial seawater | GC[a] | Søgaard et al., 2010 |
| $CO_2$ | Bulk ice | Thaw/freeze cycling | GC[a] | Matsuo and Miyake, 1966 |
| | | Dry head-space equilibration | GC[a] | Gosink, 1978 |
| | | | | Geilfus et al., 2012b; 2014a;b |
| | | Dry crushing | GC[a] | Tison et al., 2002 |
| | | *In situ* | NDIR[b] | Miller et al., 2011a |
| | | | GC[a] | Miller et al., 2011b |
| | Brine | Sackholes | NDIR[b] | Geilfus et al., 2012a;b; 2014a;b |
| | Snow | Syringe | GC[a] | Gosink and Kelley, 1985[c] |
| $CH_4$ | Bulk ice | Purge and trap | GC[a] | Gosink, 1980[c] |
| | | Thaw/freeze cycling | GC[a] | Zhou et al., 2014a |
| CO | Bulk ice | Melt head-space equilibration | GC[a] | Gosink, 1980[c] |
| | | | | Xie and Gosselin, 2005 |
| | | | | Song et al., 2011 |
| $N_2$ | Bulk ice | Dry crushing | GC[a] | Tison et al., 2002 |
| Ar | Bulk ice | Dry crushing | GC[a] | Zhou et al., 2013 |
| $N_2O$ | Bulk ice | Purge and trap | GC[a] | Kelley and Gosink, 1979 |
| | | | | Gosink, 1980[c] |
| | | | | Randall et al., 2012 |

| Gas | Samples | Extraction method | Analysis | References |
|---|---|---|---|---|
| DMS | Bulk ice | Melting in base[d] | GC[a] | Turner et al., 1995 |
| | | Melting in acid | GC[a] | Trevena and Jones, 2006 |
| | | Melting in brine | GC[a]; PTR-MS[e] | Stefels et al., 2012 |
| | | Dry crushing | GC[a]; PTR-MS[e] | Tison et al., 2010 |
| | | | | Stefels et al., 2012 |
| | Brine | Sackholes | GC[a] | Delille et al., 2007 |
| | | | | Asher et al., 2011 |
| Halocarbons | Bulk ice | Purge and trap | GC[a] | Kelley and Gosink, 1979[c] |
| | | | | Sturges et al., 1997 |
| | | | | Granfors et al., 2013a |
| | Brine | Purge and trap | GC[a] | Mattson et al., 2012 |
| | | | | Granfors et al., 2013a;b |
| | Snow/Frost flowers | Purge and trap | GC[a] | Sturges et al., 1997 |
| | | | | Granfors et al., 2013a;b |

[a]GC: Gas Chromatography with suitable detectors
[b]NDIR: Non-dispersive infrared spectroscopy
[c]Methods not specified
[d]Determined total DMS+DMSP
[e]Proton-transfer-reaction mass spectrometry

most obvious difficulty in sampling sea ice for gases is the potential for exchange with the air. In most cases, gases are lost from the ice to the atmosphere, but ice samples can also be contaminated by contact with air, particularly by pollutant volatile organics and hydrocarbons. In addition, gases occur in sea ice both as solutes and as bubbles; some methods will extract both fractions, while others collect only the fraction dissolved in the brines. In this section, we discuss general concerns with analyzing gases in sea ice and summarize the specific methods used to date (Table 6). The details of measuring dimethylsulfide and carbon dioxide are discussed in respective sections 4.5 and 4.6, elemental mercury is covered in section 4.3, halocarbons are discussed in section 5.2, and studies of other volatile organic compounds are included in section 4.2.4.

The oldest methods for extracting gases from sea ice are wet extractions, involving sequential melting and refreezing of ice samples (Matsuo and Miyake, 1966). For insoluble gases at high concentrations in the ice or for which analyses with low detection limits exist, the refreezing step is not required. With knowledge of the gas partitioning coefficient between the aqueous melt and air, the initial melt can simply be equilibrated with a volume of ambient air, which is then analyzed. This approach has been used successfully for analyses of total gas content (Tison et al., 2002), CO (Xie and Gosselin, 2005; Song et al., 2011), $N_2O$ (Kelley and Gosink, 1979; Randall et al., 2012), methane (Zhou et al., 2014a), and organohalides (section 5.2).

Dry-extraction (or dry crushing), involving crushing an ice sample with steel balls in a vacuum chamber, has proven effective for determining $O_2$, $N_2$ (Tison et al., 2002), and Ar (Zhou et al., 2013) in sea ice. The size of the crushed ice sample mainly depends on the concentration of the target gas in the ice and the sensitivity of the gas chromatography detector. For trace gases (such as DMS, section 4.5), the gas may need to be preconcentrated before injection into the gas chromatograph. Crushing the ice has one intrinsic problem: contamination by methane released by the metal-metal friction between the stainless steel balls and the container during the crushing process (Higaki et al., 2006). Therefore, other extraction methods are used for analyses of carbon-containing species.

Sea-ice brines for gas analyses are usually collected from sackholes, although Nomura et al. (2009) used full-core gravity drainage (Figure 4c) to collect brines for $O_2$ analysis. In general, extracting brines for gas analyses is only satisfactory for relatively soluble gases, such as $O_2$ and $CO_2$, and even so, only if temperatures are high enough for the brines to accumulate quickly and the sackhole is capped (Papadimitriou et al., 2007). In contrast, the bulk of the insoluble gases in sea ice is probably located in bubbles within brine pockets and channels; when only the brines are analyzed, such gas bubbles within the ice are lost. Zhou et al. (2014a) found that brine $CH_4$ concentrations deduced from measurements in bulk ice can be up to 10 times higher than the concentrations directly measured in brine samples, a difference almost certainly due to exchange with the atmosphere during brine percolation.

*In situ* probes hold great potential for determining gas concentrations in sea ice (McMinn et al., 2009). Probes based on photochemical detection (*i.e.*, optodes) are particularly promising for sea-ice applications; oxygen has been measured successfully in sea ice using commercially available optodes (Figure 12a; Mock

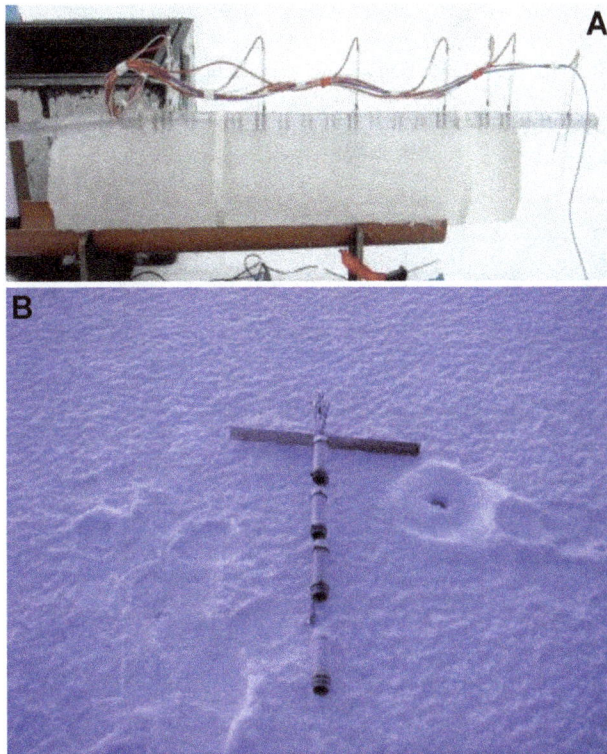

Figure 12

*In situ* probes for measuring gases in sea ice.

(a) Oxygen optodes deployed in an ice core, prior to being replaced into the original core hole. Photo: A. Krell. (b) Silicone chamber 'peeper' array ready for deployment through the adjacent hole in the sea ice. Amundsen Gulf, December 2007. Photo: N. Sutherland.

et al., 2002; Rysgaard et al., 2008), although reaction times are slow and calibrations are potentially complicated by both high salinities and organic matter concentrations. Another potential drawback of microprobe measurements is uncertainty in the specific microenvironment sampled. For example, an oxygen optode might only sense the liquid phase, missing the (often dominant) gas phase. In addition, any *in situ* probe will change the thermodynamic environment of the ice to at least some extent; particularly under sunny conditions, objects frozen into the ice absorb heat, causing localized excess melt. Gas-permeable silicone chambers developed by the soil science community have been deployed for measuring *in situ* $CO_2$ mole fractions in sea ice (Figure 12b; section 4.6.4; Miller et al., 2011a;b) and theoretically could be used for analyses of other gases, including $O_2$, $CH_4$, and $N_2O$ (*e.g.*, Holter, 1990; Kammann et al., 2001). Electrochemical probes are generally unsuitable for sea-ice applications, because their internal electrolyte solutions freeze, but have potential for further development. The primary drawbacks to *in situ* sensors, in general, are that the study site must be occupied for an extended period and, unless the sensors can be deployed at freeze-up, their installation requires disturbing the ice cover.

## 4.5. Sulfur species

Dimethylsulfide (DMS) is one of the most abundant volatile sulfur compounds in the ocean and accounts for more than half of the global biogenic sulfur flux to the atmosphere (*e.g.*, Liss et al., 1997). Sea ice usually contains larger amounts of DMS and its precursor dimethylsulfonioproprionate (DMSP) than does the under-ice water (up to two orders of magnitude higher), although the amounts in sea ice are highly variable (*e.g.*, Trevena and Jones, 2006). Therefore, seasonal ice melting introduces elevated DMS concentrations to surface waters from the release of sea-ice DMS and DMSP (*e.g.*, Tison et al., 2010). The sulfur compounds studied in sea ice to date are DMS, DMSP (which occurs in both particulate and dissolved fractions), and dimethylsulfoxide (DMSO). Other sulfur species important in the sulfur cycle that have not yet been investigated in sea ice include sulfur dioxide ($SO_2$), hydrogen sulfide ($H_2S$), carbonyl sulfide (COS), and carbon disulfide ($CS_2$).

Total and dissolved DMSP have been measured in ice brines recovered from sackholes (Trevena and Jones, 2006; Asher et al., 2011), although it is not clear how well dissolved DMSP measurements from brines distinguish the *in situ* partitioning between dissolved and particulate fractions within the brine network of undisturbed ice (sections 2.2.2, 4.2.1). Total DMSP is often analyzed in bulk ice melts (Levasseur et al., 1994; Curran and Jones, 2000; Trevena et al., 2000; Trevena and Jones, 2006), whether obtained by melting in filtered seawater (Levasseur et al., 1994), in acidified filtered seawater (Trevena et al., 2000; Trevena and Jones, 2006), or in concentrated brine (Stefels et al., 2012). However, in a direct comparison of DMSP analyses from twin cores, one melted in concentrated filtered seawater and one dry-crushed (section 4.4), Stefels et al. (2012) found that large amounts of DMSP can be converted to DMS during the melting process.

Because of its very low solubility, accurately measuring DMS concentrations in ice samples has been a challenge. Although brines have been analyzed for DMS (Table 6), it is generally considered too insoluble to be recovered confidently from brines collected from sackholes. Trevena and Jones (2006) melted ice samples directly into acid within purge chambers. Small ice samples have also been directly crushed to recover and

Table 7. $CO_2$ system analyses in sea ice

| Parameter[a] | Samples | Processing | References |
|---|---|---|---|
| DIC and TIC | Bulk ice | Melting in distilled water | Rysgaard et al., 2007 |
| | | | Søgaard et al., 2013 |
| | | Direct melting | Fransson et al., 2011; 2013 |
| | | | Miller et al., 2011a;b |
| | | | Geilfus et al., 2012a; 2014a |
| | | | Hawes et al., 2012 |
| | Brine | Sackholes | Garrison et al., 2003 |
| | | | Papadimitriou et al., 2004; 2007; 2009; 2012 |
| | | | Munro et al., 2010 |
| | | | Fransson et al., 2011; 2013 |
| | | | Miller et al., 2011a |
| | | | Geilfus et al., 2012a; 2014a |
| | | | Nomura et al., 2010b; 2013b |
| | | Gravity drainage | Nomura et al., 2009 |
| | | Centrifugation | Munro et al., 2010 |
| | Frost flowers | Direct melting | Miller et al., 2011a |
| | | | Fransson et al., 2013 |
| $A_T$ | Bulk ice | Direct melting | Lyakhin, 1970 |
| | | | Anderson and Jones, 1985 |
| | | | Nedashkovsky et al., 2009 |
| | | | Fransson et al., 2011; 2013 |
| | | | Miller et al., 2011a;b |
| | | | Geilfus et al., 2012a;b; 2013; 2014a |
| | | | Hare et al., 2013 |
| | | | Nomura et al., 2013a |
| | | Melting in distilled water | Ryssgaard et al., 2007 |
| | | | Søgaard et al., 2013 |
| | | Melting in seawater | Nedashkovsky and Shvetsova, 2010 |
| | Brine | Sackholes | Gleitz et al., 1995 |
| | | | Kennedy et al., 2002 |
| | | | Delille et al., 2007 |
| | | | Papadimitriou et al., 2007; 2009; 2012 |
| | | | Fransson et al., 2011; 2013 |
| | | | Geilfus et al., 2012a;b; 2014a;b |
| | | | Nomura et al., 2010b; 2013a |
| | | Gravity drainage | Nomura et al., 2009 |
| | Snow | Melting in seawater | Nedashkovsky and Shvetsova, 2010 |
| | Frost flowers | Direct melting | Miller et al., 2011a |
| | | | Douglas et al., 2012 |
| | | | Fransson et al., 2013 |
| | | | Geilfus et al., 2013 |

| Parameter[a] | Samples | Processing | References |
|---|---|---|---|
| pH | Brine | Sackholes | Gleitz et al., 1995 |
| | | | Kennedy et al., 2002 |
| | | | Papadimitriou et al., 2004 |
| | | | Delille et al., 2007 |
| | | | Miller et al., 2011a |
| | | | Hare et al., 2013 |
| | | | Geilfus et al., 2014b |
| PIC | Brine | Centrifugation | Tison et al., 2002 |
| | Bulk ice | Pipetting from melts | Dieckmann et al., 2008 |
| | | | Miller et al., 2011b |
| | | | Fischer et al., 2012 |
| | | | Geilfus et al., 2013 |
| | | | Nomura et al., 2013a |
| | | Melt filtration | Dieckmann et al., 2010 |
| | | | Fischer et al., 2012 |
| | | | Søgaard et al., 2013 |
| | | Microscopy | Rysgaard et al., 2013 |
| | | | Geilfus et al., 2014a |
| | Frost flowers | Pipetting from melts | Geilfus et al., 2013 |

[a]See Table 6 for $pCO_2$ analyses.

analyze DMS (Tison et al., 2010; Stefels et al., 2012). Fluxes of DMS out of sea ice have been estimated using both chamber (section 5.1.1; Nomura et al., 2012) and micrometeorological (section 5.1.4; Zemmelink et al., 2008) techniques. The analyses generally use gas chromatography with either flame photometric or mass spectrometric detection.

Analyses of DMSO in sea ice are still scarce and complicated by the need to eliminate interference from DMSP (Brabant et al., 2011). Lee et al. (2001) measured DMSO associated with the particulate (algal) fraction in sea ice, Brabant et al. (2011) determined total DMSO in bulk ice, and Asher et al. (2011) measured dissolved DMSO in sea-ice brines. All of these studies analyzed DMSO as DMS, after purging and reduction.

The rates at which sulfur species are formed and degraded is one of the largest uncertainties in the global sulfur cycle (*e.g.*, Ayers and Cainey, 2007); in only one study have natural rates of interconversion between the various sulfur compounds in sea ice brines been determined directly (using short-term incubations with $^2$H- and $^{13}$C-labelled DMS, DMSP, and DMSO tracers; Asher et al., 2011). Stefels et al. (2012) described a method for spiking samples with deuterated DMS and DMSP before melting, in order to document and then correct for DMSP degradation. Although used to examine sample storage and processing artifacts (Stefels et al., 2012), this tracer method also has potential for investigating conversion rates by the *in situ* community. As for primary production and bacterial production (sections 3.4.1 and 3.4.3), a satisfactory method for determining sulfur cycling rates in undisturbed ice environments has not been reported, making continued method development a high priority. In general, we encourage immediate processing and analysis of sea-ice samples for sulfur cycle studies to limit interconversion between sulfur species and loss of insoluble DMS.

## 4.6. The carbon dioxide system

The inorganic carbon system in sea ice is controlled by complex biogeochemical processes that transform carbon between phases (gas bubbles, brine, and particles) and between inorganic and organic forms, while also exchanging with the atmosphere and the underlying water (*e.g.*, Thomas et al., 2010; Loose et al., 2011). A number of methods have been used to collect and process samples for determining $CO_2$ system parameters in sea ice (Table 7), but sample storage and analyses have generally followed standard protocols for seawater (Dickson et al., 2007); methodological intercalibrations are desperately needed. For example, in the one study that determined $pCO_2$ within sea ice by multiple methods (calculated from TIC and alkalinity in brines, section 4.6.1; calculated from TIC and alkalinity in bulk ice melts, section 4.6.3; and with *in situ* peepers, section 4.6.4), poor agreement was found between them (Miller et al., 2011a).

Interpretations of inorganic carbon data from sea ice have also been heavily influenced by our understanding of the seawater inorganic carbon system (*e.g.*, Zeebe and Wolf-Gladrow, 2001). This reliance on the seawater model is perilous, because the seawater methods have been optimized for seawater and thus calibrated for very narrow concentration and salinity ranges; both sea-ice brines and bulk melt samples almost always fall outside those ranges. In addition, pH is defined analytically on a number of different scales (*e.g.*, Dickson, 1993). The standard scale used for seawater (the total hydrogen ion scale, $pH_T$), and standard buffers certified on that scale, cannot be applied rigorously to sea-ice brines (*e.g.*, Miller et al., 2011a). Finally, the conditional stability constants used to convert between $CO_2$ partial pressure ($pCO_2$), dissolved inorganic carbon (DIC), total alkalinity ($A_T$), and pH in seawater are only rigorously valid for temperatures above 0 °C and salinities between 5 and 50 g kg$^{-1}$. Studies in spring ice (Delille et al., 2007) indicated that seawater thermodynamic relationships may be acceptable in warm, low-salinity sea ice, but in sea-ice brines at even moderate brine salinities of 80 g kg$^{-1}$, Brown et al. (2014) found that measured and calculated values of the $CO_2$ system parameters can differ by as much as 40%. On the other hand, because the $CO_2$ system parameters are much more variable in sea ice than in seawater, sea-ice measurements demand less precision than those in seawater.

As indicated by Tables 6 and 7, most carbonate system parameters have been measured in both bulk ice and brines. Attempts to use those data to understand sea-ice biogeochemistry always involves uncomfortable assumptions about the validity of the sample handling and the *in situ* behavior of the $CO_2$ system. For example, whereas brine samples are compromised by gas exchange during sampling, interpreting measurements in bulk sea ice requires assumptions about the presence of gaseous and solid inorganic carbon. Therefore, the best way to sample the ice and the best parameters to measure depend on both the conditions and specific questions targeted by the study.

Nonetheless, we can make some recommendations. In general, samples for $CO_2$ system analyses in sea ice should be collected upwind from any ship, camp, or generator to avoid contamination by $CO_2$ or soot from fossil fuel combustion. Hand-coring is preferable, although electric auger motors can be used, as long as the generator is located a substantial distance downwind. Some investigators have filtered their samples for DIC analyses (Papadimitriou et al., 2004; 2007; 2012) using specialized techniques to avoid significant gas exchange (McCorkle et al., 1985), but vacuum filtration is not recommended. Because the concentration of particulate inorganic carbon (PIC) can be high in sea ice (Dieckmann et al., 2008; 2010; Rysgaard et al., 2013), the results from DIC analyses of unfiltered sea-ice samples are properly termed total inorganic carbon (TIC).

### 4.6.1. Brines

Samples for $pCO_2$, TIC, and pH are sensitive to gas exchange and need to be isolated from the atmosphere during sampling. Therefore, analyses of these parameters in brines, which are difficult to sample without exposing them to the air, can be problematic (see also sections 2.2.2 and 4.4).

Electrochemical pH measurements are particularly challenging in ice brines, because the high sample salinities result in large liquid junction potentials and severely slow electrode response times. In addition, stable, certified standard buffers are not available for brine solutions, compromising electrode calibration. Although the first sea-ice brine pH measurements were made electrochemically (Gleitz et al., 1995), spectrophotometric measurements are becoming more common (Miller et al., 2011a; Hare et al., 2013). Wren and Donaldson (2012) have developed a spectrophotometric method for analyzing pH in surface brine films that may have potential for *in situ* applications. Particular care is needed in spectrophotometric analyses to use optical and thermodynamic parameters for the dyes that have been defined for appropriate temperature and salinity ranges (*e.g.*, Millero et al., 2009).

### 4.6.2. Gas bubbles

The standard method of crushing ice under vacuum to retrieve gases trapped in bubbles within an ice sample, as developed for glacial ice, may give artificially high $pCO_2$ values when applied to sea ice (*e.g.*, Tison et al., 2002). The vacuum likely disrupts the $CO_2$ system equilibria within the brines, causing $CO_2$ outgassing from the brine solution and possibly also precipitating $CaCO_3$ (Geilfus et al., 2012b). Therefore, Geilfus et al. (2012b) developed a method to accurately measure $CO_2$ in gas bubbles and brines in sea ice by equilibration with a headspace of known volume and $CO_2$ mole fraction (dry head-space equilibration). The headspace must be as small as possible to assure that the $CO_{2(g)}$ released from the ice dominates the $CO_2$ signal, with only a small contribution from the standard headspace gas. Gosink (1978) described an *in situ* head-space equilibration technique that involved sealing sampling cuvettes to the ice surface.

### 4.6.3. Bulk ice melts

Total alkalinity of bulk ice melts is a relatively uncomplicated analysis that has been performed for decades (*e.g.*, Lyakhin, 1970; Anderson and Jones, 1985; Nedashkovsky et al., 2009). In a standard potentiometric titration (*i.e.*, Dickson et al., 2007), the measured $A_T$ will include not only that which was in the brines *in situ*, but also a contribution from any particulate inorganic carbon (*e.g.*, $CaCO_{3(s)}$) that dissolves as the ice melts or when the sample is acidified during the titration.

On the other hand, because TIC is impacted by gas exchange, its analysis in bulk ice is more complicated; a method for confidently melting ice samples without allowing interaction with ambient $CO_2$ has not yet been devised. In the field, ice cores need to be retrieved, sectioned, and isolated from the atmosphere as quickly as possible. The most common approach to melting ice samples for TIC analysis is to use gas-impermeable bags (made from fluoropolymers such as ALTEF® or Kynar®; Rysgaard et al., 2009; Fransson et al., 2011; 2013; Miller et al., 2011a). After sealing the sample in the bag, the headspace should be removed using a hand pump, to assure that the sample is not exposed to an excessive vacuum. As long as the container in which the sample is melted is sealed and the headspace (after melting) is less than 2% of the total volume, the TIC concentration in the solution should be correct to within 0.01% (Dickson et al., 2007); in fact, because the $p\mathrm{CO_2}$ of ice melts is generally low, the melt solution should also absorb essentially all of the gaseous $\mathrm{CO_2}$ initially present as bubbles trapped in the ice. Alternatively, the ice sample can be melted without a headspace in water of known TIC concentration (Rysgaard et al., 2007; 2009).

Unless $A_T$ and TIC samples can be analyzed within 1–2 hours of collection, they generally should be poisoned with small quantities of $\mathrm{HgCl_2}$ (Dickson et al., 2007). Although straightforward and relatively safe for samples that are initially aqueous, like seawater and brines, poisoning is more complicated for sea-ice melts, which cannot be bottled for long-term storage until after melting is complete. Some researchers have added the $\mathrm{HgCl_2}$ directly to the bag with the melting sample (Rysgaard et al., 2009; Fransson et al., 2013). However, because ice melts are poorly buffered, the $\mathrm{Hg(OH)_2}$ complexes formed from the added mercury may impact the carbonate system chemistry (Fransson et al., 2013). In addition, it is difficult to completely contain the mercury when working with gas-impermeable bags (during cleaning between samples, but also because the bags can fail at low temperatures, developing small holes and leaking), creating an exposure risk for all personnel using the core-processing laboratory. Therefore, $\mathrm{HgCl_2}$ is often added to samples only after they are transferred from the bags into bottles for long-term storage. We still lack a satisfactory method for safely preserving sea-ice samples for carbonate system analyses during melting.

Unlike TIC and $A_T$ (in units of mol kg$^{-1}$), which are total quantities unaffected by the temperature and salinity changes associated with melting, pH and $p\mathrm{CO_2}$ are potentials; their values measured in sea-ice melts cannot be directly converted to the original conditions in the solid ice/brine matrix. Therefore, $p\mathrm{CO_2}$ and pH measurements in ice melts do not provide information about the initial, *in situ* conditions of the ice, although the measurements can be used to calculate other $\mathrm{CO_2}$ system parameters or to derive the theoretical characteristics of the melt that will influence the surface waters in summer (Nedashkovsky and Shvetsova, 2010, Fransson et al., 2011; Geilfus et al., 2013). Samples for $p\mathrm{CO_2}$ and pH analyses have the same issues with potential degradation during melt as TIC and $A_T$ samples: standard seawater protocols (*e.g.*, Dickson et al., 2007) indicate that $p\mathrm{CO_2}$ samples that cannot be analyzed within a couple of hours should be poisoned with $\mathrm{HgCl_2}$; and, presently, pH samples cannot be stored for they change significantly within hours of collection (*i.e.*, within the time it takes for an ice sample to melt).

### 4.6.4. *In situ* sensors

Ideally, *in situ* sensors would provide the most meaningful $p\mathrm{CO_2}$ and pH measurements in sea ice, particularly if the sensors could be deployed at freeze-up, so that installation would not disrupt an established ice cover. To date, available pH microelectrodes are still unsuitable for deployment in sea ice, because their electrolyte solutions freeze at low temperatures, in addition to the calibration and response-time issues discussed in section 4.6.1. Although *in situ* silicone gas exchange chambers ("peepers") have been used for $p\mathrm{CO_2}$ measurements (section 4.4; Miller et al., 2011a;b), gas diffusion rates in silicone decrease dramatically with temperature, and peepers require long equilibration times, limiting deployments to extended occupations of a single site. In addition, peepers have not yet been fully tested or calibrated under controlled conditions. As noted in section 4.4, *in situ* $\mathrm{CO_2}$ or pH probes will modify their local thermodynamic environment within the ice to at least some extent.

### 4.6.5. Particulate inorganic carbon (PIC)

A number of carbonate salts are known to precipitate from brines in sea ice, including calcium carbonate (generally in the form of ikaite, $\mathrm{CaCO_3 \cdot 6H_2O}$; *e.g.*, Dieckmann et al., 2008) and magnesium- and mixed magnesium-calcium-carbonates (*e.g.*, Assur, 1958). Precipitation of any of these minerals likely has a strong influence on $p\mathrm{CO_2}$ and the entire $\mathrm{CO_2}$ system within the ice. Ikaite has been identified in both Antarctic and Arctic sea ice (Delille, 2006; Dieckmann et al., 2008; 2010; Rysgaard et al. 2012; 2013; Geilfus et al., 2013; Nomura et al., 2013a), but solid salts are not recovered from all samples (*e.g.*, Nomura et al., 2013a). The lack of observable PIC in some samples is likely due to natural variability in sea ice (possibly related to phosphate concentrations, thermal history of the ice, and/or sea-ice permeability; Nomura et al., 2013a; Rysgaard et al., 2013; Papadimitriou et al., 2013; 2014) and not methodological differences, as the presence and absence of precipitates are observed by the same groups. Recent laboratory studies have defined the thermodynamics and kinetics of ikaite precipitation in abiotic ice brines (Papadimitriou et al., 2013; 2014), but the conditions controlling the formation and preservation of solid $\mathrm{CaCO_3}$ in natural sea ice are still largely unknown.

Because carbonate salts are water soluble to varying extents at temperatures above freezing, the ice samples must be melted at a low temperature and processed as soon as melting is completed. Although the melted samples can be filtered to collect the solid precipitate (Dieckmann et al., 2010; Fischer et al., 2012; Søgaard et al., 2013), the quantity of particulate organic matter present in the sea ice can interfere with subsequent visualization and analysis of the inorganic salts, so the precipitate is often collected from sea-ice melt samples using a pipette (Dieckmann et al., 2008; Miller et al., 2011b; Geilfus et al., 2013; Nomura et al., 2013a). Rysgaard et al. (2013) have developed a microscopic method to visually identify and quantify ikaite crystals from the ice as it melts. Even when dry, ikaite is unstable at temperatures above 4 °C; if confirmation of the specific calcium carbonate mineralogy (*i.e.*, ikaite versus calcite, aragonite, or vaterite) is required, the sample must be kept below 4 °C throughout sample recovery, melting, storage, transport, and analysis. Analysis is usually by x-ray diffraction spectrometry, but facilities able to keep a sample cold throughout the analysis are rare. If the specific mineralogy of the salt is not required, the precipitate sample can be stored indefinitely at room temperature and analyzed on any x-ray diffraction instrument (Dieckmann et al., 2008; Miller et al., 2011b) or with a standard calcium assay (Fischer et al., 2012).

The question of whether carbonate minerals that precipitate within sea ice are mobile with the brines has not been completely resolved. Despite observations that particulates are under-represented in percolating sea-ice brines (section 2.2.2), solid $CaCO_3$ has been recovered from centrifuged brines (Tison et al., 2002), and circumstantial evidence from the Sea of Okhotsk (Lyakhin, 1970) and the Beaufort Sea (Fransson et al., 2013) has indicated that abiotic $CaCO_3$ precipitates from sea ice may be released to the water column.

## 4.7. Photochemistry: CDOM, hydrogen peroxide, and ozone

Photochemical processes are likely to be very important in many sea-ice biogeochemical cycles, but while numerous studies have examined the transmission of electromagnetic radiation through sea ice (*e.g.*, Perovich, 2009), there has been little research on photochemistry within the ice (Belzile et al., 2000). Colored dissolved organic matter (CDOM, usually measured spectrophotometrically and reported as absorption coefficients, in units of $m^{-1}$) likely represents the most photochemically active fraction of the non-living sea-ice components and has been measured in sea ice by a number of investigators (Belzile et al., 2000; Scully and Miller, 2000; Granskog et al., 2005b; 2015; Uusikivi et al., 2010; Norman et al., 2011). Fluorescence has also been used to measure and differentiate the components of CDOM in natural and laboratory sea ice and in frost flowers (Stedmon et al., 2007; 2011; Müller et al., 2013; Granskog et al., 2015). Although generally assumed to be less prone to contamination than bulk DOC, samples of CDOM, particularly those measured by fluorescence, can easily be contaminated. In addition, the effects of melting protocols on the absorbance of organic matter from sea ice has not been investigated explicitly. Hydrogen peroxide and other photochemically produced oxidizers, such as ozone, are likely important players in any photochemical reactions occurring in sea ice (Klánová et al., 2003; King et al., 2005), but no one has reported direct sea-ice measurements of these compounds.

## 4.8. Stable isotopes: $^{18}O$, $^{2}H$, $^{13}C$, $^{15}N$, $^{30}Si$

The stable oxygen isotope ratio ($^{18}O$ relative to $^{16}O$) in the water molecules of a sea-ice sample is controlled by many of the processes that influence sea-ice biogeochemistry, including freezing, melting, flooding, and snowfall, making $\delta^{18}O$ a powerful tool for sea-ice studies (*e.g.*, Eicken, 1998; Granskog et al., 2003; Tison et al., 2008; Nomura et al., 2009; 2011). Particularly in the Arctic Ocean, where $\delta^{18}O$ measurements can also help distinguish between riverine versus sea-ice melt sources of freshwater in the surface ocean (*e.g.*, Macdonald et al., 1989), $\delta^{18}O$ is often considered a mandatory parameter, along with salinity.

Sampling $\delta^{18}O$ is simple and inexpensive; the main concern is evaporation during sample storage. Although evaporation during melting could also be an issue, if the container is open or the headspace is large, evaporation is no more of a problem for $\delta^{18}O$ than for salinity. Mass spectrometric analyses of $\delta^{18}O$ require as little as a few milliliters of sample; both bulk sea-ice melts and brines have been analyzed for $\delta^{18}O$ (*e.g.*, Zhou et al., 2013). Glass containers with caps forming a tight seal are preferred, particularly if the samples are likely to be stored for more than a few months before analysis (*e.g.*, McLaughlin et al., 2012). Parafilm can also be wrapped around the outside of the cap, to further protect against leakage (Miller et al., 2011b).

Sea-ice measurements of the stable isotopes of other elements are summarized in Table 8. Deuterium fractionation in the water molecules in sea ice is greater than that of $^{18}O$, providing additional information on brine convection and ice melt. Sampling and handling of $^{2}H$ samples for mass spectrometric analysis should follow those for $^{18}O$ samples (Zhou et al., 2013). Stable isotope ratios of carbon ($^{13}C$ relative to $^{12}C$), nitrogen ($^{15}N$ relative to $^{14}N$), and silicon ($^{30}Si$ relative to $^{28}Si$) are proving to be useful tools in investigations of sea-ice biogeochemical cycles, including interpretations of sedimentary records in the polar oceans. In sea ice, stable isotope measurements can potentially help distinguish between the origins (*i.e.*, land, seawater, or sea ice) of the particulate organic matter, as well as between biogeochemical cycling pathways. Samples for stable isotope analyses are typically collected from bulk sea ice and sackhole brines following the methodologies for POC, PON, $bSiO_2$ (section 3.3.1), $NO_3^-$, $Si(OH)_4$ (section 4.1), TDN (section 4.2.2), and TIC (section 4.6); the analyses have used standard mass-spectrometric techniques (*e.g.*, McCorkle et al., 1985; Kennedy and Robertson, 1995; Sigman et al., 2001; Cardinal et al., 2003).

Table 8. Stable isotope measurements in sea ice[a]

| Element | Isotope measured | Samples | References |
|---|---|---|---|
| Hydrogen | $\delta^2H$-$H_2O$ | Snow | Zhou et al., 2013 |
| | | Bulk ice | Zhou et al., 2013 |
| | | | Geilfus et al., 2014a |
| | | Brine | Zhou et al., 2013 |
| | | | Geilfus et al., 2014a |
| Carbon | $\delta^{13}C$-DIC | Platelet ice/interstitial waters | Thomas et al., 2001b |
| | | Gap Layer Water | Kennedy et al., 2002 |
| | | | Papadimitriou et al., 2009 |
| | | Brine | Papadimitriou et al., 2004; 2007 |
| | | | Munro et al., 2010 |
| | $\delta^{13}C$-POC | Bulk ice | Gibson et al., 1999 |
| | | | Schubert and Calvert, 2001 |
| | | | Arrigo et al., 2003 |
| | | | Tremblay et al., 2006 |
| | | | Pineault et al., 2013 |
| | | Platelet ice/ interstitial waters | Thomas et al., 2001b |
| | | Gap layer water | Kennedy et al., 2002 |
| | | | Papadimitriou et al., 2009 |
| Nitrogen | $\delta^{15}N$-PON | Bulk ice | Rau et al., 1991 |
| | | | Schubert and Calvert, 2001 |
| | | | Tremblay et al., 2006 |
| | | | Pineault et al., 2013 |
| | | | Fripiat et al., 2014a |
| | $\delta^{15}N$-$NO_3^-$ | Bulk ice | Fripiat et al., 2014a |
| | $\delta^{15}N$-TDN[b] | Bulk ice | Fripiat et al., 2014a |
| Silicon | $\delta^{30}Si$-$Si(OH)_4$ | Brine | Fripiat et al., 2007 |
| | | | Fripiat et al., 2014b |
| | $\delta^{30}Si$-$bSiO_2$ | Bulk ice | Fripiat et al., 2007 |

[a]$\delta^{18}O$ analyses are routine and, therefore, not included.
[b]TDN = $NO_3^-$ + $NO_2^-$ + $NH_4^+$ + DON and DON = dissolved organic N.

# 5. Ice-atmosphere and ice-ocean fluxes

Because sea ice is porous, it exchanges material with both the overlying atmosphere and the underlying water. Brine rejection from sea ice has long been recognized as a primary driver of oceanic deepwater formation and global circulation (*e.g.*, Chu and Gascard, 1991), with likely implications for the biogeochemical cycles of many elements. Likewise, the atmosphere, both in the boundary layer directly above the sea ice and at higher altitudes (*e.g.*, Begoin et al., 2010), is strongly influenced by sea-ice biogeochemistry.

Qualitative information about the direction of fluxes can be derived from measurements of concentration gradients between the ice and the atmosphere or the water: the larger the gradient, the larger the flux might be. However, confirming and quantifying those presumed fluxes requires more sophisticated methods. In particular, when material fluxes above sea ice are measured directly, the fluxes estimated from the measured ice-air gradients can be wrong both in magnitude and direction, because of reactions occurring at the interface, in the surface brines, frost flowers, and snow cover, that produce or consume gases and aerosols.

## 5.1. Air-ice gas fluxes

A number of methods have been developed to estimate gas fluxes above sea ice, but no systematic intercomparisons between various gas flux techniques over ice surfaces have been published. In particular, micrometeorological (sections 5.1.2–5.1.6) and chamber (section 5.1.1) methods measure fluxes on very different temporal and spatial scales, which has confounded efforts to compare the resulting flux estimates. Comparisons between eddy covariance and chamber methods have been carried out over terrestrial surfaces

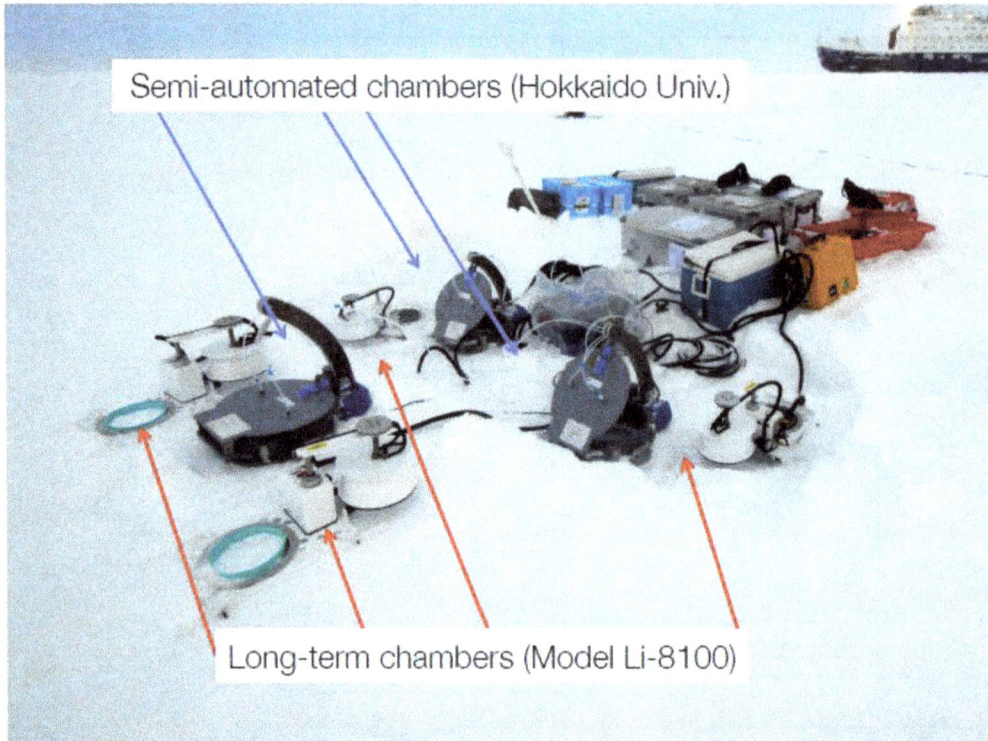

**Figure 13**

**Gas flux chambers deployed to measure air-sea ice $CO_2$ exchange.**

Configured for an intercalibration experiment between two types of chamber systems: 1) semi-automated $CO_2$ chambers originally developed at Hokkaido University for soil $CO_2$ flux measurements; and 2) long-term chambers (Li-8100) manufactured by LI-COR Biosciences, USA. Weddell Sea, July 2013. Photo: D. Nomura.

Table 9. Sea ice-air flux measurements

| Gas | Method | References |
|---|---|---|
| $CO_2$ | Enclosure | Gosink, 1978 |
| | | Semiletov et al., 2004 |
| | | Delille, 2006 |
| | | Nomura et al., 2010a;b; 2013b |
| | | Sejr et al., 2011 |
| | | Geilfus et al., 2012a; 2013; 2014a;b |
| | | Fischer, 2013 |
| | Eddy covariance | Semiletov et al., 2004 |
| | | Zemmelink et al., 2006 |
| | | Else et al., 2011 |
| | | Miller et al., 2011b |
| | | Papakyriakou & Miller, 2011 |
| DMS | Eddy accumulation | Zemmelink et al., 2008 |
| | Enclosure | Nomura et al., 2012 |
| CO | Mass balance | Gosink and Kelley, 1979 |
| | | Kelley and Gosink, 1979 |
| $O_3$ | Eddy covariance | Muller et al., 2012 |
| Iodated organics | Mass balance | Shaw et al., 2011 |

(Wang et al., 2013; Riederer et al., 2014), but carefully designed intercalibration experiments over sea ice are still needed to resolve remaining important questions about how each type of flux measurement performs in the sea-ice environment.

### 5.1.1. Flux chambers

Enclosure methods, which were widely used in early biosphere-atmosphere exchange studies (*e.g.*, Mosier, 1989), are still common for some applications, including sea-ice biogeochemistry (Figure 13; Table 9). The method is based on the rate of increase (or decrease) of the trace gas concentration with time within a chamber placed directly on the ice (or snow) surface providing the flux (McMinn et al., 2009).

All enclosure methods are subject to potential artifacts in the measured flux (*e.g.*, Winston et al., 1995; Fowler et al., 2001; Nomura et al., 2012), because the enclosure itself introduces:

- changes in the radiation balance (both short and long wave);
- changes in the temperatures of the air and the surface (*i.e.*, the ice or snow);
- changes in turbulence, wind speed, and the vertical density profile;
- a pressure gradient between inside and outside the chamber; and
- a surface-atmosphere gas concentration gradient, including changes in the gas concentration inside the chamber resulting from the flux.

To overcome these issues, flux measurements should be completed quickly, before changing conditions introduce artifacts. Although the optimum length of time for a measurement is dependent on the situation (*e.g.*, weather, gas concentration gradient), 20 to 30 minutes is generally recommended. However, over snow, even a perfect chamber deployment may underestimate the flux, because in an undisturbed system, wind-driven pressure pumping within a snow cover can enhance the transport beyond molecular diffusion by up to 40% (Bowling and Massman, 2011).

Two fundamentally different types of chamber systems are available (Mosier, 1989; Luo and Zhou, 2006): in closed systems, the change in the gas concentration inside the chamber is measured directly, usually by connecting the interior to an on-line gas analyzer in a closed loop; in open systems, ambient air is pumped through the chamber, and the flux is calculated from the air flow rate and the difference in gas concentrations between the inlet and outlet. Changes in the trace gas concentration within the enclosure are a larger concern with a closed system, but open chamber methods are more susceptible to artifacts arising from interior-to-exterior pressure gradients. To date, only closed-chamber systems have been used to measure gas fluxes above sea ice; because of the difficulty in controlling the pressure differential in open systems, we recommend using closed systems, particularly if they incorporate dampening to equalize the pressure (as described by Xu et al., 2006). Current generation closed-chamber systems (*e.g.*, the Li-COR 8100-104) include such dampening mechanisms, making them probably the best chambers for measuring gas fluxes over sea ice (Fischer, 2013).

The greatest advantages and disadvantages to using enclosure methods are both due to spatial variability. Chamber enclosures only integrate the signal from the area they cover (generally, a few hundred cm$^2$); if the exchange is governed by factors that vary on larger horizontal scales (*i.e.*, the thickness and wetness of the snow cover, melt ponds, leads, under-ice hydrology, *etc.*), a prohibitive number of individual chamber measurements over a large area may be required to estimate the flux accurately (section 2.1). On the other hand, the method is ideal for studying specific, small-scale processes influencing variations in the flux (*i.e.*, brine channel distributions, ice algae respiration, *etc.*), and enclosure methods are the only technique available to determine fluxes on the same scale as most sea-ice biogeochemical measurements. In contrast, the micrometeorological techniques (sections 5.1.2–5.1.6) cover areas several orders of magnitude larger than chambers, integrating fluxes from different ice types and any open water in the footprint; micrometeorological results can, therefore, be difficult to interpret over heterogeneous surfaces.

### 5.1.2. Micrometeorological methods: general

Micrometeorological techniques, such as eddy covariance (EC), can be used to measure fluxes of gases, as well as of momentum, sensible heat, and latent heat (*e.g.*, Vihma et al., 2009) above sea ice. Unlike enclosure techniques, micrometeorological methods do not modify the observed environment, and they integrate processes occurring over relatively large spatial areas (up to several hundred m$^2$). This integration can be a problem if the surface is heterogeneous (*i.e.*, sea ice, leads, or other open water features may contribute to the observed fluxes), but typically the issue can be addressed by calculating the "footprint" of the flux measurement (*e.g.*, Vesala et al., 2008). The EC method has been used widely to examine $CO_2$ fluxes over all types of surfaces, but its application to many other chemical compounds has been hampered by a lack of fast-response sensors and by small signal levels. These limitations have led to development of a variety of alternative methods, including eddy accumulation (Businger and Oncley, 1990) and gradient techniques (*e.g.*, Businger et al., 1971). However, EC is, by definition, a direct flux measurement method (*e.g.*, Swinbank, 1951), while the others are based on several assumptions that break down over heterogeneous surfaces and in stably stratified atmospheres, such as often occur over sea ice.

The basic framework for measuring and interpreting micrometeorological flux data, including those from EC systems, is based on simultaneous measurements of the gas concentration and vertical, turbulent motion in the atmosphere (Figure 14). Ideal conditions for micrometeorological flux measurements include a horizontal and homogeneous surface, no source or sink in the atmosphere that can alter the concentration above the surface, and consistent atmospheric conditions (*i.e.*, air temperature, wind velocity, and mean gas concentration).

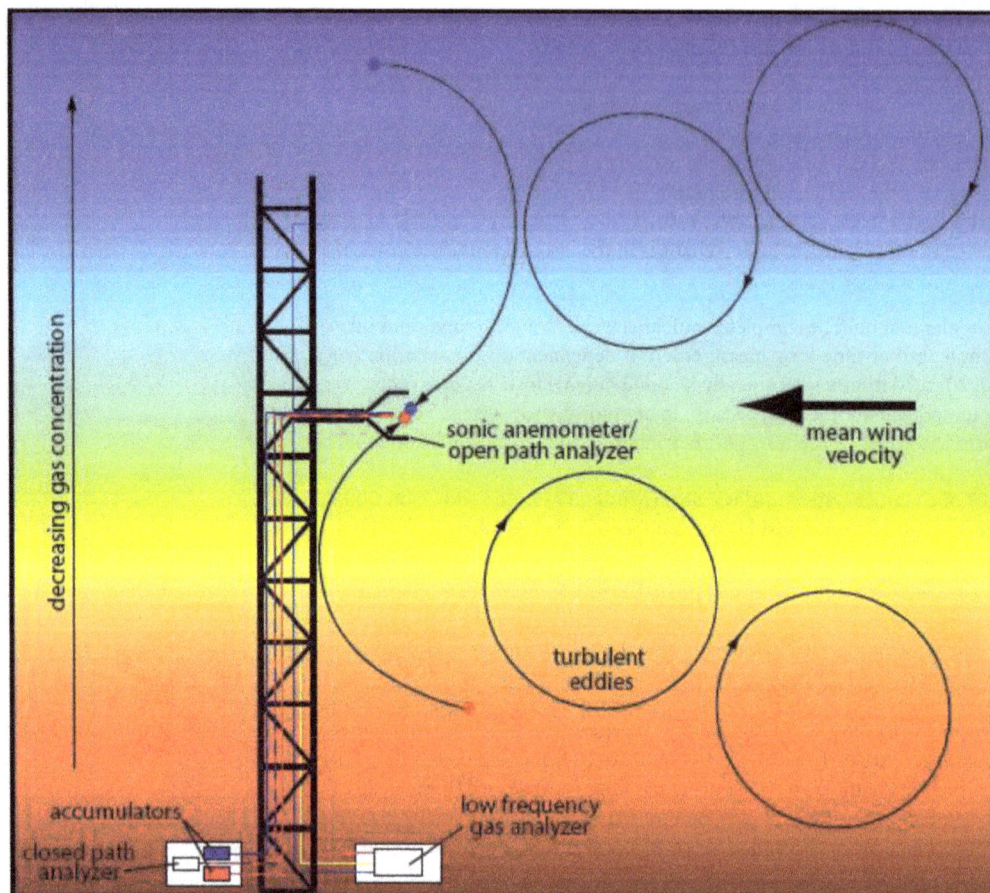

Figure 14

Principles behind micrometeorological methods for measuring
gas exchange.

In this example, the surface is
releasing a gas to the atmosphere,
resulting in a vertical concentration
gradient, with higher concentrations
(red colors) near the surface, and
lower concentrations (blue colors)
away from the surface. Upward
moving eddies will have higher
gas concentrations than downward
moving eddies, and by sampling
these concentrations (either *in situ*
or with a sampling tube) along with
the vertical wind velocity, fluxes can
be calculated via eddy covariance.
For eddy accumulation, the
samples are collected conditionally,
depending on whether the eddy is
moving upwards or downwards,
and the samples accumulate in a
bag or chamber for later laboratory
analyses. The g radient m ethod
measures the vertical concentration
gradient (usually by transporting
sample air to a low-frequency
gas analyzer), and then estimates
the rate of transport across that
gradient by parameterization or
comparison to measured transport
rates of heat or momentum.

### 5.1.3. Eddy covariance

A general description of EC methods is given by Lee et al. (2004), but briefly, both the three-dimensional motion field and the analyte must be measured at the same place and at the same high frequency. Sonic anemometers are used for the motion field, while the analyte can either be measured *in situ* by an instrument exposed to the atmosphere (an "open path" system) or by transporting air in a constant stream to an analyzer (a "closed path" system). Non-dispersive infrared (NDIR) analyzers are generally used to measure $CO_2$ fluxes in EC systems (*e.g.*, Baldocchi, 2003), but other types of detectors are also available for $CO_2$ and $CH_4$ (Crosson, 2008; Detto et al., 2011) and for volatile organic compounds (VOCs; Müller et al., 2010). Fluxes are derived from the covariance between the vertical velocity and the concentration of the species of interest (Figure 14); the EC approach is, of course, only valid when the fluctuations in the concentration are caused by the vertical turbulence, not by sources or sinks within the boundary layer. In order to resolve the turbulence and measure a sufficient fraction of the vertical transport, the sampling frequency must be higher than 10 Hz.

Over sea ice, the EC method has primarily been used to measure $CO_2$ exchanges (Table 9) through deployments either on ships or directly on the ice (Figure 15). All published studies, with the possible exception of Semiletov et al. (2004) who did not specify their instrumentation, have used the LI-7500 open-path $CO_2/H_2O$ analyzer, manufactured by LI-COR Biosciences. This instrument has been used widely at lower latitudes, but it suffers from some biases under cold conditions and in the marine environment. In cold weather, the instrument may significantly heat the column of air it is sampling, which lowers the measured gas concentration and biases flux measurements towards $CO_2$ uptake (Burba et al., 2008). In addition, salt deposited on the lens of the instrument interferes with the infrared absorption measurement, resulting in flux overestimations (Prytherch et al., 2010). In most cases, researchers have taken steps to minimize or correct for these biases, but results obtained using these open-path sensors should still be interpreted with a healthy dose of skepticism.

Closed-path EC systems potentially avoid many of these problems associated with the open-path sensors, because the temperature of the air sample can be controlled and instrument lenses can be protected by filters. However, the closed-path systems also suffer from their own shortcomings. Most importantly, closed-path systems inevitably attenuate gas concentration fluctuations and thus degrade the flux signal (Leuning and King, 1992; Lee et al., 2004), making it difficult to confidently identify gas exchanges between sea ice and the atmosphere, which are often much smaller than those observed over terrestrial or open-water surfaces.

Figure 15

Eddy covariance systems for measuring $CO_2$ exchange over sea ice.

Clockwise from left: a micrometeorological tower installed on the CCGS *Amundsen*; a close-up view of the flux instrumentation on the *Amundsen*; and a portable micrometeorological tower deployed on a sled directly on the sea ice. Amundsen Gulf, March–June, 2008. Photos: B. Else.

This difficulty may be why no sea ice-air $CO_2$ flux measurements using closed-path EC systems have yet been reported, despite several attempted deployments. However, Muller et al. (2012) have reported ozone fluxes over sea ice measured using a closed-path EC system.

Eddy covariance measurements can be subject to significant biases and random errors (Businger, 1986; Finkelstein and Sims, 2001), but there are no straightforward ways to calibrate or validate EC flux measurements in the field. These errors are exacerbated over sea ice, where the observed fluxes are often small. Particularly when temperatures and, therefore, sea-ice permeability are low, the fluxes are likely to be close to eddy covariance detection limits (estimated to be 1 μg C m$^{-2}$ s$^{-1}$ for $CO_2$; Wang et al., 2013), and the uncertainty in the measured flux can be over 200% in some cases (Sørensen et al., 2014). However, evaluating the frequency spectra of the sampled EC data can eliminate some sources of error (*e.g.*, advection, noise), and coupling fluxes estimated from EC and by spectral techniques gives more robust results (Kaimal et al., 1972; Sørensen and Larsen, 2010; Norman et al., 2012; Sørensen et al., 2014). Flux estimations based on the spectral methods use the same instrumental configuration as EC, but corrections for atmospheric stability are also required.

### 5.1.4. Eddy accumulation

For many trace gases, fast-response sensors, as required for EC flux measurements, are not available; relaxed eddy accumulation (REA) provides an alternate method for estimating the fluxes of these gases. Measurements by REA rely on conditional sampling (Hicks and McMillen, 1984; Businger and Oncley, 1990) of the gas into separate reservoirs depending on whether the bulk air movement is upward or downward. The "relaxation" refers to the fact that samples are taken with a constant flow rate and are not weighted according to the vertical wind speed; the data consequently lack information on the vertical wind speed. To date, only DMS fluxes have been measured over sea ice by REA (Zemmelink et al., 2008), but the method has potential for application to fluxes of other gases.

### 5.1.5. Gradient techniques

Gradient techniques provide another alternative for measuring fluxes of trace species for which no fast response sensor is available. Basically, the gradient technique requires measurements of gas concentrations from at least two levels (one typically within 1 m of the air-ice surface, the other 1–10 m from the surface), along with some estimate of how rapidly the gas is transported between those levels. This transport rate estimate is generally based on observations or parameterizations of atmospheric turbulence, which are usually derived from wind velocity and temperature measurements (preferably made at the same two levels). The method is indirect, can require a number of empirical functions to account for thermal stratification of the atmosphere, and is based on the assumption that turbulent transfer is analogous to molecular diffusion. The most common gradient technique is the aerodynamic method, which is based on the momentum flux equation and the wind speed-gradient relationship (Businger et al., 1971; Businger, 1986; Baldocchi et al., 1988; Sørensen et al., 2005). To date, the gradient technique has only been applied to measure nitric acid (Beine et al., 2003) and ozone (Bocquet et al., 2011) fluxes over terrestrial snow, but the method has potential for applications over sea ice.

### 5.1.6. Best practices for micrometeorological techniques

Several textbooks and review articles (we recommend Lee et al., 2004) have dealt extensively with issues of best practice for micrometeorological techniques, particularly with respect to eddy covariance. The interested reader should consult those texts, but we address several points here that are unique to the sea-ice environment.

The first important consideration is the installation of the meteorological tower, especially if the measurements are to last through a portion of the melt season. A melting sea-ice surface is inherently unstable; towers can tilt and even topple in such conditions. We recommend one of two approaches to combat such tower instability. First, the tower can be mounted on a qamutiq (sled), which can be periodically moved or repositioned to keep the tower level (Figure 15). Alternatively, the tower should be frozen into the ice by drilling holes (to a depth of about 50 cm), in which the ends of the tower posts are anchored with frozen fresh water. This system can be improved by also passing the tower posts through a plywood base on the ice surface, as this will shade the posts and prevent localized melt. The tower can also be directly fixed to a sheet of plywood (preferably with rigid insulation underneath), but freezing the tower legs into the ice is preferred.

The orientation of the tower is also important. Even a low-profile lattice-style tower can create significant snow drifts; it is best to place the meteorological equipment on the side of the tower that faces the prevailing wind direction, so that it will measure fluxes predominately over an undisturbed, upwind surface. Most eddy covariance installations over sea ice are by necessity close to the ground (less than 5 m above the surface), leading us to recommend sampling at a rate of 20 Hz to capture the small-scale eddies that exist near the surface. Finally, it is important to choose instruments that have low-temperature ratings and are well sealed against blowing snow. As discussed previously, particular attention should be paid to the choice of infrared gas analyzer to ensure that it operates properly in cold environments.

## 5.2. Aerosols, frost flowers, and saline snow

The formation of atmospheric particulate aerosols is another important interface flux in the sea-ice system. The polar regions frequently receive aerosols from lower latitudes, but wintertime sea-ice formation is also associated with vertical particle fluxes. Whereas under open-water conditions, the sea-to-air transport of both solid and liquid aerosols is controlled by breaking waves, in polar regions brine-wetted saline snow and frost flowers formed on new sea ice provide additional sources of atmospheric aerosols. In particular, frost flowers may be broken up by blowing wind, and the submicron fraction may have lifetimes up to a week. Both frost flowers and saline snow contain NaCl, other ocean salts from sea-ice brines, and organic components, including both organisms and their products (Alvarez-Aviles et al., 2008; Bowman and Deming, 2010; Douglas et al., 2012; Ewert et al., 2013). Particles from frost flowers and saline snow can be lofted into the atmosphere and transported long distances by wind, eventually deposited in new places under dry or wet conditions. Therefore, the size and composition of such particles should be characterized in studies of atmospheric aerosols in polar regions (Yang et al., 2008; Obbard et al., 2009; Roscoe et al., 2011).

Several methods exist to measure aerosols: in-line with mass spectrometers and ion chromatographs, as well as off-line by gas chromatography-mass spectrometry and Fourier transform infrared spectroscopy (e.g., Russell, 2014). Each of these recommended methods characterizes different, complementary chemical qualities. However, comparability between methods requires well-characterized inlets and careful techniques for minimizing artifacts during collection and storage. New advances in quantifying aerosol fluxes require simultaneous analyses of sea-ice components, with careful attention to the local meteorology, so that the measured aerosols can be linked to their upwind source regions.

Of particular interest is the role of halogens in atmospheric chemistry of the polar regions (Simpson et al., 2007; Abbatt et al., 2012), including the function of Br in tropospheric ozone and mercury depletion and the possible importance of iodine in new particle formation. High concentrations of BrO observed in satellite datasets during the polar spring, in particular, suggest that halogens originate at the sea-ice surface; the available studies of halocarbons in sea ice, to date, have confirmed variable and often high concentrations of biogenic halocarbons in ice, brines, and overlying snow (Sturges et al., 1997; Simpson et al., 2005; Atkinson et al., 2012; Mattson et al., 2012; Granfors et al., 2013a;b). Inorganic halides are easily sampled and analyzed (after melting and filtering) by, for example, ion chromatography for bromide, cyclic voltammetry for iodide, and spectrophotometry for iodate (e.g., Atkinson et al., 2012). On the other hand, sampling sea ice for halocarbons, which are volatile, should follow methods to limit gas exchange (section 4.4), including melting samples in gas-impermeable bags with minimal headspace (as for inorganic carbon species, section 4.6.3). Halocarbon samples are usually analyzed by gas chromatography, requiring aqueous (i.e., melted, see section 3.2) or gas samples. Shaw et al. (2011) measured iodated organics in artificial sea ice from laboratory tank studies by melting ice samples directly in sealed syringes to limit gas exchange. New methods to confidently analyze the halogen chemistry of frost flowers are particularly needed.

## 5.3. Ice-water fluxes

Most efforts to quantify biogeochemical exchanges at the ice-water interface have used mass balance and budgeting tactics. This approach has been most successful in identifying water-to-ice fluxes of nutrients (Cota et al., 1987; 1990; Rahm et al., 1995; Nishi and Tabeta, 2008) and of trace metals (Granskog and Kaartokallio, 2004; Lannuzel et al., 2010; 2011; van der Merwe et al., 2011a). Recent models of ice-brine dynamics have confirmed that seawater pumping into the brine network of growing sea ice could cycle enough surface seawater through the lower parts of the ice cover to account for observations of the nutrient and iron distributions in the ice (Vancoppenolle et al., 2010). On the other hand, efforts to estimate inorganic carbon fluxes from sea ice into the underlying surface waters have been confounded by the need to identify small changes in large, variable concentrations (e.g., Miller et al., 2011b; Fransson et al., 2013).

Attempts to use the gradient method to derive sea ice-water chemical fluxes have been limited by the difficulty of collecting high-vertical resolution water samples from under the undisturbed ice sheet. Although Dieckmann et al. (1992) developed a promising device for detailed under-ice sampling, it has not been utilized widely, and efforts by divers to collect biogeochemical samples from under the ice have not produced consistent results. *In situ* microsensors, such as those deployed for measuring oxygen by Rysgaard et al. (2001) and McMinn et al. (2000), show some promise for measuring gradients near the ocean-ice interface, but deployment of such instruments (usually by divers) remains difficult.

Eddy covariance shows promise for measuring ice-water, as well as ice-air, fluxes. Although to date only oxygen fluxes have been determined (section 3.4.1; Long et al., 2012), the method should be applicable to any analyte for which an *in situ* sensor is available with a response time less than 1 second. Sensors for measuring fluxes of nitrate (Johnson et al., 2011) and hydrogen sulfide (McGinnis et al., 2011) have been employed in benthic studies and could be adapted to the sea-ice environment.

Sinking particle fluxes from the sea ice can be measured using particle interceptor traps tethered below the ice (Michel et al., 1996; Fortier et al., 2002; Juul-Pedersen et al., 2008; Nishi and Tabeta, 2008); time series from such "sediment" traps have proven to be a valuable tool in estimating carbon budgets of sea-ice primary production, export, and transfers to pelagic and benthic grazers (e.g., Michel et al., 2002; Renaud et al., 2007). The traps are typically deployed at shallow depths specifically to capture particles exported from the ice (as opposed to produced within the water column) and to avoid excessive drag on the trap line. However, the traps can also be deployed at deeper depths, thereby providing insights into sea ice-pelagic-benthic coupling (Fortier et al., 2002). The extensive literature pertaining to sediment trap methodology does not deal specifically with under-ice deployments, although standard protocols for particle trap deployments should be followed as much as possible (i.e., Gardner, 2000). Particular challenges associated with under-ice deployments include over- and under-trapping if the traps are deployed under moving ice floes or under fast ice in high current regimes.

# 6. The future

Biogeochemists working in the sea-ice environment have made tremendous strides over recent years in learning how to measure and then understand the biogeochemical processes occurring in sea ice. Nonetheless, we still have much work ahead of us to resolve the uncertainties in the measurements we are making and identify which of the methods we are using are most suitable under various conditions (Table 10).

Dedicated, collaborative, field and laboratory studies are required to compare and intercalibrate a number of methods, most notably those for primary production and the $CO_2$ system, but also for nutrients, biomass, cell abundance, and size fractionation of organic matter. Time on icebreakers and in ice camps is expensive; it is often difficult to rationalize, within interdisciplinary process studies, the kind of redundant sampling and analyses that are required for rigorous intercalibrations. Therefore, to resolve methodological discrepancies, we will need expeditions and experiments that are focused first on methods. This kind of approach is difficult for many sea-ice scientists, as well as our funding agencies, who are motivated to use every opportunity in remote polar environments to address questions on how sea ice impacts the functioning of our planet. However, in the end, the confidence in our measurements provided by focused methodological studies will make these efforts worthwhile. In the meantime, to the extent possible, we encourage interdisciplinary process studies to include methodological intercalibrations.

For the most part, large icebreakers are not essential (and may be overkill) for the intercalibration and method validation experiments required (Table 10). Rather, a number of coastal laboratory facilities with ready access to fast ice, such as those in Barrow (Alaska, USA), Svalbard (Norway), McMurdo Sound (Antarctica), Saroma-ko (Hokkaido, Japan), and Tvärminne (Hanko, Finland), could be very useful sites for this work. Laboratory studies, including those in large-scale ice-tank facilities, will also be a critical component of our efforts to understand our methods and their limitations. The spatial and temporal variability in natural sea ice

Table 10. Priorities for sea-ice biogeochemical method development, validation, and intercalibration[a]

| Method development | *In situ* probes for $CO_2$ system parameters and gases |
|---|---|
| | Analyses of small-volume and high-salinity samples |
| | Sampling and studying ridged and deformed ice |
| | Sulfur cycling rates in undisturbed ice environments |
| | Net community production using $O_2$: Ar ratios |
| | Gross primary production using stable oxygen isotope ratios |
| | Halogen chemistry in frost flowers |
| | Inoculation of laboratory sea ice with representative sympagic communities |
| | Sea ice-seawater exchange processes |
| | Characterization of EPS polymer composition, structure, and molecular size |
| | Sampling and studying surface brine skims and frost flowers |
| | Rates of biogenic silica cycling using Si isotope methods |
| | Preserving bulk ice samples for $CO_2$ system parameters during melting |
| | Improving precision of gas extraction procedures |
| Validation | Errors in sackhole brine measurements of particulates and soluble versus insoluble gases |
| | Tracer-to-carbon conversion ratios for bacterial production measurements |
| | $CO_2$ system thermodynamics in ice brines |
| | Sensitivity of sea-ice TOC samples to contamination |
| | Accuracy of flow cytometric analyses in sea-ice samples with high organic matter content |
| | Impact of high EPS concentrations on nominal pore sizes of different filters |
| | EPS interference in analyses of salts, trace metals, and other organic compounds |
| | Certified reference materials over sea-ice concentration and salinity ranges for DOC, macro-nutrients, salinity, $A_T$, and DIC |
| | Incubation temperature impacts on measured microbial metabolic rates |
| | Accuracy of sea-ice equations of state in the presence of high organic matter concentrations |
| | Effects of melting method on measurements of macronutrients, EPS, and other organic compounds |
| | Precision of fluxes determined by eddy covariance |
| Intercalibration | Melting methods for determining biomass and community composition |
| | Primary and secondary production measurements |
| | Melting and filtration methods for macronutrient and organic matter analyses |
| | $pCO_2$ analysis |
| | Chamber and micrometeorological methods for determining ice-air $CO_2$ fluxes |
| | Aerosol production |
| | PIC extraction and analyses |
| | Nucleic acid extraction |
| | EPS analyses |

[a]The order in which items are listed does not necessarily imply priority.

often prevents us from accurately constraining our methods; in controlled laboratory tanks can we determine the true methodological precision and accuracy of chemical measurements in sea ice. On the other hand, to date, laboratory tanks have not been successfully inoculated with a natural sea-ice community, which, when coupled with universal challenges to laboratory-based biological experiments (*i.e.,* maintaining nutrient supply, *etc.*), severely limits the utility of laboratory experiments for studying the complexity of biological processes in sea ice.

The field of sea-ice biogeochemistry is also ripe for new technological developments; we have only begun by trying to adapt methods we know from other disciplines. As the field matures, we hope that entirely new approaches and methods will be developed. These efforts could be facilitated by recruiting more students with degrees in analytical chemistry, biotechnology, bioengineering, and electronics, as well as through new collaborations. In particular, *in situ* probes are critically needed to answer many of our questions effectively. Electrochemical and optical technologies show the most promise; throughout this paper, we have highlighted a number of biogeochemical parameters that we believe are particularly well-suited for sensor development, but there will certainly also be others. Robust ice buoys on which such chemical sensors could be deployed received a substantial boost from the 2007–08 International Polar Year (*e.g.*, Knepp et al., 2010) but require additional development to recover data dependably throughout the entire cycle of freeze-up and melt. Recent advances in under-water vehicle technology, including successful under-ice deployments of instrumented remotely operated vehicles (ROVs) and autonomous underwater vehicles (AUVs) also promise new tools to study sea-ice physical-ecological-biogeochemical interactions on scales from meters to kilometers (Wadhams and Doble, 2008; Williams et al., 2013).

Finally, a key challenge for the future is integrating interdisciplinary measurements on different scales to tell a more complete story of sea-ice biogeochemistry. For example, measurements of biogenic gas fluxes paired with analyses of bacterial gene expression and *in situ* nutrient concentrations are far more valuable than any of these observations, alone.

In closing, we have described many of the problems with existing methods for studying sea-ice biogeochemistry, while also noting the successes, which have been significant. This important and exciting research field is now beginning to mature. We hope that the insights presented here will provide inspiration for new scientists to boldly tackle some of these challenging methodological problems, because we must solve these problems, if we are to understand how sea ice impacts the Earth's biogeochemical cycles.

# References

Abbatt JPD, Thomas JL, Abrahamsson K, Boxe C, Granfors A, et al. 2012. Halogen activation via interactions with environmental ice and snow in the polar lower troposphere and other regions. *Atmos Chem Phys* **12**: 6237–6271. doi: 10.5194/acp-12-6237-2012.

Aguilar-Islas AM, Rember RD, Mordy CW, Wu J. 2008. Sea ice-derived dissolved iron and its potential influence on the spring algal bloom in the Bering Sea. *Geophys Res Lett* **35**: L24601. doi: 10.1029/2008GL035736.

Alou-Font E, Mundy C-J, Roy S, Gosselin M, Agustí S. 2013. Snow cover affects ice algal pigment composition in the coastal Arctic Ocean during spring. *Mar Ecol-Prog Ser* **474**: 89–104. doi: 10.3354/meps10107.

Alvarez-Aviles L, Simpson WR, Douglas TA, Sturm M, Perovich D, et al. 2008. Frost flower chemical composition during growth and its implications for aerosol production and bromine activation. *J Geophys Res* **113**: D21304. doi: 10.1029/2008JD010277.

Anderson LG, Jones EP. 1985. Sea ice melt water, a source of alkalinity, calcium and sulfate? Results from the CESAR Ice Station. *Rit Fiskideildar* **9**: 90–96.

Arar EJ, Collins GB. 1997. Method 445.0: In Vitro Determination of Chlorophyll *a* and Pheophytin *a* in Marine and Freshwater Algae by Fluorescence. Cincinnati: Office of Research and Development, U.S. Environmental Protection Agency. #445.0.

Arrigo KR, Dieckmann G, Gosselin M, Robinson DH, Fritsen CH, et al. 1995. High resolution study of the platelet ice ecosystem in McMurdo Sound, Antarctica: Biomass, nutrient, and production profiles within a dense microalgal bloom. *Mar Ecol-Prog Ser* **127**: 255–268.

Arrigo KR, Mock T, Lizotte MP. 2010. Primary producers in sea ice, in Thomas DN, Dieckmann GS, eds., *Sea Ice*. 2nd ed. Oxford: Wiley-Blackwell. pp. 283–325.

Arrigo KR, Robinson DH, Dunbar RB, Leventer AR, Lizotte MP. 2003. Physical control of chlorophyll *a*, POC, and TPN distributions in the pack ice of the Ross Sea, Antarctica. *J Geophys Res* **108**: 3316. doi:10.1029/2001JC001138.

Arrigo KR, Sullivan CW. 1992. The influence of salinity and temperature covariation on the photophysiological characteristics of Antarctic sea ice microalgae. *J Phycol* **28**: 746–756.

Asher EC, Dacey JWH, Mills MM, Arrigo KR, Tortell PD. 2011. High concentrations and turnover rates of DMS, DMSP and DMSO in Antarctic sea ice. *Geophys Res Lett* **38**: L23609. doi: 10.1029/2011GL049712.

Aslam SN, Underwood GJC, Kaartokallio H, Norman L, Autio R, et al. 2012. Dissolved extracellular polymeric substances (dEPS) dynamics and bacterial growth during sea ice formation in an ice tank study. *Polar Biol* **35**: 661–676. doi: 10.1007/s00300-011-1112-0.

Assur A. 1958. Composition of sea ice and its tensile strength, in, *Arctic Sea Ice*. Easton, Maryland: National Academy of Sciences, National Research Council: pp. 106–138.

Atkinson HM, Huang R-J, Chance R, Roscoe HK, Hughes C, et al. 2012. Iodine emissions from the sea ice of the Weddell Sea. *Atmos Chem Phys* **12**: 11,229–11,244. doi: 10.5194/acp-12-11229-2012.

Ayers GP, Cainey JM. 2007. The CLAW hypothesis: a review of the major developments. *Environ Chem* **4**: 366–374. doi: 10.1071/EN07080.

Baldocchi DD. 2003. Assessing the eddy covariance technique for evaluating carbon dioxide exchange rates of ecosystems: Past, present and future. *Glob Change Biol* **9**: 479–492.

Baldocchi DD, Hicks BB, Meyers TP. 1988. Measuring biosphere-atmosphere exchanges of biologically related gases with micrometerological methods. *Ecology* **69**(5): 1331–1340.

Barber DG, Ehn JK, Pućko M, Rysgaard S, Deming JW, et al. 2014. Frost flowers on young Arctic sea ice: The climatic, chemical and microbial significance of an emerging ice type. *J Geophys Res-Atmos*. doi: 10.1002/2014JD021736.

Becquevort S, Dumont I, Tison J-L, Lannuzel D, Sauvée M-L, et al. 2009. Biogeochemistry and microbial community composition in sea ice and underlying seawater off East Antarctica during early spring. *Polar Biol* **32**: 879–895. doi: 10.1007/s00300-009-0589-2.

Begoin M, Richter A, Weber M, Kaleschke L, Tian-Kunze X, et al. 2010. Satellite observations of long range transport of a large BrO plume in the Arctic. *Atmos Chem Phys* **10**: 6515–6526. doi: 10.5194/acp-10-6515-2010.

Beine HJ, Dominé F, Ianniello A, Nardino M, Allegrini I, et al. 2003. Fluxes of nitrates between snow surfaces and the atmosphere in the European high Arctic. *Atmos Chem Phys* **3**: 335–346.

Belt ST, Brown TA, Ringrose AE, Cabedo-Sanz P, Mundy CJ, et al. 2013. Quantitative measurement of the sea ice diatom biomarker IP$_{25}$ and sterols in Arctic sea ice and underlying sediments: Further considerations for palaeo sea ice reconstruction. *Org Geochem* **62**: 33–45. doi: 10.1016/j.orggeochem.2013.07.002.

Belzile C, Johannessen SC, Gosselin M, Demers S, Miller WL. 2000. Ultraviolet attenuation by dissolved and particulate constituents of first-year ice during late spring in an Arctic polynya. *Limnol Oceanogr* **45**: 1265–1273.

Bender M, Grande K, Johnson K, Marra J, Williams PJL, et al. 1987. A comparison of four methods for determining planktonic community production. *Limnol Oceanogr* **32**: 1085–1098.

Bennington KO. 1963. Some chemical composition studies on Arctic sea ice, in Kingery WD, ed, *Ice and Snow: Properties, Processes, and Applications*. Cambridge, Massachusetts: MIT Press: pp. 248–257.

Bidigare RR, Van Heukelem L, Trees CC. 2005. Analysis of algal pigments by high-performance liquid chromatography, in Andersen RA, ed, *Algal Culturing Techniques*. Burlington: Elsevier Academic Press: pp 327–345.

Bitter T, Muir HM. 1962. A modified uronic acid carbazole reaction. *Anal Biochem* **4**: 330–334.

Bocquet F, Helmig D, Van Dam BA, Fairall CW. 2011. Evaluation of the flux gradient technique for measurement of ozone surface fluxes over snowpack at Summit, Greenland. *Atmos Meas Tech* **4**: 2305–2321. doi: 10.5194/amt-4-2305-2011.

Bowling DR, Massman WJ. 2011. Persistent wind-induced enhancement of diffusive $CO_2$ transport in a mountain forest snowpack. *J Geophys Res* **116**: G04006. doi: 10.1029/2011JG001722.

Bowman JS, Deming JW. 2010. Elevated bacterial abundance and exopolymers in saline frost flowers and implications for atmospheric chemistry and microbial dispersal. *Geophys Res Lett* **37**: L13501. doi: 10.1029/2010GL043020.

Bowman JS, Larose C, Vogel TM, Deming JW. 2013. Selective occurrence of *Rhizobiales* in frost flowers on the surface of young sea ice near Barrow, Alaska and distribution in the polar marine rare biosphere. *Environ Microbiol Rep* **5**: 575–582. doi: 10.1111/1758-2229.12047.

Bowman JS, Rasmussen S, Blom N, Deming JW, Rysgaard S, et al. 2012. Microbial community structure of Arctic multiyear sea ice and surface seawater by 454 sequencing of the 16S RNA gene. *ISME J* **6**: 11–20. doi: 10.1038/ismej.2011.76.

Boye M, van den Berg CMG, de Jong JTM, Leach H, Croot P, et al. 2001. Organic complexation of iron in the Southern Ocean. *Deep-Sea Res Pt I* **48**: 1477–1497.

Brabant F, El Amri S, Tison J-L. 2011. A robust approach for the determination of dimethylsulfoxide in sea ice. *Limnol Oceanogr Methods* **9**: 261–274. doi: 10:4319/lom.2011.9.261.

Brakstad OG, Nonstad I, Faksness L-G, Brandvik PJ. 2008. Responses of microbial communities in Arctic sea ice after contamination by crude petroleum oil. *Microb Ecol* **55**: 540–552. doi: 10.1007/s00248-007-9299-x.

Brinkmeyer R, Knittel K, Jürgens J, Weyland H, Amann R, et al. 2003. Diversity and structure of bacterial communities in Arctic versus Antarctic pack ice. *Appl Environ Microb* **69**: 6610–6619. doi: 10.1128/AEM.69.11.6610-6619.2003.

Bronk DA, Lomas MW, Glibert PM, Schukert KJ, Sanderson MP. 2000. Total dissolved nitrogen analysis: comparisons between the persulfate, UV and high temperature oxidation methods. *Mar Chem* **69**: 163–178.

Brown KA, Miller LA, Davelaar M, Francois R, Tortell PD. 2014. Over-determination of the carbonate system in natural sea ice brine and assessment of carbonic acid dissociation constants under low temperature, high salinity conditions. *Mar Chem* **165**: 36–45. doi: 10.1016/j.marchem.2014.07.005.

Brown MV, Bowman JP. 2001. A molecular phylogenetic survey of sea-ice microbial communities (SIMCO). *FEMS Microbiol Ecol* **35**: 267–275.

Bruland KW, Rue EL. 2001. Analytical methods for the determination of concentrations and speciation of iron, in Turner DR, Hunter KA, eds., *The Biogeochemistry of Iron in Seawater*. West Sussex: John Wiley & Sons Ltd.: pp 255–289.

Burba GG, McDermitt DK, Grelle A, Anderson DJ, Xu L. 2008. Addressing the influence of instrument surface heat exchange on the measurements of $CO_2$ flux from open-path gas analyzers. *Glob Change Biol* **14**: 1854–1876. doi: 10.1111/j.1365-2486.2008.01606.x.

Burkholder PR, Mandelli EF. 1965. Productivity of microalgae in Antarctic sea ice. *Science* **149**: 872–874.

Burt A, Wang F, Pućko M, Mundy C-J, Gosselin M, et al. 2013. Mercury uptake within an ice algal community during the spring bloom in first-year Arctic sea ice. *J Geophys Res-Oceans* **118**: 4746–4754. doi: 10.1002/jgrc.20380.

Businger JA. 1986. Evaluation of the accuracy with which dry deposition can be measured with current micrometeorological techniques. *J Clim Appl Meteorol* **25**: 1100–1124.

Businger JA, Oncley SP. 1990. Flux measurement with conditional sampling. *J Atmos Ocean Tech* **7**: 349–352.

Businger JA, Wyngaard JC, Isumi Y, Bradley EF. 1971. Flux-profile relationships in the atmospheric surface layer. *J Atmos Sci* **28**: 181–189.

Campbell JA, Yeats PA. 1982. The distribution of manganese, iron, nickel, copper and cadmium in the waters of Baffin Bay and the Canadian Arctic Archipelago. *Oceanol Acta* **5**(2): 161–168.

Cardinal D, Alleman LY, de Jong J, Ziegler K, André L. 2003. Isotopic composition of silicon measured by multicollector plasma source mass spectrometry in dry plasma mode. *J Anal Atom Spectrom* **18**: 213–218. doi: 10.1039/b210109b.

Chaulk A, Stern GA, Armstrong D, Barber DG, Wang F. 2011. Mercury distribution and transport across the ocean-sea-ice-atmosphere interface in the Arctic Ocean. *Environ Sci Technol* **45**: 1866–1872. doi: 10.1021/es103434c.

Chu PC, Gascard GC, eds. 1991. *Deep Convection and Deep Water Formation in the Oceans*. Amsterdam: Elsevier.

Collins RE, Carpenter SD, Deming JW. 2008. Spatial heterogeneity and temporal dynamics of particles, bacteria, and pEPS in Arctic winter sea ice. *J Marine Syst* **74**: 902–917. doi: 10.1016/j.jmarsys.2007.09.005.

Collins RE, Rocap G, Deming JW. 2010. Persistence of bacterial and archaeal communities in sea ice through an Arctic winter. *Environ Microbiol* **12**: 1828–1841. doi: 10.1111/j.1462-2920.2010.02179.x.

Comiso JC. 2010. Variability and trends of the global sea ice cover, in Thomas DN, Dieckmann GS, eds., *Sea Ice*. 2nd ed. Oxford: Wiley-Blackwell: pp. 205–246.

Conover RJ, Mumm N, Bruecker P, MacKenzie S. 1999. Sources of urea in arctic seas: seasonal fast ice? *Mar Ecol-Prog Ser* **179**: 55–69.

Cossa D, Heimbürger L-E, Lannuzel D, Rintoul SR, Butler ECV, et al. 2011. Mercury in the Southern Ocean. *Geochim Cosmochim Acta* **75**: 4037–4052. doi: 10.1016/j.gca.2011.05.001.

Cota GF, Anning JL, Harris LR, Harrison WG, Smith REH. 1990. Impact of ice algae on inorganic nutrients in seawater and sea ice in Barrow Strait, NWT, Canada, during spring. *Can J Fish Aquat Sci* **47**: 1402–1415.

Cota GF, Prinsenberg SJ, Bennett EB, Loder JW, Lewis MR, et al. 1987. Nutrient fluxes during extended blooms of Arctic ice algae. *J Geophys Res* **92**: 1951–1962.

Cottier F, Eicken H, Wadhams P. 1999. Linkages between salinity and brine channel distribution in young sea ice. *J Geophys Res* **104**(C7): 15,859–15,871.

Cowie ROM. 2011. Bacterial Community Structure, Function and Diversity in Antarctic Sea Ice [Ph.D. thesis], Wellington: Victoria University of Wellington, Ecology and Biodiversity.

Cox GFN, Weeks WF. 1983. Equations for determining the gas and brine volumes in sea-ice samples. *J Glaciol* **29**: 306–316.

Cozzi S. 2008. High-resolution trends of nutrients, DOM and nitrogen uptake in the annual sea ice at Terra Nova Bay, Ross Sea. *Antarct Sci* **20**: 441–454. doi: 10.1017/S0954102008001247.

Craig H, Hayward T. 1987. Oxygen supersaturation in the ocean: Biological versus physical contributions. *Science* **235**: 199–202.

Crosson ER. 2008. A cavity ring-down analyzer for measuring atmospheric levels of methane, carbon dioxide, and water vapor. *Appl Phys B* **92**: 403–408. doi: 10.1007/s00340-008-3135-y.

Curran MAJ, Jones GB. 2000. Dimethyl sulfide in the Southern Ocean: Seasonality and flux. *J Geophys Res* **105**(D16): 20,451–20,459.

Cutter G, Andersson P, Codispoti L, Croot P, Francois R, et al. 2010. Sampling and Sample-handling Protocols for GEOTRACES Cruises.

de Jong J, Schoemann V, Maricq N, Mattielli N, Langhorne P, et al. 2013. Iron in land-fast sea ice of McMurdo Sound derived from sediment resuspension and wind-blown dust attributes to primary productivity in the Ross Sea, Antarctica. *Mar Chem* **157**: 24–40. doi: 10.1016/j.marchem.2013.07.001.

Delille B. 2006. Inorganic carbon dynamics and air-ice-sea $CO_2$ fluxes in the open and coastal waters of the Southern ocean [thesis], Liège: Université de Liège, Faculté des Sciences.

Delille B, Jourdain B, Borges AV, Tison J-L, Delille D. 2007. Biogas ($CO_2$, $O_2$, dimethylsulfide) dynamics in spring Antarctic fast ice. *Limnol Oceanogr* **52**: 1367–1379.

Deming JW. 2010. Sea ice bacteria and viruses, in Thomas DN, Dieckmann GS, eds., *Sea Ice*. 2nd ed. Oxford: Wiley-Blackwell: pp. 247–282.

Detto M, Verfaillie J, Anderson F, Xu L, Baldocchi D. 2011. Comparing laser-based open- and closed-path gas analyzers to measure methane fluxes using the eddy covariance method. *Agr Forest Meteorol* **151**: 1312–1324. doi: 10.1016/j.agrformet.2011.05.014.

Dickson AG. 1993. The measurement of sea water pH. *Mar Chem* **44**: 131–142.

Dickson AG, Sabine CL, Christian JR. 2007. Guide to Best Practices for Ocean $CO_2$ Measurements. Sidney: North Pacific Marine Science Organization. PICES Special Publication 3.

Dieckmann GS, Arrigo K, Sullivan CW. 1992. A high-resolution sampler for nutrient and chlorophyll a profiles of the sea ice platelet layer and underlying water column below fast ice in polar oceans: Preliminary results. *Mar Ecol-Prog Ser* **80**: 291–300.

Dieckmann GS, Eicken H, Haas C, Garrison DL, Gleitz M, et al. 1998. A compilation of data on sea ice algal standing crop from the Bellingshausen, Amundsen and Weddell Seas from 1983 to 1994, in Lizotte MP, Arrigo KR, eds., *Antarctic Sea Ice: Biological Processes, Interactions and Variability*. Washington: American Geophysical Union: pp. 85–92.

Dieckmann GS, Lange MA, Ackley SF, Jennings Jr JC. 1991a. The nutrient status in sea ice of the Weddell Sea during winter: Effects of sea ice texture and algae. *Polar Biol* **11**: 449–456.

Dieckmann GS, Nehrke G, Papadimitriou S, Göttlicher J, Steininger R, et al. 2008. Calcium carbonate as ikaite crystals in Antarctic sea ice. *Geophys Res Lett* **35**: L08501. doi: 10.1029/2008GL033540.

Dieckmann GS, Nehrke G, Uhlig C, Göttlicher J, Gerland S, et al. 2010. Ikaite ($CaCO_3 \cdot 6H_2O$) discovered in Arctic sea ice. *The Cryosphere* **4**: 227–230. doi: 10.5194/tc-4-227-2010.

Dieckmann GS, Spindler M, Lange MA, Ackley SF, Eicken H. 1991b. Antarctic sea ice: A habitat for the foraminifer *Neogloboquadrina pachyderma*. *J Foraminiferal Res* **21**: 182–189.

Domine F, Sparapani R, Ianniello A, Beine HJ. 2004. The origin of sea salt in snow on Arctic sea ice and in coastal regions. *Atmos Chem Phys* **4**: 2259–2271.

Douglas TA, Domine F, Barret M, Anastasio C, Beine HJ, et al. 2012. Frost flowers growing in the Arctic ocean-atmosphere-sea ice-snow interface: 1. Chemical composition. *J Geophys Res* **117**: D00R09. doi:10.1029/2011JD016460.

Douglas TA, Sturm M, Simpson WR, Blum JD, Alvarez-Aviles L, et al. 2008. Influence of snow and ice crystal formation and accumulation on mercury deposition to the Arctic. *Environ Sci Technol* **42**: 1542–1551. doi: 10.1021/es070502d.

Douglas TA, Sturm M, Simpson WR, Brooks S, Lindberg SE, et al. 2005. Elevated mercury measured in snow and frost flowers near Arctic sea ice leads. *Geophys Res Lett* **32**: L04502. doi: 10.1029/2004GL022132.

Dubois M, Gilles KA, Hamilton JK, Rebers PA, Smith F. 1956. Colorimetric method for determination of sugars and related substances. *Anal Chem* **28**: 350–356.

Ducklow H. 2000. Bacterial production and biomass in the oceans, in Kirchman DL, ed., *Microbial Ecology of the Oceans*. Toronto: John Wiley & Sons, Inc.: pp. 85–120.

Dugdale RC, Goering JJ. 1967. Uptake of new and regenerated forms of nitrogen in primary productivity. *Limnol Oceanogr* **12**: 196–206.

Dumont I, Schoemann V, Lannuzel D, Chou L, Tison J-L, et al. 2009. Distribution and characterization of dissolved and particulate organic matter in Antarctic pack ice. *Polar Biol* **32**: 733–750. doi: 10.1007/s00300-008-0577-y.

Edwards R, Sedwick P. 2001. Iron in East Antarctic snow: Implications for atmospheric iron deposition and algal production in Antarctic waters. *Geophys Res Lett* **28**(20): 3907–3910.

Ehrhardt M, Koeve W. 1999. Determination of particulate organic carbon and nitrogen, in Grasshoff K, Kremling K, Ehrhardt M, eds., *Methods of Seawater Analysis*. Toronto: Wiley-VCH: pp. 437–444.

Eicken H. 1992. The role of sea ice in structuring Antarctic ecosystems. *Polar Biol* **12**: 3–13.

Eicken H. 1998. Deriving modes and rates of ice growth in the Weddell Sea from microstructural, salinity and stable-isotope data, in Jeffries MO, ed., *Antarctic Sea Ice: Physical Processes, Interactions and Variability*. Washington: American Geophysical Union: pp. 89–122.

Eicken H. 2009. Chapter 3.3, Ice sampling and basic sea ice core analysis, in Eicken H, Gradinger R, Salganek M, Shirasawa K, Perovich D, Leppäranta M, eds., *Field Techniques for Sea Ice Research*. Fairbanks: University of Alaska Press: pp. 117–140.

Eicken H, Gradinger R, Salganek M, Shirasawa K, Perovich D, et al. eds. 2009. *Field Techniques for Sea Ice Research*. Fairbanks: University of Alaska Press.

Eicken H, Lange MA, Dieckmann GS. 1991. Spatial variability of sea-ice properties in the northwestern Weddell Sea. *J Geophys Res* **96**(C6): 10,603–10,615.

Else BGT, Papakyriakou TN, Galley RJ, Drennan WM, Miller LA, et al. 2011. Wintertime $CO_2$ fluxes in an Arctic polynya using eddy covariance: Evidence for enhanced air-sea gas transfer during ice formation. *J Geophys Res* **116**: C00G03. doi:10.1029/2010JC006760.

EPA. 2002. Method 1631, Revision E: Mercury in Water by Oxidation, Purge and Trap, and Cold Vapor Atomic Fluorescence Spectrometry. Washington, DC: Environmental Protection Agency.

Ewert M, Carpenter SD, Colangelo-Lillis J, Deming JW. 2013. Bacterial and extracellular polysaccharide content of brine-wetted snow over Arctic winter first-year sea ice. *J Geophys Res-Oceans* **118**: 726–735. doi:10.1002/jgrc.20055.

Ewert M, Deming JW. 2013. Sea ice microorganisms: Environmental constraints and extracellular responses. *Biology* **2**(2): 603–628.

Falkowski PG, Raven JA. 2007. Photosynthesis and primary production in nature, in, *Aquatic Photosynthesis*. Princeton: Princeton University Press: pp. 319–363.

Finkelstein PL, Sims PF. 2001. Sampling error in eddy correlation flux measurements. *J Geophys Res* **106**(D4): 3503–3509.

Fischer M. 2013. Sea ice and the air-sea exchange of $CO_2$ [thesis], Bremen: Universität Bremen, Biologie/Chemie.

Fischer M, Thomas DN, Krell A, Nehrke G, Göttlicher J, et al. 2012. Quantification of ikaite in Antarctic sea ice. *Antarct Sci*. doi:10.1017/S0954102012001150.

Fortier M, Fortier L, Michel C, Legendre L. 2002. Climatic and biological forcing of the vertical flux of biogenic particles under seasonal Arctic sea ice. *Mar Ecol-Prog Ser* **225**: 1–16.

Fowler D, Coyle M, Flechard C, Hargreaves K, Nemitz E, et al. 2001. Advances in micrometeorological methods for the measurement and interpretation of gas and particle nitrogen fluxes. *Plant Soil* **228**: 117–129.

Frache R, Abelmoschi ML, Grotti M, Ianni C, Magi E, et al. 2001. Effects of ice melting on Cu, Cd and Pb profiles in Ross Sea waters (Antarctica). *Int J Environ An Ch* **79**: 301–313.

Fransson A, Chierici M, Miller LA, Carnat G, Shadwick E, et al. 2013. Impact of sea-ice processes on the carbonate system and ocean acidification at the ice-water interface of the Amundsen Gulf, Arctic Ocean. *J Geophys Res-Oceans* **118**: 7001–7023. doi: 10.1002/2013JC009164.

Fransson A, Chierici M, Yager PL, Smith Jr WO. 2011. Antarctic sea ice carbon dioxide system and controls. *J Geophys Res* **116**: C12035. doi:10.1029/2010JC006844.

Fripiat F, Cardinal D, Tison J-L, Worby A, André L. 2007. Diatom-induced silicon isotopic fractionation in Antarctic sea ice. *J Geophys Res* **112**: G02001. doi: 10.1029/2006JG000244.

Fripiat F, Corvaisier R, Navez J, Elskens M, Schoemann V, et al. 2009. Measuring production-dissolution rates of marine biogenic silica by $^{30}Si$-isotope dilution using a high-resolution sector field inductively coupled plasma mass spectrometer. *Limnol Oceanogr Methods* **7**: 470–478.

Fripiat F, Sigman DM, Fawcett SE, Rafter PA, Weigand MA, et al. 2014a. New insights into sea ice nitrogen biogeochemical dynamics from the nitrogen isotopes. *Glob Biogeochem Cy* **28**: 115–130. doi: 10.1002/2013GB004729.

Fripiat F, Tison J-L, André L, Notz D, Delille B. 2014b. Biogenic silica recycling in sea ice inferred from Si-isotopes: constraints from Arctic winter first-year sea ice. *Biogeochemistry* **119**: 25–33. doi: 10.1007/s10533-013-9911-8.

Fritsen CH, Memmott JC, Ross RM, Quetin LB, Vernet M, et al. 2011. The timing of sea ice formation and exposure to photosynthetically active radiation along the Western Antarctic Peninsula. *Polar Biol* **34**: 683–692. doi: 10.1007/s00300-010-0924-7.

Garbarino JR, Snyder-Conn E, Leiker TJ, Hoffman GL. 2002. Contaminants in Arctic snow collected over Northwest Alaskan sea ice. *Water Air Soil Poll* **139**: 183–214.

Gardner WD. 2000. Sediment trap sampling in surface waters, in Hanson RB, Ducklow HW, Field JG, eds., *The Changing Ocean Carbon Cycle: A Midterm Synthesis of the Joint Global Ocean Flux Study*. Cambridge: Cambridge University Press: pp. 240–281.

Garrison DL, Buck KR. 1986. Organism losses during ice melting: A serious bias in sea ice community studies. *Polar Biol* **6**: 237–239.

Garrison DL, Jeffries MO, Gibson A, Coale SL, Neenan D, et al. 2003. Development of sea ice microbial communities during autumn ice formation in the Ross Sea. *Mar Ecol-Prog Ser* **259**: 1–15.

Geilfus N-X, Carnat G, Dieckmann GS, Halden N, Nehrke G, et al. 2013. First estimates of the contribution of $CaCO_3$ precipitation to the release of $CO_2$ to the atmosphere during young sea ice growth. *J Geophys Res-Oceans* **118**: 244–255. doi: 10.1029/2012JC007980.

Geilfus N-X, Carnat G, Papakyriakou T, Tison J-L, Else B, et al. 2012a. Dynamics of $pCO_2$ and related air-ice $CO_2$ fluxes in the Arctic coastal zone (Amundsen Gulf, Beaufort Sea). *J Geophys Res* **117**: C00G10. doi:10.1029/2011JC007118.

Geilfus N-X, Delille B, Verbeke V, Tison J-L. 2012b. Towards a method for high vertical resolution measurements of the partial pressure of $CO_2$ within bulk sea ice. *J Glaciol* **58**: 287–300. doi: 10.3189/2012JoG11J071.

Geilfus N-X, Galley RJ, Crabeck O, Papakyriakou T, Landy J, et al. 2014a. Inorganic carbon dynamics of melt pond-covered first year sea ice in the Canadian Arctic. *Biogeosciences Discuss* **11**: 7485–7519. doi: 10.5194/bgd-11-7485-2014.

Geilfus N-X, Tison J-L, Ackley SF, Galley RJ, Rysgaard S, et al., 2014b. Sea ice $pCO_2$ dynamics and air-ice $CO_2$ fluxes during the Sea Ice Mass Balance in the Antarctic (SIMBA) experiment – Bellingshausen Sea, Antarctica. *The Cryosphere* **8**: 2395–2407. doi: 10.5194/tc-8-2395-2014.

Gerdes B, Brinkmeyer R, Dieckmann G, Helmke E. 2005. Influence of crude oil on changes of bacterial communities in Arctic sea-ice. *FEMS Microbiol Ecol* **53**: 129–139. doi: 10.1016/j.femsec.2004.11.010.

Giannelli V, Thomas DN, Haas C, Kattner G, Kennedy H, et al. 2001. Behaviour of dissolved organic matter and inorganic nutrients during experimental sea-ice formation. *Ann Glaciol* **33**: 317–321.

Gibson JAE, Trull T, Nichols PD, Summons RE, McMinn A. 1999. Sedimentation of $^{13}$C-rich organic matter from Antarctic sea-ice algae: A potential indicator of past sea-ice extent. *Geology* **27**: 331–334.

Gleitz M, Rutgers vd Loeff M, Thomas DN, Dieckmann GS, Millero FJ. 1995. Comparison of summer and winter inorganic carbon, oxygen and nutrient concentrations in Antarctic sea ice brine. *Mar Chem* **51**: 81–91.

Glud RN, Rysgaard S, Kühl M. 2002. A laboratory study on O$_2$ dynamics and photosynthesis in ice algal communities: quantification by microsensors, O$_2$ exchange rates, $^{14}$C incubations and a PAM fluorometer. *Aquat Microb Ecol* **27**: 301–311.

Golden KM, Ackley SF, Lytle VI. 1998. The percolation phase transition in sea ice. *Science* **282**: 2238–2241.

Gosink T. 1978. The Arctic: A significant source-sink of carbon dioxide, in Gosink TA, Kelley JJ, eds., *Gases in the Sea Ice*. Fairbanks, AK: Institute of Marine Science, University of Alaska: pp. 79–91.

Gosink TA. 1980. Atmospheric trace gases in association with sea ice. *Antarct J US* **15**: 82–83.

Gosink TA, Kelley JJ. 1979. Carbon monoxide evolution from Arctic surfaces during spring thaw. *J Geophys Res* **84**: 7041.

Gosink T, Kelley JJ. 1985. Final Report: Carbon Dioxide in Arctic and Subarctic Regions. University of Alaska, 215 pp.

Gosselin M, Legendre L, Therriault J-C, Demers S, Rochet M. 1986. Physical control of the horizontal patchiness of sea-ice microalgae. *Mar Ecol-Prog Ser* **29**: 289–298.

Gosselin M, Levasseur M, Wheeler PA, Horner RA, Booth BC. 1997. New measurements of phytoplankton and ice algal production in the Arctic Ocean. *Deep-Sea Res Pt II* **44**: 1623–1644.

Gradinger R. 1999. Vertical fine structure of the biomass and composition of algal communities in Arctic pack ice. *Mar Biol* **133**: 745–754.

Gradinger R. 2009. Sea-ice algae: Major contributors to primary production and algal biomass in the Chukchi and Beaufort Seas during May/June 2002. *Deep-Sea Res Pt II* **56**: 1201–1212.

Gradinger R, Bluhm B, Iken K. 2010. Arctic sea-ice ridges — Safe heavens for sea-ice fauna during periods of extreme ice melt? *Deep-Sea Res Pt II* **57**: 86–95. doi: 10.1016/j.dsr2.2009.08.008.

Gradinger R, Ikävalko J. 1998. Organism incorporation into newly forming Arctic sea ice in the Greenland Sea. *J Plankton Res* **20**: 871–886.

Granfors A, Andersson M, Chierici M, Fransson A, Gårdfeldt K, et al. 2013a. Biogenic halocarbons in young Arctic sea ice and frost flowers. *Mar Chem* **155**: 124–134. doi: 10.1016/j.marchem.2013.06.002.

Granfors A, Karlsson A, Mattsson E, Smith Jr. WO, Abrahamsson K. 2013b. Contribution of sea ice in the Southern Ocean to the cycling of volatile halogenated organic compounds. *Geophys Res Lett* **40**: 3950–3955. doi: 10.1002/grl.50777.

Granskog MA, Kaartokallio H. 2004. An estimation of the potential fluxes of nitrogen, phosphorus, cadmium and lead from sea ice and snow in the northern Baltic Sea. *Water Air Soil Poll* **154**: 331–347.

Granskog MA, Kaartokallio H, Kuosa H, Thomas DN, Ehn J, et al. 2005a. Scales of horizontal patchiness in chlorophyll *a*, chemical and physical properties of landfast sea ice in the Gulf of Finland (Baltic Sea). *Polar Biol* **28**: 276–283. doi: 10.1007/s00300-004-0690-5.

Granskog MA, Kaartokallio H, Shirasawa K. 2003. Nutrient status of Baltic Sea ice: Evidence for control by snow-ice formation, ice permeability, and ice algae. *J Geophys Res* **108**(C8). doi: 10.1029/2002JC001386.

Granskog MA, Kaartokallio H, Thomas DN, Kuosa H. 2005b. Influence of freshwater inflow on the inorganic nutrient and dissolved organic matter within coastal sea ice and underlying waters in the Gulf of Finland (Baltic Sea). *Estuar Coast Shelf S* **65**: 109–122. doi: 10.1016/j.ecss.2005.05.011.

Granskog MA, Nomura D, Müller S, Krell A, Toyota T, et al. 2015. Evidence for significant protein-like dissolved organic matter accumulation in Sea of Okhotsk sea ice. *Ann Glaciol* **56**(69): 1–8. doi: 10.3189/2015AoG69A002.

Granskog MA, Virkanen J. 2001. Observations on sea-ice and surface-water geochemistry — Implications for importance of sea ice in geochemical cycles in the northern Baltic Sea. *Ann Glaciol* **33**: 311–316.

Granskog MA, Virkkunen K, Thomas DN, Ehn J, Kola H, et al. 2004. Chemical properties of brackish water ice in the Bothnian Bay, the Baltic Sea. *J Glaciol* **50**: 292–302.

Grossi SM, Kottmeier ST, Moe RL, Taylor GT, Sullivan CW. 1987. Sea ice microbial communities. VI. Growth and primary production in bottom ice under graded snow cover. *Mar Ecol-Prog Ser* **35**: 153–164.

Grossmann S, Dieckmann GS. 1994. Bacterial standing stock, activity, and carbon production during formation and growth of sea ice in the Weddell Sea, Antarctica. *Appl Environ Microbiol* **60**(8): 2746–2753.

Grotti M, Soggia F, Ianni C, Frache R. 2005. Trace metals distributions in coastal sea ice of Terra Nova Bay, Ross Sea, Antarctica. *Antarct Sci* **17**(2): 289–300. doi: 10.1017/S0954102005002695.

Hansell DA, Carlson CA, Repeta DJ, Schlitzer R. 2009. Dissolved organic matter in the ocean: A controversy stimulates new insights. *Oceanography* **22**: 202–211.

Hare AA, Wang F, Barber D, Geilfus N-X, Galley RJ, et al. 2013. pH evolution in sea ice grown at an outdoor experimental facility. *Mar Chem* **154**: 46–54. doi: 10.1016/j.marchem.2013.04.007.

Harrison WG, Cota GF, Smith REH. 1990. Nitrogen utilization in ice algal communities of Barrow Strait, Northwest Territories, Canada. *Mar Ecol-Prog Ser* **67**: 275–283.

Hawes I, Lund-Hansen LC, Sorrell BK, Nielsen MH, Borzák R, et al. 2012. Photobiology of sea ice algae during initial spring growth in Kangerlussuaq, West Greenland: insights from imaging variable chlorophyll fluorescence of ice cores. *Photosynth Res* **112**: 103–115. doi: 10.1007/s11120-012-9736-7.

Helmke E, Weyland H. 1995. Bacteria in sea ice and underlying water of the eastern Weddell Sea in midwinter. *Mar Ecol-Prog Ser* **117**: 269–287.

Hendry KR, Meredith MP, Measures CI, Carson DS, Rickaby REM. 2010a. The role of sea ice formation in cycling of aluminium in northern Marguerite Bay, Antarctica. *Estuar Coast Shelf S* **87**: 103–113. doi: 10.1016/j.ecss.2009.12.017.

Hendry KR, Rickaby REM, de Hoog JCM, Weston K, Rehkamper M. 2010b. The cadmium-phosphate relationship in brine: biological versus physical control over micronutrients in sea ice environments. *Antarct Sci* **22**(1): 11–18. doi:10.1017/S0954102009990381.

Herborg L-M, Thomas DN, Kennedy H, Haas C, Dieckmann GS. 2001. Dissolved carbohydrates in Antarctic sea ice. *Antarct Sci* **13**: 119–125.

Hicks BB, McMillen RT. 1984. A simulation of the eddy accumulation method for measuring pollutant fluxes. *J Clim Appl Meteorol* **23**: 637–643.

Higaki S, Oya Y, Makide Y. 2006. Emission of methane from stainless steel surface investigated by using tritium as a radioactive tracer. *Chem Lett* **35**: 292–293. doi: 10.1246/cl.2006.292.

Hilmer T, Bate GC. 1989. Filter types, filtration and post-filtration treatment in phytoplankton production studies. *J Plankton Res* **11**(1): 49–63.

Hobbie JE, Daley RJ, Jasper S. 1977. Use of Nuclepore filters for counting bacteria by fluorescence microscopy. *Appl Environ Microbiol* **33**: 1225–1228.

Hölemann JA, Schirmacher M, Kassens H, Prange A. 1999. Geochemistry of surficial and ice-rafted sediments from the Laptev Sea (Siberia). *Estuar Coast Shelf S* **49**: 45–59.

Holmes RM, Aminot A, Kérouel R, Hooker BA, Peterson BJ. 1999. A simple and precise method for measuring ammonium in marine and freshwater ecosystems. *Can J Fish Aquat Sci.* **56**: 1801–1808.

Holter P. 1990. Sampling air from dung pats by silicone rubber diffusion chambers. *Soil Biol Biochem* **22**: 995–997.

Horner R. 1985. History of ice algal investigations, in Horner RA ed., *Sea Ice Biota*. Boca Raton, Florida: CRC Press. pp. 1–19.

Horner R, Ackley SF, Dieckmann GS, Gulliksen B, Hoshiai T, et al. 1992. Ecology of sea ice biota: 1. Habitat, terminology, and methodology. *Polar Biol* **12**: 417–427.

Horner R, Schrader GC. 1982. Relative contributions of ice algae, phytoplankton, and benthic microalgae to primary production in nearshore regions of the Beaufort Sea. *Arctic* **35**: 485–503.

Hunke EC, Notz D, Turner AK, Vancoppenolle M. 2011. The mulitphase physics of sea ice: a review for model developers. *The Cryosphere* **5**: 989–1009. doi: 10.5194/tc-5-989-2011.

Hydes DJ, Aoyama M, Aminot A, Bakker K, Becker S, et al. 2010. Determination of dissolved nutrients (N, P, Si) in seawater with high precision and inter-comparability using gas-segmented continuous flow analysers. *The GO-SHIP Repeat Hydrography Manual: A Collection of Expert Reports and Guidelines*.

Johnson KS, Barry JP, Coletti LJ, Fitzwater SE, Jannasch HW, et al. 2011. Nitrate and oxygen flux across the sediment-water interface observed by eddy correlation measurements on the open continental shelf. *Limnol Oceanogr Methods* **9**: 543–553. doi: 10.4319/lom.2011.9.543.

Juhl AR, Krembs C, Meiners KM. 2011. Seasonal development and differential retention of ice algae and other organic fractions in first-year Arctic sea ice. *Mar Ecol-Prog Ser* **436**: 1–16. doi: 10.3354/meps09277.

Junge K, Eicken H, Deming JW. 2004. Bacterial activity at -2 to -20°C in Arctic wintertime sea ice. *Appl Environ Microbiol* **70**(1): 550–557.

Junge K, Krembs C, Deming J, Stierle A, Eicken H. 2001. A microscopic approach to investigate bacteria under in situ conditions in sea-ice samples. *Ann Glaciol* **33**: 304–310.

Juul-Pedersen T, Michel C, Gosselin M, Seuthe L. 2008. Seasonal changes in the sinking export of particulate material under first-year sea ice on the Mackenzie Shelf (western Canadian Arctic). *Mar Ecol-Prog Ser* **353**: 13–25. doi: 10.3354/meps07165.

Kaartokallio H. 2001. Evidence for active microbial nitrogen transformation in sea ice (Gulf of Bothnia, Baltic Sea) in midwinter. *Polar Biol* **24**: 21–28.

Kaartokallio H. 2004. Food web components, and physical and chemical properties of Baltic Sea ice. *Mar Ecol-Prog Ser* **273**: 49–63.

Kaartokallio H, Søgaard DH, Norman L, Rysgaard S, Tison J-L, et al. 2013. Short-term variability in bacterial abundance, cell properties, and incorporation of leucine and thymidine in subarctic sea ice. *Aquat Microb Ecol* **71**: 57–73. doi: 10.3354/ame01667.

Kaartokallio H, Tuomainen J, Kuosa H, Kuparinen J, Martikainen PJ, et al. 2008. Succession of sea-ice bacterial communities in the Baltic Sea fast ice. *Polar Biol* **31**: 783–793. doi: 10.1007/s00300-008-0416-1.

Kaimal JC, Wyngaard JC, Izumi Y, Coté OR. 1972. Spectral characteristics of surface-layer turbulence. *Q J Roy Meteor Soc* **98**: 563–589.

Kammann C, Grünhage L, Jäger H-J. 2001. A new sampling technique to monitor concentrations of $CH_4$, $N_2O$ and $CO_2$ in air at well-defined depths in soils with varied water potential. *Eur J Soil Sci* **52**: 297–303.

Kattner G. 1999. Storage of dissolved inorganic nutrients in seawater: poisoning with mercuric chloride. *Mar Chem* **67**: 61–66.

Kelley JJ, Gosink TA. 1979. Gases in Sea Ice: 1975–1979. Fairbanks, Alaska: University of Alaska, Fairbanks. Final report to ONR N000 14-76C-0331.

Kelley JJ, Gosink TA. 1985. Sources and sinks of carbon dioxide in the Arctic regions, in *Final Report: Carbon Dioxide in Arctic and Subarctic Regions*. Fairbanks: University of Alaska: 131–165.

Kennedy H, Robertson J. 1995. Variations in the isotopic composition of particulate organic carbon in surface waters along an 88°W transect from 67°S to 54°S. *Deep-Sea Res Pt II* **42**: 1109–1122.

Kennedy H, Thomas DN, Kattner G, Haas C, Dieckmann GS. 2002. Particulate organic matter in Antarctic summer sea ice: concentration and stable isotopic composition. *Mar Ecol-Prog Ser* **238**: 1–13.

King MD, France JL, Fisher FN, Beine HJ. 2005. Measurement and modelling of UV radiation penetration and photolysis rates of nitrate and hydrogen peroxide in Antarctic sea ice: An estimate of the production rate of hydroxyl radicals in first-year sea ice. *J Photochem Photobiol A* **176**: 39–49. doi: 10.1016/j.jphotochem.2005.08.032.

Klánová J, Klán P, Heger D, Holoubek I. 2003. Comparison of the effects of UV, $H_2O_2$/UV and γ-irradiation processes on frozen and liquid water solutions of monochlorophenols. *Photochem Photobiol Sci* **2**: 1023–1031. doi: 10.1039/b303483f.

Knefelkamp B, Carstens K, Wiltshire KH. 2007. Comparison of different filter types on *chlorophyll-a* retention and nutrient measurements. *J Exp Mar Biol Ecol* **345**: 61–70. doi: 10.1016/j.jembe.2007.01.008.

Knepp TN, Bottenheim J, Carlsen M, Carlson D, Donohoue D, et al. 2010. Development of an autonomous sea ice tethered buoy for the study of ocean-atmosphere-sea ice-snow pack interactions: the O-buoy. *Atmos Meas Tech* **3**: 249–261.

Koh EY, Atamna-Ismaeel N, Martin A, Cowie ROM, Beja O, et al. 2010. Proteorhodopsin-bearing bacteria in Antarctic sea ice. *Appl Environ Microbiol* **76**: 5918–5925. doi: 10.1128/AEM.00562-10.

Krembs C, Eicken H, Deming JW. 2011. Exopolymer alteration of physical properties of sea ice and implications for ice habitability and biogeochemistry in a warmer Arctic. *Proc Natl Acad Sci* **108**: 3653–3658.

Krembs C, Eicken H, Junge K, Deming JW. 2002. High concentrations of exopolymeric substances in Arctic winter sea ice: implications for the polar ocean carbon cycle and cryoprotection of diatoms. *Deep-Sea Res Pt I* **49**: 2163–2181.

Kristiansen S, Farbrot T, Kuosa H, Myklestad S, Quillfeldt CHv. 1998. Nitrogen uptake in the infiltration community, and ice algal community in Antarctic pack-ice. *Polar Biol* **19**: 307–315.

Kristiansen S, Syvertsen EE, Farbrot T. 1992. Nitrogen uptake in the Weddell Sea during late winter and spring. *Polar Biol* **12**: 245–251.

Kühl M, Glud RN, Borum J, Roberts R, Rysgaard S. 2001. Photosynthetic performance of surface-associated algae below sea ice as measured with a pulse-amplitude-modulated (PAM) fluorometer and $O_2$ microsensors. *Mar Ecol-Prog Ser* **223**: 1–14.

Lannuzel D, Bowie AR, van der Merwe PC, Townsend AT, Schoemann V. 2011. Distribution of dissolved and particulate metals in Antarctic sea ice. *Mar Chem* **124**: 134–146. doi:10.1016/j.marchem.2011.01.004.

Lannuzel D, de Jong J, Schoemann V, Trevena A, Tison J-L, et al. 2006. Development of a sampling and flow injection analysis technique for iron determination in the sea ice environment. *Anal Chim Acta* **556**: 476–483. doi:10.1016/j.aca.2005.09.059.

Lannuzel D, Schoemann V, de Jong J, Chou L, Delille B, et al. 2008. Iron study during a time series in the western Weddell pack ice. *Mar Chem* **108**(1–2): 85–95. doi: 10.1016/j.marchem.2007.10.006.

Lannuzel D, Schoemann V, de Jong J, Pasquer B, van der Merwe P, et al. 2010. Distribution of dissolved iron in Antarctic sea ice: Spatial, seasonal, and inter-annual variability. *J Geophys Res.* **115**: G03022. doi: 10.1029/2009JG001031.

Lannuzel D, Schoemann V, de Jong J, Tison J-L, Chou L. 2007. Distribution and biogeochemical behaviour of iron in the East Antarctic sea ice. *Mar Chem* **106**: 18–32. doi:10.1016/j.marchem.2006.06.010.

Lannuzel D, van der Merwe PC, Townsend AT, Bowie AR. 2014. Size fractionation of iron, manganese and aluminium in Antarctic fast ice reveals a lithogenic origin and low iron solubility. *Mar Chem* **161**: 47–56. doi: 10.1016/j.marchem.2014.02.006.

Latasa M. 2007. Improving estimations of phytoplankton class abundances using CHEMTAX. *Mar Ecol-Prog Ser* **329**: 13–21.

Laws E, Sakshaug E, Babin M, Dandonneau Y, Falkowski P, et al. 2002. Photosynthesis and Primary Productivity in Marine Ecosystems: Practical Aspects and Application of Techniques. Bergen: Joint Global Ocean Flux Study. JGOFS Report No. 36.

Lee PA, de Mora SJ, Gosselin M, Levasseur M, Bouillon R-C, et al. 2001. Particulate dimethylsulfoxide in Arctic sea-ice algal communities: The cryoprotectant hypothesis revisited. *J Phycol* **37**: 488–499.

Lee X, Massman W, Law B, eds. 2004. *Handbook of Micrometeorology: A Guide for Surface Flux Measurement and Analysis.* Boston: Kluwer Academic Publishers.

Leppäranta M, Manninen T. 1988. *The Brine and Gas Content of Sea Ice with Attention to Low Salinities and High Temperatures.* Helsinki: Finnish Institute of Marine Research.

Leuning R, King KM. 1992. Comparison of eddy-covariance measurements of $CO_2$ fluxes by open- and closed-path $CO_2$ analysers. *Bound-Lay Meteorol* **59**: 297–311.

Levasseur M, Gosselin M, Michaud S. 1994. A new source of dimethylsulfide (DMS) for the arctic atmosphere: Ice diatoms. *Mar Biol* **121**: 381–387.

Liss PS, Hatton AD, Malin G, Nightingale PD, Turner SM. 1997. Marine sulphur emissions. *Philos T Roy Soc B* **352**: 159–169.

Long MH, Koopmans D, Berg P, Rysgaard S, Glud RN, et al. 2012. Oxygen exchange and ice melt measured at the ice-water interface by eddy correlation. *Biogeosciences* **9**: 1957–1967. doi: 10.5194/bg-9-1957-2012.

Loose B, Miller LA, Elliott S, Papakyriakou T. 2011. Sea ice biogeochemistry and material transport across the frozen interface. *Oceanography* **24**: 202–218. doi: 10.5670/oceanog.2011.72.

Löscher BM, deBaar HJW, de Jong JTM, Veth C, Dehairs F. 1997. The distribution of Fe in the Antarctic Circumpolar current. *Deep-Sea Res Pt II* **44**(1–2): 143–187.

Luo Y, Zhou X. 2006. *Soil Respiration and the Environment.* San Francisco: Academic Press.

Lyakhin YI. 1970. Saturation of water of the Sea of Okhotsk with calcium carbonate. *Oceanology* **10**: 789–795.

Maas EW, Simpson AM, Martin A, Thompson S, Koh EY, et al. 2012. Phylogenetic analyses of bacteria in sea ice at Cape Hallett, Antarctica. *New Zeal J Mar Fresh* **46**: 3–12.

Macdonald RW, Carmack EC, McLaughlin FA, Iseki K, Macdonald DM, et al. 1989. Composition and modification of water masses in the Mackenzie shelf estuary. *J Geophys Res* **94**(C12): 18,057–18,070.

Macdonald RW, Carmack EC, Paton DW. 1999. Using the $\delta^{18}O$ composition in landfast ice as a record of arctic estuarine processes. *Mar Chem* **65**: 3–24.

Manes SS, Gradinger R. 2009. Small scale vertical gradients of Arctic ice algal photophysiological properties. *Photosynth Res* **102**: 53–66. doi: 10.1007/s11120-009-9489-0.

Mantoura RFC, Wright SW, Jeffrey SW, Barlow RG, Cummings DE. 1997. Filtration and storage of pigments from microalgae, in Jeffrey SW, Mantoura RFC, Wright SW, eds., *Phytoplankton Pigments in Oceanography.* UNESCO Publishing: 283–305.

Martin A, Anderson MJ, Thorn C, Davy SK, Ryan KG. 2011. Response of sea-ice microbial communities to environmental disturbance: an *in situ* transplant experiment in the Antarctic. *Mar Ecol-Prog Ser* **424**: 25–37. doi: 10.3354/meps08977.

Martiny AC, Vrugt JA, Primeau FW, Lomas MW. 2013. Regional variation in the particulate organic carbon to nitrogen ratio in the surface ocean. *Glob Biogeochem Cy* **27**: 723–731. doi: 10.1002/gbc.20061.

Massom RA, Eicken H, Haas C, Jeffries MO, Drinkwater MR, et al. 2001. Snow on Antarctic sea ice. *Rev Geophys* **39**: 413–445. doi: 10.1029/2000RG000085.

Massom RA, Stammerjohn SE. 2010. Antarctic sea ice change and variability - Physical and ecological implications. *Polar Sci* **4**: 149–186. doi: 10.1016/j.polar.2010.05.001.

Matsuo S, Miyake Y. 1966. Gas composition in ice samples from Antarctica. *J Geophys Res* **71**: 5235–5241.

Mattson E, Karlsson A, Smith Jr. WO, Abrahamsson K. 2012. The relationship between biophysical variables and halocarbon distributions in the waters of the Amundsen and Ross Seas, Antarctica. *Mar Chem* **140-1**: 1–9. doi: 10.1016/j.marchem.2012.07.002.

McCorkle DC, Emerson SR, Quay PD. 1985. Stable carbon isotopes in marine porewaters. *Earth Planet Sc Lett* **74**: 13–26.

McGinnis DF, Cherednichenko S, Sommer S, Berg P, Rovelli L, et al. 2011. Simple, robust eddy correlation amplifier for aquatic dissolved oxygen and hydrogen sulfide flux measurements. *Limnol Oceanogr Methods* **9**: 340–347. doi:10.4319/lom.2011.9.340.

McLaughlin F, Proshutinsky A, Carmack EC, Shimada K, Brown K, et al. 2012. Physical, Chemical and Zooplankton Data from the Canada Basin and Canadian Arctic Archipelago, July 20 to September 14, 2006. Sidney: Institute of Ocean Sciences. Canadian Data Report of Hydrography and Ocean Sciences 186.

McMinn A, Ashworth C. 1998. The use of oxygen microelectrodes to determine the net production by an Arctic sea ice algal community. *Antarct Sci* **10**: 39–44.

McMinn A, Ashworth C, Ryan KG. 2000. *In situ* net primary productivity of an Antarctic fast ice bottom algal community. *Aquat Microb Ecol* **21**: 177–185.

McMinn A, Gradinger R, Nomura D. 2009. Chapter 3.8, Biogeochemical properties of sea ice, in Eicken H, Gradinger R, Salganek M, Shirasawa K, Perovich D, et al., eds., *Field Techniques for Sea Ice Research*. Fairbanks: University of Alaska Press: pp. 259–82.

McMinn A, Hegseth EN. 2007. Sea ice primary productivity in the northern Barents Sea, spring 2004. *Polar Biol* **30**: 289–294. doi: 10.1007/s00300-006-0182-x.

McMinn A, Ryan K, Gademann R. 2003. Diurnal changes in photosynthesis of Antarctic fast ice algal communities determined by pulse amplitude modulation fluorometry. *Mar Biol* **143**: 359–367. doi: 10.1007/s00227-003-1052-5.

McMinn A, Ryan KG, Ralph PJ, Pankowski A. 2007. Spring sea ice photosynthesis, primary productivity and biomass distribution in eastern Antarctica, 2002–2004. *Mar Biol* **151**: 985–995. doi: 10.1007/s00227-006-0533-8.

Meiners K, Brinkmeyer R, Granskog MA, Lindfors A. 2004. Abundance, size distribution and bacterial colonization of exopolymer particles in Antarctic sea ice (Bellingshausen Sea). *Aquat Microb Ecol* **35**: 283–296.

Meiners K, Gradinger R, Fehling J, Civitarese G, Spindler M. 2003. Vertical distribution of exopolymer particles in sea ice of the Fram Strait (Arctic) during autumn. *Mar Ecol-Prog Ser* **248**: 1–13.

Meiners K, Krembs C, Gradinger R. 2008. Exopolymer particles: microbial hotspots of enhanced bacterial activity in Arctic fast ice (Chukchi Sea). *Aquat Microb Ecol* **52**: 195–207. doi: 10.3354/ame01214.

Meiners KM, Papadimitriou S, Thomas DN, Norman L, Dieckmann GS. 2009. Biogeochemical conditions and ice algal photosynthetic parameters in Weddell Sea ice during early spring. *Polar Biol* **32**: 1055–1065. doi: 10.1007/s00300-009-0605-6.

Meiners KM, Vancoppenolle M, Thanassekos S, Dieckmann GS, Thomas DN, et al. 2012. Chlorophyll *a* in Antarctic sea ice from historical ice core data. *Geophys Res Lett* **39**: L21602. doi: 10.1029/2012GL053478.

Michel C, Legendre L, Ingram RG, Gosselin M, Levasseur M. 1996. Carbon budget of sea-ice algae in spring: Evidence of a significant transfer to zooplankton grazers. *J Geophys Res* **101**(C8): 18,345–18,360.

Michel C, Nielsen TG, Nozais C, Gosselin M. 2002. Significance of sedimentation and grazing by ice micro- and meiofauna for carbon cycling in annual sea ice (northern Baffin Bay). *Aquat Microb Ecol* **30**: 57–68.

Michel C, Niemi A. 2009. Field and laboratory methods for biogeochemical analyses of sea ice, seawater and particle interceptor trap samples. Winnipeg: Fisheries and Oceans Canada. Canadian Technical Report of Fisheries and Aquatic Sciences # 2852.

Mikkelsen DM, Witkowski A. 2010. Melting sea ice for taxonomic analysis: a comparison of four melting procedures. *Polar Res* **29**: 451–454.

Miller LA, Carnat G, Else BGT, Sutherland N, Papakyriakou TN. 2011a. Carbonate system evolution at the Arctic Ocean surface during autumn freeze-up. *J Geophys Res* **116**: C00G04. doi: 10.1029/2011JC007143.

Miller LA, Papakyriakou TN, Collins RE, Deming JW, Ehn JK, et al. 2011b. Carbon dynamics in sea ice: A winter flux time series. *J Geophys Res* **116**: C02028. doi: 10.1029/2009JC006058.

Millero FJ, DiTrolio B, Suarez AF, Lando G. 2009. Spectroscopic measurements of the pH in NaCl brines. *Geochim Cosmochim Acta* **73**: 3109–3114. doi: 10.1016/j.gca.2009.01.037.

Mock T. 2002. *In situ* primary production in young Antarctic sea ice. *Hydrobiologia* **470**: 127–132.

Mock T, Dieckmann GS, Haas C, Krell A, Tison J-L, et al. 2002. Micro-optodes in sea ice: A new approach to investigate oxygen dynamics during sea ice formation. *Aquat Microb Ecol* **29**: 297–306.

Mock T, Gradinger R. 1999. Determination of Arctic ice algal production with a new *in situ* incubation technique. *Mar Ecol-Prog Ser* **177**: 15–26.

Mock T, Kruse M, Dieckmann GS. 2003. A new microcosm to investigate oxygen dynamics at the sea ice water interface. *Aquat Microb Ecol* **30**: 197–205.

Mock T, Meiners KM, Giesenhagen HC. 1997. Bacteria in sea ice and underlying brackish water at 54° 26'50" N (Baltic Sea, Kiel Bight). *Mar Ecol-Prog Ser* **158**: 23–40.

Moeseneder MM, Winter C, Herndl GJ. 2001. Horizontal and vertical complexity of attached and free-living bacteria of the eastern Mediterranean Sea, determined by 16S rDNA and 16S rRNA fingerprints. *Limnol Oceanogr* **46**: 95–107.

Mosier AR. 1989. Chamber and isotope techniques, in Andreae MO, Schimel DS, eds., *Exchange of Trace Gases between Terrestrial Ecosystems and the Atmosphere*. Toronto: John Wiley & Sons: pp. 175–187.

Muller JBA, Dorsey JR, Flynn M, Gallagher MW, Percival CJ, et al. 2012. Energy and ozone fluxes over sea ice. *Atmos Environ* **47**: 218–225. doi: 10.1016/j.atmosenv.2011.11.013.

Müller M, Graus M, Ruuskanen TM, Schnitzhofer R, Bamberger I, et al. 2010. First eddy covariance flux measurements by PTR-TOF. *Atmos Meas Tech* **3**: 387–395.

Müller S, Vähätalo AV, Stedmon CA, Granskog MA, Norman L, et al. 2013. Selective incorporation of dissolved organic matter (DOM) during sea ice formation. *Mar Chem* **155**: 148–157. doi: 10.1016/j.marchem.2013.06.008.

Mundy CJ, Barber DG, Michel C. 2005. Variability of snow and ice thermal, physical and optical properties pertinent to sea ice algae biomass during spring. *J Marine Syst* **58**: 107–120.

Mundy CJ, Ehn JK, Barber DG, Michel C. 2007. Influence of snow cover and algae on the spectral dependence of transmitted irradiance through Arctic landfast first-year sea ice. *J Geophys Res* **112**: C03007. doi: 10.1029/2006JC003683.

Mundy CJ, Gosselin M, Ehn JK, Belzile C, Poulin M, et al. 2011. Characteristics of two distinct high-light acclimated algal communities during advanced stages of sea ice melt. *Polar Biol* **34**: 1869–1886. doi: 10.1007/s00300-011-0998-x.

Munro DR, Dunbar RB, Mucciarone DA, Arrigo KR, Long MC. 2010. Stable isotope composition of dissolved inorganic carbon and particulate organic carbon in sea ice from the Ross Sea, Antarctica. *J Geophys Res* **115**: C09005. doi: 10.1029/2009JC005661.

Myklestad SM, Skånøy E, Hestmann S. 1997. A sensitive and rapid method for analysis of dissolved mono- and polysaccharides in seawater. *Mar Chem.* **56**: 279–286.

Nedashkovskii AP. 2002. Cadmium and lead in the ice of Amur Bay (Sea of Japan). *Oceanology* **42**: 344–349.

Nedashkovsky AP, Khvedynich SV, Petrovsky TV. 2009. Alkalinity of sea ice in the high-latitudinal Arctic according to the surveys performed at North Pole Drifting Station 34 and characterization of the role of the Arctic ice in the $CO_2$ exchange. *Oceanology* **49**: 55–63. doi: 10.1134/S000143700901007X.

Nedashkovsky AP, Shvetsova MG. 2010. Total inorganic carbon in sea ice. *Oceanology* **50**: 861–868. doi: 10.1134/S0001437010060056.

Nelson KH, Thompson TG. 1954. Deposition of salts from sea water by frigid concentration. *J Mar Res* **13**: 166–182.

Nishi Y, Tabeta S. 2008. Relation of material exchange between sea ice and water to a coupled ice-ocean ecosystem at the Hokkaido coastal region of the Okhotsk Sea. *J Geophys Res* **113**: C01003. doi: 10.1029/2006JC004077.

Noble RT, Fuhrman JA. 1998. Use of SYBR Green I for rapid epifluorescence counts of marine viruses and bacteria. *Aquat Microb Ecol* **14**: 113–118.

Nomura D, Assmy P, Nehrke G, Granskog MA, Fischer M, et al. 2013a. Characterization of ikaite ($CaCO_3 \cdot 6H_2O$) crystals in first-year Arctic sea ice north of Svalbard. *Ann Glaciol* **54**: 125–131. doi: 10.3189/2013AoG62A034.

Nomura D, Eicken H, Gradinger R, Shirasawa K. 2010a. Rapid physically driven inversion of the air-sea ice $CO_2$ flux in the seasonal landfast ice off Barrow, Alaska after onset of surface melt. *Cont Shelf Res* **30**: 1998–2004. doi: 10.1016/j.csr.2010.09.014.

Nomura D, Granskog MA, Assmy P, Simizu D, Hashida G. 2013b. Arctic and Antarctic sea ice acts as a sink for atmospheric $CO_2$ during periods of snowmelt and surface flooding. *J Geophys Res-Oceans* **118**: 6511–6524. doi: 10.1002/2013JC009048.

Nomura D, Koga S, Kasamatsu N, Shinagawa H, Simizu D, et al. 2012. Direct measurements of DMS flux from Antarctic fast sea ice to the atmosphere by a chamber technique. *J Geophys Res* **117**: C04011. doi:10.1029/2010JC006755.

Nomura D, McMinn A, Hattori H, Aoki S, Fukuchi M. 2011. Incorporation of nitrogen compounds into sea ice from atmospheric deposition. *Mar Chem* **127**: 90–99. doi: 10.1016/j.marchem.2011.08.002.

Nomura D, Takatsuka T, Ishikawa M, Kawamura T, Shirasawa K, et al. 2009. Transport of chemical components in sea ice and under-ice water during melting in the seasonally ice-covered Saroma-ko Lagoon, Hokkaido, Japan. *Estuar Coast Shelf S* **81**: 201–209. doi: 10.1016/j.ecss.2008.10.012.

Nomura D, Yoshikawa-Inoue H, Toyota T, Shirasawa K. 2010b. Effects of snow, snowmelting and refreezing processes on air-sea-ice $CO_2$ flux. *J Glaciol* **56**: 262–270.

Norman L, Thomas DN, Stedmon CA, Granskog MA, Papadimitriou S, et al. 2011. The characteristics of dissolved organic matter (DOM) and chromophoric dissolved organic matter (CDOM) in Antarctic sea ice. *Deep-Sea Res Pt II* **58**: 1075–1091. doi: 10.1016/j.dsr2.2010.10.030.

Norman M, Rutgersson A, Sørensen LL, Sahlée E. 2012. Methods for estimating air-sea fluxes of $CO_2$ using high-frequency measurements. *Bound-Lay Meteorol* **144**: 379–400. doi: 10.1007/s10546-012-9730-9.

Not C, Brown K, Ghaleb B, Hillaire-Marcel C. 2012. Conservative behavior of uranium vs. salinity in Arctic sea ice and brine. *Mar Chem* **130–1**: 33–39. doi:10.1016/j.marchem.2011.12.005.

Notz D, Wettlaufer JS, Worster MG. 2005. A non-destructive method for measuring the salinity and solid fraction of growing sea ice in situ. *J Glaciol* **51**(172): 159–166.

Notz D, Worster MG. 2009. Desalination processes of sea ice revisited. *J Geophys Res* **114**: C05006. doi: 10.1029/2008JC004885.

Obbard RW, Roscoe HK, Wolff EW, Atkinson HM. 2009. Frost flower surface area and chemistry as a function of salinity and temperature. *J Geophys Res* **114**: D20305. doi: 10.1029/2009JD012481.

Papadimitriou S, Kennedy H, Kattner G, Dieckmann GS, Thomas DN. 2004. Experimental evidence for carbonate precipitation and $CO_2$ degassing during sea ice formation. *Geochim Cosmochim Acta* **68**: 1749–1761. doi: 10.1016/j.gca.2003.07.004.

Papadimitriou S, Kennedy H, Kennedy P, Thomas DN. 2013. Ikaite solubility in seawater-derived brines at 1 atm and sub-zero temperatures to 265 K. *Geochim Cosmochim Acta* **109**: 241–253. doi:10.1016/j.gca.2013.01.044.

Papadimitriou S, Kennedy H, Kennedy P, Thomas DN. 2014. Kinetics of ikaite precipitation and dissolution in seawater-derived brines at sub-zero temperatures to 265 K. *Geochim Cosmochim Acta* **140**: 199–211. doi: 10.1016/j.gca.2014.05.031.

Papadimitriou S, Kennedy H, Norman L, Kennedy DP, Dieckmann GS, et al. 2012. The effect of biological activity, $CaCO_3$ mineral dynamics, and $CO_2$ degassing in the inorganic carbon cycle in sea ice in late winter-early spring in the Weddell Sea, Antarctica. *J Geophys Res* **117**: C08011. doi:10.1029/2012JC008058.

Papadimitriou S, Thomas DN, Kennedy H, Haas C, Kuosa H, et al. 2007. Biogeochemical composition of natural sea ice brines from the Weddell Sea during early austral summer. *Limnol Oceanogr* **52**: 1809–1823.

Papadimitriou S, Thomas DN, Kennedy H, Kuosa H, Dieckmann GS. 2009. Inorganic carbon removal and isotopic enrichment in Antarctic sea ice gap layers during early austral summer. *Mar Ecol-Prog Ser* **386**: 15–27. doi: 10.3354/meps08049.

Papakyriakou T, Miller L. 2011. Springtime $CO_2$ exchange over seasonal sea ice in the Canadian Arctic Archipelago. *Ann Glaciol* **52**: 215–224.

Passow U. 2002. Transparent exopolymer particles (TEP) in aquatic environments. *Prog Oceanogr* **55**: 287–333.

Paterson H, Laybourn-Parry J. 2012. Sea ice microbial dynamics over an annual ice cycle in Prydz Bay, Antarctica. *Polar Biol* **35**: 993–1002. doi: 10.1007/s00300-011-1146-3.

Perovich D. 2009. Chapter 3.6, Sea ice optics measurements, in Eicken H, Gradinger R, Salganek M, Shirasawa K, Perovich D, et al., eds., *Field Techniques for Sea Ice Research*. Fairbanks: University of Alaska Press: pp. 215–229.

Perovich DK, Richter-Menge JA. 2009. Loss of sea ice in the Arctic. *Annu Rev Mar Sci* **1**: 417–441. doi: 10.1146/annurev. marine.010908.163805.

Petrich C, Eicken H. 2010. Growth, structure and properties of sea ice, in Thomas DN, Dieckmann GS, eds., *Sea Ice*. 2nd ed., Oxford: Wiley-Blackwell: 23–77.

Pineault S, Tremblay J-É, Gosselin M, Thomas H, Shadwick E. 2013. The isotopic signature of particulate organic C and N in bottom ice: Key influencing factors and applications for tracing the fate of ice-algae in the Arctic Ocean. *J Geophys Res-Oceans* **118**: 287–300. doi:10.1029/2012JC008331.

Piwosz K, Wiktor JM, Niemi A, Tatarek A, Michel C. 2013. Mesoscale distribution and functional diversity of picoeukaryotes in the first-year sea ice of the Canadian Arctic. *ISME J* **7**: 1461–1471. doi: 10.1038/ismej.2013.39.

Porter KG, Feig YS. 1980. The use of DAPI for identifying and counting aquatic microflora. *Limnol Oceanogr* **25**: 943–948.

Post E, Bhatt US, Bitz CM, Brodie JF, Fulton TL, et al. 2013. Ecological consequences of sea-ice decline. *Science* **341**: 519–524. doi: 10.1126/science.1235225.

Poulain AJ, Garcia E, Amyot M, Campbell PGC, Ariya PA. 2007. Mercury distribution, partitioning and speciation in coastal vs. inland high Arctic snow. *Geochim Cosmochim Acta* **71**: 3419–3431. doi: 10.1016/j.gca.2007.05.006.

Price NM, Harrison PJ. 1987. Comparison of methods for the analysis of dissolved urea in seawater. *Mar Biol* **94**: 307–317.

Prytherch J, Yelland MJ, Pascal RW, Moat BI, Skjelvan I, et al. 2010. Direct measurements of the $CO_2$ flux over the ocean: Development of a novel method. *Geophys Res Lett* **37**: L03607. doi:10.1029/2009GL041482.

Pućko M, Stern GA, Barber DG, Macdonald RW, Rosenberg B. 2010a. The International Polar Year (IPY) Circumpolar Flaw Lead (CFL) System Study: The importance of brine processes for α- and γ-hexachlorocyclohexane (HCH) accumulation or rejection in sea ice. *Atmos Ocean* **48**: 244–262. doi: 10.3137/OC318.2010.

Pućko M, Stern GA, Macdonald RW, Barber DG. 2010b. α- and γ-hexachlorocyclohexane measurements in the brine fraction of sea ice in the Canadian high Arctic using a sump-hole technique. *Environ Sci Technol* **44**: 9258–9264. doi: 10.1021/es102275b.

Qian J, Mopper K. 1996. Automated high-performance, high-temperature combustion total organic carbon analyzer. *Anal Chem* **68**: 3090–3097.

Ragueneau O, Savoye N, Del Amo Y, Cotten J, Tardiveau B, et al. 2005. A new method for the measurement of biogenic silica in suspended matter of coastal waters: using Si:Al ratios to correct for the mineral interference. *Cont Shelf Res* **25**: 697–710. doi: 10.1016/j.csr.2004.09.017.

Rahm L, Håkansson B, Larsson P, Fogelqvist E, Bremle G, et al. 1995. Nutrient and persistent pollutant deposition on the Bothnian Bay ice and snow fields. *Water Air Soil Poll* **84**: 187–201.

Ralph PJ, Gademann R. 2005. Rapid light curves: A powerful tool to assess photosynthetic activity. *Aquat Bot* **82**: 222–237. doi: 10.1016/j.aquabot.2005.02.006.

Ralph PJ, Ryan KG, Martin A, Fenton G. 2007. Melting out of sea ice causes greater photosynthetic stress in algae than freezing in. *J Phycol* **43**: 948–956. doi: 10.1111/j.1529-8817.2007.00382.x.

Randall K, Scarratt M, Levasseur M, Michaud S, Xie H, et al. 2012. First measurements of nitrous oxide in Arctic sea ice. *J Geophys Res* **117**: C00G15. doi: 10.1029/2011JC007340.

Rau GH, Sullivan CW, Gordon LI. 1991. $\delta^{13}C$ and $\delta^{15}N$ variations in Weddell Sea particulate organic matter. *Mar Chem* **35**: 355–369.

Raymond B, Meiners K, Fowler CW, Pasquer B, Williams GD, et al. 2009. Cumulative solar irradiance and potential large-scale sea ice algae distribution off East Antarctica (30°E–150°E). *Polar Biol* **32**: 443–452. doi: 10.1007/s00300-008-0538-5.

Renaud PE, Riedel A, Michel C, Morata N, Gosselin M, et al. 2007. Seasonal variation in benthic community oxygen demand: A response to an ice algal bloom in the Beaufort Sea, Canadian Arctic? *J Marine Syst* **67**: 1–12. doi: 10.1016/j.jmarsys.2006.07.006.

Riedel A, Michel C, Gosselin M. 2006. Seasonal study of sea-ice exopolymeric substances on the Mackenzie shelf: Implications for transport of sea-ice bacteria and algae. *Aquat Microb Ecol* **45**: 195–206.

Riedel A, Michel C, Gosselin M. 2007a. Grazing of large-sized bacteria by sea-ice heterotrophic protists on the Mackenzie Shelf during the winter-spring transition. *Aquat Microb Ecol* **50**: 25–38. doi:10.3354/ame01155.

Riedel A, Michel C, Gosselin M, LeBlanc B. 2007b. Enrichment of nutrients, exopolymeric substances and microorganisms in newly formed sea ice on the Mackenzie shelf. *Mar Ecol-Prog Ser* **342**: 55–67.

Riedel A, Michel C, Gosselin M, LeBlanc B. 2008. Winter-spring dynamics in sea-ice carbon cycling in the coastal Arctic Ocean. *J Marine Syst* **74**: 918–932. doi: 10.1016/j.jmarsys.2008.01.003.

Riederer M, Serafimovich A, Foken T. 2014. Net ecosystem $CO_2$ exchange measurements by the closed chamber method and the eddy covariance technique and their dependence on atmospheric conditions. *Atmos Meas Tech* **7**: 1057–1064. doi:10.5194/amt-7-1057-2014.

Ringer WE. 1928. Ueber die Veränderungen in der Zusammensetzung des Meereswassersalzes beim Ausfrieren. Copenhagen: Conseil Permanent International Pour L'Exploration de la Mer. Volume 47.

Rivkin RB, Legendre L. 2001. Biogenic carbon cycling in the upper ocean: Effects of microbial respiration. *Science* **291**: 2398–2400.

Robinson DH, Arrigo KR, Kolber Z, Gosselin M, Sullivan CW. 1998. Photophysiological evidence of nutrient limitation of platelet ice algae in McMurdo Sound, Antarctica. *J Phycol* **34**: 788–797.

Roscoe HK, Brooks B, Jackson AV, Smith MH, Walker SJ, et al. 2011. Frost flowers in the laboratory: Growth, characteristics, aerosol, and the underlying sea ice. *J Geophys Res* **116**: D12301. doi:10.1029/2010JD015144.

Roy S, Llewellyn CA, Egeland ES, Johnsen G. 2011. *Phytoplankton Pigments: Characterization, Chemotaxonomy and Applications in Oceanography*. Cambridge: Cambridge University Press.

Rusch DB, Halpern AL, Sutton G, Heidelberg KB, Williamson S, et al. 2007. The *Sorcerer II* global ocean sampling expedition: Northwest Atlantic through eastern Tropical Pacific. *PLoS Biology* 5(3): e77. doi: 10.1371/journal.pbio.0050077.

Russell LM. 2014. Carbonaceous particles: Source-based characterization of their formation, composition, and structures, in Holland HD and Turekian KK, eds., *Treatise on Geochemistry*. Second ed. Oxford: Elsevier: pp. 291–316.

Ryan KG, McMinn A, Mitchell KA, Trenerry L. 2002. Mycosporine-like amino acids in Antarctic sea ice algae, and their response to UVB radiation. *Z Naturforsch* 54c: 471–477.

Rysgaard S, Bendtsen J, Delille B, Dieckmann GS, Glud RN, et al. 2011. Sea ice contribution to the air-sea $CO_2$ exchange in the Arctic and Southern Oceans. *Tellus B* 63: 823–830. doi: 10.1111/j.1600-0889.2011.00571.x.

Rysgaard S, Bendtsen J, Pedersen LT, Ramløv H, Glud RN. 2009. Increased $CO_2$ uptake due to sea ice growth and decay in the Nordic Seas. *J Geophys Res* 114: C09011. doi:10.1029/2008JC005088.

Rysgaard S, Glud RN. 2004. Anaerobic $N_2$ production in Arctic sea ice. *Limnol Oceanogr* 49: 86–94.

Rysgaard S, Glud RN, Lennert K, Cooper M, Halden N, et al. 2012. Ikaite crystals in melting sea ice - implications for $pCO_2$ and pH levels in Arctic surface waters. *The Cryosphere* 6: 901–908. doi:10.5194/tc-6-901-2012.

Rysgaard S, Glud RN, Sejr MK, Bendtsen J, Christensen PB. 2007. Inorganic carbon transport during sea ice growth and decay: A carbon pump in polar seas. *J Geophys Res* 112: C03016. doi:10.1029/2006JC003572.

Rysgaard S, Glud RN, Sejr MK, Blicher ME, Stahl HJ. 2008. Denitrification activity and oxygen dynamics in Arctic sea ice. *Polar Biol* 31: 527–537. doi: 10.1007/s00300-007-0384-x.

Rysgaard S, Kühl M, Glud RN, Hansen JW. 2001. Biomass, production and horizontal patchiness of sea ice algae in a high-Arctic fjord (Young Sound, NE Greenland). *Mar Ecol-Prog Ser* 223: 15–26.

Rysgaard S, Søgaard DH, Cooper M, Pućko M, Lennert K, et al. 2013. Ikaite crystal distribution in winter sea ice and implications for $CO_2$ system dynamics. *The Cryosphere* 7: 707–718. doi: 10.5194/tc-7-707-2013.

Satoh H, Watanabe K. 1988. Primary productivity in the fast ice area near Syowa Station, Antarctica, during spring and summer 1983/84. *J Oceanogr Soc Japan* 44: 287–292.

Schubert CJ, Calvert SE. 2001. Nitrogen and carbon isotopic composition of marine and terrestrial organic matter in Arctic Ocean sediments: implications for nutrient utilization and organic matter composition. *Deep-Sea Res Pt I* 48: 789–810.

Scully NM, Miller WL. 2000. Spatial and temporal dynamics of colored dissolved organic matter in the North Water Polynya. *Geophys Res Lett* 27: 1009–1011.

Sejr MK, Krause-Jensen D, Rysgaard S, Sørensen LL, Christensen PB, et al. 2011. Air-sea flux of $CO_2$ in arctic coastal waters influenced by glacial melt water and sea ice. *Tellus B* 63: 815–822. doi: 10.1111/j.1600-0889.2011.00540.x.

Semiletov I, Makshtas A, Akasofu S-I, Andreas EL. 2004. Atmospheric $CO_2$ balance: The role of Arctic sea ice. *Geophys Res Lett* 31: L05121. doi:10.1029/2003GL017996.

Shaw MD, Carpenter LJ, Baeza-Romero MT, Jackson AV. 2011. Thermal evolution of diffusive transport of atmospheric halocarbons through artificial sea-ice. *Atmos Environ* 45: 6393–6402. doi: 10.1016/j.atmosenv.2011.08.023.

Sherman LS, Blum JD, Douglas TA, Steffen A. 2012. Frost flowers growing in the Arctic ocean-atmosphere-sea ice-snow interface: 2. Mercury exchange between the atmosphere, snow, and frost flowers. *J Geophys Res* 117: D00R10. doi: 10.1029/2011JD016186.

Sigman DM, Casciotti KL, Andreani M, Barford C, Galanter M, et al. 2001. A bacterial method for the nitrogen isotopic analysis of nitrate in seawater and freshwater. *Anal Chem* 73: 4145–4153.

Sime-Ngando T, Juniper SK, Demers S. 1997. Ice-brine and planktonic microheterotrophs from Saroma-ko Lagoon, Hokkaido (Japan): quantitative importance and trophodynamics. *J Marine Syst* 11: 149–161.

Simpson WR, Alvarez-Aviles L, Douglas TA, Sturm M, Domine F. 2005. Halogens in the coastal snow pack near Barrow, Alaska: Evidence for active bromine air-snow chemistry during springtime. *Geophys Res Lett* 32: L04811. doi: 10.1029/2004GL021748.

Simpson WR, von Glasow R, Riedel K, Anderson P, Ariya P, et al. 2007. Halogens and their role in polar boundary-layer ozone depletion. *Atmos Chem Phys* 7: 4375–4418.

Smedsrud LH, Skogseth R. 2006. Field measurements of Arctic grease ice properties and processes. *Cold Reg Sci Technol* 44: 171–183. doi: 10.1016/j.coldregions.2005.11.002.

Smith REH, Clement P. 1990. Heterotrophic activity and bacterial productivity in assemblages of microbes from sea ice in the high Arctic. *Polar Biol* 10: 351–357.

So A, Pel J, Rajan S, Marziali A. 2010. Efficient genomic DNA extraction from low target concentration bacterial cultures using SCODA DNA extraction technology. *Cold Spring Harb Protoc* 2010: 1150–1153, 1185–1198. doi:10.1101/pdb.prot5506.

Søgaard DH, Kristensen M, Rysgaard S, Glud RN, Hansen PJ, et al. 2010. Autotrophic and heterotrophic activity in Arctic first-year sea ice: seasonal study from Malene Bight, SW Greenland. *Mar Ecol-Prog Ser* 419: 31–45. doi: 10.3354/meps08845.

Søgaard DH, Thomas DN, Rysgaard S, Glud RN, Norman L, et al. 2013. The relative contributions of biological and abiotic processes to carbon dynamics in subarctic sea ice. *Polar Biol* 36: 1761–1777. doi:10.1007/s00300-013-1396-3.

Song G, Xie H, Aubry C, Zhang Y, Gosselin M, et al. 2011. Spatiotemporal variations of dissolved organic carbon and carbon monoxide in first-year sea ice in the western Canadian Arctic. *J Geophys Res* 116: C00G05. doi: 10.1029/2010JC006867.

Sørensen LL, Jensen B, Glud RN, McGinnis DF, Sejr MK, et al. 2014. Parameterization of atmosphere-surface exchange of $CO_2$ over sea ice. *The Cryosphere* 8: 853–866. doi:10.5194/tc-8-853-2014.

Sørensen LL, Larsen SE. 2010. Atmosphere-surface fluxes of $CO_2$ using spectral techniques. *Bound-Layer Meteorol* 136: 59–81. doi: 10.1007/s10546-010-9499-7.

Sørensen LL, Pryor SC, de Leeuw G, Schulz M. 2005. Flux divergence of nitric acid in the marine atmospheric surface layer. *J Geophys Res* 110: D15306. doi: 10.1029/2004JD005403.

Spindler M, Dieckmann GS. 1986. Distribution and abundance of the planktic foraminifer *Neogloboquadrina pachyderma* in sea ice of the Weddell Sea (Antarctica). *Polar Biol* 5: 185–191.

Spyres G, Nimmo M, Worsfold PJ, Achterberg EP, Miller AEJ. 2000. Determination of dissolved organic carbon in seawater using high temperature catalytic oxidation techniques. *TrAC Trends Anal Chem* 19(8): 498–506.

Stedmon CA, Thomas DN, Granskog M, Kaartokallio H, Papadimitriou S, et al. 2007. Characteristics of dissolved organic matter in Baltic coastal sea ice: Allochthonous or autochthonous origins? *Environ Sci Technol* **41**: 7273–7279. doi: 10.1021/es071210f.

Stedmon CA, Thomas DN, Papadimitriou S, Granskog MA, Dieckmann GS. 2011. Using fluorescence to characterize dissolved organic matter in Antarctic sea ice brines. *J Geophys Res* **116**: G03027. doi: 10.1029/2011JG001716.

Steele JH, Turekian KK, Thorpe SA, eds. 2001. Appendix 2, Useful Values, in, *Encyclopedia of Ocean Sciences*. 2nd ed. Oxford: Academic Press: 376.

Stefels J, Carnat G, Dacey JWH, Goossens T, Elzenga JTM, et al. 2012. The analysis of dimethylsulfide and dimethyl-sulfoniopropionate in sea ice: Dry-crushing and melting using stable isotope additions. *Mar Chem* **128–129**: 34–43. doi: 10.1016/j.marchem.2011.09.007.

Steffens M, Granskog MA, Kaartokallio H, Kuosa H, Luodekari K, et al. 2006. Spatial variation of biogeochemical proper-ties of landfast sea ice in the Gulf of Bothnia, Baltic Sea. *Ann Glaciol* **44**: 80–87.

Sturges WT, Cota GF, Buckley PT. 1997. Vertical profiles of bromoform in snow, sea ice, and seawater in the Canadian Arctic. *J Geophys Res* **102**(C11): 25,073–25,083.

Sturm M. 2009. Field techniques for snow observations on sea ice, in Eicken H, Gradinger R, Salganek M, Shirasawa K, Perovich D, et al., eds., *Field Techniques for Sea Ice Research*. Fairbanks: University of Alaska Press: pp. 25–47.

Suggett DJ, Oxborough K, Baker NR, Macintyre HL, Kana TM, et al. 2003. Fast repetition rate and pulse amplitude modulation chlorophyll *a* fluorescence measurements for assessment of photosynthetic electron transport in marine phytoplankton. *Eur J Phycol* **38**: 371–384. doi: 10.1080/09670260310001612655.

Swadling KM, Gibson JAE, Ritz DA, Nichols PD. 1997. Horizontal patchiness in sympagic organisms of the Antarctic fast ice. *Antarct Sci* **9**: 399–406.

Swinbank WC. 1951. The measurement of vertical transfer of heat and water vapor by eddies in the lower atmosphere. *J Meteorol* **8**: 135–145.

Tebbe CC, Vahjen W. 1993. Interference of humic acids and DNA extracted directly from soil in detection and transfor-mation of recombinant DNA from bacteria and a yeast. *Appl Environ Microb* **59**: 2657–2665.

Thomas DN, Lara RJ, Eicken H, Kattner G, Skoog A. 1995. Dissolved organic matter in Arctic multi-year sea ice during winter: Major components and relationship to ice characteristics. *Polar Biol* **15**: 477–483.

Thomas DN, Lara RJ, Haas C, Schnack-Schiel SB, Dieckmann GS, et al. 1998. Biological soup within decaying summer sea ice in the Amundsen Sea, Antarctica, in Lizotte MP, Arrigo KR, eds., *Antarctic Sea Ice: Biological Processes, Interac-tions and Variability*. Washington: American Geophysical Union: pp. 161–171.

Thomas DN, Kattner G, Engbrodt R, Giannelli V, Kennedy H, et al. 2001a. Dissolved organic matter in Antarctic sea ice. *Ann Glaciol* **33**: 297–303.

Thomas DN, Kennedy H, Kattner G, Gerdes D, Gough C, et al. 2001b. Biogeochemistry of platelet ice: Its influence on particle flux under fast ice in the Weddell Sea, Antarctica. *Polar Biol* **24**: 486–496.

Thomas DN, Papadimitriou S, Michel C. 2010. Biogeochemistry of sea ice, in Thomas DN, Dieckmann GS, eds., *Sea Ice*. 2nd ed., Chichester: Wiley-Blackwell: pp. 425–467.

Tison J-L, Brabant F, Dumont I, Stefels J. 2010. High-resolution dimethyl sulfide and dimethylsulfoniopropionate time series profiles in decaying summer first-year sea ice at Ice Station Polarstern, western Weddell Sea, Antarctica. *J Geophys Res* **115**: G04044. doi: 10.1029/2010JG001427.

Tison J-L, Haas C, Gowing MM, Sleewaegen S, Bernard A. 2002. Tank study of physico-chemical controls on gas content and composition during growth of young sea ice. *J Glaciol* **48**: 177–191.

Tison J-L, Worby A, Delille B, Brabant F, Papdimitriou S, et al. 2008. Temporal evolution of decaying summer first-year sea ice in the Western Weddell Sea, Antarctica. *Deep-Sea Res Pt II* **55**: 975–987. doi: 10.1016/j.dsr2.2007.12.021.

Tovar-Sánchez A, Duarte CM, Alonso JC, Lacorte S, Tauler R, et al. 2010. Impacts of metals and nutrients released from melting multiyear Arctic sea ice. *J Geophys Res* **115**: C07003. doi: 10.1029/2009JC005685.

Tremblay J-É, Michel C, Hobson KA, Gosselin M, Price NM. 2006. Bloom dynamics in early opening waters of the Arctic Ocean. *Limnol Oceanogr* **51**(2): 900–912.

Trevena AJ, Jones GB. 2006. Dimethylsulphide and dimethylsulphoniopropionate in Antarctic sea ice and their release during sea ice melting. *Mar Chem* **98**: 210–222. doi: 10.1016/j.marchem.2005.09.005.

Trevena AJ, Jones GB, Wright SW, van den Enden RL. 2000. Profiles of DMSP, algal pigments, nutrients and salinity in pack ice from eastern Antarctica. *J Sea Res* **43**: 265–273.

Tsurikov VL. 1965. Formation of the ionic composition and salinity of sea ice. *Oceanology* **5**: 59–66.

Turner SM, Nightingale PD, Broadgate W, Liss PS. 1995. The distribution of dimethyl sulphide and dimethylsulphonio-propionate in Antarctic waters and sea ice. *Deep-Sea Res Pt II* **42**: 1059–1080.

Tütken T, Eisenhauer A, Wiegand B, Hansen BT. 2002. Glacial-interglacial cycles in Sr and Nd isotopic composition of Arctic marine sediments triggered by the Svalbard/Barents Sea ice sheet. *Mar Geol* **182**: 351–372.

Underwood GJC, Aslam SN, Michel C, Niemi A, Norman L, et al. 2013. Broad-scale predictability of carbohydrates and exopolymers in Antarctic and Arctic sea ice. *P Natl Acad Sci* **100**(39): 15734–15739. doi:10.1073/pnas.1302870110.

Underwood GJC, Fietz S, Papadimitriou S, Thomas DN, Dieckmann GS. 2010. Distribution and composition of dissolved extracellular polymeric substances (EPS) in Antarctic sea ice. *Mar Ecol-Prog Ser* **404**: 1–19. doi:10.3354/meps08557.

Uusikivi J, Vähätalo AV, Granskog MA, Sommaruga R. 2010. Contribution of mycosporine-like amino acids and colored dissolved and particulate matter to sea ice optical properties and ultraviolet attenuation. *Limnol Oceanogr* **55**: 703–713.

van der Merwe P, Lannuzel D, Bowie AR, Mancuso Nichols CA, Meiners KM. 2011a. Iron fractionation in pack and fast ice in East Antarctica: Temporal decoupling between the release of dissolved and particulate iron during spring melt. *Deep-Sea Res Pt II* **58**: 1222–1236. doi: 10.1016/j.dsr2.2010.10.036.

van der Merwe P, Lannuzel D, Bowie AR, Meiners KM. 2011b. High temporal resolution observations of spring fast ice melt and seawater iron enrichment in East Antarctica. *J Geophys Res* **116**: G03017. doi:10.1029/2010JG001628.

van der Merwe P, Lannuzel D, Mancuso Nichols CA, Meiners K, Heil P, et al. 2009. Biogeochemical observations during the winter-spring transition in East Antarctic sea ice: Evidence of iron and exopolysaccharide controls. *Mar Chem* **115**: 163–175. doi:10.1016/j.marchem.2009.08.001.

Vancoppenolle M, Goosse H, de Montety A, Fichefet T, Tremblay B, et al. 2010. Modeling brine and nutrient dynamics in Antarctic sea ice: The case of dissolved silica. *J Geophys Res* **115**: C02005. doi:10.1029/2009JC005369.

Vancoppenolle M, Meiners KM, Michel C, Bopp L, Brabant F, et al. 2013. Role of sea ice in global biogeochemical cycles: emerging views and challenges. *Quat Sci Rev* **79**: 207–230. doi: 10.1016/j.quascirev.2013.04.011.

Vesala T, Kljun N, Rannik U, Rinne J, Sogachev A, et al. 2008. Flux and concentration footprint modelling: State of the art. *Environ Pollut* **152**(3): 653–666. doi:10.1016/j.envpol.2007.06.070.

Vihma T, Johansson MM, Launiainen J. 2009. Radiative and turbulent surface heat fluxes over sea ice in the western Weddell Sea in early summer. *J Geophys Res* **114**: C04019. doi:10.1029/2008JC004995.

Wadhams P, Doble MJ. 2008. Digital terrain mapping of the underside of sea ice from a small AUV. *Geophys Res Lett* **35**: L01501. doi:10.1029/2007GL031921.

Wang K, Liu C, Zheng X, Pihlatie M, Li B, et al. 2013. Comparison between eddy covariance and automatic chamber techniques for measuring net ecosystem exchange of carbon dioxide in cotton and wheat fields. *Biogeosciences* **10**: 6865–6877. doi: 10.5194/bg-10-6865-2013.

Wang Y, Hammes F, Boon N, Egli T. 2007. Quantification of the filterability of freshwater bacteria through 0.45, 0.22, and 0.1 μm pore size filters and shape-dependent enrichment of filterable bacterial communities. *Environ Sci Technol* **41**: 7080–7086. doi: 10.1021/es0707198.

Wangersky PJ. 1993. Dissolved organic carbon methods: a critical review. *Mar Chem* **41**: 61–74.

Weissenberger J. 1992. The environmental conditions in the brine channels of Antarctic sea-ice. *Ber Polarfosch* **111**: 1–159.

Welch HE, Bergmann MA, Jorgenson JK, Burton W. 1988. A subice suction corer for sampling epontic ice algae. *Can J Fish Aquat Sci* **45**: 562–568.

Wells LE, Deming JW. 2006. Modelled and measured dynamics of viruses in Arctic winter sea-ice brines. *Environ. Microbiol* **8**: 1115–1121. doi: 10.1111/j.1462-2920.2005.00984.x.

Westerlund S, Öhman P. 1991. Iron in the water column of the Weddell Sea. *Mar Chem* **35**: 199–217.

Wiese W. 1930. Zur Kenntnis der Salze des Meereises. *Annalen der Hydrographie und Maritimen Meteorologie* **58**: 282–286.

Williams GD, Maksym T, Kunz C, Kimball P, Singh H, et al. 2013. Beyond point measurements: Sea ice floes characterized in 3-D. *EOS* **94**: 69–70. doi:10.1002/2013EO070002.

Winston GC, Stephens BB, Sundquist ET, Hardy JP, Davis RE. 1995. Seasonal variability in $CO_2$ transport through snow in a boreal forest, in Tonnessen KA, Williams MW, Tranter M, eds., *Biogeochemistry of Seasonally Snow-Covered Catchments*. Boulder: IAHS: pp. 61–70.

Winton VHL, Dunbar GB, Bertler NAN, Millet M-A, Delmonte B, et al. 2014. The contribution of aeolian sand and dust to iron fertilization of phytoplankton blooms in southwestern Ross Sea, Antarctica. *Glob Biogeochem Cy* **28**: 423–436. doi: 10.1002/2013GB004574.

Wren SN, Donaldson DJ. 2012. Laboratory study of pH at the air-ice interface. *J Phys Chem C* **116**: 10,171–10,180. doi: 10.1021/jp3021936.

Wright SW, Jeffrey SW. 2006. Pigment markers for phytoplankton production, in Volkman JK, ed., *Marine Organic Matter: Biomarkers, Isotopes and DNA*. Heidelberg: Springer-Verlag: pp. 71–104.

Xie H, Gosselin M. 2005. Photoproduction of carbon monoxide in first-year sea ice in Franklin Bay, southeastern Beaufort Sea. *Geophys Res Lett* **32**: L12606. doi: 10.1029/2005GL022803.

Xu L, Furtaw MD, Madsen RA, Garcia RL, Anderson DJ, et al. 2006. On maintaining pressure equilibrium between a soil $CO_2$ flux chamber and the ambient air. *J Geophys Res* **111**: C08S10. doi: 10.1029/2005JD006435.

Yang X, Pyle JA, Cox RA. 2008. Sea salt aerosol production and bromine release: Role of snow on sea ice. *Geophys Res Lett* **35**: L16815. doi: 10.1029/2008GL034536.

Zeebe RE, Wolf-Gladrow D. 2001. *$CO_2$ in Seawater: Equilibrium, Kinetics, Isotopes*. San Francisco: Elsevier.

Zemmelink HJ, Dacey JWH, Houghton L, Hintsa EJ, Liss PS. 2008. Dimethylsulfide emissions over the multi-year ice of the western Weddell Sea. *Geophys Res Lett* **35**: L06603. doi: 10.1029/2007GL031847.

Zemmelink HJ, Delille B, Tison JL, Hintsa EJ, Houghton L, et al. 2006. $CO_2$ deposition over the multi-year ice of the western Weddell Sea. *Geophys Res Lett* **33**: L13606. doi: 10.1029/2006GL026320.

Zhou J, Delille B, Brabant F, Tison J-L. 2014b. Insights into oxygen transport and net community production in sea ice from oxygen, nitrogen and argon concentrations. *Biogeosciences* **11**: 5007–5020. doi: 10.5194/bg-11-5007-2014.

Zhou J, Delille B, Eicken H, Vancoppenolle M, Brabant F, et al. 2013. Physical and biogeochemical properties in landfast sea ice (Barrow, Alaska): Insights on brine and gas dynamics across seasons. *J Geophys Res-Oceans* **118**: 3172–3189. doi: 10.1002/jgrc.20232.

Zhou J, Tison J-L, Carnat G, Geilfus N-X, Delille B. 2014a. Physical controls on the storage of methane in landfast sea ice. *The Cryosphere* **8**: 1019–1029. doi: 10.5194/tc-8-1-2014.

## Contributions

- LAM wrote the first draft and coordinated and synthesized the contributions from the other authors.
- FF and BGTE contributed material and coordinated and synthesized contributions for sections 3 and 5, respectively.
- All other authors contributed material.

## Acknowledgments

This manuscript is a product of SCOR working group 140 on Biogeochemical Exchange Processes at Sea-Ice Interfaces (BEPSII); we thank BEPSII chairs Jacqueline Stefels and Nadja Steiner and SCOR executive director Ed Urban for their practical and moral support of this endeavour. This manuscript was first conceived at an EU COST Action 735 workshop held in Amsterdam in April 2011; in addition to COST 735, we thank the other participants of the "methods" break-out group at that meeting, namely Gerhard Dieckmann, Christoph Garbe, and Claire Hughes. Our editors, Steve Ackley and Jody Deming, and our reviewers, Mats Granskog and two anonymous reviewers, provided invaluable advice that not only identified and helped fill in some gaps, but also suggested additional ways to make what is by nature a rather dry subject (methods) at least a bit more interesting and accessible. We also thank the librarians at the Institute of Ocean Sciences for their unflagging efforts to track down the more obscure references we required. Finally, and most importantly, we thank everyone who has braved the unknown and made the new measurements that have helped build sea-ice biogeochemistry into the robust and exciting field it has become.

## Competing interest

To the best of our knowledge, none of the authors have a competing interest in the publication of this manuscript.

## Data accessibility statement

This manuscript does not present original data.

# 4

# Climate change and ice hazards in the Beaufort Sea

D. G. Barber[1*] • G. McCullough[1] • D. Babb[1] • A. S. Komarov[1] • L. M. Candlish[1] • J.V. Lukovich[1] •
M. Asplin[1] • S. Prinsenberg[2] • I. Dmitrenko[1] • S. Rysgaard[1,3,4]

[1]Centre for Earth Observation Science, University of Manitoba, Winnipeg, Manitoba, Canada
[2]Coastal Ocean Science Bedford Institute of Oceanography, Department of Fisheries and Oceans Canada,
 Dartmouth, Nova Scotia, Canada
[3]Greenland Climate Research Centre, Greenland Institute of Natural Resources, Nuuk, Greenland
[4]Arctic Research Centre (ARC), Aarhus University, Aarhus, Denmark
*David.Barber@umanitoba.ca

**Domain Editor-in-Chief**
Jody W. Deming, University of Washington

**Associate Editor**
Edward C. Carmack, Fisheries and Oceans Canada

**Knowledge Domain**
Ocean Science

## Abstract

Recent reductions in the summer extent of sea ice have focused the world's attention on the effects of climate change. Increased $CO_2$-derived global warming is rapidly shrinking the Arctic multi-year ice pack. This shift in ice regimes allows for increasing development opportunities for large oil and gas deposits known to occur throughout the Arctic. Here we show that hazardous ice features remain a threat to stationary and mobile infrastructure in the southern Beaufort Sea. With the opening up of the ice pack, forecasting of high-frequency oscillations or local eddy-driven ice motion will be a much more complex task than modeling average ice circulation. Given the observed reduction in sea ice extent and thickness this rather counterintuitive situation, associated with a warming climate, poses significant hazards to Arctic marine oil and gas development and marine transportation. Accurate forecasting of hazardous ice motion will require improved real-time surface wind and ocean current forecast models capable of ingesting local satellite-derived wind data and/or local, closely-spaced networks of anemometers and improved methods of determining high-frequency components of surface ocean current fields 'up-stream' from drilling and extraction operations.

## Introduction

It is becoming increasingly apparent that climate change is having a significant impact on sea ice conditions in the northern hemisphere. The multiyear sea ice (MYI) pack is rapidly shrinking (Polyakov et al., 2012; Comiso, 2012). This loss of thick ice has significant implications throughout the ocean-sea ice-atmosphere interface and, through teleconnections, to temperate and even tropical parts of our planet (e.g., Budikova, 2009; Francis and Vavrus, 2012). The impacts of this abrupt change extend throughout physical, biological, and biogeochemical aspects of the icescape due to the control sea ice has on mass, gas and energy fluxes. There are also many impacts of a practical nature affecting Inuit use of the icescape and the safe operation of maritime industries now working to develop newly accessible Arctic resources.

Shipping and the oil and gas industry are particularly susceptible to the changing icescape as they require information on ice hazards and the oceanic and atmospheric forcing of these hazards. Drill ships in particular will require improved forecasting of large, dense ice features that may cause unmanageable collisions with stationary platforms. Ironically, the decrease in MYI has actually increased the frequency and relative velocities of certain ice hazards (Galley et al., 2013) as global climate change affects both local and regional wind fields (Overland et al., 2012; Ogi and Wallace, 2012). Ocean forcing of sea ice remains poorly understood, relative to wind forcing, partly because of a paucity of *in situ* under-ice ocean current data.

The high dependence of ice floe drift velocity on geostrophic winds was established as early as the 1970s by investigators such as Thorndike and Colony (1982), who reported sea ice motion to wind velocity correlations of 0.95 and 0.85 in summer and winter, respectively. Recent studies have pointed to increases in drift velocity that can be attributed to increased storminess in the Arctic and regional increases in wind stress in summer, with implications for vertical mixing and erosion of the halocline (Hakkinen et al., 2008; Bourgain and Gascard, 2011). Spreen et al. (2011) ascribe accelerated drift to a thinner ice cover, while Rampal et al. (2009) also attribute changes in ice drift and deformation primarily to sea ice conditions compared to atmospheric forcing, based on an assessment of the International Arctic Buoy Programme data from 1979 to 2007. Investigations of sea ice drift, using beacons launched during the International Polar Year Circumpolar Flaw Lead system study in winter 2007–2008 (Barber et al., 2010) showed seasonality in ice drift-to-wind ratios and angles that corresponded to expected variation of internal ice stress in the Beaufort Sea, with the most responsive ice-atmosphere coupling from mid-November to January, prior to annual compacting of the seasonal ice zone along the coast of the Canadian Arctic Archipelago (CAA) and the adjacent perennial ice pack (Richter-Menge et al., 2002; Lukovich et al., 2011).

Recent studies have highlighted the fact that ice shelves along the NW flank of Ellesmere Island have lost significant mass and area (e.g., Copland et al., 2007). Three-quarters of the 7500–8900 km² total shelf area, estimated from records of Robert Peary's 1906 expedition (Spedding, 1977, as quoted in Jefferies, 1987; Vincent et al., 2001) had disintegrated by 1960 (England et al., 2008); by 2008 only 720 km² remained (Mueller et al., 2008). Although the loss represents an annualized disintegration rate of 29 km² a$^{-1}$ since 1960, in fact, most has occurred in massive events, sometimes grouped in a single year. For example, in 2008, a total of 214 km² of shelf ice was lost, including the entire Markham Shelf, and large blocks calved from the Serson and Ward Hunt Shelves (50, 122 and 22 km², respectively; Mueller et al., 2008; England et al., 2008). Ice islands originating from such events typically enter either the CAA and drift through the inter-island straits, or they become entrained within the Beaufort Gyre (BG) where in the past they have drifted for years. A prime example is ice island T-3 that was detected in 1950 roughly 650 km northwest of Barrow, Alaska (Koenig et al., 1952). T-3 drifted in the BG for 27 years, completing two revolutions before exiting the Arctic through Fram Strait (Jefferies, 1992). Jefferies (1992) estimated that between 1946 and 1992 there were 600 ice islands identified in the Arctic Ocean.

The region NW of Ellesmere Island also represents the origin of extreme MYI features that, like the ice islands, become entrained within the BG and circulate towards regions of potential offshore development. These extreme MYI features arise due to the convergence of sea ice within the BG and the Transpolar Drift Stream against the NW flank of the CAA (Bourke and Garrett, 1987). This region has been described as the most dynamic region of the Arctic marine icescape. It created the thickest (Wadhams et al., 2011) and most heavily deformed sea ice in the world (Bourke and Garrett, 1987; Melling, 2002). Both the ice islands and the highly deformed first year and thick MYI created here pose a significant potential hazard to industrial operations as they drift south and southwest through the southern Beaufort Sea.

The objective of this research is to illustrate some of the current ice-related hazards in the southern Beaufort Sea in order to inform ice management planning for ship navigation and for the oil and gas industry in the area, and potentially wherever industrial development is planned or underway in other Arctic marine areas. We also present data which demonstrate that in the marginal ice zone, over short periods (hours to days, at least) thick MYI and glacial ice can travel significant distances (km) in any direction relative to the general ice flow field, thereby posing a particular challenge to forecasting paths of hazardous ice features in the southern Beaufort Sea.

## Methods

Data presented in this study were collected between August 12 and August 20, 2011, in the eastern Beaufort Sea pack ice, at and near stations 140–150 km WNW of Banks Island (75.0°N 128.9°W) during Leg 2A of the 2011 ArcticNet mission of the CCGS *Amundsen*. The ice here is a mix of MYI floes interspersed with glacial ice, embedded in or among younger first and second year ice floes. This complex icescape is created by pack ice carried by the BG advecting against the western and northern coast of the CAA. We selected this area to study ice hazards because it lies upstream of license areas for oil and gas exploration to the south and southwest (Figure 1). Two on-ice stations were intensively investigated: Site 1 (August 16, 2011) and Site 2 (August 18, 2011), denoted as S1 and S2 respectively, on MYI with an average thickness of 6.6 m and 4.1 m respectively.

Sea ice thickness was surveyed using a helicopter-mounted electromagnetic induction (EMI) system. Four surveys were flown in the vicinity of sites S1 and S2, each comprising 4–6 parallel transects, with each transect 10–15 km in length. The overall system comprises the EMI instrument, a laser profiler, a nadir-facing video system, and support navigation, control and archive systems. The EMI method for measurements of sea ice thickness over seawater is well established, as are its limitations (e.g., Prinsenberg and Holladay, 1993). At our typical survey altitude of 2–4 m, the footprint of the system was roughly 20–35 m in diameter over 5–10 m thick ice. Further details are reported by Holladay (2006). We compared EMI ice thickness data with direct measurements in 58 coincident augured holes at S1 and S2. Although the helicopter-borne EMI system underestimated drill hole thickness in 8–10 m thick ice by as much as 2–3 m, in 5–7 m thick ice the average difference was insignificant (mean difference = 0.06 m, s.d. = 0.68 m, n = 14).

**Figure 1**

**Photographs of thick MYI, glacial ice and a map showing the location of industry exploration.**

Example photographs of: A) thick MYI, B) glacial ice and C) location of significant license areas for oil and gas exploration in the Southern Beaufort Sea. The southern limb of the Beaufort Gyre advects ice features, such as in A and B, into the lease areas depicted in C.

Fifteen position-only ice beacons were deployed by helicopter and air-ice boat on ice islands and MYI ice floes to track ice motion in the region. Ten were Canatec Associates International Ltd beacons which transmitted at 15-min intervals, and five were Oceanetics model 703 iridium ice tracking buoys, transmitting at 2-h intervals. Two main components of ice motion in the region, parallel shift and rotation, were measured using a RADARSAT sea ice motion tracking system (Komarov and Barber, 2013) applied to sequential Synthetic Aperture Radar (SAR) images. The sea ice tracking approach employs a phase-correlation technique to detect both the translational and the rotational components of sea ice motion. Further details of this ice motion tracking system are reported elsewhere (Komarov 2009; Komarov and Barber, 2013). In this study we derive sea ice motion (and surface wind fields—see below) both from sequential RADARSAT-1 and from RADARSAT-2 ScanSAR, with images at daily intervals. In addition, selected floes were identified visually in successive RADARSAT images and overlay maps were created to illustrate local, anomalous ice motion.

Ship-based wind records, Canadian Meteorological Centre (CMC) forecast winds, and the North American Regional Reanalysis (NARR) winds were used to assess atmospheric forcing of ice motion. A micrometeorological tower located on the front deck of the *Amundsen* provided continuous monitoring of wind speed. Wind speed data used in this paper are corrected to 10 m above ocean surface. Additional details on the ship winds are available elsewhere (Else et al., 2012), NARR winds (Mesinger et al., 2006) and CMC winds (Laroche et al., 1999). CMC and NARR forecasts were generated at three-hourly intervals on August 16 and 18, 2011, for comparison with the ship's winds. Local and regional over-water wind speeds were also calculated from RADARSAT-2 dual-polarization HH-HV ScanSAR images as a function of HH normalized radar cross section (NRCS), HV NRCS, incidence angle and noise floor of the SAR instrument. For RADARSAT-1 and 2 single-polarization HH images the model predictors are HH NRCS, and the incidence angle only. Further details are provided in Komarov et al. (2013).

A 600 kHz Z-cell Nortek Aquadopp acoustic Doppler current profiler (ADCP) was installed to measure upper ocean currents at both stations. Current speed and direction were recorded at 1-min intervals for 17.5 and 19.0 h at S1 and S2, respectively, in thirty 2-m vertical cells extending from 0.5–60.5 m below the bottom surface of the ice. The instrument (and ice floe) track was logged at 10-sec intervals with a Garmin Etrex GPS. Currents were first corrected to true north (Natural Resources Canada at http://geomag.nrcan.gc.ca/calc/mdcal-eng.php) and then further corrected by subtraction of ice drift velocity. Due to the proximity of the magnetic pole, it is not unreasonable to doubt the accuracy of magnetic compass records. However, the GPS record of instrument drift provides a means to test this record. Given the proximity of numerous keels, some to at least 30-m depth, we expect ocean currents in the shallowest bins, sheltered by these keels, to be dragged in the direction of drift. In fact, in the 19-h record at S2, the mean of current directions reported 4 and 6 m under the ice were on average within 2° of the mean of ice drift directions calculated from GPS tracks. At S1, currents at 4 and 6 m tracked an average of 22–25° to the right of ice drift. Hence, we can be reasonably confident that the record of current directions at S2 is accurate to a few degrees, and at S1, to within about 25°. Even at S1, the rotation, at least, of shallow currents is real, and the general direction—moving to the WSW, with rotation to the right—of currents in shallow sheltered water does tend to follow the track of the ice floe (Figure 2).

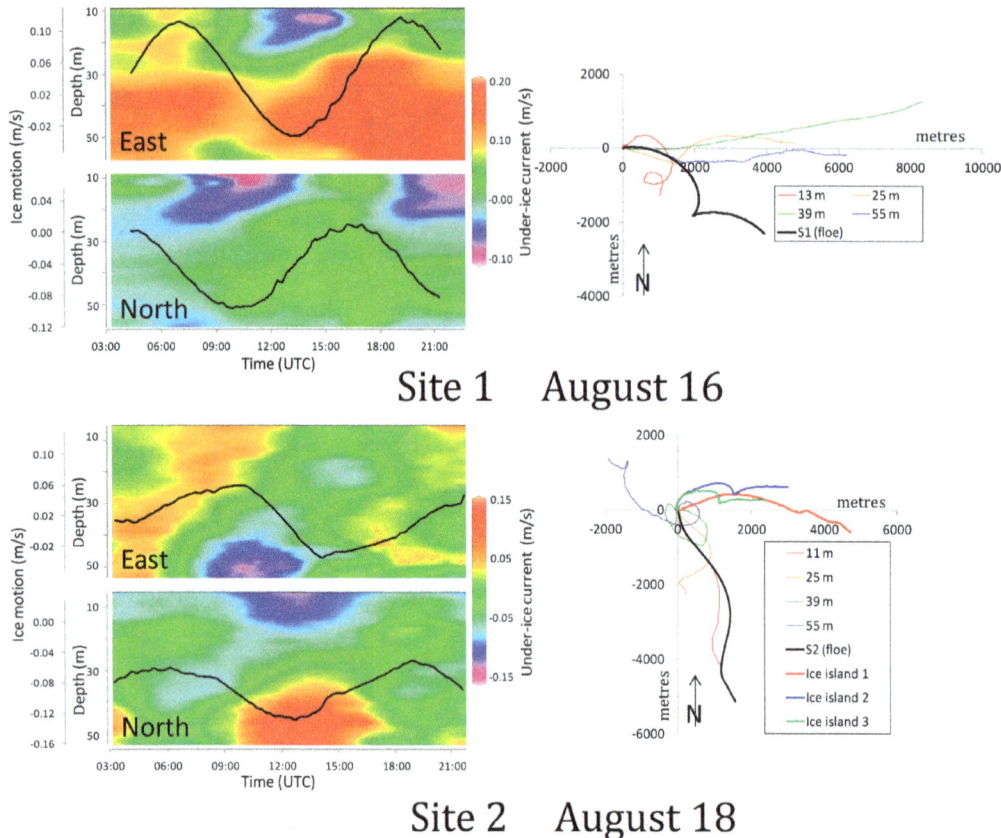

Site 1     August 16

Site 2     August 18

**Figure 2**

**Ice motion and under-ice currents at Site 1 and Site 2.**

Panels at left are current velocity time-depth cross-sections, with speeds indicated by the colour scale at right. Superimposed black lines show ice floe velocity, with positive values indicating eastward (northward) motion. Charts at right show relative ice motion and water circulation at four depths, starting at coordinates 0,0. Simultaneous tracks of three ice islands are also shown (for S2 only). Water circulation tracks were determined by summing incremental velocities. See Figure 6 and 7 for site locations.

## Results and discussion

### Field observations on multi-year ice

In preparation for the Amundsen mission, multiyear floes of potential interest were identified by a Canatec Consulting (Calgary) RADARSAT analysis team. We note here that none of the features that they identified were actually observed to be unmanageable ice features (e.g., in the sense that they would pose an unmanageable hazard to drill ships). In consequence, we used the *Amundsen's* helicopter to conduct visual surveys and select study sites within a 200-km radius of the ship. Although, the region appeared to be dominated by first and second year ice, nonetheless, embedded MYI was common. Sea ice thickness distributions at both S1 and S2 showed an open water mode, 0 m thick, and within the ice, modes at 1, 3 and 6–7 m thickness (Figure 3). In the four EMI surveys flown, 20–40% of the ice was > 4 m thick. Several observations were ≥ 10 m, the saturation limit of the EMI. At S1 and S2, both in MYI, mean thicknesses measured by boreholes were of the order of 3–5 m, but a few holes were 10–15 m deep. In air-ice boat surveys, we identified several thicker floes in the region. Figure 1A shows one example of a floe with maximum sail height more than 10 m above water and a visible keel reaching more than 40 m below the water surface.

During the helicopter surveys we identified 17 glacial ice features that ranged in size from hundreds to thousands of meters in diameter. All had drafts greater than the surrounding sea ice (by visual inspection) and most were embedded within the multiyear and/or first-year sea ice (Figure 1b). Some had gravel on

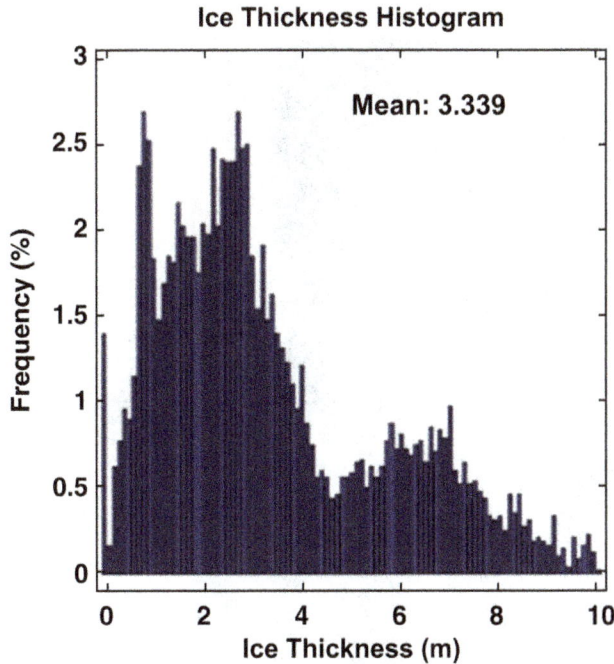

## Ice Thickness Histogram

**Mean: 3.339**

*Frequency (%)* vs *Ice Thickness (m)*

**Figure 3**

**Sea ice thickness distributions derived from the helicopter-borne EMI induction system in the study area.**

Values of 8–10 m underestimate thicknesses measured at coincident augur holes. The EMI system saturated at 10-m thickness so that maximum values should be considered as ≥ 10 m.

their surface (anywhere from small stones and gravel to boulders the size of a small car), which Jeffries (1992) described as a characteristic of ice islands with glacial origins. These features were visually obvious from the helicopter, due to the deformation of the sea ice surrounding the glacial floes and to dark ice surfaces (when rocks and gravel were present). None could be distinguished on SAR.

These glacial ice features probably derive from the collapse of the ice shelves along the NW flank of the CAA (Jeffries, 1992; Copland et al., 2007). If the Arctic continues to warm, it is likely that this loss of ice mass flux will continue at the present rate at least as long as suitable shelf ice material remains. Historically, such ice islands have survived for several years to decades as they drift around the BG (e.g., ice island T-3 described above; Koenig et al., 1952; Jeffries, 1992; Rigor et al., 2002; Rigor and Wallace, 2004), but statistics of MYI age need to be revisited now that there has been such a dramatic reduction in MYI in the BG (Comiso, 2012) and an increase in summer melt (Perovich et al., 2008) and velocity (Galley et al., 2013) of the BG icescape.

A key feature of any ice management system is the use of reliable local surface winds for wave height and ice motion modeling. Although the wind speed recorded at the ship was modeled reasonably well by CMC, and modestly overestimated by NARR, neither predicted wind direction adequately (Figure 4). Even the light breeze on August 18 (~ 3 m s$^{-1}$) was mispredicted by roughly 60°. Moderate (at best) to poor agreement between forecast and observed winds is an ongoing issue (e.g., Garrett, 1985; Bromwich et al., 2009) that requires new approaches to surface wind forecasting.

One such approach may incorporate local wind speed estimates using RADARSAT. We applied the method of Komarov et al. (2013) using RADARSAT-2 HH-HV and RADARSAT-1 HH imagery over the study

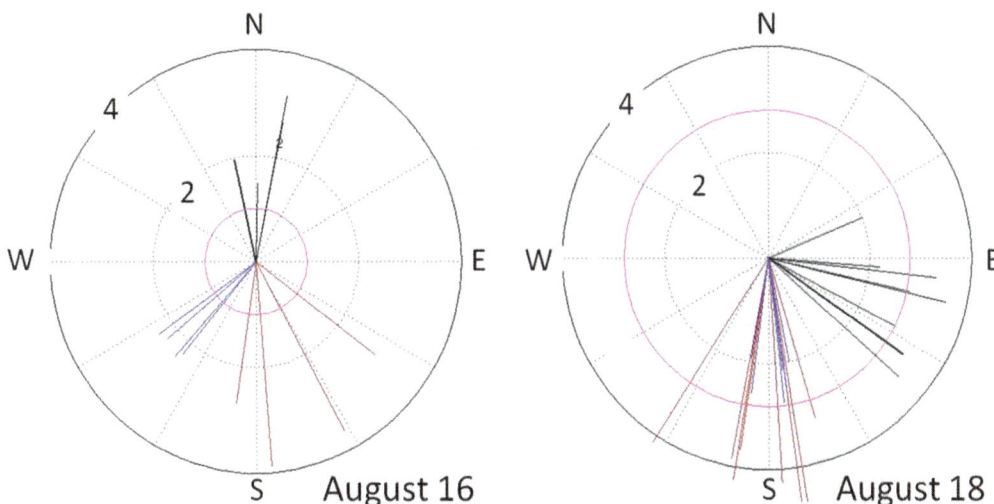

August 16

August 18

**Figure 4**

**Comparisons of surface winds.**

Wind roses showing ship-based (black), NARR (red), and CMC (blue) wind speed and direction at 3-h intervals on August 16 (06–15 h, left panel) and August 18 (00–21 h, right panel), 2011. Retrieved SAR wind speed values for August 16 (2.97 m s$^{-1}$) and August 18 (5.76 m s$^{-1}$) are shown as magenta circles in each panel.

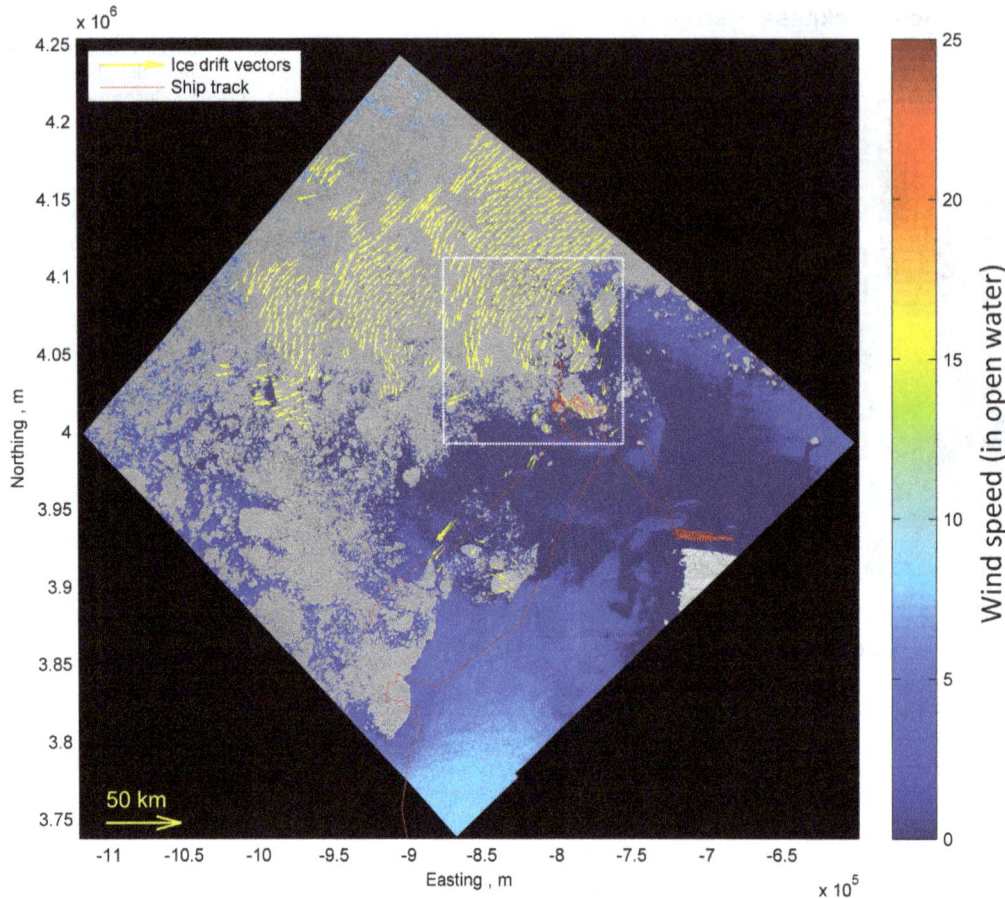

Figure 5
Surface wind speed, in open water, derived from a model.

Surface wind speed, in open water, derived from a model (Komarov *et al.*, 2013) relating RADARSAT-2 backscatter to surface winds through their effect on wave height. Wave estimates are valid for open water areas outside of and within the sea ice pack. Wind speed is extracted from a RADARSAT-2 HH-HV Image acquired on Aug 16, 15:46, 2011. Color bar indicates wind speed in m s⁻¹. Yellow vectors denote sea ice motion derived from a SAR image pair August 15, 15:23—August 16, 15:46, 2011. M'Clure Strait and the NW corner of Banks Island are visible near the right corner of the image.

area. A surface wind product derived from a RADARSAT-2 HH-HV image acquired on August 16, 2011, is shown in Figure 5. Owing to calm wind conditions on August 16, retrieved SAR wind speeds underestimated ship wind speeds, although they were within the error reported for the method (1 m s⁻¹; Komarov et al., 2013), while on August 18 the average ship wind speed (2.8 m s⁻¹; Figure 4) was very well predicted by SAR. From the low wind speeds for regional (CMC and NARR) and local (ship and SAR retrieved) winds on August 16 and 18, we infer that local atmospheric forcing played a minor role in the drift of ice features recorded at stations S1 and S2.

Differences in wind direction between the ship-based, reanalysis and forecast data may be attributed to the coarse resolution of the numerical models unable to capture small-scale wind and ice drift variability. The advantage of the SAR retrieval algorithm over the numerical model output resides in its ability to capture ice drift and wind effects over spatial scales on the order of hundreds of meters. Furthermore, for wind speeds in excess of 2 m s⁻¹—values for which accurate wind representation in forecasting ice hazards and drift are increasingly important—retrieved SAR wind speeds are shown to outperform model and reanalysis data. Komarov et al. (2013) report that SAR-derived wind speeds fit buoy data with a root-mean-square error (RMSE) = 1.6 m s⁻¹ (for HH-VV). Bromwich et al. (2009) reported that the Polar Weather Research and Forecasting model (PWRF; considered state-of-the-art for arctic weather modeling) predicted SHEBA (Surface Heat Budget of the Arctic Ocean, 1997–1998) ice camp data with RMSE = 1.5 (January) to 1.6 (August) m s⁻¹. However, the PWRF model output retained a bias of –0.6 m s⁻¹; here the Komarov method outperformed the PWRF model, with a reported bias for the prediction of the validation set of only 0.1 m s⁻¹.

Ice beacon trajectories and ice motion tracking from a SAR image pair (15:17 August 17 and 15:35 August 18, 2011) are shown in Figure 6, overlaid on the first image of the pair. The SAR-derived vectors agree well with the beacon recording motion at S2 over the same period. We observed a similarly close fit of the SAR-derived ice track with trajectories of beacons placed at S1 two days earlier (not shown).

Variability in drift direction among floes, and between ice islands and floes, was first observed as different displacements, between helicopter flights, of readily distinguishable ice features. It was confirmed by beacon tracks over the next few days of the study. While S1 and S2 moved coherently southward with the surrounding ice pack, several ice islands, on which we also placed beacons, did not. From August 16 to 17 three of these drifted first westward and then northwestward, across and then opposite to the general flow of the sea ice (Figures 7 and 8); through the late afternoon and evening of August 17 they continued to drift northward before, on the morning of August 18, turning southeastward (Figure 6C).

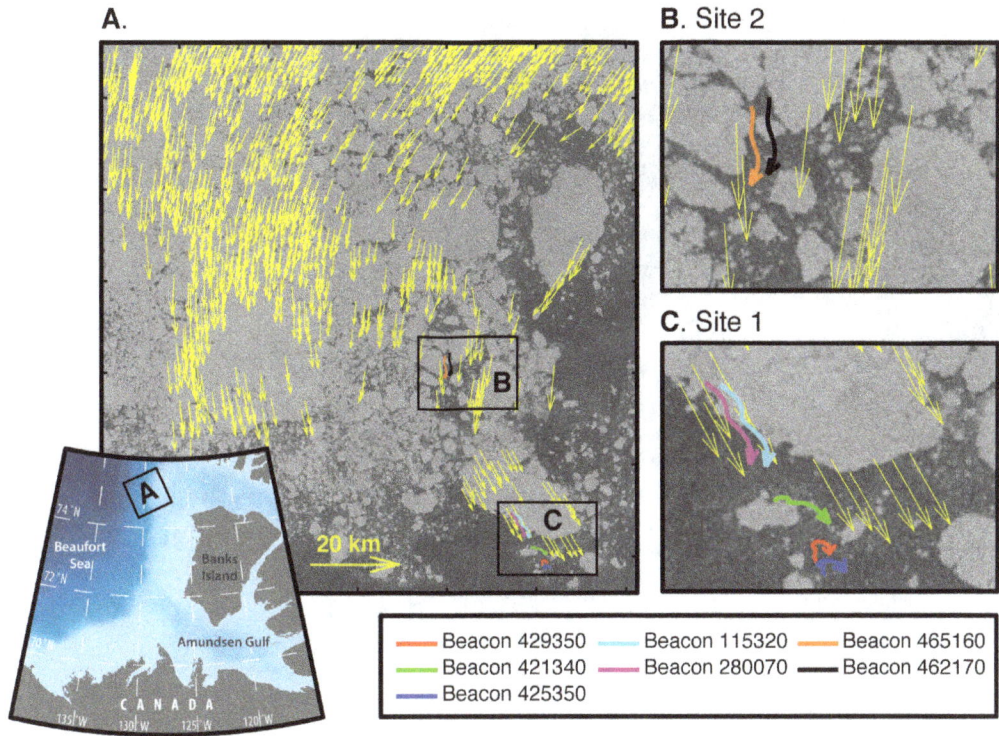

**Figure 6**

**RADARSAT derived ice motion.**

RADARSAT derived ice motion between images acquired on August 17, 15:17, and August 18, 15:35, 2011, showing the average ice motion as yellow vectors. Beacon data, shown in other colours, in general indicate good agreement between the general ice motion and the motion of particular floes. This is general agreement breaks down, however, when one examines the motion of thick ice features (e.g., red and blue lines in C. Site 1) which were marine glacial features with drafts > 40 m.

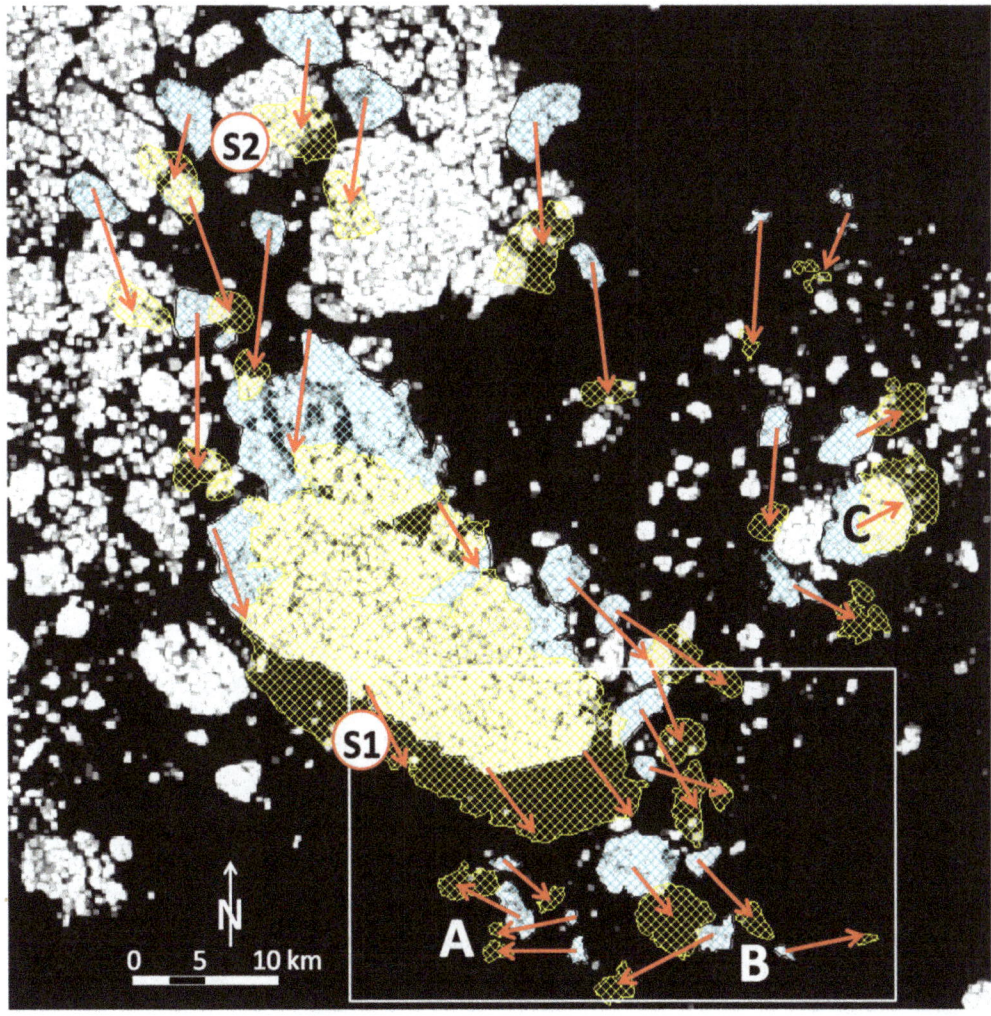

**Figure 7**

**Displacement of selected floes over 24 h.**

RADARSAT image acquired at 15:46 UTC 16 August, 2011 (white = ice, black = open water), with selected floes indicated by blue cross-hatch. Yellow cross-hatch indicates positions of same floes at 15:17 on 17 August; pale yellow indicates overlap. Red arrows indicate displacement of selected floes over the ~ 24 h between images. Letters highlight three regions where the daily motion varied greatly, both locally and relative to the average pack motion (*c.f.* Figure 6). 'S1' and 'S2' indicate ADCP locations. The area within the white rectangle is enlarged in Figure 8.

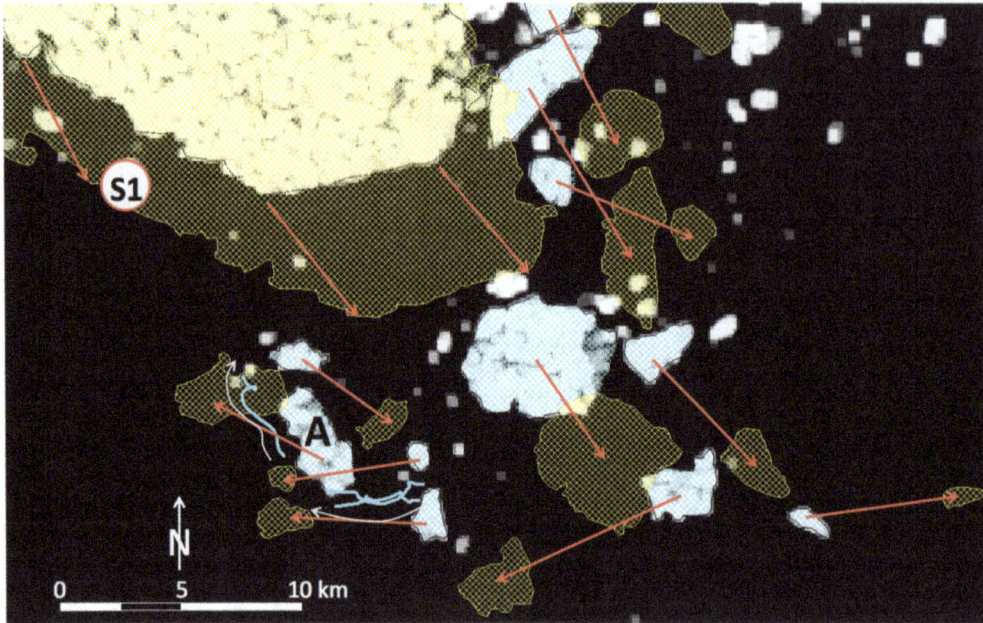

Figure 8
Displacement of ice floes with ice beacon tracking data.

See Figure 7 for context. Here, cyan lines indicate tracks of ice beacons on three ice islands over the period between retrieval of RADARSAT images on August 16 and 17 (with adjacent arrows indicating the direction of travel). The ice islands are not visible at the resolution of the RADARSAT 2 data. Note the clockwise rotation of floe "A" which, together with displacements of surrounding floes and beacon tracks, suggests a local clockwise eddy.

Stations S1 and S2 drifted 4–5 km to the southeast during the ADCP deployments following the regional motion of the pack (Figure 2). A periodic element in the motion of both drift trajectories (semi-diurnal) is roughly consistent with either inertial or tidal oscillations; the length of each record, less than two full periods of the oscillation, is too brief to distinguish between these two forcings with any certainty. In ADCP bins nearest the under-ice surface, ocean currents tended to follow the ice (as would be expected if water within or near the range of keel depths was dragged with the ice); deeper currents tended more eastward (at S1) or northwestward, opposite to the ice motion (at S2). The ADCP range of 60 m reached to the upper halocline (from local CTD profile observations; not shown) and hence to the upper boundary of Pacific water flow (Melling, 1998; McLaughlin et al., 2002). The vertical velocity gradients described can be explained as a reconciliation between the drag of the ice floes, following the atmospherically forced southerly flow of the BG in this region, and Pacific water, which here flows prevailingly northward along the coast of the CAA, or eastward towards M'Clure Strait (McLaughlin et al., 2002).

The large ice floe on which we established station S1 drifted southeastward from August 16 to 17, following the general motion of the regional pack. Over the same period, two nearby ice islands moved westward, across the paths of vectors describing general pack motion, and a third drifted to the northwest, opposite to the pack (Figure 8). The base of the ice islands reached deeper (greater than 70 m, given the freeboard in excess of 10 m) than most keels under the MYI floes. Nonetheless, the westward and northwestward travel of the ice islands was not explained by interaction of the relatively thick ice islands with the deep currents recorded at nearby S1 (Figure 2), which were eastward at the time. On closer examination, the ice islands and several nearby floes moved coherently. The largest of these floes, "A" in Figure 8, clearly rotated clockwise and, while two ice islands and two floes just to the south drifted westward, a third floe just to the north drifted to the southeast. Together, these motions are best explained by a clockwise eddy structure centered near the north end of floe "A".

These deviations from the general motion of the pack were not unique. In the same image, we note that two nearby floes at "B" in Figure 7 drifted perpendicular to the path of the pack, but in opposite directions. Similarly, several floes at "C" drifted to the east-northeast, again perpendicular to the average flow field of the ice pack. The average speed of these floes across the path of the pack was of the same order as the speed of the pack itself, in the range of 4–8 km d$^{-1}$. None of these anomalous tracks are adequately explained by wind, whether measured at the ship or modeled, or by local under-ice current records, measured only 10–20 km away, or forcing by the regional motion of the pack (all floes described were separated by at least 1–5 km of open water). In fact, without very local surface wind and ocean current observations, iceberg trajectory models fail to report motion due to high-frequency current structure (inertial or tidal motion, eddies) with sufficient precision to forecast local motion of hazardous ice features (e.g., Gaskill and Rochester, 1983). More specifically, it has long been accepted that currents or current profiles recorded at a well rig are not adequate indicators of currents at icebergs as little as a few kilometers distant (e.g., Garrett, 1985). Near-field near-future velocity is as well forecasted by near-field antecedent velocity, with the error estimate for the forecast trajectory estimated by the high-frequency variability in the antecedent velocity. Using this and related approaches, the uncertainty in the forecast trajectory, over periods of hours, typically exceeds the full trajectory length

(Garrett, 1985). Wind-driven iceberg trajectory models, without ocean current forcing, develop significant errors very quickly. Using such a model in a study of 28 iceberg trajectories in the Hibernia oil field, Finlayson et al. (1992) reported a median error of 6 km (and a maximum error of 25 km) after only 12 h.

Given that both marine glacial ice and thick MYI are significant hazards to mobile or stationary infrastructure, the modeling and prediction of this highly dynamic floe field remains a significant challenge to ice management in the southern Beaufort Sea and in other locations with a similar icescape (e.g. the CAA, Baffin Bay, Fram Strait, and along the coasts of Greenland).

## Hazardous ice features

With increasing oil and gas development planned in the southern Beaufort Sea, it is important that we learn whether the ice islands and the hazardous MYI that we have described are likely to persist given the well-known reduction in extent and average thickness of the BG icescape. Here, we use evidence from relevant literature to discuss potential changes in glacial and multiyear ice mass, velocity and circulation as they may constitute hazards to Arctic marine industries.

Sea ice along the northwest flank of the CAA experiences great internal stresses as it is compressed against the CAA and is thus described as the most dynamically active and thickest sea ice in the world (Bourke and Garrett, 1987; Melling, 1999). These characteristics are ascribed to the large-scale drift patterns of Arctic sea ice which act to transport and pile ice up along the northwest coasts of the CAA and Greenland. Due to the stresses and geographical constraints of the area, ice can remain there for years, growing thicker and more heavily ridged (Wadhams et al., 2011). Hence, this region has been referred to as a "redoubt" of Arctic MYI, where MYI coverage has remained relatively constant from 1980 through to 2011 (Wadhams et al., 2011; Maslanik et al. 2011). Although Maslanik et al. (2011) note a reduction in the spatial coverage of 5+ year old ice within this region from a fairly stable ~ 400 000 km$^2$ between 1980 and 2005 to a minimum of ~ 110 000 km$^2$ in 2011, with significant losses in 2008 after the 2007 sea ice minimum, this MYI remains a reflection of the imbalance between the export and import of MYI from this region. It continues to be exported southwards along the coast of the CAA into the Beaufort Sea via the Beaufort Gyre (Maslanik et al., 2011) with a possible recent increase in the quantity of ice transported southwards out of the region north of the CAA (Stroeve et al., 2011). On the other hand, faster melt of MYI during its transition through the southern BG (e.g., Perovich et al., 2008) has reduced the amount of MYI transported back towards the CAA on the northern limb of the BG (Maslanik et al., 2011).

This reduction of MYI is accompanied by a reduction in the average thickness of sea ice (Haas et al., 2008). Measurements of ice thickness are sparse in the eastern Beaufort Sea; hence we refer to the helicopter-borne EMI ice thickness surveys of Haas et al. (2010) over the North Pole and submarine thickness surveys of Wadhams (1990), Wadhams et al. (2011), Rothrock et al. (1999) and Rothrock et al. (2008). Haas et al. (2008) compared ice thickness data from 1991, 2001, 2004 and 2007 for the region over the North Pole and found a regime shift from earlier MYI dominance to first year ice dominance by 2007. As part of this regime shift a decrease in mean ice thickness of up to 44% occurred between 2001 and 2007, to a mean thickness of 1.27 ± 0.77 m in 2007. Wadhams et al. (2011) compared submarine data from 1976, 1987, 2004 and 2007 from north of Greenland between 20°W and 30°W and found a reduction in mean drafts from 5.5–6.5 m (1976) to 4.5–6.0 m (1987) down to 3.5–4.0 m (2004 and 2007). Historically, submarine-based ice thickness data from along the CAA is sparse; however, sea ice within the northern portion of the nearby region of declassified submarine sonar data (the 'SCICEX box'; Rothrock et al., 2008) was shown to decrease by as much as 2.3 m between the periods of 1958–1976 and 1993–1997 (Rothrock et al., 1999). In another analysis the same area decreased from thicknesses of 4–5 m in 1988 (Rothrock et al., 2008) to thicknesses of 2.5–4 m in 2008 (Kwok and Rothrock, 2009). Clearly, in these regions, there has been a dramatic reduction in ice thickness. However, significant thick MYI remains. From their 2004 and 2007 surveys, Wadhams et al. (2011) also reported that in spite of the widespread trend to thinner MYI, much unmanageably thick MYI remained along the northern coast of the CAA. West of 50°W off the north coast of Ellesmere Island, they reported mean drafts of 5–6 m. Haas et al. (2010) found similarly thick MYI along the coast of Ellesmere Island during aerial surveys in April 2009. In the region we studied in August 2011, 20–40% of the ice was > 4 m thick. Clearly, enough of the Arctic's old, thick MYI persists off the coast of the CAA to remain a significant threat to potential oil and gas infrastructure.

Is this also true of the CAA ice shelves, the source of ice islands to the Beaufort Gyre (BG)? Certainly, since 2008, there have been further calving events from the remaining shelves. Based on comparison of two MODIS satellite images (August 29, 2008, and August 13, 2013), only a few small remnants of the Serson Ice Shelf remain. An 80-km$^2$ section of the Ward Hunt Ice Shelf was lost between August 2008 and August 2013, leaving two large segments separated by 5–6 km of open sea and very exposed to erosion or further weakening by marine forces. Only the Milne Ice Shelf appears to be largely intact. The remaining shelf ice survives in a weakened condition. Calving from the Ward Hunt Ice Shelf in 2008 was accompanied by widespread fracturing throughout the shelf, preconditioning it for future deterioration (Mueller et al., 2008). Moreover, it may have thinned by as much as 15 m between 1981 and 2002 (Braun, 2012, as quoted in Mortimer et al., 2012). The Milne Ice Shelf, too, has thinned; it lost 8 m in average thickness in the 28 years from 1981 to 2009 (Mortimer et al, 2012). Such thinning has further contributed to weakening of the remaining shelf ice. Although sustained warming probably accounts for thinning and thereby weakening of the shelves, the more proximate cause of calving events is likely to have been processes associated with increasingly frequent, longer periods of exposure to open water and loose pack ice along shelf margins. Historically there has been a broad band of multiyear landfast sea ice located along the seaward face of the ice shelves that held it in place and protected it from external forces—mainly direct thermal and mechanical erosion by waves, and direct collisions with mobile pack ice (Copland et al., 2007; Pope et al. 2012). Major calving events from 2005 to 2008 all occurred during periods when the protective band of multiyear landfast sea ice was not present (Mueller et al., 2008; Mortimer et al., 2012). The consequence is that the ice shelves of Ellesmere Island are expected to continue to deteriorate and collapse (Braun et al., 2004; Copland et al., 2007; Mueller et al., 2008; Mortimer et al., 2012). In the last 5 years, at least 130 km² have been lost from the 720 km² of ice shelf extent in 2008. At this rate of disintegration, the last of the shelf ice will deteriorate in the next two decades. Given the increased exposure of the two remaining parts of the Ward Hunt Ice Shelf and the weakened state of both the Ward Hunt and the Milne Ice Shelves, it is unlikely that they will survive even that long. Meanwhile, until the remaining ice shelves are completely gone, the continued calving of the shelves will maintain or even, for a brief time, increase the population of ice islands in the Beaufort Sea.

## Conclusions and recommendations

Arctic climate change would lead us to believe that sea ice hazards are decreasing in the Beaufort Sea as we increase the proportion of annual ice and decrease the proportion of perennial ice. Our results show that, in fact, significant challenges remain in terms of the detection and prediction of motion of sea and glacial ice hazards. Each of the thick ice features identified during our field program (Figure 1) were of sufficient size and mass that a drill ship would have to detach to evade them or icebreakers would have to tow or break them up (difficult to impossible depending on mass). The many hazardous glacial and multiyear ice features that remain in the southern Beaufort Sea are difficult or impossible to distinguish by satellite remote sensing. We expect these hazards to exist for at least the next several decades, because the BG will continue to carry glacial and MYI southward from along the CAA into hydrocarbon exploration areas and to circulate at least some of this ice back to the MYI-generating NW flank of the CAA, making development of an ice management system for oil and gas development challenging.

Our observation of a highly variable circulation of small but very thick ice islands and MYI floes illustrates a number of key challenges: 1) we need better remote sensing detection methods to be able to distinguish marine glacial ice entrained in sea ice; 2) we need improved surface wind direction forecasting if we are to describe the average flow field of the ice with useful accuracy; 3) we need high resolution measurements of near-surface currents since high frequency oscillations and eddies force deviations in sea ice motion away from the average flow field. We summarize recommendations into three areas required for management of ice hazards:

*1. Hazardous ice feature detection:* Current active microwave satellite sensors have limited ability to detect hazardous ice features. Future research is required to exploit polarimetry and the higher temporal resolution data which will be available through constellation missions (e.g., Radarsat Constellation and the ESA missions). Given the limitations of current state-of-the-art satellite hazardous ice detection, dedicated aircraft surveillance may continue to be necessary for the foreseeable future in ice management systems. Because hazardous ice features may be hidden among manageable floes, the required density of flight surveillance will be higher than, for instance, in the managing for icebergs in the north Atlantic Ocean. However, this solution is inevitably limited by weather conditions. Very high resolution optical satellite data may be used to supplement airborne surveillance, but they are limited to cloud-free days.

*2. Wind field observations and forecasting:* Local wind estimates from satellite data (e.g., Komarov and Barber, 2013) may provide improved forecasting of wind speed; with the advent of the proposed SAR constellation system by the European and Canadian Space Agencies, repeat coverage would be reduced to 3 h, making this approach a good candidate for estimating the local surface wind field. We recommend further research into the potential for estimating wind direction by SAR. For now, however, improved forecasting of wind direction will require local wind observations, at the very least including winds recorded at oil or gas extraction support vessels and permanent platforms. One further option may be installation of recoverable or disposable anemometers, with real-time telemetry, on upstream floes, a solution that would require parallel development of modeling routines that self-correct semi-continuously using real-time, local wind data from a Lagrangian network of observatories.

*3. Surface current observations:* It is possible that High Frequency Coastal RADAR (sometimes called CODAR, i.e. Coastal Ocean Dynamics Applications RADAR) could be used to measure the surface current field updrift of installations. It can be effective over a range of 30–50 km. However, since it measures only radial components of velocity, multiple stations would be required to characterize current vector fields precisely enough to model high-frequency ice drift velocities (local deviations from the average pack motion) in a field updrift of a drill rig or other structure. An alternative would be an array of upward looking Acoustic Doppler Current Profilers installed as a real-time cable network upstream of any drill ship location. Such an ADCP network would need a spatial resolution on the order of 10 km to adequately represent or model the highly variable surface currents illustrated in Figure 2. An ice motion forecasting model would also benefit from data on pack compression rates, due to ice-ice and ice-shore interactions, telemetered from stress sensors in the updrift pack. We realize that these recommendations may seem impractical at this time, for to implement them fully would require development of sea-ice-capable observational instruments and of forecast models capable of ingesting moving arrays of real-time observational data. We believe, however, that, in the long term, developing such a system is possible and that the benefits of an accurate forecasting model could outweigh the costs of the system.

The results that we have presented here illustrate a very complex circulation regime with forcing by both the ocean and the atmosphere combining in ways that make precise prediction of ice motion a significant challenge yet paramount to avoiding unwanted interactions of ice and industrial operations. We conclude that development of potential oil reserves in the Arctic, particularly in the Beaufort Sea, will require significant investments in technology and modeling capability if a fully functioning operational ice management system is to be developed.

# References

Barber DG, Asplin M, Gratton Y, Lukovich J, Galley R, Raddatz R, Leitch D. 2010. The International Polar Year (IPY) Circumpolar Flaw Lead (CFL) system study: Introduction and physical system. *Atmos Ocean* 48(4): 225–243. doi:10.3137/OC317.2010

Bourgain P, Gascard JC. 2011. The Arctic Ocean halocline and its interannual variability from 1997 to 2008. *Deep-Sea Res Part I* 58(7): 745–756.

Bourke RH, Garrett RP. 1987. Sea ice thickness distribution in the Arctic Ocean. *Cold Regions Sci Tech* 13: 259–280.

Braun C, Hardy DR, Bradley RS, Sahanatien V. 2004. Surface mass balance of the Ward Hunt Ice Rise and Ward Hunt Ice Shelf, Ellesmere Island, Nunavut, Canada. *J Geophys Res* 109: D22110. doi:10.1029/2004JD004560

Braun C. 2012. The surface mass balance of the Ward Hunt Ice Shelf and the Ward Hunt Ice Rise, Ellesmere Island, Nunavut, Canada, in Copland L, Mueller DR, eds. *Arctic Ice Shelves and Ice Islands*, Netherlands: Springer, Dordrecht (in press).

Bromwich DH, Hines KM, Bai L-S. 2009. Development and testing of Polar Weather Research and Forecasting model: 2. Arctic Ocean. *J Geophys Res* 114: D08122. doi:10.1029/208JD010300

Budikova D. 2009. Role of Arctic sea ice in global atmospheric circulation: A review. *Global Planet Change* 68(3): 149–163. doi:10.1016/j. gloplacha.2009.04.001

Comiso J. 2012. Large decadal decline of Arctic MYI cover. *J Climate* 25: 1176–1193. doi:10.1175/JCLI-D-11-00113.1

Copland L, Mueller DR, Weir L. 2007. Rapid loss of the Ayles Ice Shelf, Ellesmere Island, Canada. *Geophys Res Lett* 34: L21501. doi:10.1029/2007GL031809

Else B, Galley R, Papakyriakou T, Miller LA, Mucci A, Barber DG. 2012. Sea surface $pCO_2$ cycles and $CO_2$ fluxes at landfast sea ice edges in Amundsen Gulf, Canada. *J Geophys Res Oceans* (in press).

England JH, Lakemen TR, Lemmen DS, Bednarski JM, Stewart TG, Evans DJA. 2008. A millennial-scale record of Arctic Ocean sea ice variability and the demise of the Ellesmere Island ice shelves. *Geophys Res Lett* 35: L19502. doi:10.1029/2008GL034470

Finlayson DJ, Bobbitt J, Rudkin P, Jordaan IJ. 1992. Iceberg Trajectory Model – real-time verification. *Environmental Studies Research Funds Report* No. 113, Calgary ix +47 p.

Francis JA, Vavrus SJ. 2012. Evidence linking Arctic amplification to extreme weather in mid-latitudes. *Geophys Res Lett* 39: L06801. doi:10.1029/2012GL051000, 2012

Galley RJ, Else BGT, Prinsenberg SJ, Barber DG. 2013. Sea ice concentration, extent, age, motion and thickness in regions of proposed offshore oil and gas development near the Mackenzie Delta - Canadian Beaufort Sea. *Arctic* 66(1): 105–116.

Garrett C. 1985. Statistical prediction of iceberg trajectories. *Cold Regions Sci Tech* 11: 255–266.

Gaskill HS, Rochester J. 1983. A new technique for iceberg drift prediction. *Cold Regions Sci Tech* 8(3): 223–234.

Hakkinen S, Proshutinsky A, Ashik I. 2008. Sea ice drift in the Arctic since the 1950s, *Geophys Res Lett* 35: L19704. doi:10.1029/2008GL034791

Haas C, Pfaffling A, Hendricks S, Rabenstein L, Etienne JL, Rigor I. 2008. Reduced ice thickness in Arctic Transpolar Drift favors ice retreat. *Geophys Res Lett* 35: L17501. doi:10.1029/2008GL034457

Haas C, Hendricks S, Eicken H, Herber A. 2010. Synoptic airborne thickness surveys reveal state of Arctic sea ice cover. *Geophys Res Lett* 37: L09501. doi:10.1029/2010GL042652

Holladay JS. 2006. *Application of Airborne and Surface-based EM/laser Measurements to Ice/water/sediment Models at Mackenzie Delta Sites*. Ocean Sciences Division, Maritimes Region, Fisheries and Oceans Canada, Bedford Institute of Oceanography. http://www.dfo-mpo.gc.ca/Library/324817.pdf

Jeffries MO. 1987. The growth, structure and disintegration of Arctic ice shelves. *Polar Record* 23(147): 631–649.

Jeffries MO. 1992. Arctic ice shelves and ice islands: Origin, growth and disintegration, physical characteristics, structural-stratigraphic variability and dynamics. *Rev Geophys* 30: 245–267.

Koenig LS, Greenway KR, Dunbar M, Haatersly-Smith G. 1952. Arctic Ice Islands. *Arctic* 5: 67–103.

Komarov A. 2009. Sea ice tracking system – technical documentation, *CIS Internal Report*, Canadian Ice Service, Ottawa, 33 pp.

Komarov A, Barber D. 2013. Sea ice motion tracking from sequential dual-polarization RADARSAT-2 images. *IEEE Trans on Geoscience and Remote Sensing* (in press). doi:10.1109/TGRS.2012.2236845

Komarov A, Zabeline V, Barber D. 2013. Ocean surface wind speed retrieval from C-band SAR images without wind direction input. *IEEE Transactions on Geoscience and Remote Sensing* (in press). doi:10.1109/TGRS.2013.2246171

Kwok R, Rothrock DA. 2009. Decline in Arctic sea ice thickness from submarine and ICESat records: 1958–2008. *Geophys Res Lett* **36**: L15501. doi:10.1029/2009GL039035

Laroche S, Gauthier P, St-James J, Morneau J. 1999. Implementation of a 3D variational data assimilation system at the Canadian Meteorological Centre. Part II: The regional analysis. *Atmosphere-Ocean* **37.3**: 281–307.

Lukovich JV, Babb DG, Barber DG. 2011. On the scaling laws derived from ice beacon trajectories in the southern Beaufort Sea during the International Polar Year — Circumpolar Flaw Lead study, 2007–2008. *J Geophys Res* **116**: C00G07.

Maslanik J, Stroeve J, Fowler C, Emery W. 2011. Distribution and trends in Arctic sea ice age through spring 2011. *Geophys Res Lett* **38**:L13502. doi: 10.1029/2011GL047735

McLaughlin F, Carmack E, Macdonald R, Weaver AJ, Smith J. 2002. The Canada Basin, 1989–1995: Upstream events and far-field effects of the Barents Sea. *J Geophys Res* **107**(C7): 3082.

Melling H. 1998. Hydrographic changes in the Canada Basin of the Arctic Ocean: 1979–1996. *J Geophys Res* **103**(C4): 7637–7645.

Melling H. 1999. Exchanges of freshwater through the shallow straits of the North American Arctic, in Lewis EL, Jones EP, eds. *The Freshwater Budget of the Arctic Ocean*. Netherlands: Kluwer Academic Publishers: p. 479–502.

Melling H. 2002. Sea ice of the northern Canadian Arctic Archipelago. *J Geophys Res* **107**(C11): 3181. doi:10.1029/2001JC001102

Mesinger F, DiMego G, Kalnay E, Mitchell K, Shafran PC, et al. 2006. North American regional reanalysis. *Bull Amer Meteor Soc* **87**(3): 343–360. doi:http://dx.doi.org/10.1175/BAMS-87-3-343

Mortimer CA, Copland L, Mueller DR. 2012. Volume and area changes of the Milne Ice Shelf, Ellesmere Island, Nunavut, Canada, since 1950. *J Geophys Res* **117**: F04011. doi:10.1029/2011JF002074

Mueller DR, Copland L, Hamilton A, Stern D. 2008. Examining Arctic ice shelves prior to the 2008 breakup. *EOS* **89**(49): 502–503.

Overland JE, Francis JA, Hanna E, Wang M. 2012. The recent shift in early summer Arctic atmospheric circulation. *Geophys Res Lett* **39**: L19804. doi:10.1029/2012GL053268

Ogi M, Wallace JM. 2012. The role of summer surface wind anomalies in the summer Arctic sea ice extent in 2010 and 2011. *Geophys Res Lett* **39**: L09704. doi:10.1029/2012GL051330

Perovich DK, Richter-Menge JA, Jones KF, Light B. 2008. Sunlight, water, and ice: Extreme Arctic sea ice melt during the summer of 2007. *Geophys Res Lett* **35**: L11501. doi:10.1029/2008GL034007

Polyakov IV, Walsh JE, Kwok R. 2012. Recent changes of Arctic MYI coverage and the likely causes. *Bull Amer Meteor Soc* **93**(2): 145–151. doi:10.1175/BAMS-D-11-00070.1

Pope S, Copland L, Mueller D. 2012. Loss of multiyear landfast sea ice from Yelverton Bay, Ellesmere Island, Nunavut, Canada. *Arctic Antarctic Alpine Res* **44**(2): 210–221.

Prinsenberg SJ, Holladay JS. 1993. Using air-borne electromagnetic ice thickness sensor to validate remotely sensed marginal ice zone properties. *Port and Ocean Engineering under Arctic Conditions (POAC 93)*: 936–948.

Rampal P, Weiss J, Marsan D. 2009. Positive trend in the mean speed and deformation rate of Arctic sea ice, 1979–2007. *J Geophys Res* **114**: C05013. doi:10.1029/2008JC005066

Richter-Menge JA, McNutt SL, Overland JE, Kwok R. 2002. Relating Arctic pack ice stress and deformation under winter conditions. *J Geophys Res* **107**(C10): 8040. doi:10.1029/2000JC000477

Rigor IG, Wallace JM, Colony RL. 2002. Response of sea ice to the Arctic Oscillation. *J Climate* **15**(18): 2648–2663.

Rigor IG, Wallace JM. 2004. Variations in the age of Arctic sea-ice and summer sea-ice extent. Geophys Res Lett **31**(9).

Rothrock DA, Yu Y, Maykut GA. 1999. Thinning of the Arctic Sea-Ice Cover. *Geophys Res Lett* **26**(23): 3469–3472.

Rothrock DA, Percival DB, Wensnahan M. 2008. The decline in arctic sea-ice thickness: separating the spatial, annual, and interannual variability in a quarter century of submarine data. *J Geophys Res* **113**: C05003. doi:10.1029/JC004252

Spedding LG. 1977. Ice Island Count, Southern Beaufort Sea, 1976. Report IPRT-13ME-77 of Arctic Petroleum Operations Association Project 99-3.

Spreen G, Kwok R, Menemenlis D. 2011. Trends in Arctic sea ice drift and role of wind forcing: 1992–2009. *Geophys Res Lett* **38**: L19501. doi:10.1029/2011GL048970

Stroeve JS, Maslanik J, Serreze MC, Rigor I, Meier W, Fowler C. 2011. Sea ice response to an extreme negative phase of the Arctic Oscillation during winter 2009/2010. *Geophys Resh Lett* **38**: L0502. doi: 10.1029/2010GL045662

Thorndike AS, Colony R. 1982. Sea ice motion in response to geostrophic winds. *J Geophys Res Oceans (1978–2012)* **87**(C8): 5845–5852.

Vincent WF, Gibson JAE, Jeffries MO. 2001. Ice shelf collapse, climate change, and habitat loss in the Canadian high Arctic. *Polar Record* **37**(201): 133–142.

Wadhams P. 1990. Evidence for thinning of the Arctic ice cover north of Greenland. *Nature* **345**: 795–797.

Wadhams P, Hughes N, Rodrigues J. 2011. Arctic sea ice thickness characteristics in winter 2004 and 2007 from submarine sonar transects. *J Geophys Res* **116**: C00E02. doi:10.1029/2011JC006982

**Contributions**

- Contributed to conception and design: DGB
- Contributed to acquisition of data: DGB, GM, DB, ASK, LMC, JVL, SP
- Contributed to analysis and interpretation of data: DGB, GM, DB, ASK, LMC, JVL, SP, MA, SR
- Drafted and/or revised the article: DGB, GM, DB, ASK, LMC, JVL, MA, ID, SR
- Approved the submitted version for publication: DGB, GM, LMC

**Acknowledgments**

This work is a contribution to the Arctic Science Partnership (ASP) and the ArcticNet Networks of Centres of Excellence. We thank Dmitri Matskevich (Exxon Mobile Upstream Research) for useful discussions during the field program and Martin Fortier (ArcticNet), Jim Hawkins and Neal Darlow (Imperial Oil) for project management. NCEP Reanalysis data was provided by the NOAA/OAR/ESRL PSD, Boulder, Colorado, USA, from their Web site at http://www.esrl.noaa.gov/psd/. We also thank the Canadian Ice Service, Environment Canada, for providing the RADARSAT data, and the officers and crew of the NGCC Amundsen for excellence in field support. This work benefitted greatly from the contributions of two anonymous reviewers. It is dedicated to our colleague, Dr. Klaus Hochheim, who tragically lost his life while conducting sea ice research in northern Canada in September 2013.

#### Funding information

This work was conducted under a joint ArcticNet — Oil and Gas Industry Partnership program with funding from both Imperial Oil and BP. Funding was also provided by ArcticNet, the Natural Sciences and Engineering Research Council (NSERC), the Canada Research Chairs (CRC) program, the University of Manitoba, and the Canada Excellence Research Chairs (CERC) program. Data from this program are freely available to the public through the Polar Data Catalogue (University of Waterloo) or by contacting the lead author.

#### Competing interests

There are no competing interests with regards to this manuscript.

#### Data accessibility statement

The following publically available datasets were used:
*   National Centers for Environmental Prediction, National Weather Service, NOAA, U.S. Department of Commerce. 2005, updated monthly. *NCEP North American Regional Reanalysis (NARR)*. Research Data Archive at the National Center for Atmospheric Research, Computational and Information Systems Laboratory. http://rda.ucar.edu/datasets/ds608.0.

#### The following datasets were generated:
*   GPS Ice motion beacon data - ArcticNet 2011 - South Beaufort Sea - www.polardata.ca CCIN #11726
*   Snow and Sea Ice thickness using a Helicopter EM Induction System - ArcticNet 2011 - Beaufort Sea - www.polardata.ca CCIN #11725

# 5

# Fe availability drives phytoplankton photosynthesis rates during spring bloom in the Amundsen Sea Polynya, Antarctica

Anne-Carlijn Alderkamp[1]* • Gert L. van Dijken[1] • Kate E. Lowry[1] • Tara L. Connelly[2,6] • Maria Lagerström[3,7] • Robert M. Sherrell[3,4] • Christina Haskins[3] • Emily Rogalsky[3] • Oscar Schofield[3] • Sharon E. Stammerjohn[5] • Patricia L. Yager[2] • Kevin R. Arrigo[1]

[1]Department of Environmental Earth System Science, Stanford University, Stanford, California, United States
[2]Department of Marine Sciences, University of Georgia, Athens, Georgia, United States
[3]Department of Marine and Coastal Sciences, Rutgers University, New Brunswick, New Jersey, United States
[4]Department of Earth and Planetary Sciences, Rutgers University, New Brunswick, New Jersey, United States
[5]Institute of Arctic and Alpine Research, University of Colorado, Boulder, Colorado, United States
[6]Marine Science Institute, The University of Texas at Austin, Port Aransas, Texas, United States
[7]Department of Applied Environmental Science (ITM), Stockholm University, Stockholm, Sweden

*alderkamp@stanford.edu

**Domain Editor-in-Chief**
Jody W. Deming, University of Washington

**Associate Editor**
Jean-Éric Tremblay, Université Laval

**Knowledge Domain**
Ocean Science

## Abstract

To evaluate what drives phytoplankton photosynthesis rates in the Amundsen Sea Polynya (ASP), Antarctica, during the spring bloom, we studied phytoplankton biomass, photosynthesis rates, and water column productivity during a bloom of *Phaeocystis antarctica* (Haptophyceae) and tested effects of iron (Fe) and light availability on these parameters in bioassay experiments in deck incubators. Phytoplankton biomass and productivity were highest (20 µg chlorophyll $a$ L$^{-1}$ and 6.5 g C m$^{-2}$ d$^{-1}$) in the central ASP where sea ice melt water and surface warming enhanced stratification, reducing mixed layer depth and increasing light availability. In contrast, maximum photosynthesis rate ($P^*_{max}$), initial light-limited slope of the photosynthesis–irradiance curve ($\alpha^*$), and maximum photochemical efficiency of photosystem II ($F_v / F_m$) were highest in the southern ASP near the potential Fe sources of the Dotson and Getz ice shelves. In the central ASP, $P^*_{max}$, $\alpha^*$, and $F_v / F_m$ were all lower. Fe addition increased phytoplankton growth rates in three of twelve incubations, and at a significant level when all experiments were analyzed together, indicating Fe availability may be rate-limiting for phytoplankton growth in several regions of the ASP early in the season during build-up of the spring bloom. Moreover, Fe addition increased $P^*_{max}$, $\alpha^*$, and $F_v / F_m$ in almost all experiments when compared to unamended controls. Incubation under high light also increased $P^*_{max}$, but decreased $F_v / F_m$ and $\alpha^*$ when compared to low light incubation. These results indicate that the lower values for $P^*_{max}$, $\alpha^*$, and $F_v / F_m$ in the central ASP, compared to regions close to the ice shelves, are constrained by lower Fe availability rather than light availability. Our study suggests that higher Fe availability (e.g., from higher melt rates of ice shelves) would increase photosynthesis rates in the central ASP and potentially increase water column productivity 1.7-fold, making the ASP even more productive than it is today.

## Introduction

Antarctic shelf waters are strong sinks for atmospheric $CO_2$ due to high biological productivity, intense winds, high air-sea gas exchange, formation of bottom water and extensive winter ice cover. These factors make these regions important for the biogeochemical cycling of elements, particularly of carbon (C) (Sarmiento et al., 2004; Arrigo et al., 2008). Specifically, coastal polynyas (areas of open water surrounded by ice) are hot spots for energy and C transfer between the atmosphere and ocean (Smith and Barber, 2007, Mu et al., 2014). The reduced ice cover increases air-sea gas exchange and results in enhanced light availability in the water column in early spring, thereby increasing primary productivity through phytoplankton photosynthesis. In addition to its importance for the global C cycle, phytoplankton productivity supports the biota occupying higher trophic levels including krill, penguins, and whales (Arrigo et al., 2003; Ainley et al., 2006).

Phytoplankton productivity in the Southern Ocean is often limited by the availability of iron (Fe) (Boyd et al., 2007, and references therein), although light limitation may also occur due to deep vertical mixing below the critical depth (Mitchell et al., 1991; De Baar et al., 2005). The Fe supply for phytoplankton growth in polynyas is enhanced compared to the open ocean due to input of dissolved Fe (DFe) from melting sea ice (Sedwick and DiTullio, 1997; Lannuzel et al., 2010), shelf sediments (Hatta et al., 2013; De Jong et al., 2013), icebergs (Raiswell et al., 2008; Raiswell, 2011; Shaw et al., 2011), upwelling Circumpolar Deep Water (CDW) (Klunder et al., 2011), remineralized Fe in winter water (WW), and melting glaciers (Raiswell et al., 2006; Gerringa et al., 2012, Sherrell et al., 2015). Despite these sources, phytoplankton growth is often still seasonally Fe-limited in later stages of blooms in polynyas such as in the Ross Sea Polynya (Sedwick and DiTullio, 1997; Sedwick et al., 2000; Tagliabue and Arrigo, 2005) and the Weddell Sea (Buma et al., 1991). Bioassay experiments in the Ross Sea Polynya revealed that phytoplankton growth was Fe-limited later in the season, but not early (Sedwick and DiTullio, 1997; Sedwick et al., 2000). It is believed that phytoplankton blooms gradually draw down a "winter stock" of DFe in the WW that eventually limits the bloom, especially away from DFe sources such as sea ice (Hopkinson et al., 2013). However, recent measurements show low and potentially limiting concentrations of DFe early in the season (Sedwick et al., 2011; Marsay et al 2014), implying that significant sources of new DFe are required to sustain a phytoplankton bloom throughout the season.

Satellite estimates of primary productivity in Antarctic polynyas reveal the highest productivity per surface area in the Amundsen Sea (Arrigo and Van Dijken, 2003). The Amundsen Sea contains two polynyas, the Pine Island Polynya (PIP) in the east and the Amundsen Sea Polynya (ASP) in the west. The Amundsen Sea is located in the western Antarctic, where rates of ice sheet thinning are the highest in all of Antarctica (Pritchard et al., 2009; Rignot et al., 2013). Several fast-flowing glaciers that are rapidly thinning drain into the Amundsen Sea where they form floating ice shelves or glacier tongues that are also thinning (Randall-Goodwin et al., 2015). These are the Pine Island Glacier in the PIP and the Getz and Dotson glaciers in the ASP. The Thwaites and Crosson glaciers form ice tongues between the PIP and ASP and may affect both polynyas (Pritchard et al., 2009; Rignot et al., 2013).

The thinning of the ice shelves is mainly attributed to regional bathymetry and oceanography. As the Antarctic Circumpolar Current (ACC) flows close to the continental shelf break, Circumpolar Deep Water (CDW) intrudes southward through deep troughs onto the Antarctic continental shelf (Jacobs et al., 1996, 2011; Arneborg et al., 2012). On the shelf, the CDW mixes with WW and becomes modified CDW (mCDW) that is warm (> 0.6°C) and salty (> 34.5) relative to WW. Near the coast, mCDW has access to ice shelf cavities and drives basal melting of floating ice shelves (Jenkins et al., 2010; Jacobs et al., 2011, 2013; Dutrieux et al., 2014). The resulting mCDW mixed with glacial melt water (meltwater-laden mCDW) becomes fresher (< 34.0), colder (-1.1 to -0.5°C), and more buoyant. At the surface of the ASP, Antarctic Surface Water (AASW) shows a range in salinity (33.6 to 34.1) and temperature (-1.8 to > 0°C), depending on length of time since sea ice melt, degree of solar warming, and wind- or buoyancy-induced mixing with the underlying waters (Yager et al., 2012; Ha et al., 2014). The meltwater-laden mCDW outflowing at subsurface depths from under the Dotson ice shelf (DIS) appears to be a major source of DFe and Particulate Fe (PFe) to the phytoplankton bloom in the ASP (Gerringa et al., 2012; Sherrell et al., 2015). This subsurface source is likely made available to AASW at the surface through horizontal diffusivity (Gerringa et al., 2012), advective eddy transport (e.g., Årthun et al., 2013), mixing along the Dotson trough (e.g., St-Laurent et al., 2013), and by wind- and iceberg-induced mixing (Randall-Goodwin et al., 2015). Similarly, the meltwater-laden mCDW from the Pine Island Glacier was found to be a major source of DFe for the phytoplankton bloom in the PIP (Gerringa et al., 2012). These Fe sources support phytoplankton blooms with high biomass and productivity in both the PIP and ASP (Alderkamp et al., 2012a; Yager et al., 2012). Moreover, Fe addition bioassay experiments at the peak and during the decline of the phytoplankton bloom revealed that Fe was not limiting phytoplankton growth in either polynya at these times (Mills et al., 2012), suggesting relatively high Fe availability to the phytoplankton.

Light availability for phytoplankton in Antarctic polynyas is temporally and spatially variable. Early in the season, the shrinking sea ice cover, in combination with increasing day length and solar elevation, results in greater light availability in surface waters. After the polynya opens up, light availability is determined by cloud cover, mixed layer depth (MLD), and attenuation in the water column that is controlled primarily by phytoplankton biomass. In general, melting sea ice introduces fresh water at the surface, which stabilizes the water column and creates a well-lit shallow mixed layer. Conversely, basal ice shelf melt introduces meltwater-laden mCDW into the water column at the base of the ice shelf, at 150-400 m depth, depending both on the draft of the ice shelf (Jacobs et al., 2012; Mankoff et al., 2012) and the degree of buoyancy driven upwelling. Introducing buoyant water at depth destabilizes the water column and may increase MLD and decrease light availability to the phytoplankton, as was observed at the face of the Pine Island ice shelf (Alderkamp et al., 2012a).

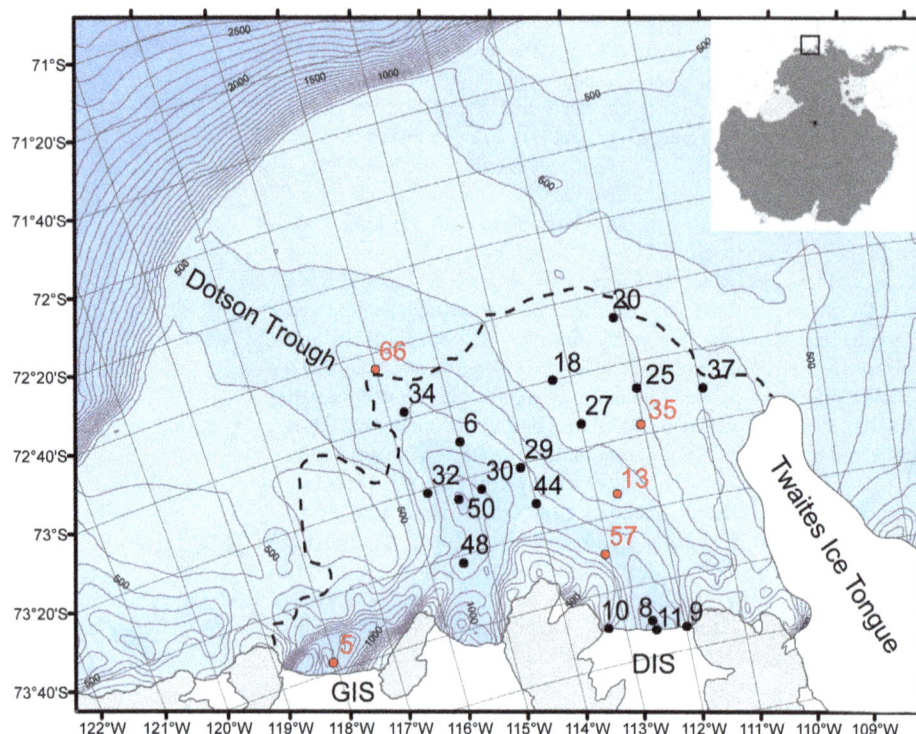

Figure 1
**ASPIRE stations in the Amundsen Sea Polynya.**

Hydrographic stations sampled during the ASPIRE cruise are projected on the Amundsen Sea Polynya (ASP) bathymetry. Fe addition bioassay experiments were performed at red stations; the water column was sampled at all stations. The dashed line shows the sea ice edge on 1 January 2011; the Getz Ice Shelf (GIS), Dotson Ice Shelf (DIS), and Twaites Ice Tongue are shown in white. Inset shows the location of the ASP in Antarctica.

Phytoplankton photoacclimate to low light by increasing their cellular pigment concentration to maximize the capture of photons (Falkowski and LaRoche, 1991; MacIntyre et al., 2002). However, Fe availability affects photoacclimation because biosynthesis of pigments requires Fe and the photosynthetic apparatus has a high Fe content (Raven, 1990; Greene et al., 1992). Thus, phytoplankton in low Fe regions may be impaired in their ability to photoacclimate (Greene et al., 1992; Vassiliev et al., 1995) or may have adapted mechanisms to increase the capture of photons without increasing Fe requirements (Strzepek and Harrison, 2004; Strzepek et al., 2012).

In this study, we investigated the effects of Fe additions on phytoplankton growth and photosynthesis rates during the build-up phase of a phytoplankton bloom in the ASP and assessed how Fe availability affected phytoplankton photoacclimation. We measured phytoplankton biomass, photosynthetic parameters, and productivity rates in surface waters of the ASP (Figure 1) and performed Fe addition bioassay experiments throughout the ASP to test effects of Fe additions on these parameters. In addition, effects of Fe addition on bacterial productivity were evaluated in the experiments. Moreover, interactions between Fe and light availability were tested in the experiments by including incubations at different light levels. Experimental results were then used to draw conclusions about Fe limitation of phytoplankton photosynthesis and productivity in the ASP.

## Methods

### Sampling

Seawater samples were collected during the NBP 10-05 cruise on the RVIB *Nathaniel B. Palmer* in the ASP during the austral spring and summer, 13 December 2010 to 12 January 2011. Water for analysis of parameters listed in the section on *Analytical methods* was sampled from discrete depths in the upper 300 m of the water column at 22 stations (Figure 1) during the middle of the day within three hours of solar noon. Water was sampled with trace metal clean (TMC) techniques using externally-closing 12 L Niskin bottles (Model 110BES, Ocean Test Equipment, Ft. Lauderdale, FL, USA) mounted on a GEOTRACES-style non-contaminating CTD-Rosette deployed on a coated aramid cable (see Sherrell et al., 2015, for details). Continuous vertical profiles of temperature, salinity, irradiance, fluorescence, and suspended particle abundance were obtained from the water column using a SeaBird 911+ CTD, a Chelsea fluorometer, photosynthetically active radiation (PAR) sensor (Biospherical), and a 25-cm WETLabs transmissometer, respectively, mounted on a TMC rosette.

## Fe addition bioassay experiments

At five stations (Figure 1), large volumes of seawater (~ 100 L) were collected at select depths (see below) within the upper 50 m of the water column for bioassay experiments to study the effects of Fe addition under different light conditions, using TMC techniques throughout the experiments. Acid-washed polycarbonate bottles (2 L) were rinsed three times with MilliQ water and once with seawater from the same station before being filled to the brim with unfiltered seawater. Triplicate bottles for each treatment were incubated at in-situ water temperature in transparent deck incubators under incident irradiance shaded with different levels of neutral transmission screening. $FeCl_3$ was added to the Fe treatments from a 1000x stock in weakly acidified, 0.2 μm filtered seawater, in a final concentration of 4 nmol $L^{-1}$ (Mills et al. 2012). Nothing was added to the control treatments. Bottles were capped and caps were wrapped with parafilm to prevent contamination by water from the incubator. All treatments were sampled within three hours of solar noon at the beginning of the experiment (T = 0 days), at 4 days, and at 7 or 8 days for all parameters listed in the section on *Analytical methods*. At three of the five stations (Stations 5, 13, and 35), parallel experiments were conducted with water from two different depths, the surface (S: 8–12 m) and subsurface (deep, D: 35–50 m). Water from each depth was incubated at 10% of incident irradiance. At the two remaining stations (Stations 57 and 66), only surface water was incubated in three parallel experiments under different light conditions: high (50%), medium (10%), and low (1%) incident irradiance.

## Analytical methods

### Nutrients and iron concentrations

Concentrations of nitrate ($NO_3$), ammonium ($NH_4$), phosphate ($PO_4$) and silicate ($Si(OH)_4$) from seawater samples at 22 stations and bioassay experiments were determined by flow injection analysis using a Lachat Instruments Quickchem 8000 Autoanalyzer according to standard protocols (see Vernet et al., 2011). Samples were collected directly from the experimental bottles, filtered through 0.2 μm Acrodisc® filters and stored at 4°C until analysis on the same day.

Concentrations of DFe from seawater samples at 16 stations and T = 0 samples of the bioassay experiments were determined by preconcentration and isotope dilution ICP-MS and are described in full in Sherrell et al. (2015). Concentrations of particulate Fe (PFe) from seawater samples at 11 stations and T = 0 samples of the bioassay experiments were determined by ICP-MS by methods following Planquette and Sherrell (2012) and have been reported elsewhere (Harazin et al., 2014).

### Particulate organic carbon (POC) and particulate organic nitrogen (PON)

Duplicate samples from the bioassay experiments (100–1000 ml) were filtered onto precombusted (450°C for 4 h) 25 mm Whatman GF/F filters and dried at 60°C for analysis of POC and PON on a Carlo-Erba NA-1500 elemental analyzer using acetanilide as a calibration standard.

The POC concentrations at T = 0 and at 4 days were used to calculate phytoplankton growth rates (μ: $d^{-1}$) in the bioassay experiments using the equation:

$$POC_{T4} = POC_{T0}e^{\mu T}. \tag{1}$$

### Pigment analysis

Seawater samples at 19 stations and bioassay experiment samples (50–500 mL) for chlorophyll *a* (Chl *a*) were filtered onto 25 mm Whatman GF/F filters, extracted overnight at 4°C in 5 mL of 90% acetone, and analyzed on a Turner Model 10AU fluorometer before and after acidification (Holm-Hansen et al., 1965).

The full pigment composition was analyzed for seawater samples at 12 stations and T = 0 samples of the bioassay experiments by High Performance Liquid Chromatography (HPLC). Samples (100–2000 mL) were filtered onto 25 mm Whatman GF/F filters, flash-frozen in liquid $N_2$, and stored at –80°C until analysis. Filters were extracted for two hours in 98% methanol: 2% ammonium acetate [vol:vol] in the dark at –20°C after disruption by sonication. Pigments were separated on a SPD-M10AVP HPLC system (Shimadzu, Inc.) using a Agilent 4.6 x 250 mm C18 column kept at 30°C according to Wright et al. (1991), using standards for Chl *a*, chlorophyll *b* (Chl *b*), chlorophyll $c_3$ (Chl $c_3$), peridinin (Per), 19′-butanoyloxylfucoxanthin (19′-But), fucoxanthin (Fuc), 19′-hexanoyloxyfucoxanthin (19′-Hex), neoxanthin (Neo), prasinoxanthin (Pras), violaxanthin (Viol), alloxanthin (Allo), lutein (Lut), antheraxanthin (Anth), diadinoxanthin (Diad), diatoxanthin (Diat), and β-carotene (β-Car). The Chl *a* breakdown product chlorophillide *a* was detected in few samples at < 2% of Chl *a* concentrations.

The ratios of the first 12 of these pigments were used to determine the phytoplankton class abundance using the CHEMTAX analysis package, version 1.95 (Mackey et al., 1996; Wright et al., 1996). The initial input ratios (Table 1A) consisted of specific pigment ratios for eight phytoplankton classes that generally dominate Antarctic waters (Wright et al., 2010), including prasinophytes, chlorophytes, cryptophytes, diatoms (with a separate class for Chl $c_3$ containing diatoms such as *Pseudonitzschia*), dinoflagellates, and two classes of

Table 1A. Initial pigment: Chlorophyll *a* (Chl *a*) ratios used in the CHEMTAX analysis of pigment data[a]

| Phytoplankton class | Pigment used in ratio with Chl *a* | | | | | | | | | | |
|---|---|---|---|---|---|---|---|---|---|---|---|
| | Chl $c_3$ | Lut | Per | 19′-But | Fuc | 19′-Hex | Neo | Pras | Allo | Viol | Chl *b* |
| Prasinophytes | 0 | 0.006 | 0 | 0 | 0 | 0 | 0.030 | 0.315 | 0 | 0.056 | 0.620 |
| Chlorophytes | 0 | 0.220 | 0 | 0 | 0 | 0 | 0.062 | 0 | 0 | 0.031 | 0.180 |
| Cryptophytes | 0 | 0 | 0 | 0 | 0 | 0 | 0 | 0 | 0.220 | 0 | 0 |
| Diatoms | 0 | 0 | 0 | 0 | 0.520 | 0 | 0 | 0 | 0 | 0 | 0 |
| *Pseudonitzschia* | 0.033 | 0 | 0 | 0 | 0.610 | 0 | 0 | 0 | 0 | 0 | 0 |
| *P. antarctica* H | 0.130 | 0 | 0 | 0.010 | 0.080 | 0.400 | 0 | 0 | 0 | 0 | 0 |
| *P. antarctica* L | 0.270 | 0 | 0 | 0.001 | 0.010 | 1.100 | 0 | 0 | 0 | 0 | 0 |
| Dinoflagellates | 0 | 0 | 1.060 | 0 | 0 | 0 | 0 | 0 | 0 | 0 | 0 |

Table 1B. Optimized pigment: Chl *a* ratios after CHEMTAX analysis[a]

| Phytoplankton class | Pigment used in ratio with Chl *a* | | | | | | | | | | |
|---|---|---|---|---|---|---|---|---|---|---|---|
| | Chl $c_3$ | Lut | Per | 19′-But | Fuc | 19′-Hex | Neo | Pras | Allo | Viol | Chl *b* |
| Prasinophytes | 0 | 0.006 | 0 | 0 | 0 | 0 | 0.030 | 0.094 | 0 | 0.056 | 0.620 |
| Chlorophytes | 0 | 0.220 | 0 | 0 | 0 | 0 | 0.062 | 0 | 0 | 0.031 | 0.180 |
| Cryptophytes | 0 | 0 | 0 | 0 | 0 | 0 | 0 | 0 | 0.220 | 0 | 0 |
| Diatoms | 0 | 0 | 0 | 0 | 0.520 | 0 | 0 | 0 | 0 | 0 | 0 |
| *Pseudonitzschia* | 0.033 | 0 | 0 | 0 | 0.624 | 0 | 0 | 0 | 0 | 0 | 0 |
| *P. antarctica* H | 0.201 | 0 | 0 | 0.010 | 0.080 | 0.284 | 0 | 0 | 0 | 0 | 0 |
| *P. antarctica* L | 0.098 | 0 | 0 | 0.001 | 0.010 | 1.493 | 0 | 0 | 0 | 0 | 0 |
| Dinoflagellates | 0 | 0 | 1.060 | 0 | 0 | 0 | 0 | 0 | 0 | 0 | 0 |

[a]Abbreviations: Chl $c_3$ = chlorophyll $c_3$, Lut = lutein, Per = peridinin, 19′-But = 19′-butanoyloxyfucoxanthin, Fuc = fucoxanthin, 19′-Hex = 19′-hexanoyloxyfucoxanthin, Neo = neoxanthin, Pras = prasinoxanthin, Allo = alloxanthin, Viol = violaxanthin, Chl *b* = chlorophyll *b*.

*P. antarctica*. The two classes of *P. antarctica* account for variations in pigment ratios between strains (Zapata et al., 2004) and changes in response to Fe limitation (Van Leeuwe and Stefels, 2007, DiTullio et al., 2007, Alderkamp et al., 2012b). The pigment ratios in the output matrix (Table 1B) were within those reported in the literature (Zapata et al., 2004, Van Leeuwe and Stefels, 2007, DiTullio et al., 2007, Wright et al., 2010, Alderkamp et al., 2012b). Diatoms and *Pseudonitzschia* are presented together as diatoms and the two *P. antarctica* classes are presented together as *P. antarctica*.

## Phytoplankton photosynthesis rates

Photosynthesis vs irradiance (*P-E*) relationships were determined in surface water samples (2–10 m depth) at 12 stations and in one sample from pooled replicates for each incubation treatment. *P-E* relationships were determined using the $^{14}$C-bicarbonate incorporation technique by incubating 2 mL aliquots of seawater in a photosynthetron for two hours over a range of 20 different light intensities ranging from 3 to 542 μmol photons m$^{-2}$ s$^{-1}$ at 0°C (Lewis and Smith, 1983; the full method is outlined in Arrigo et al., 2010). $CO_2$ incorporation normalized by Chl *a* concentration was calculated from radioisotope incorporation and the data were fit by least squares nonlinear regression to the equation of Webb et al. (1974):

$$P^* = P^*_{max} \left( 1 - exp \left( -\alpha^* \frac{E}{P^*_{max}} \right) \right) \tag{2}$$

where $P^*_{max}$ is the maximum rate of photosynthesis ($CO_2$ incorporation in g C g$^{-1}$ Chl *a* hr$^{-1}$) and $\alpha^*$ is the initial slope of the *P-E* curve (g C g$^{-1}$ Chl *a* hr$^{-1}$ [μmol photons m$^{-2}$ s$^{-1}$]$^{-1}$) where photosynthesis rates are light-limited. The photoacclimation parameter, $E_k$, was calculated as $P^*_{max}/\alpha^*$. *P-E* data were also fitted to the model of Platt et al. (1980), which contains the photoinhibition parameter ß* (g C g$^{-1}$ Chl *a* hr$^{-1}$ [μmol photons m$^{-2}$ s$^{-1}$]$^{-1}$). However, ß* was not significantly different from zero in any of the *P-E* curves and, therefore, this model was disregarded.

## Phytoplankton optical absorption ($\bar{a}^*$)

The spectrally averaged optical absorption cross section ($\bar{a}^*$, $m^2$ $mg^{-1}$ Chl $a$) was determined in surface water samples (2–10 m depth) at 12 stations and in one sample from pooled replicates for each bioassay incubation treatment. Aliquots of the seawater sample (100–1000 mL) were filtered onto 25 mm Whatman GF/F filters for measurement of particulate absorption spectra ($a_p$, 300–800 nm) and detrital absorption ($a_{det}$, 300–800 nm) on a Perkin-Elmer Lambda 35 spectrophotometer equipped with an integrating sphere (Labsphere) using the filter pad method and optical corrections in Mitchell and Kiefer (1988) and the coefficients of Bricaud and Stramski (1990). Detrital absorption ($a_{det}$, 300–800 nm) was assayed after methanol extraction according to the method of Kishino et al. (1985). Chl $a$-specific optical absorption cross sections ($a^*_{ph}$) at each wavelength ($\lambda$) were calculated as:

$$a^*_{ph}(\lambda) = \frac{a_p(\lambda) - a_{det}(\lambda)}{[Chla]} \tag{3}$$

where $[Chla]$ is the Chl $a$ concentration of the sample.

Spectrally averaged Chl $a$-specific optical absorption cross sections ($\bar{a}^*$, $m^2$ $mg^{-1}$ Chl $a$) were calculated using the equation:

$$\bar{a}^* = \frac{\sum_{\lambda=400}^{\lambda=700} a^*_{ph} E(\lambda)}{\sum_{\lambda=400}^{\lambda=700} E(\lambda)} \tag{4}$$

where $E(\lambda)$ ($\mu$mol photons $m^{-2}$ $s^{-1}$) is the spectral irradiance of the photosynthetron light source.

## Quantum yield of photosynthesis

The quantum yield of photosynthesis ($\Phi_m$ in C $mol^{-1}$ photons) was calculated as:

$$\Phi_m = \frac{\alpha^*}{43.2\bar{a}^*} \tag{5}$$

after first confirming that $\Phi_m$ was maximal at the lowest light level used in each of the assays (Johnson and Barber, 2003).

## Variable fluorescence

A Satlantic Fluorescence Induction and Relaxation (FIRe) system was used to determine the maximum photochemical efficiency ($F_v/F_m$) and the functional absorption cross section ($\sigma_{PSII}$) ($Å^2$ $photon^{-1}$) of photosystem II (Gorbunov et al., 1999) for water samples at 11 stations and bioassay experiment samples. Prior to analysis, the FIRe was blanked with GF/F-filtered seawater from the same station. After sampling from the sample bottles, samples were acclimated in the dark at 2°C for 30 min to fully oxidize the photosynthetic reaction centers.

## Bacterial productivity

Bacterial productivity was estimated in the bioassay experiments from incorporation of $^3H$-leucine into protein (Williams et al., 2015). Each of the incubation bottles served as a replicate in the measurements.

## *Water column analysis*

### Diffuse attenuation coefficient

The diffuse attenuation coefficient of downwelling PAR ($K_d$) in the water column was determined by fitting the equation:

$$E_z = E_0\, e^{-K_d z} \tag{6}$$

to each PAR profile, where $E_z$ is the irradiance at depth $z$ and $E_0$ is the irradiance just below the sea surface.

### Mixed layer depth (MLD)

The MLD was determined from each CTD profile as the shallowest depth at which the density ($\sigma_T$) was 0.02 kg $m^{-3}$ greater than at the surface (Cisewski et al., 2008; Alderkamp et al., 2012a).

**Mean light level in the mixed layer ($E_{UML}$)**

To calculate the mean daily PAR in the upper mixed layer ($E_{UML}$, mol photons m$^{-2}$ day$^{-1}$), we used the equation of Riley (1957):

$$E_{UML} = \frac{\overline{E}_{surf}\, T\left(1 - e^{-K_d z_{UML}}\right)}{K_d z_{UML}} \tag{7}$$

where $\overline{E}_{surf}$ is the total daily surface PAR averaged over five days and $T$ is the mean transmittance through the sea surface (0.85 for open water, 0.20 for grey ice and nilas, and 0.05 for snow covered and multiyear ice).

**Water column productivity**

Phytoplankton productivity throughout the water column was estimated at each station from the Chl $a$ concentrations, light availability in the water column, and $P$-$E$ parameters, as described in Alderkamp et al. (2012a). Briefly, at depth intervals of 1 m, Chl $a$ concentrations were estimated from continuous vertical fluorescence profiles that were calibrated to the measured Chl $a$ concentrations at similar depths (Chl $a$ = 0.71 fluorescence; R$^2$ = 0.69). The daily light cycle was binned in 10-min intervals and the mean over the previous five days was used to estimate the sinusoidal light cycle at 1-m depth intervals at each station based on the measured $K_d$ of that station. These light levels were then used to calculate the phytoplankton productivity at each depth using $P$-$E$ parameters of the phytoplankton collected at 10 m depth. These $P$-$E$ parameters were assumed to be representative of phytoplankton in the upper mixed layer (UML) where > 99% of the phytoplankton productivity occurred (virtually no light penetrated below the MLD because the high phytoplankton biomass levels resulted in high $K_d$ in all stations).

## Statistical analysis

Effects of Fe addition were tested by comparing Fe addition treatments to unamended controls in each bioassay experiment using one-way ANOVA analysis. All parameters for phytoplankton photosynthesis and cellular composition, as well as bacterial productivity, were tested at day four to eliminate potential effects of NO$_3$ limitation (see *Phytoplankton biomass response to Fe addition*). Interactions between Fe addition and original sample depth, as well as interactions between Fe addition and light availability, were tested using two-way ANOVA analysis. Differences were considered significant at $p < 0.05$. Simple linear regression was used to test relationships between phytoplankton parameters in seawater samples.

# Results

## Phytoplankton bloom characteristics

The physical properties of the upper water column, which affected the Fe and light availability to the phytoplankton, varied markedly within the ASP (Randall-Goodwin et al., 2015). In the southern ASP, near the ice sheets, AASW was relatively salty, due to little overall sea ice or near surface glacial melt, and/or enhanced mixing with WW, as observed in the southwest ASP close to the Getz Ice Shelf (GIS) (e.g., Station 5, Figure 2), or due to enhanced mixing with meltwater-laden mCDW waters as observed in front of the Dotson Ice Shelf (DIS) and at Station 57 (Figure 2). In the central ASP, AASW was moderately fresh, due to less recent sea ice melt and/or wind mixing (e.g., Stations 13 and 35, Figure 2). In the northern ASP, AASW was very fresh due to recent sea ice melt, as observed along the sea ice edge and in the MIZ (e.g., Station 66, Figure 2). These surface water characteristics affected the phytoplankton throughout our sampling during the buildup of a dense spring-summer phytoplankton bloom that typically peaks in mid-January (Arrigo et al., 2012). The phytoplankton biomass increased over the course of the cruise (Yager et al., 2012), conforming to this typical bloom development.

**Near the Getz Ice Shelf**

The salty AASW with the largest contribution of WW was observed in the southwest ASP close to the GIS (Station 5; Table 2A) and had a MLD of ~ 35 m (Figure 3C), resulting in moderate light levels in the UML ($E_{UML}$ 158 µmol photons m$^{-2}$ s$^{-2}$). DFe in these surface waters were > 0.18 nmol L$^{-1}$ (Figure 3E), whereas PFe was two orders of magnitude higher than DFe at > 18 nmol L$^{-1}$ (Figure 3F). Phytoplankton biomass was 2.3 mg Chl $a$ m$^{-3}$ in surface waters (Figure 3G) and 177 mg m$^{-2}$ integrated over depth (Figure 3H). The phytoplankton community was dominated by *Phaeocystis antarctica*, but diatoms and prasinophytes were also present and constituted up to 12% of the phytoplankton community at different depths (Table 2B, Figures 3I, 3J). Of the five stations sampled for the Fe addition bioassay experiments (see *Phytoplankton biomass response to Fe addition*), Station 5 is the most representative of salty AASW with WW influence (Figure 2).

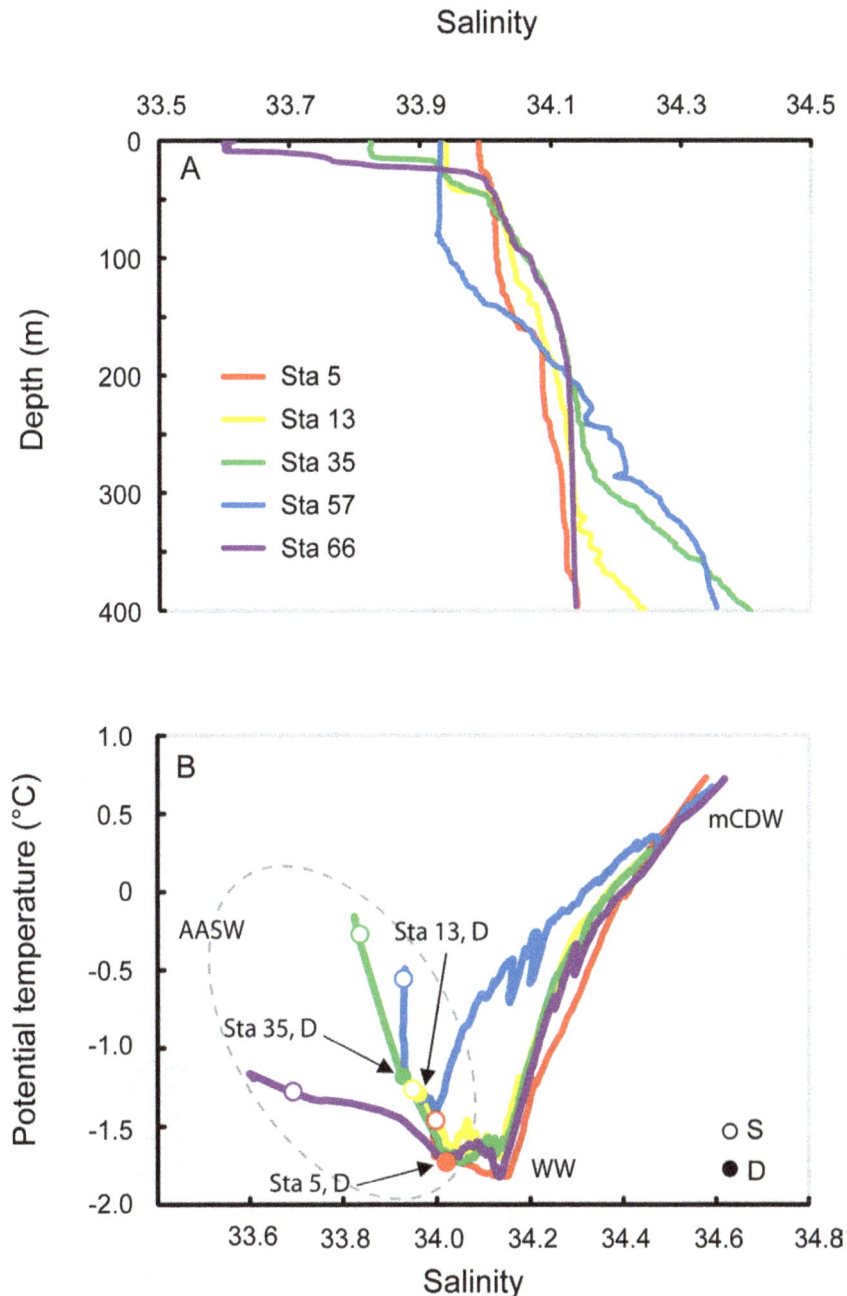

Figure 2

**Water properties of Fe addition bioassay experiment stations.**

Depth profiles of salinity are shown in (A), and temperature and salinity properties in (B), where AASW is Antarctic Surface Water, WW is Winter Water, and mCDW is modified Circumpolar Deep Water. Open symbols indicate surface waters (10 m depth); solid symbols indicate deeper subsurface waters (D, 35–50 m depth).

**Near the Dotson Ice Shelf**

Salty AASW with the largest contribution of mCDW was observed in the southern ASP close to the DIS (Stations 8, 9, 10, 11, and 57). The buoyancy-driven upwelling of mCDW destabilized the upper water column resulting in deep MLDs (40 to 70 m, Figure 3C). The deep MLD, in combination with variable light attenuation due to variable phytoplankton biomass, resulted in variable $E_{UML}$ (50 to 220 μmol photons $m^{-2}$ $s^{-2}$, Figure 3D). The meltwater-laden mCDW flowing from under the DIS appears to be the main source of DFe and PFe to the ASP (Sherrell et al., 2015) and the salty AASW contained 0.11–1.31 nmol $L^{-1}$ DFe (Figure 3E) and 15–50 nmol $L^{-1}$ PFe (Figure 3F). Phytoplankton biomass was <2.0 mg Chl $a$ $m^{-3}$ in surface waters (Figure 3G) and < 200 mg Chl $a$ $m^{-2}$ integrated over depth near the face of the DIS where the mCDW signature was strongest. At Station 57, approximately 50 km away from the DIS, surface Chl $a$ increased to 8.6 mg $m^{-3}$ (Figure 3G) and depth-integrated Chl $a$ to 618 mg $m^{-2}$ (Figure 3H).

**Table 2A. Oceanographic data on water samples used to initiate bioassay experiments**

| Type of data | Station-specific data | | | | | | | |
|---|---|---|---|---|---|---|---|---|
| Station | 5 | | 13 | | 35 | | 57 | 66 |
| Latitude (°S) | 73°57.72 | | 73°34.22 | | 73°16.78 | | 73°43.69 | 72°44.45 |
| Longitude (°W) | 118°01.18 | | 112°40.03 | | 112°06.28 | | 113°15.20 | 116°01.21 |
| Date | 14 Dec 2010 | | 19 Dec 2010 | | 26 Dec 2010 | | 1 Jan 2011 | 5 Jan 2011 |
| Mixed layer depth (m) | 28 | | 42 | | 22 | | 80 | 25 |
| Mean irradiance during experiment ($\mu$mol photons m$^{-2}$ s$^{-1}$) | 1243 | | 932 | | 1098 | | 1082 | 776 |
| Sample type (surface, S; deeper subsurface, D) | S | D | S | D | S | D | S | S |
| Sample depth (m) | 8.8 | 50.3 | 9.9 | 40.0 | 12.0 | 35.0 | 9.9 | 10.2 |
| Salinity | 34.01 | 34.05 | 33.94 | 33.94 | 33.83 | 33.93 | 33.93 | 33.69 |
| Temperature (°C) | −1.46 | −1.71 | −1.25 | −1.25 | −0.26 | −1.23 | −0.55 | −1.33 |
| NO$_3$ ($\mu$mol L$^{-1}$) | 26.3 | 26.7 | 21.1 | 24.8 | 12.5 | 22.0 | 24.9 | 15.8 |
| PO$_4$ ($\mu$mol L$^{-1}$) | 1.76 | 1.78 | 1.33 | 1.61 | 0.88 | 1.49 | 1.67 | 1.18 |
| Si(OH)$_4$ ($\mu$mol L$^{-1}$) | 88.0 | 93.0 | 77.9 | 83.8 | 70.8 | 72.5 | 96.5 | 74.1 |
| DFe (nmol L$^{-1}$) | 0.18 | 0.30 | 0.09 | 0.22 | 0.32 | 0.20 | 0.10 | 0.116 |
| PFe (nmol L$^{-1}$) | 18.88 | 12.54 | ND | ND | 7.28 | 6.78 | 21.07 | 15.43 |

**Table 2B. Biological data on water samples used to initiate bioassay experiments[a]**

| Type of data | Station 5 | | Station 13 | | Station 35 | | Station 57 | Station 66 |
|---|---|---|---|---|---|---|---|---|
| | S | D | S | D | S | D | S | S |
| Chl $a$ | 3.8 | 3.7 | 8.4 | 6.5 | 14.0 | 7.4 | 5.6 | 10.8 |
| POC | 244 | 179 | 386 | 287 | 903 | 471 | 237 | 544 |
| POC/Chl $a$ | 65.4 | 48.5 | 46.4 | 44.4 | 66.8 | 63.7 | 42.7 | 51.1 |
| POC/PON | 7.05 | 4.50 | 5.66 | 6.20 | 7.79 | 6.02 | 5.32 | 5.66 |
| $F_v/F_m$ | 0.49 | 0.46 | 0.35 | 0.39 | 0.27 | 0.39 | 0.40 | 0.36 |
| $\sigma_{PSII}$ | 525 | 561 | 737 | 677 | 608 | 648 | 544 | 638 |
| $P^*_{max}$ | 3.52 | 3.00 | 2.97 | 2.61 | 2.27 | 2.92 | 3.18 | 2.48 |
| $\alpha^*$ | 0.080 | 0.065 | 0.051 | 0.058 | 0.034 | 0.075 | 0.032 | 0.067 |
| $E_k$ | 44 | 46 | 58 | 45 | 67 | 39 | 99 | 37 |
| $\overline{a}^*$ | 0.015 | 0.008 | 0.013 | 0.011 | 0.014 | 0.014 | 0.013 | 0.010 |
| $\Phi_m$ | 0.127 | 0.191 | 0.091 | 0.123 | 0.057 | 0.128 | 0.060 | 0.119 |
| Dominant phytoplankton (fraction of total) | *Phaeocystis antarctica* (0.80) | *P. antarctica* (0.85) | *P. antarctica* (0.87) | *P. antarctica* (0.88) | *P. antarctica* (0.84) | *P. antarctica* (0.84) | *P. antarctica* (0.87) | *P. antarctica* (0.84) |
| 2$^{nd}$ most abundant phytoplankton | Prasinophytes (0.11) | Diatoms (0.11) | Prasinophytes (0.09) | Prasinophytes (0.08) | Prasinophytes (0.14) | Prasinophytes (0.15) | Diatoms (0.05) | Diatoms (0.11) |
| BP | 6.4 | 5.6 | 63.6 | 55.6 | 74.1 | 61.7 | 49.5 | 89.6 |

[a] Abbreviations (and units): S = surface water sample, D = deeper subsurface sample, Chl $a$ = Chlorophyll $a$ ($\mu$g L$^{-1}$), POC = particulate organic carbon ($\mu$g L$^{-1}$), POC/Chl $a$ (wt/wt), PON = particulate organic nitrogen, POC/PON (mol/mol), $F_v/F_m$ = maximum photochemical efficiency of photosystem II (no units), $\sigma_{PSII}$ = functional absorption cross section (Å photon$^{-1}$), $P^*_{max}$ = maximum rate of photosynthesis (g C g$^{-1}$ Chl $a$ h$^{-1}$), $\alpha^*$ the initial slope of the photosynthesis versus irradiance curve (g C g$^{-1}$ Chl $a$ h$^{-1}$ [$\mu$mol photons m$^{-2}$ s$^{-1}$]$^{-1}$), $E_k$ = photoacclimation parameter ($\mu$mol photons m$^{-2}$ s$^{-1}$), $\overline{a}^*$ = spectrally averaged Chl $a$-specific optical absorption cross section (photons m$^{-2}$), $\Phi_m$ = quantum yield of photosynthesis (mol C mol$^{-1}$ photons), BP = bacterial productivity (pmol Leu uptake L$^{-1}$ h$^{-1}$).

### Central Amundsen Sea Polynya

In the center of the polynya, at increasing distance from the Dotson and Getz Ice Shelves (Stations 6, 18–32, 35–50), solar warming increased the temperature of AASW (-0.7 °C, Figure 3B) and recent sea ice melt lowered the salinity (< 33.9, Figure 3A). Both processes increase water column stratification, which decreased the MLD to 10–30 m (Figure 3C). The highest phytoplankton biomass was observed in the central ASP with

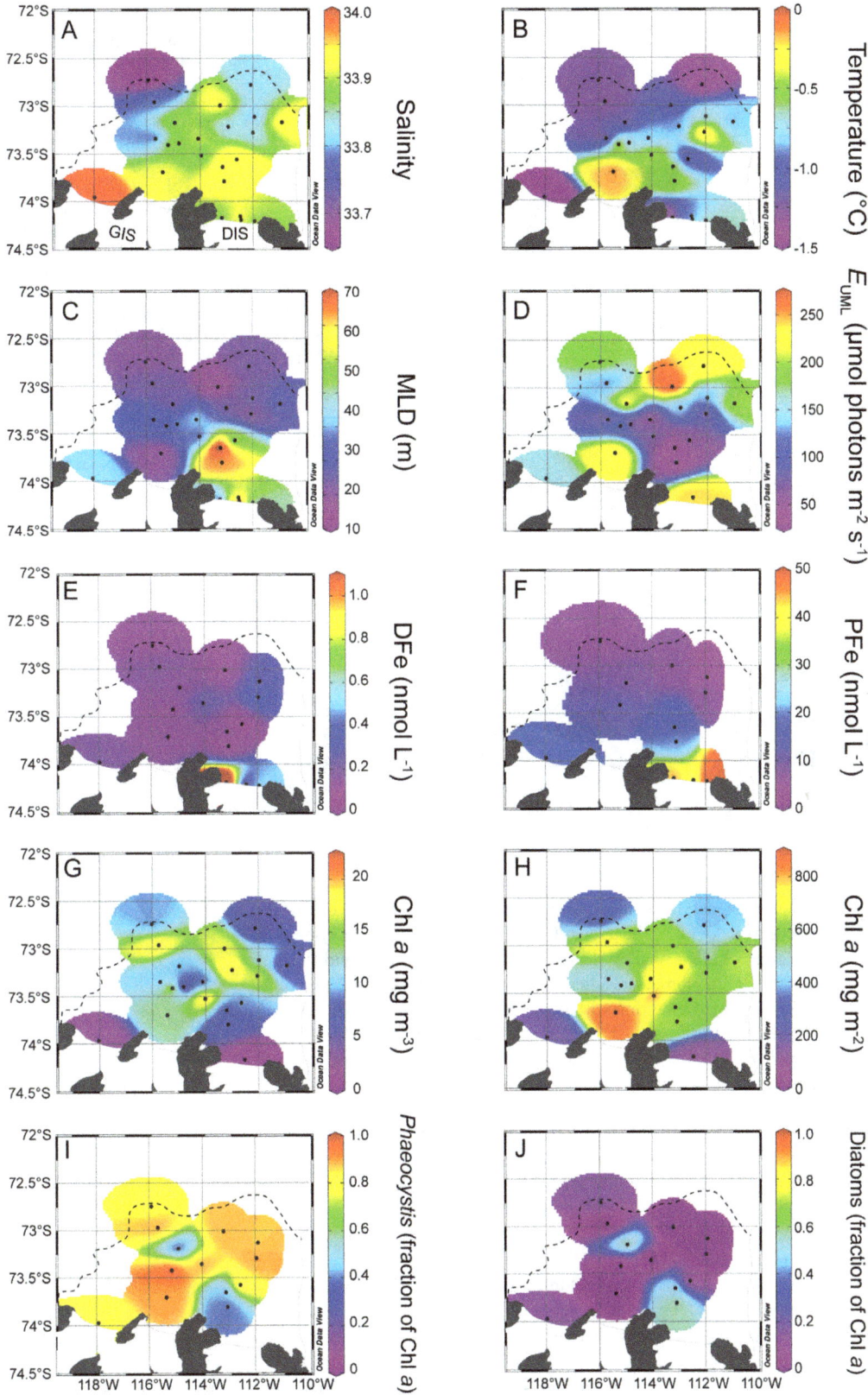

**Figure 3**

**Surface water properties in the Amundsen Sea Polynya.**

Properties shown for surface waters (2–10 m depth) are: salinity (A); temperature (B); mixed layer depth (MLD), the depth where density $\sigma_t$ increased 0.02 from surface waters (C); $E_{UML}$, mean light in the upper mixed layer (D); dissolved Fe (DFe) concentrations in surface water (data from Sherrell et al., 2015) (E); particulate Fe (PFe) concentrations in surface waters (data from Harazin et al., 2014) (F); chlorophyll (Chl) $a$ concentrations (G); depth-integrated Chl $a$ (H); fraction of phytoplankton community as *Phaeocystis antarctica* (I); and fraction of phytoplankton community as diatoms (J). In each panel, the dashed line shows the sea ice edge on 1 January 2011 and the Getz Ice Shelf (GIS) and Dotson Ice Shelf (DIS) are shown in white.

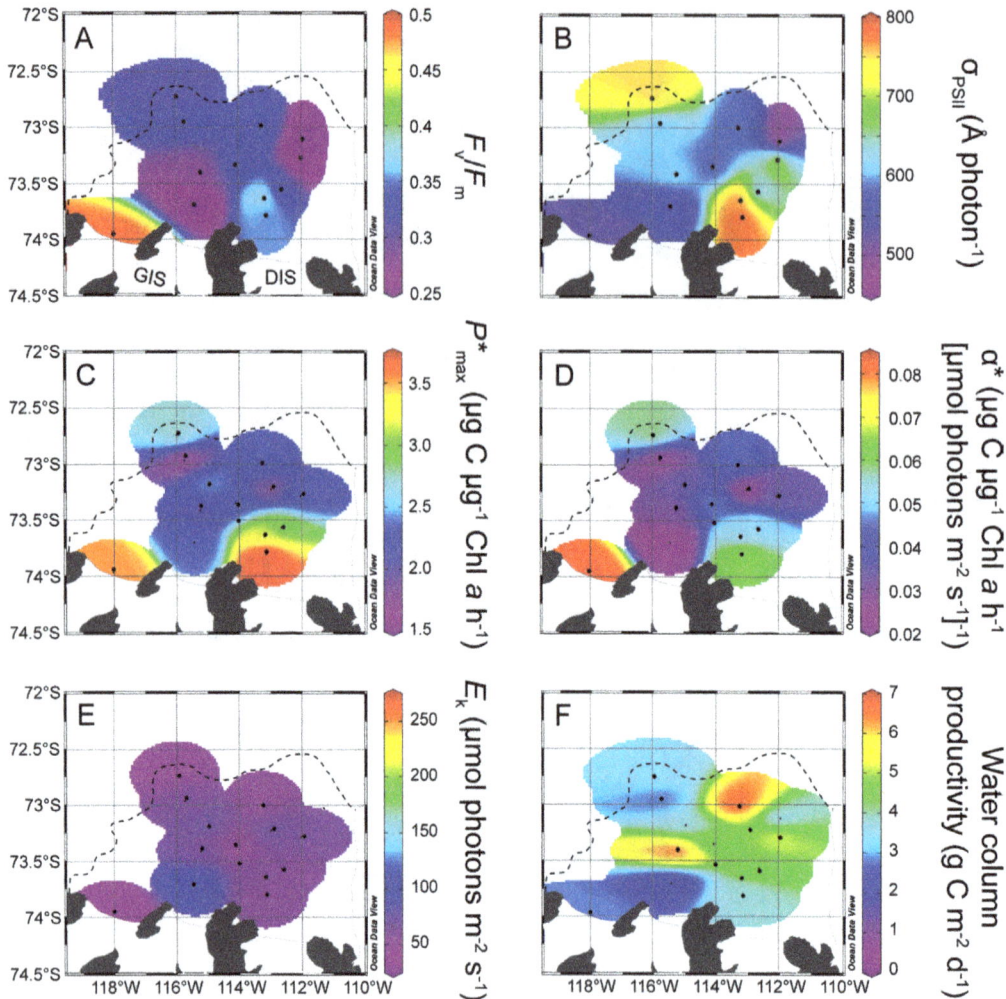

**Figure 4**

**Photosynthetic parameters of surface phytoplankton.**

Photosynthetic parameters shown for phytoplankton in surface waters (2–10 m depth) are: maximum photochemical efficiency of Photosystem (PS) II ($F_v/F_m$) (A), functional cross section of PS II ($\sigma_{PSII}$) (B), maximum photosynthesis rates normalized to chlorophyll $a$ (Chl $a$) ($P^*_{max}$) (C), initial light-limited slope of the photosynthesis–irradiance normalized to Chl $a$ ($\alpha^*$) (D), and photoacclimation parameter ($E_k$) (E). Water column productivity is shown in (F). In each panel, the sea ice edge on 1 January 2011, is shown by dashed line and the ice shelves are white: Getz Ice Shelf (GIS) and Dotson Ice Shelf (DIS).

mean surface Chl $a$ of 12.6 ± 5.7 mg m$^{-3}$ (Figure 3G) and mean depth-integrated Chl $a$ of 604 ± 150 mg m$^{-2}$ (Figure 3H). The highest Chl $a$ concentrations were found in the upper 20 m of the water column, dropping to 5–10 mg m$^{-3}$ at a depth of 20–50 m (Figure 5A). The shallow MLD, in combination with rapid light attenuation with depth due to the high phytoplankton biomass, resulted in a relatively low $E_{UML}$ (< 50 μmol photons m$^{-2}$ s$^{-1}$, Figure 3D). Moreover, the rapid light attenuation with depth resulted in a shallow euphotic zone with a 1% light depth shallower than the MLD. DFe and PFe in surface waters of the central polynya were on average 0.18 ± 0.07 nmol L$^{-1}$ (Figure 3E) and 8.7 ± 3.6 nmol L$^{-1}$ (Figure 3F), respectively. The phytoplankton community was dominated by $P.$ $antarctica$ at almost all stations throughout the central polynya. Diatoms contributed up to 60% to the phytoplankton biomass at two stations (Figures 3I, 3J), but contributed less or were almost absent at most stations. Prasinophytes were present in low numbers at most stations (< 10%) and contributed up to 14% to the phytoplankton community at some. The bioassay experiments at Stations 13 and 35 were conducted in the central ASP, where sea ice melt influence was stronger at Station 35 than at Station 13 (Figure 2).

### Northern Amundsen Sea Polynya and Marginal Ice Zone (MIZ)

Very fresh AASW that was strongly affected by sea ice melt was found in the MIZ (Station 66) and in the northern waters of the ASP near the sea ice (Station 34). Sea ice melt waters were relatively fresh (< 33.85) and cold (< –1.5°C) (Figures 3A, 3B), but warmed over time. Freshening of surface waters by sea ice melt water increased stratification, which resulted in a shallow MLD (< 20 m, Figure 3C) and a relatively high $E_{UML}$ > 150 μmol photons m$^{-2}$ s$^{-2}$ (Figure 3D). DFe in very fresh AASW was relatively low (0.11 ± 0.03 nmol L$^{-1}$; Figure 3E), as was PFe (6.9 ± 1.6 nmol L$^{-1}$; Figure 3F) when compared to other regions in the ASP. Phytoplankton biomass in very fresh AASW ranged from 6 to 10 mg Chl $a$ m$^{-3}$ (Figure 3G) and depth-integrated Chl $a$ exceeded 400 mg m$^{-2}$ (Figure 3H). All very fresh AASW stations were dominated by $P.$ $antarctica$, although diatoms and prasinophytes were also present and constituted up to 4% and 13% of the phytoplankton assemblages, respectively. The bioassay experiment at Station 66 was situated in the MIZ where phytoplankton biomass was high, even though waters were still partially ice covered (Table 2B).

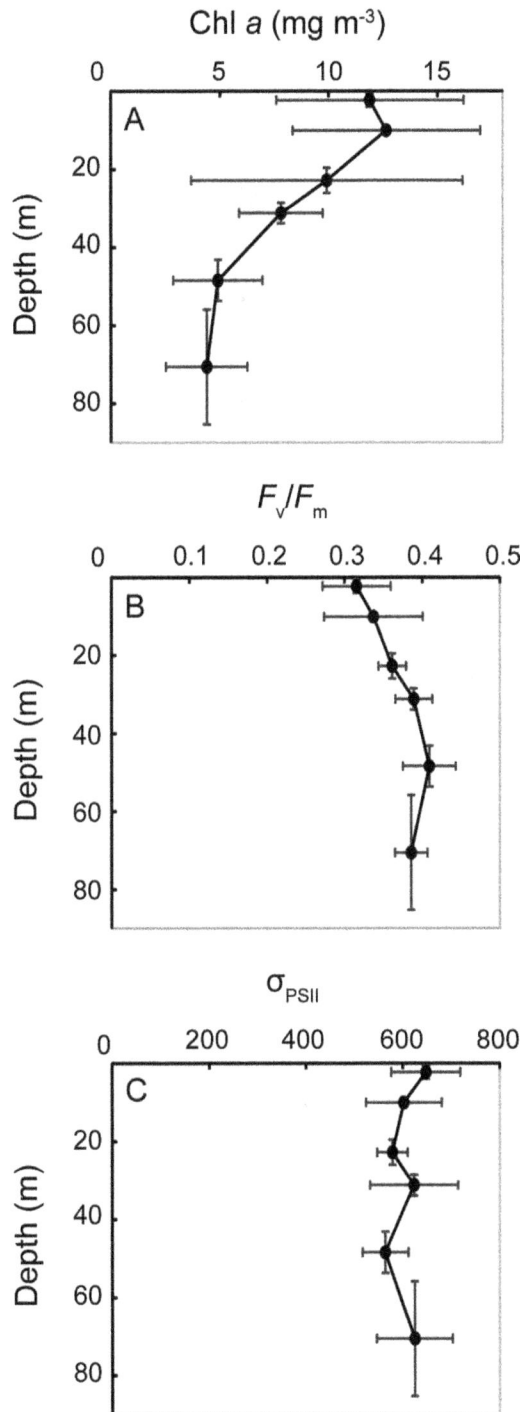

Figure 5

Depth profiles of phytoplankton variable fluorescence.

Mean and standard deviation of chlorophyll $a$ (Chl $a$, $n = 18$) (A), maximum photochemical efficiency of Photosystem (PS) II ($F_v/F_m$, $n = 14$) (B), and functional cross section of PS II ($\sigma_{PSII}$, $n = 14$) (C) are shown with depth in the water column.

## Phytoplankton photosynthesis in the ASP

### Variable fluorescence parameters

In general, the $F_v/F_m$ of phytoplankton was low in surface waters of the central ASP (< 0.35), and higher (> 0.4) in the salty AASW of the southern ASP near the DIS and GIS, at stations with a relatively strong WW and mCDW influence, respectively (Figure 4A). Phytoplankton in the very fresh AASW of the northern exhibited $F_v/F_m$ values of 0.35. In general, $F_v/F_m$ increased with depth in the upper 80 m of the water column (Figure 5B). The lowest $F_v/F_m$ was observed in surface waters in the central ASP with the highest phytoplankton biomass. The relationship between $F_v/F_m$ and Chl $a$ in surface waters was negative (Table 3).

Table 3. Simple linear regressions for phytoplankton photosynthesis parameters in surface waters of the Amundsen Sea Polynya[a]

| Variable 1 | Variable 2 | Equation | $n$ | $R^2$ | $p$ |
|---|---|---|---|---|---|
| $F_v/F_m$ | Chl $a$ | $F_v/F_m = -0.008$ Chl $a + 0.433$ | 14 | 0.594 | < 0.01 |
| $\sigma_{PSII}$ | Chl $a$ | | 14 | 0.005 | 0.820 |
| $P^*_{max}$ | $F_v/F_m$ | $P^*_{max} = 8.84 F_v/F_m - 0.13$ | 13 | 0.411 | < 0.05 |
| $P^*_{max}$ | DFe | | 14 | 0.154 | 0.208 |
| $P^*_{max}$ | PFe | $P^*_{max} = 0.09$ PFe $+ 1.67$ | 8 | 0.745 | < 0.01 |
| $\alpha^*$ | $F_v/F_m$ | $\alpha^* = 0.26 F_v/F_m - 0.035$ | 13 | 0.678 | < 0.001 |
| $\alpha^*$ | $P^*_{max}$ | $\alpha^* = 0.02 P^*_{max} - 0.009$ | 14 | 0.820 | < 0.001 |
| $\alpha^*$ | DFe | | 14 | 0.081 | 0.370 |
| $\alpha^*$ | PFe | $\alpha^* = 0.002$ PFe $+ 0.024$ | 8 | 0.506 | < 0.05 |
| WCP | Chl $a$ | | 11 | 0.001 | 0.928 |
| WCP | Depth-integr Chl $a$ | | 11 | 0.041 | 0.552 |
| WCP | MLD | | 11 | 0.002 | 0.906 |
| WCP | $E_{UML}$ | | 11 | 0.059 | 0.473 |
| WCP | $P^*_{max}$ | | 11 | 0.043 | 0.538 |
| WCP | $\alpha^*$ | | 11 | 0.037 | 0.571 |
| WCP | DFe | | 11 | 0.005 | 0.829 |
| WCP | PFe | | 8 | 0.063 | 0.548 |

[a] Abbreviations (and units): $F_v/F_m$ = maximum photochemical efficiency of photosystem II (no units), Chl $a$ = Chlorophyll $a$ ($\mu$g L$^{-1}$), depth-integrated Chl $a$ (mg m$^{-2}$), $\sigma_{PSII}$ = functional absorption cross section (Å photon$^{-1}$), $P^*_{max}$ = maximum rate of photosynthesis (g C g$^{-1}$ Chl $a$ h$^{-1}$), $\alpha^*$ = initial slope of photosynthesis versus irradiance curve (g C g$^{-1}$ Chl $a$ h$^{-1}$ [$\mu$mol photons m$^{-2}$ s$^{-1}$]$^{-1}$), DFe = dissolved iron (nmol L$^{-1}$), PFe = particulate iron (nmol L$^{-1}$), WCP = water column productivity (g C m$^{-2}$ d$^{-1}$), MLD = mixed layer depth (m), $E_{UML}$ = mean daily PAR in the upper mixed layer (mol photons m$^{-2}$ day$^{-1}$).

The $\sigma_{PSII}$ of phytoplankton was variable (500–900 Å photon$^{-1}$) throughout surface waters in the ASP (Figure 4B). The highest $\sigma_{PSII}$ (800–900 Å photon$^{-1}$) was observed in salty AASW near the DIS, whereas the lowest (500 Å photons$^{-1}$) was observed close to the GIS (Figure 4B). The $\sigma_{PSII}$ in the central polynya ranged from 473 to 938 Å photon$^{-1}$ and was similar to that in waters affected by sea ice melt. There was no trend in $\sigma_{PSII}$ with depth (Figure 5C); moreover, there was no significant relationship between $\sigma_{PSII}$ and Chl $a$ (Table 3).

## P-E parameters

The spatial distribution of $P^*_{max}$ resembled that of $F_v/F_m$. $P^*_{max}$ was highest in the salty AASW in the southern ASP near the DIS and GIS (Figure 4C), exceeding 3.0 $\mu$g C $\mu$g$^{-1}$ Chl $a$ h$^{-1}$. The $P^*_{max}$ was lowest (< 2.0 $\mu$g C $\mu$g$^{-1}$ Chl $a$ h$^{-1}$) in the central ASP in the areas with the highest phytoplankton biomass. In fresh AASW waters affected by sea ice melt water, $P^*_{max}$ was intermediate. There was no trend in $P^*_{max}$ with depth (Table 1B). The relationship between $P^*_{max}$ and $F_v/F_m$ was positive (Table 3). There was no relationship between $P^*_{max}$ and DFe, but a positive relationship between $P^*_{max}$ and PFe was detected (Table 3).

The spatial distribution of $\alpha^*$ resembled that of $P^*_{max}$, with high $\alpha^*$ (> 0.06 $\mu$g C $\mu$g$^{-1}$ Chl $a$ h$^{-1}$ [umol photons m$^{-2}$ s$^{-1}$]) in the southern ASP near the DIS and GIS (Figure 4D) and low $\alpha^*$ (< 0.03 $\mu$g C $\mu$g$^{-1}$ Chl $a$ h$^{-1}$ [umol photons m$^{-2}$ s$^{-1}$]) in the central ASP. The parameter $\alpha^*$ was intermediate in fresh AASW affected by sea ice melt water in the northern ASP (Figure 4D). In general, $\alpha^*$ increased with depth (Table 2B). There were strong positive relationships between $\alpha^*$ and $F_v/F_m$, and between $\alpha^*$ and $P^*_{max}$ (Table 3). There was no relationship between $\alpha^*$ and DFe, but the relationship between $\alpha^*$ and PFe was positive (Table 3).

The $E_k$ ranged from 40 to 100 $\mu$mol photons m$^{-2}$ s$^{-1}$ (Figure 4E) and was generally lower than $E_{UML}$ (Figure 4E). The strong positive relationship between $P^*_{max}$ and $\alpha^*$ resulted in no spatial pattern in $E_k$, although $E_k$ did decrease with depth (Table 2B).

Water column productivity in the ASP ranged from 2.0 to 6.5 g C m$^{-2}$ d$^{-1}$ (Figure 4F). In general, productivity was highest (> 5.0 g C m$^{-2}$ d$^{-1}$) in the central polynya in waters with relatively shallow MLD (< 30 m) that resulted in high values for $E_{UML}$ (> 130 $\mu$mol photons m$^{-2}$ s$^{-2}$) and Chl $a$ concentrations (> 7 mg m$^{-3}$ and > 400 mg m$^{-2}$). Water column productivity depends on phytoplankton biomass, light availability, and $P$-$E$ parameters of phytoplankton. However, productivity showed no relationship with any of these parameters, such as surface Chl $a$, depth-integrated Chl $a$, MLD, $E_{UML}$, $P^*_{max}$, or $\alpha^*$ (Table 3). Similarly, there was no relationship between productivity and DFe in surface waters or between productivity and PFe (Table 3).

*Phytoplankton biomass response to Fe addition*

In general, phytoplankton biomass increased significantly over the course of the incubation in all experiments at 10% and 50% irradiance (Figure 6, left panels). The increase in biomass was accompanied by a rapid drawdown of $NO_3$ (Figure 6, right panels), which was depleted by the end (7 or 8 days) in all incubations at 10% and 50% irradiance, both in the control and the Fe treatments (Figure 6, right panels). Moreover, all $NO_3$ was depleted by day 4 in both control and Fe treatments of surface water from Station 35, such that these incubations were likely $NO_3$-limited (Figure 6F).

Fe addition resulted in higher biomass compared to controls in four of the bioassay incubations in the ASP, whereas there was no significant effect of Fe in eight incubations (Figure 6). In the incubations of surface water from Station 5, in the salty AASW near the face of the GIS, Fe addition enhanced phytoplankton biomass 1.37-fold by day 4, which increased growth rate to 0.29 $d^{-1}$ compared to 0.26 $d^{-1}$ in the control (one-way ANOVA, $p < 0.05$, Table 4). By day 7, when $NO_3$ was depleted, biomass in the Fe treatment was still 1.27-fold higher than in the control treatment (one-way ANOVA, $p < 0.05$, Figure 6A). In the incubations of subsurface water from Station 35, in the moderately fresh AASW in the central ASP, Fe addition enhanced phytoplankton biomass 1.32-fold by day 4 (one-way ANOVA $p < 0.05$, Figure 6E), which increased the phytoplankton growth rates to 0.26 $d^{-1}$ compared to 0.12 $d^{-1}$ in the control (one-way ANOVA, $p < 0.05$, Table 4). By day 8, when $NO_3$ was depleted, biomass in the Fe treatment was still 1.13-fold higher than in the control treatment (one-way ANOVA, $p < 0.05$, Figure 6E). In the incubations of subsurface water from Station 57, in the salty AASW affected by mCDW, Fe addition enhanced biomass 1.14-fold in the 10% irradiance incubation by day 4 (one-way ANOVA, $p < 0.05$), which increased phytoplankton growth rates to 0.31 $d^{-1}$ compared to 0.29 $d^{-1}$ in the control (one-way ANOVA, $p < 0.05$, Table 4). By day 7, when $NO_3$ was depleted, Fe addition enhanced the biomass 1.42-fold in the 50% irradiance treatment (one-way ANOVA, $p < 0.05$, Figure 6G). Fe addition also enhanced $NO_3$ drawdown when compared to controls in two subsurface water incubations (Stations 13 and 35), whereas there was no significant Fe effect in the ten other incubations.

Interactions between effects of Fe and sample depth on phytoplankton growth rates and $NO_3$ drawdown were studied by analyzing data from day 4 of Stations 5, 13 and 35 bioassay experiments together (Table 5). Both growth rates and $NO_3$ drawdown were higher in the subsurface than surface water incubations over these three experiments (one-way ANOVA, $p < 0.05$). These results are likely due to $NO_3$-limitation in the incubations of surface water from Station 35, resulting in lower growth rates and $NO_3$ drawdown compared to subsurface water incubations that were not $NO_3$-depleted. There was no interaction between Fe and sample depth on phytoplankton growth rates, nor on $NO_3$ drawdown (Table 5).

Data from day 4 of the Stations 57 and Station 66 incubations were analyzed together to study interactions between Fe and light on phytoplankton growth rates and $NO_3$ drawdown (Table 5). Whereas both higher Fe and light availability individually increased the phytoplankton growth rates in the incubations (one-way ANOVA $p < 0.05$), there was no interaction between these factors. Similarly, there were no interactions between Fe and light effects on $NO_3$ drawdown (Table 5).

*Phytoplankton photosynthesis response to Fe additions*

**Variable fluorescence responses**

All photosynthesis responses to Fe additions were studied on day 4 of the incubations, when $NO_3$ was not depleted, except for the surface water incubation from Station 35. The $F_v/F_m$ ranged from 0.25 to 0.48 (Figures 7A, 7B) and was in the same range as the initial $F_v/F_m$ of surface phytoplankton (Figure 4A). Fe addition increased the $F_v/F_m$ in almost all of the 10% and 50% irradiance incubations when compared to control treatments (one-way ANOVA, $p < 0.05$, Figures 7A, 7B) by an average of 1.17-fold, whereas there was no effect in the 1% irradiance incubations. Despite this trend, the interaction between Fe and light was not significant (Table 5) when experiments at Stations 57 and 66 were analyzed together, likely due to the small but significant decrease in $F_v/F_m$ after Fe addition in the 10% irradiance incubation at Station 57 (Figure 7B). In addition, the original sampling depth did not affect the $F_v/F_m$ in the incubations, nor was there an interaction between Fe and sample depth (Table 5).

The $\sigma_{PSII}$ by day 4 of the incubations ranged from 442 to 765 Å photon$^{-1}$, which was within the range of the initial $\sigma_{PSII}$ of the incubations (Table 2B) and that of surface phytoplankton in the ASP (Figure 4B). Fe effects on $\sigma_{PSII}$ were inconsistent (Figures 7C, 7D), reducing $\sigma_{PSII}$ in two incubations (Stations 5, S incubation; Station 57, 10% irradiance incubation), increasing $\sigma_{PSII}$ in another incubation (Station 13, D incubation), and not affecting $\sigma_{PSII}$ in the remaining incubations (Figures 7C, 7D). Analyzing all incubations together revealed no overall Fe effect on $\sigma_{PSII}$ (Table 5). The $\sigma_{PSII}$ of the surface water incubations was slightly higher than the subsurface water incubations, but this effect was not significant (Table 5). Moreover, there was no interaction between effects of Fe and sample depth (Table 5). Light did not affect $\sigma_{PSII}$, but there was a significant interaction between Fe and light (two-way ANOVA, $p < 0.05$), where Fe addition slightly decreased $\sigma_{PSII}$ at low light, but did not affect $\sigma_{PSII}$ at high light (Figure 7D).

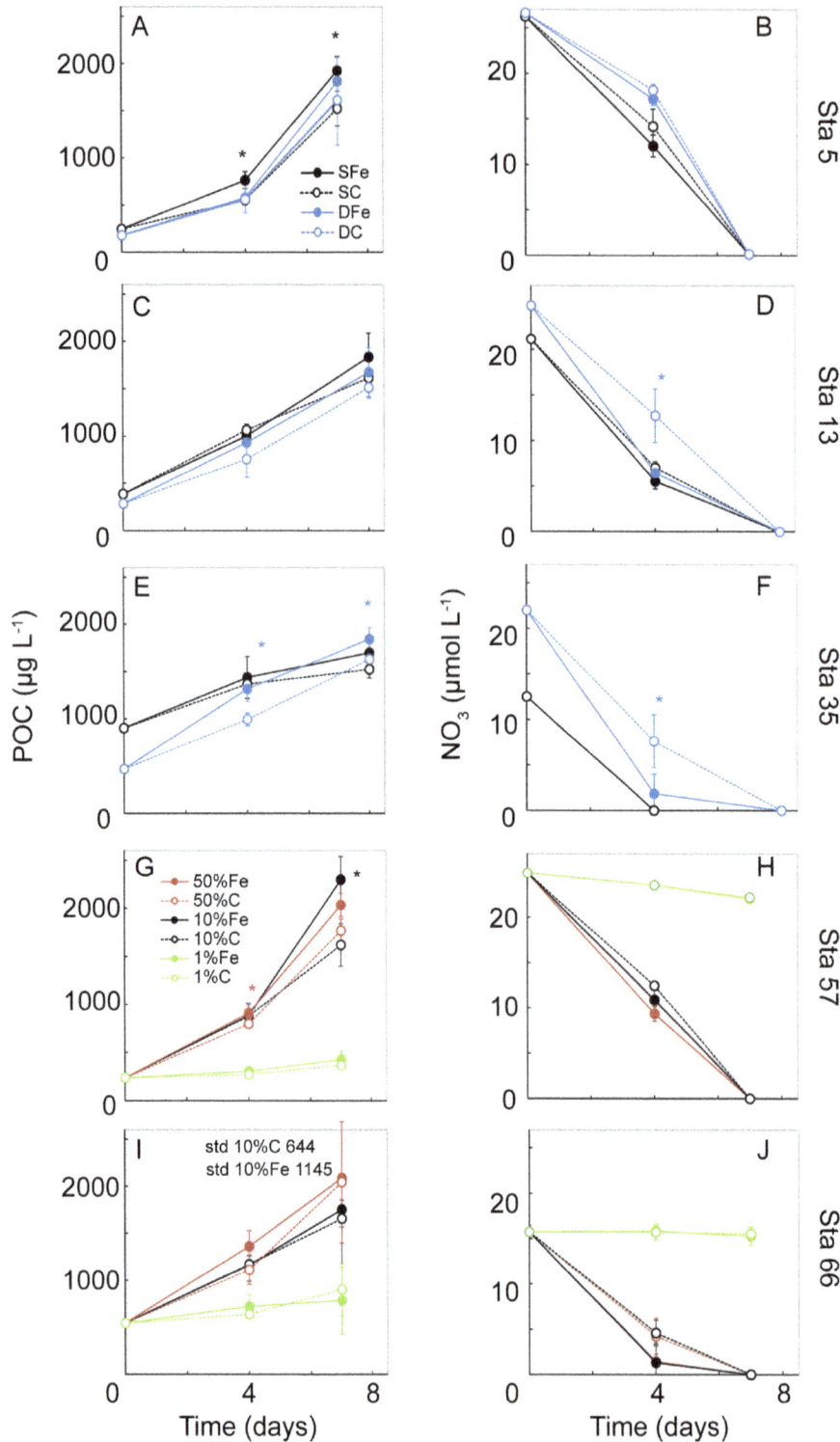

**Figure 6**

**Phytoplankton responses to iron addition.**

Mean and standard deviation of the measured parameters are shown for triplicate incubations of Fe additions (solid symbols) and unamended controls (open symbols). Left panels indicate change in biomass (particulate organic carbon: POC) over time of incubation; right panels indicate change in nitrate ($NO_3$). The upper three rows of panels show responses during incubation of surface (black) and deeper subsurface (blue) waters from Station 5 (A, B) near the Dotson glacier, Station 13 (C, D) in the central Amundsen Sea Polynya (ASP), and Station 35 (E, F) in the central ASP. The lower two rows of panels show responses at different levels of incident irradiance (1%, green; 10%, black; 50%, red) for surface waters from Station 57 (G, H) close to the Dotson Ice Shelf and Station 66 (I, J) in the marginal ice zone (MIZ). An asterisk indicates significant difference between Fe addition and control of the same color at the time of sampling (one-way ANOVA, $p < 0.05$). In (F), data for SC are identical to data for SFe.

**Photosynthesis vs irradiance (P-E) parameters**

High $P^*_{max}$ exceeding 2.2 μg C μg⁻¹ Chl $a$ h⁻¹ were observed in all incubations by day 4 (Figures 8A, 8B). Fe addition increased $P^*_{max}$ compared to controls in almost all incubations, on average 1.31-fold. This increase was greatest in the experiments where Fe also affected phytoplankton growth (S for Station 5, D for Station 35), with $P^*_{max}$ increasing 1.75-fold and 1.56-fold, respectively. $P^*_{max}$ was lowest for surface water incubations from Station 35, where $NO_3$ was depleted by day 4. When experiments were analyzed together, the Fe effect on $P^*_{max}$ was significant (Table 5, one-way ANOVA, $p < 0.05$). The original sample depth did not affect $P^*_{max}$ in the incubations for Stations 5, 13, and 35, and there was no interaction between Fe and

Table 4. Physiological characteristics[a] of phytoplankton in the bioassay experiments[b]

| Characteristic | Bioassay experiment | Station-specific data by sample type | | | | | | Station-specific data by light level | | | | | |
| --- | --- | --- | --- | --- | --- | --- | --- | --- | --- | --- | --- | --- | --- |
| | | Station 5 | | Station 13 | | Station 35 | | Station 57 | | | Station 66 | | |
| | | S | D | S | D | S | D | 1% | 10% | 50% | 1% | 10% | 50% |
| Phytoplankton growth rate (d⁻¹) | C | 0.26 (0.01) | 0.31 (0.01) | 0.25 (0.01) | 0.24 (0.02) | 0.10 (0.01) | 0.12 (0.01) | 0.06 (0.01) | 0.29 (0.00) | 0.28 (0.01) | 0.07 (0.01) | 0.18 (0.01) | 0.19 (0.01) |
| | Fe | 0.29* (0.00) | 0.33 (0.01) | 0.24 (0.01) | 0.30 (0.00) | 0.19 (0.01) | 0.26* (0.01) | 0.08 (0.01) | 0.31* (0.00) | 0.33 (0.00) | 0.06 (0.02) | 0.23 (0.01) | 0.19 (0.01) |
| POC/Chl a (wt/wt) | C | 58 (5) | 102 (17) | 73 (20) | 89 (17) | 75 (2) | 56 (12) | 36 (3) | 60 (10) | 89 (9) | 65 (7) | 79 (5) | 108 (7) |
| | Fe | 72 (14) | 84 (15) | 229 (212) | 67 (19) | 75 (19) | 79 (15) | 44 (3) | 64 (7) | 101 (26) | 73 (10) | 116* (15) | 131 (29) |
| POC/PON (mol/mol) | C | 5.1 (1.5) | 5.4 (0.5) | 7.3 (0.6) | 6.8 (0.2) | 6.3 (0.2) | 6.2 (0.3) | 6.3 (3.5) | 6.6 (0.2) | 4.9 (1.3) | 5.6 (0.4) | 5.8 (0.6) | 4.6 (0.3) |
| | Fe | 4.0 (0.7) | 5.1 (1.0) | 6.8 (1.1) | 6.5 (0.2) | 8.5 (1.9) | 6.5 (0.5) | 4.6 (0.3) | 5.4 (0.8) | 5.1 (1.1) | 6.2 (1.1) | 6.6 (0.6) | 5.4 (1.1) |
| $\bar{a}^*$ (photons m⁻²) | C | 0.014 | 0.016 | 0.014 | 0.017 | 0.015 | 0.013 | 0.010 | 0.012 | 0.014 | 0.011 | 0.012 | 0.012 |
| | Fe | 0.015 | 0.015 | 0.016 | 0.013 | 0.011 | 0.009 | 0.014 | 0.012 | 0.011 | ND | 0.013 | 0.013 |
| $\Phi^m$ (mol C mol⁻¹ photons) | C | 0.076 | 0.102 | 0.051 | 0.051 | 0.066 | 0.061 | 0.134 | 0.100 | 0.056 | 0.120 | 0.080 | 0.044 |
| | Fe | 0.139 | 0.119 | 0.073 | 0.081 | 0.150 | 0.173 | 0.144 | 0.099 | 0.137 | ND | 0.093 | 0.096 |

[a] Mean (standard deviation) by day 4; asterisk indicates significant effect of Fe addition (one-way ANOVA, $p < 0.05$)

[b] Abbreviations (and units): S = surface water sample, D = deeper subsurface sample, % = level of incident irradiance, C = unamended control, Fe = Fe addition, Chl $a$ = Chlorophyll $a$ (µg L⁻¹), POC = particulate organic carbon (µg L⁻¹), PON = particulate organic nitrogen, $\bar{a}^*$ = spectrally averaged Chl $a$-specific optical absorption cross section, $\Phi_m$ = quantum yield of photosynthesis

Table 5. Significance of effects[a] of Fe addition, sample depth, and light levels on phytoplankton variables[b] and bacterial productivity

| Variable | $p$ values[a] for bioassay test (number of experiments) | | | | |
| --- | --- | --- | --- | --- | --- |
| | Fe (5) | Depth (3) | Fe x Depth (3) | Light (2) | Fe x Light (2) |
| Phytoplankton growth rate | 0.018* | 0.009* | 0.668 | 0.000* | 0.723 |
| NO₃ drawdown | 0.037* | 0.034* | 0.202 | 0.000* | 0.280 |
| $F_v/F_m$ | 0.008* | 0.164 | 0.866 | 0.012* | 0.093 |
| $\sigma_{PSII}$ | 0.471 | 0.087 | 0.665 | 0.617 | 0.017* |
| $P^*_{max}$ | 0.039* | 0.500 | 0.527 | 0.217 | 0.868 |
| $\alpha^*$ | 0.006* | 0.851 | 0.473 | 0.024* | 0.322 |
| $E_k$ | 0.419 | 0.359 | 0.701 | 0.000* | 0.002* |
| $\bar{a}^*$ | 0.234 | 0.379 | 0.219 | 0.381 | 0.822 |
| $\Phi_m$ | 0.005* | 0.796 | 0.935 | 0.005* | 0.051 |
| POC/PON | 0.977 | 0.665 | 0.731 | 0.119 | 0.586 |
| POC/Chl $a$ | 0.124 | 0.440 | 0.170 | 0.000* | 0.785 |
| Bacterial productivity | 0.311 | 0.928 | 0.533 | 0.618 | 0.951 |

[a] Measured at day 4 of the experiment; $p$ values are for two-way ANOVA analysis of the bioassay experiments analyzed together; asterisk indicates significant effect at the $p < 0.05$ level.

[b] Variables: $F_v/F_m$ = maximum photochemical efficiency of photosystem II, $\sigma_{PSII}$ = functional absorption cross section, $P^*_{max}$ = maximum rate of photosynthesis, $\alpha^*$ = initial slope of photosynthesis versus irradiance curve, $E_k$ = photoacclimation parameter, $\Phi_m$ = quantum yield of photosynthesis, $\bar{a}^*$ = spectrally averaged Chl $a$-specific optical absorption cross section, POC = particulate organic carbon, PON = particulate organic nitrogen.

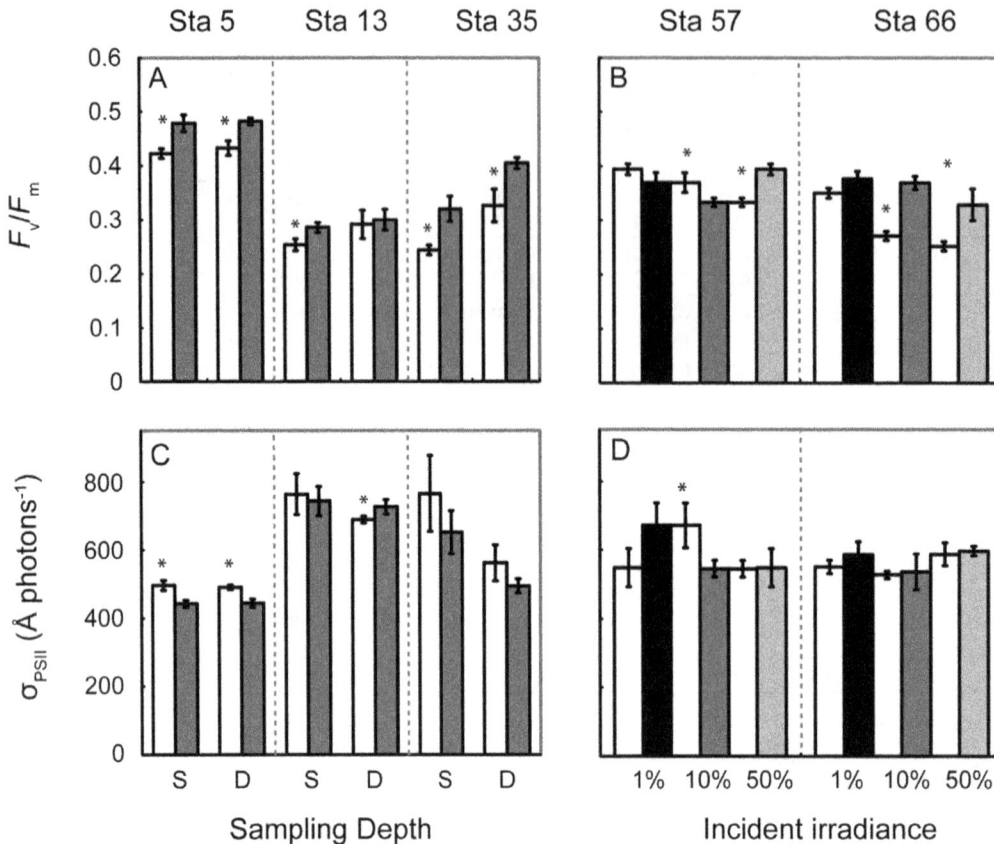

Figure 7

**Variable fluorescence in incubation experiments.**

Mean and standard deviation of variable fluorescence by day 4 of the incubation experiments are shown for triplicate incubations of Fe additions (dark bars) and unamended controls (white bars). An asterisk indicates significant difference between Fe addition and control (one-way ANOVA, $p < 0.05$). Maximum photochemical efficiency of Photosystem (PS) II ($F_v/F_m$) is shown for surface (S) and deeper subsurface (D) incubations from Stations 5, 13 and 35 (A) and for different levels of incident irradiance (1%, black; 10%, grey; 50%, light grey) for Stations 57 and 66 (B). Functional cross section of PS II ($\sigma_{PSII}$) is shown for the surface and subsurface depths (C) and levels of incident irradiance (D).

depth (Table 5). $P^*_{max}$ increased slightly with higher light, although this effect was not significant; there was no interaction between Fe and light effects (Table 5).

The $\alpha^*$ was high in all incubations (> 0.024 µg C µg$^{-1}$ Chl $a$ h$^{-1}$ [µmol photons m$^{-2}$ s$^{-1}$]$^{-1}$, Figures 8C, 8D) and generally followed the trends in $P^*_{max}$. Fe addition increased $\alpha^*$ in almost all experiments (Figures 8C, 8D), on average by 1.42-fold. This increase was greatest in the incubations where Fe additions also affected phytoplankton growth (S for Station 5, D for Station 35), where $\alpha^*$ increased 1.89-fold and 1.95-fold, respectively. Thus, in Fe-limited phytoplankton, the relative effect of Fe on $\alpha^*$ was greater than that on $P^*_{max}$. When all experiments were analyzed together, the Fe effect on $\alpha^*$ was significant (Table 5). The original sample depth did not affect $\alpha^*$ in the incubations at Stations 5, 13, and 35, and there was no interaction between Fe and sample depth (Table 5). The $\alpha^*$ decreased at higher light, but there was no interaction between Fe and light effects (Table 5).

The photoacclimation parameter ($E_k$) was high (53–104 µmol photons m$^{-2}$ s$^{-1}$) in all incubations at 10% and 50% irradiance and somewhat lower at 1% irradiance (42–46 µmol photons m$^{-2}$ s$^{-1}$) (Figures 8E, 8F). Fe addition did not affect $E_k$ because both $\alpha^*$ and $P^*_{max}$ increased relative to the controls. When all experiments were analyzed together, $E_k$ was not significantly affected by either Fe, depth, or the interaction between Fe and depth (Table 5). On the other hand, $E_k$ increased at higher light and there was an interaction between Fe and light effects (Table 5), where Fe addition at high light (50% irradiance) decreased $E_k$, but there was no effect at lower light.

The $\bar{a}^*$ in the incubations varied between 0.009 and 0.017 m$^{-2}$ mg$^{-1}$ Chl $a$ (Table 4) and was similar to initial $\bar{a}^*$ of phytoplankton in the ASP (Table 2B). Fe addition did not affect $\bar{a}^*$ in any of the incubations (Table 4). Moreover, when experiments were analyzed together, there was no effect of either Fe, sample depth, or light on $\bar{a}^*$, and there were no interactive effects (Table 5).

The $\Phi_m$ in the incubations varied between 0.051 and 0.173 mol C mol$^{-1}$ photons (Table 4) and was similar to initial $\Phi_m$ of phytoplankton in the ASP (Table 2B). Fe addition increased $\Phi_m$ in all incubations, on average 1.51-fold (Table 4, Table 5). There was no effect of original sample depth on the $\Phi_m$ in the incubations and there were no interactions between Fe and depth effects (Table 5). The $\Phi_m$ was higher at low light (Table 4, Table 5) and there was an interaction between Fe and light effects (Table 5), where the Fe effect was stronger at high light.

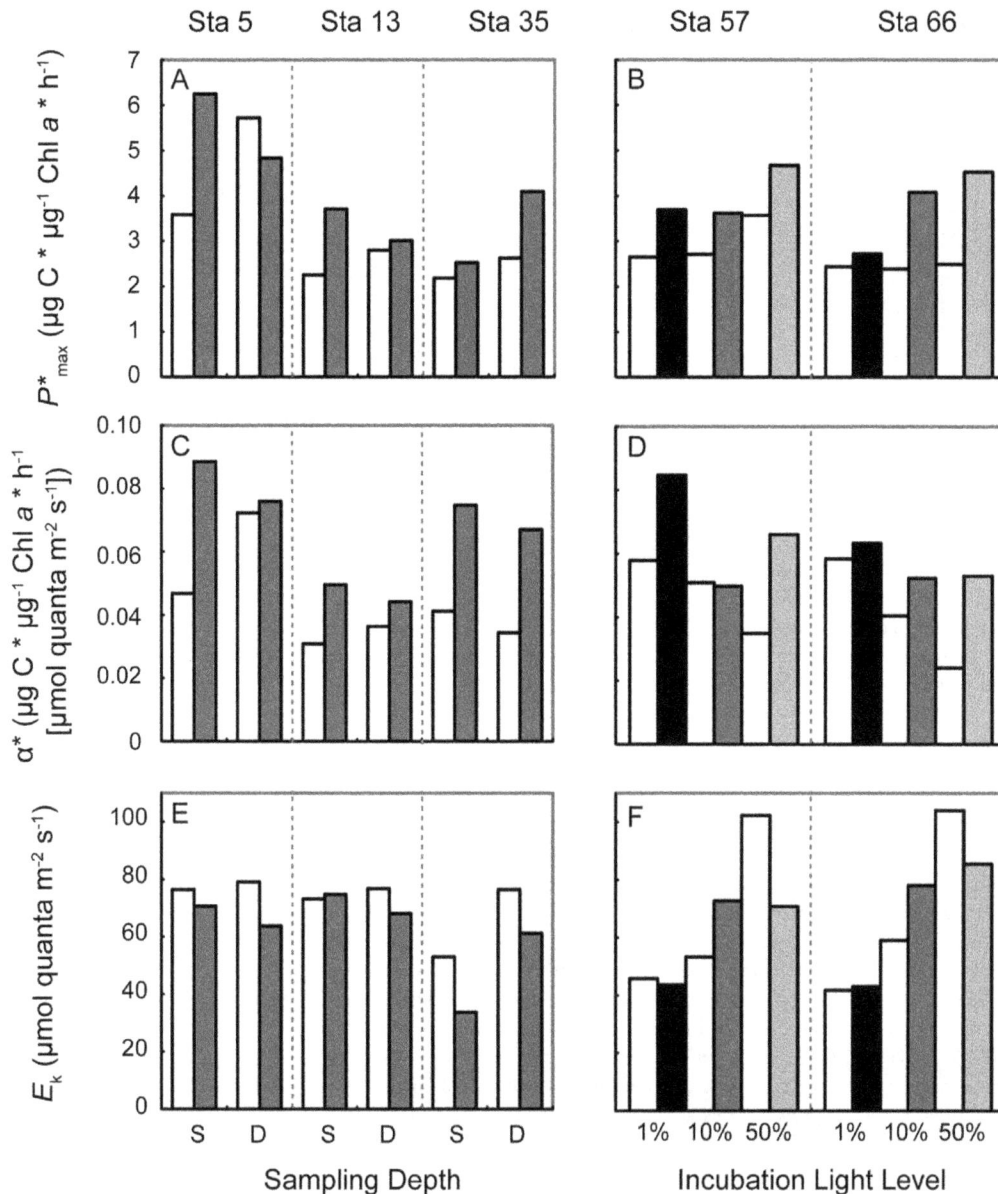

**Figure 8**

**Photosynthesis versus irradiance parameters in incubation experiments.**

Photosynthesis versus irradiance (*P-E*) parameters at day 4 of the incubation experiments are shown for incubations of Fe additions (dark bars) and unamended controls (white bars). Maximum photosynthesis rates ($P^*_{max}$) are shown for surface (S) and deeper subsurface (D) incubations from Stations 5, 13 and 35 (A) and for different levels of incident irradiance (1%, black; 10%, grey; 50%, light grey) for Stations 57 and 66 (B). Initial slope of the *P-E* curve ($\alpha^*$) is shown for the surface and subsurface depths (C) and levels of incident irradiance (D), as is photoacclimation parameter ($E_k$) (E & F).

## Cellular composition of phytoplankton

The POC/PON ratios in the incubations showed considerable variation by day 4, ranging from 3.5 to 8.5 (Table 4). The highest POC/PON ratio was found in the Fe incubations of surface waters from Station 35 that were $NO_3$-limited. Fe addition did not affect the POC/PON ratio in any of the experiments (Table 4), and the lack of Fe effect was confirmed when experiments were analyzed together (Table 5).

The POC/Chl *a* ratios in the incubations ranged from 36 to 229 g $g^{-1}$ (Table 4). Fe addition increased the POC/Chl *a* ratio in the 10% irradiance incubation at Station 66, but did not affect the POC/Chl *a* ratio in any other incubations (Table 4). When all experiments were analyzed together, Fe additions enhanced POC/Chl *a* ratios slightly (1.14-fold) compared to the controls, but this difference was not significant (Table 5). Initial sampling depth did not affect POC/Chl *a* ratios in the incubations, and there was no interaction between depth and Fe effects (Table 5). On the other hand, the POC/Chl *a* decreased at low light incubations, but there was no interaction between Fe and light effects on POC/Chl *a* (Table 5).

## Bacterial productivity

Bacterial productivity in the incubations by day 4 ranged from 36 to 289 pmol Leu uptake $L^{-1}$ $h^{-1}$ (Figure 9). It increased 2 to 17-fold over the course of 4 days in all incubations, except for Station 57 at 1% irradiance, where bacterial productivity dropped slightly from the initial value (compare Figure 9 to Table 2B). Fe addition did not affect bacterial productivity in any of the incubations (Figure 9), including the incubations where Fe addition

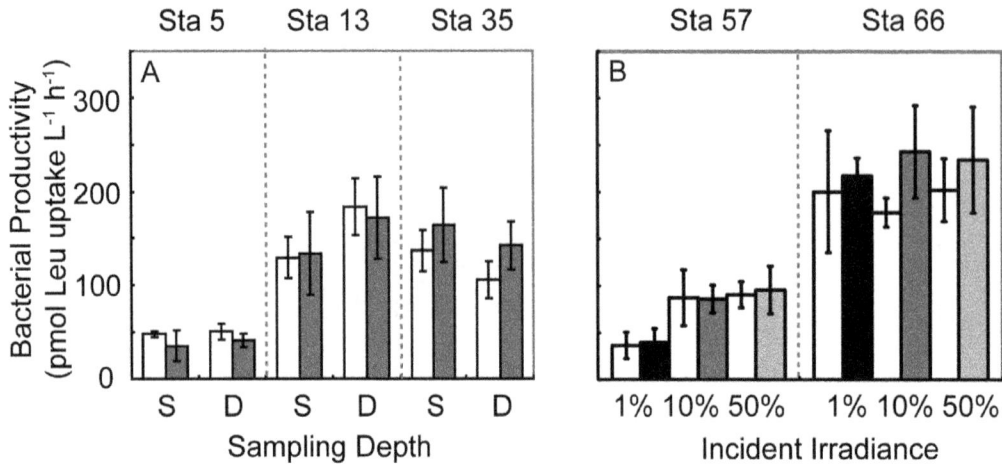

Figure 9
Bacterial productivity in incubation experiments.

Mean and standard deviation of
bacterial productivity by day 4 of
the incubation experiments are
shown for triplicate incubations
of Fe additions (dark bars) and
unamended controls (white bars).
Rates are shown for surface (S) and
deeper subsurface (D) incubations
from Stations 5, 13, and 35 (A)
and for different levels of incident
irradiance (1%, black; 10%, grey;
50%, light grey) for Stations 57
and 66 (B).

enhanced phytoplankton growth (Station 5, S; Station 35, D; Station 57, 10%), suggesting that there were no secondary effects of bacterial productivity on Fe-enhanced phytoplankton growth and photosynthesis. Moreover, original sample depth did not significantly affect bacterial productivity in the incubations, and there was no interaction between Fe and depth effects (Table 5). Light levels in the incubations did not affect bacterial productivity either, despite the big difference in phytoplankton biomass between the 1% irradiance incubations and those at higher light (10% and 50% irradiance) (Table 5).

## Discussion

### Phytoplankton response to Fe additions: carrying capacity versus rate limitation

Phytoplankton productivity in Antarctic polynyas is often assumed to be seasonally Fe limited, with a "winter reserve" of DFe that is gradually depleted over the growing season. The amount of phytoplankton biomass that is supported by this DFe has been referred to as the carrying capacity (Hopkinson et al., 2013). Since all available $NO_3$ was drawn down in both the controls and Fe addition incubation experiments with sufficient light (the 10% and 50% light incubations), the results from the incubation experiments suggest that Fe availability is not limiting the carrying capacity of waters in the ASP. Moreover, previous bioassay experiments conducted later in the growing season showed no response of the phytoplankton biomass to Fe additions in the ASP in February 2009 (Mills et al., 2012). These experiments were performed after the peak in phytoplankton bloom for that season, suggesting that Fe limitation was not the cause of the bloom demise.

In contrast, Fe addition increased phytoplankton growth rates at several stations where we performed bioassay experiments (Station 5, surface water; Station 35, subsurface water; Station 57, 10% light; Figure 6; Table 4) during the build-up of the phytoplankton spring bloom, and enhanced photosynthesis rates in almost all experiments (Figure 8). Thus, even though Fe availability is not limiting the carrying capacity of the ASP (Mills et al., 2012), the Fe effects in the bioassay experiments suggest that Fe availability may limit phytoplankton growth rates in several regions of the ASP. We could not discern physical mechanisms (e.g., MLD, water mass properties) that could explain the spatial variability of Fe effects on phytoplankton growth. The timing of the Fe effects in mid-December to early January was early in the growing season, well before the peak of the phytoplankton bloom in the ASP, which is generally in the middle of January (Arrigo et al., 2012). These findings suggest that the notion of a "winter reserve" of DFe that is gradually depleted over the growing season is an oversimplification, concurring with findings by Sedwick et al. (2011) who showed early season depletion of DFe in the Ross Sea Polynya. Instead, bioavailable Fe must be supplied throughout the growing season to support productive phytoplankton blooms such as those observed in the Amundsen and Ross Sea Polynyas.

Bacterial productivity in the incubation experiments was similar to that reported in the high productiv-ity stations in the central ASP, which was high compared to other Antarctic polynyas (see Williams et al., 2015, for a full description). In contrast to phytoplankton growth and photosynthesis, bacterial productivity did not respond to Fe addition, suggesting that bacteria are not Fe-limited during the early season in the ASP. The Fe demand of bacteria is generally lower than that of phytoplankton, as heterotrophic bacteria lack the photosynthetic apparatus that is rich in Fe (Raven et al., 1990). Moreover, bacteria are smaller, with a larger surface to volume ratio that optimizes nutrient uptake. In addition, despite the increase in bacterial productivity over the four days of incubation, there was no secondary response of bacterial productivity to either the Fe-enhanced phytoplankton growth or photosynthesis rates. Bacterial productivity in surface waters of the ASP showed a positive relationship with phytoplankton biomass (Williams et al., 2015), suggesting bacterial productivity was coupled to phytoplankton biomass and productivity. Likely, the Fe effects on phytoplankton biomass, cellular composition, and photosynthesis rates in our experiments were not sufficient to yield a detectable response in bacterial productivity during the time scale of the incubations.

## Fe availability as the main driver for phytoplankton photosynthesis rates in the ASP

The Fe addition bioassay experiments showed that greater Fe availability increased all photosynthesis parameters, including $F_v/F_m$, $P^*_{max}$, and $\alpha^*$, similar to results from culture experiments on *Phaeocystis antarctica* (Strzepek et al., 2012; Alderkamp et al., 2012b; Van Leeuwe and Stefels, 2007). On the other hand, increased light availability in the bioassay experiments increased $P^*_{max}$, but decreased $\alpha^*$ and $F_v/F_m$. Typically, photoacclimation to high light increases carbon-fixing activity, such as electron transport and the Calvin cycle, resulting in a higher cellular C/Chl $a$ ratio, whereas it decreases the photon capture and photosynthetic efficiency at low light (Falkowski and LaRoche, 1991; MacIntyre et al., 2002). The spatial distribution of $P$-$E$ parameters of surface phytoplankton (upper 10 m) in the ASP showed a positive relationship between $F_v/F_m$, $P^*_{max}$, and $\alpha^*$, indicating that phytoplankton acclimation to Fe availability, rather than light availability, is the main driver of photosynthesis rates in the ASP during the build-up of the phytoplankton bloom.

Surface phytoplankton $F_v/F_m$, $P^*_{max}$, as well as $\alpha^*$, were highest in the southern ASP near the ice shelves (Stations 5 and 57), suggesting that the WW near the GIS, as well as meltwater-laden mCDW from the DIS, are major Fe sources to phytoplankton. On the other hand, $F_v/F_m$, $P^*_{max}$ and $\alpha^*$ were all lower in the central polynya, suggesting that Fe availability limited photosynthesis rates where the phytoplankton biomass was highest. The $F_v/F_m$, $P^*_{max}$ and $\alpha^*$ in the low salinity AASW strongly affected by recent sea ice melt (Stations 34 and 66) were intermediate, suggesting that melting sea ice may be an Fe source (Sedwick and DiTullio, 1997; Lannuzel et al., 2010), albeit not as pronounced as the ice shelves. The $P^*_{max}$ at Stations 5 and 57 close to the ice shelves (3.6 g C g$^{-1}$ Chl $a$ h$^{-1}$) was 1.6-fold higher than in the central polynya (2.2 g C g$^{-1}$ Chl $a$ h$^{-1}$), whereas $\alpha^*$ close to the ice shelves (0.067 g C g$^{-1}$ Chl $a$ h$^{-1}$ [μmol photons m$^{-2}$ s$^{-1}$]$^{-1}$) was 1.9-fold higher than in the central polynya (0.035 g C g$^{-1}$ Chl $a$ h$^{-1}$ [μmol photons m$^{-2}$ s$^{-1}$]$^{-1}$). Both $P^*_{max}$ and $\alpha^*$ close to the ice shelves exceeded values in the controls of the bioassay experiments, and matched those of the Fe treatments, as well as those in *P. antarctica* cultures growing under Fe-replete and saturating light conditions (Arrigo et al., 2010; Mills et al., 2010).

Curiously, there was no direct relationship between DFe concentrations in surface waters and any $P$-$E$ parameter. High $P^*_{max}$ (> 3 g C g$^{-1}$ Chl $a$ h$^{-1}$) and high $\alpha^*$ (> 0.05 g C g$^{-1}$ Chl $a$ h$^{-1}$ [μmol photons m$^{-2}$ s$^{-1}$]$^{-1}$) were observed in waters with DFe as low as 0.11 nmol L$^{-1}$. Similarly, DFe concentrations in the initial waters of the bioassay experiments did not alter the effects of Fe addition on phytoplankton growth. For instance, Fe addition did not affect phytoplankton growth in waters with the lowest initial DFe in this study (surface waters of Station 13, 0.09 nmol L$^{-1}$ DFe), whereas initial DFe in the incubations that did show an Fe effect (Station 5 surface water, Station 35 subsurface water) were approximately twice as high at 0.18 and 0.22 nmol L$^{-1}$ DFe, respectively. Thus, DFe concentrations do not appear to be a good measure of Fe availability for phytoplankton. A lack of direct DFe effects on phytoplankton photosynthesis and growth suggests that either not all DFe was bioavailable to the phytoplankton (Visser et al., 2003) or other sources of bioavailable Fe were present besides DFe (e.g., PFe). In addition, phytoplankton cells may have stored bioavailable DFe in excess of their immediate needs, resulting in Fe-replete cells in Fe-depleted waters.

Bioavailability of at least a fraction of PFe is suggested by the positive relationship between PFe and both $P^*_{max}$ and $\alpha^*$, indicating that PFe contributes to the pool of bioavailable Fe, or becomes bioavailable at high enough rates to support high photosynthesis rates. A small fraction of this PFe may be internal or attached to phytoplankton cells (Twining and Baines, 2013). Several studies show that Fe:C ratios of Fe-replete phytoplankton are a factor of 2 to 10 higher than those of Fe-limited phytoplankton (Twining et al., 2004; Hassler and Schoemann, 2009; Strzepek et al., 2012), suggesting that phytoplankton have the ability to take up DFe in excess of their immediate requirements and store it. Studies with *P. antarctica* suggest that metals may be stored inside *P. antarctica* colonies (Lubbers et al., 1990; Schoemann et al., 2001). The fraction of PFe internal or attached to cells is likely small, given that the highest PFe was observed at the face of the DIS where phytoplankton biomass was low.

Alternatively, extracellular, inorganic PFe in the ASP, such as crustal particles and Fe hydroxides (Hazarin et al., 2014), may be made biologically available through dissolution, photoreduction, and/or biological processing (Barbeau et al., 2003; Boyd and Ellwood, 2010; Rijkenberg et al., 2006, 2008; Boyd and Ellwood, 2010). Since $E_{UML}$ was relatively high throughout the ASP (Figure 3D), photoreduction is likely an active process in surface waters of the ASP. Photoreduction can convert bound Fe(III) species to Fe(II) via ligand to metal charge transfer (Barbeau et al., 2003; Rijkenberg et al., 2006). The *P. antarctica* bloom is associated with an abundance of relatively unsaturated organic Fe-binding ligands (Thuróczy et al., 2012) that may bind the Fe (II), keep it in solution (Rijkenberg et al., 2006, 2008), and make it available to *P. antarctica*. Specifically, polysaccharides enhance Fe bioavailability to both *P. antarctica* and diatoms (Hassler et al., 2011); as the main constituent of the matrix of *Phaeocystis* colonies (Alderkamp et al., 2007), polysaccharides may thus make PFe available to *P. antarctica*.

*Effects of Fe limitation on phytoplankton photosynthesis at different light intensities*

Phytoplankton acclimation to low Fe availability in the ASP will affect their photophysiology because of the high Fe requirements for chlorophyll biosynthesis and the high Fe content of the photosynthetic apparatus and electron transport pathways (Raven, 1990; Maldonado et al., 1999; Strzepek and Price, 2000). In general, acclimation to low Fe decreases the potential to maximize photon capture (Raven, 1990; Greene et al., 1992). Despite this interaction, the incubation experiments showed no interactive effects of Fe and light availability on photosynthesis rates. Two possible reasons for this lack of interaction may be that Antarctic phytoplankton do not increase their Fe quota to photoacclimate to low light (Strzepek et al., 2012), or that light-limited phytoplankton have lower growth rates and therefore require less Fe per unit time, i.e., a lower Fe flux.

For the first explanation for the lack of interaction between Fe and light availability on phytoplankton photosynthesis in the bioassay experiments, we consider photoacclimation strategies. Photoacclimation to low light may be achieved by either increasing the number of PSII reaction centers (RCIIs) or increasing the size of the photosynthetic pigment-containing antenna associated with the RCIIs; both strategies increase the cellular photosynthetic pigment concentrations. Increasing the number of RCIIs, however, increases the Fe requirement because the RCII and downstream electron transport chain are particularly Fe-rich: photosystem II (PSII) contains two or three Fe atoms, the cytochrome $b_6f$ complex (Cyt $b_6f$) contains five Fe atoms, and PSI contains 12 Fe atoms (Raven, 1990). The $\sigma_{PSII}$ is the functional absorption cross section of PSII and can be used to distinguish between the two different photoacclimation strategies. The $\sigma_{PSII}$ is the product of absorption by the PSII antenna pigments and the probability that an exciton within the antenna will cause a photochemical reaction (Mauzerall and Greenbaum, 1989). The amount of pigment associated with each RCII will therefore determine much of the variability in $\sigma_{PSII}$ (Kolber et al., 1988; Suggett et al., 2004). Acclimation responses that alter the ratio of pigment:RCII will change the $\sigma_{PSII}$, whereas acclimation responses that alter the number of RCIIs per cell but not the amount of pigments associated with it will not affect the $\sigma_{PSII}$ (Moore et al., 2006).

Light availability did not affect $\sigma_{PSII}$ in either the incubation experiments (Figure 7D) or the water column (Figure 5C), suggesting that the *P. antarctica*-dominated phytoplankton community did not change its antenna size during photoacclimation to low light, but rather increased the number of RCIIs, which would increase the cellular Fe requirement, contradicting the notion that Antarctic phytoplankton do not increase their Fe quota at low light (Strzepek et al. 2012). Our results match field results of $\sigma_{PSII}$ from the Ross Sea, where $\sigma_{PSII}$ did not increase with depth at *P. antarctica*-dominated stations (Smith et al., 2013). In contrast, *P. antarctica* did increase $\sigma_{PSII}$ under lower growth irradiance in culture experiments (Strzepek et al., 2012). Whether different strains of *P. antarctica* vary in their ability to adjust $\sigma_{PSII}$ or the timescales of photoacclimation in the field are too long to observe changes in $\sigma_{PSII}$ during four-day incubation is unknown.

Similar to light availability, Fe availability did not affect $\sigma_{PSII}$ consistently when control and Fe treatments were compared in the bioassay experiments. Moreover, there was no spatial pattern in $\sigma_{PSII}$ of surface phytoplankton in the ASP, despite the low Fe availability in the central ASP. These observations suggest that *P. antarctica* in the ASP does not increase its antenna size under Fe limitation. These results contrast with the increase in $\sigma_{PSII}$ under Fe limitation reported in culture experiments, where the $\sigma_{PSII}$ of Fe-limited *P. antarctica* was two-fold higher than Fe-replete *P. antarctica* under similar growth irradiance (Strzepek et al., 2012). The maximum response of $\sigma_{PSII}$ to Fe limitation in our incubation experiments was more subtle, with a 1.23-fold increase in control versus Fe treatments. Again, these differences between field observations and culture experiments could be due to (unknown) variability among different strains of *P. antarctica* in their ability to adjust $\sigma_{PSII}$, or to timescales of acclimation.

The second explanation for the lack of interactions between Fe and light availability on phytoplankton photosynthesis rates is that light-limited phytoplankton have lower growth rates and therefore require less Fe per unit time. For our incubation experiments, this reduced requirement would mean that there was enough DFe in the original waters sampled to support slow-growing phytoplankton until day four of the incubation, but not enough to support fast-growing phytoplankton. In the field, if the DFe supply to a phytoplankton bloom is considered in terms of Fe flux, light-limited phytoplankton with low growth rates would require a lower Fe flux than phytoplankton with high growth rates under high light. Thus, if the DFe flux remains constant, phytoplankton under high light availability, as in the central ASP, may experience more Fe stress than light-limited phytoplankton that are growing more slowly, even if high light phytoplankton would have a lower Fe quota.

*Water column productivity in the ASP driven by both Fe and light*

Despite Fe limitation of photosynthesis rates, phytoplankton biomass and water column productivity was high throughout the ASP (2.1–6.5 g C m$^{-2}$ d$^{-1}$). These rates were similar to primary productivity rates reported by Lee et al. (2012) and Yager et al. (2014) during the same period in the ASP, and similar to rates measured later in the season in a previous year (Alderkamp et al., 2012a). Moreover, the high phytoplankton biomass and productivity match the high primary productivity rates calculated from satellite data in the ASP (Arrigo and

Van Dijken, 2003; Arrigo et al., 2012; Yager et al., 2012). In addition, these high water column productivity rates were similar to those in the Pine Island Polynya in the eastern Amundsen Sea (Alderkamp et al., 2012a).

Stations with the highest water column productivity were all located in the moderately fresh AASW layer in the central ASP, where sea ice melt signatures were found enhancing stratification (MLD < 30 m) and increasing light availability. Although we could not find any direct relationship between water column productivity and parameters affecting light availability, such as MLD and $E_{UML}$, autonomous glider observations at high spatial resolution within the central ASP found a positive relationship between phytoplankton biomass and enhanced stratification (shallower MLD) (Schofield et al., 2015), suggesting an important role for light availability in bloom formation. Stratification, resulting in enhanced light availability to phytoplankton, is especially important for bloom development early in the season (i.e., October–November) (Long et al., 2012), when incident irradiance is relatively low due to lower solar elevation and shorter days compared to the sampling period of this study (i.e. December–January).

The high water column productivity in the central ASP coincided with the stations where $P^*_{max}$ and $\alpha^*$ were limited by Fe availability. In order to estimate the effect of Fe-replete photosynthetic parameters on water column productivity in the central ASP, we used the enhanced $P$-$E$ parameters close to the ice shelves (Stations 5 and 57) for water column productivity calculations in the central polynya. This calculation showed that enhanced $P$-$E$ parameters under high Fe availability have the potential to increase water column productivity in the central polynya on average by 1.7 ±0.01-fold, assuming all other factors stay the same (e.g., light availability, temperature, grazing, phytoplankton species composition). Thus, increased Fe availability during the build-up of the bloom would significantly increase the already high rates of phytoplankton productivity in the ASP.

Many Antarctic ice shelves are thinning, including the Dotson Ice Shelf, as the glaciers that feed these ice shelves are accelerating (Pritchard et al., 2009; Rignot et al., 2013). An increase in basal ice shelf melt would most likely be driven by an increase in the amount of mCDW flowing onto the continental shelf and/or an increase in the heat content of mCDW (Jacobs et al., 2011). An increase in meltwater-laden mCDW exiting the ice shelve cavities would destabilize the water column, potentially increasing MLD and decreasing light availability. Early in the growing season when incident irradiance is relatively low, this scenario would potentially delay bloom formation and decrease water column productivity, especially near the ice shelf. However, if the DIS is a source of bioavailable Fe for the ASP, similar to that observed for the Pine Island ice shelf in the PIP (Gerringa et al., 2012), then horizontal diffusivity (Geringa et al., 2012), advective eddy transport (e.g., Årthun et al., 2013), mixing along the Dotson trough (e.g., St-Laurent et al., 2013), and wind- and iceberg-induced mixing (Randall-Goodwin et al., 2015) would ensure a bioavailable Fe flux from the DIS or neighboring ice shelves to the high phytoplankton biomass in the central ASP, where the water column is stabilized by sea ice melt water and higher Fe availability would increase the phytoplankton photosynthesis and water column productivity substantially.

The presence of mCDW has been detected regularly in the troughs of the ASP since observations began (Jacobs et al., 2012; Arneborg et al., 2012; Wåhlin et al., 2013; Ha et al., 2014), suggesting this water mass has regular access to the DIS. However, any seasonal (Thoma et al., 2008; Jacobs et al., 2012; Randall-Goodwin et al., 2015) or annual (Jacobs et al., 2013; Dutrieux et al., 2014) variability in mCDW on the continental shelf will likewise affect the quantity of meltwater-laden mCDW from the DIS, and hence the DFe flux to the ASP. Variability in the DFe flux would affect phytoplankton photosynthetic rates directly according to our results. How phytoplankton productivity across the ASP would be affected by variability in mCDW access to the DIS is unknown, but represents a key question for future investigations.

# References

Ainley DG, Ballard G, Dugger KM. 2006. Competition among penguins and cetaceans reveals trophic cascades in the western Ross Sea, Antarctica. *Ecology* **87**: 2080–2093.

Alderkamp A-C, Buma AGJ, Van Rijssel M. 2007. The carbohydrates of *Phaeocystis* and their degradation in the microbial food web. *Biogeochemistry* **83**: 99–118.

Alderkamp A-C, Kulk G, Buma AGJ, Visser RJW, Van Dijken GL, et al. 2012b. The effect of iron limitation on the photophysiology of *Phaeocystis antarctica* (Prymnesiophyceae) and *Fragilariopsis cylindrus* (Bacillariophyceae) under dynamic irradiance. *J Phycol* **48**: 45–59. doi: 10.1111/j.1529-8817.2011.01098.x.

Alderkamp A-C, Mills MM, Van Dijken GL, Laan P, Thuyróczy C-E, et al. 2012a. Iron from melting glaciers fuels phytoplankton blooms in the Amundsen Sea (Southern Ocean): Phytoplankton characteristics and productivity. *Deep-Sea Res Pt II* **71–76**: 32–48. doi: 10.1016/j.dsr2.2012.03.005.

Arneborg L, Wåhlin AK, Björk G, Liljebladh B, Orsi AH. 2012. Persistent inflow of warm water onto the central Amundsen shelf. *Nature Geosciences* **5**: 876–880. doi: 10.1038/NGEO1644.

Arrigo KR, Lowry KE, Van Dijken GL. 2012. Annual changes in sea ice and phytoplankton in polynyas of the Amundsen Sea, Antarctica. *Deep-Sea Res Pt II* **71–76**: 5–15.

Arrigo KR, Mills MM, Kropuenske LR, Van Dijken GL, Alderkamp A-C, et al. 2010. Photophysiology in two major Southern Ocean taxa: photosynthesis and growth of *Phaeocystis antarctica* and *Fragilariopsis cylindrus* under different irradiance levels. *Integrative Comp Biol* **50**: 950–966.

Arrigo KR, Van Dijken GL. 2003. Phytoplankton dynamics within 37 Antarctic coastal polynya systems. *J Geophys Res-Oceans* **108**(C8): 3271. doi: 10.1029/2002JC001739.

Arrigo KR, Van Dijken GL, Long MC. 2008. Coastal Southern Ocean: A strong anthropogenic $CO_2$ sink. *Geophys Res Lett* **35**: L21602. doi: 10.1029/2008GL035624.

Arrigo KR, Worthen DL, Robinson DH. 2003. A coupled ocean-ecosystem model of the Ross Sea: 2. Iron regulation of phytoplankton taxonomic variability and primary production. *J Geophys Res-Oceans* **108**: 3231. doi:10.1029/2001JC000856.

Årthun M, Holland PR, Nicholls KW, Feltham DL. 2013. Eddy-driven exchange between the open ocean and a sub-ice shelf cavity. *J Phys Oceanogr* **43**: 2372–2387.

Barbeau K, Rue EL, Trick CG, Bruland KW, Butler A. 2003. Photochemical reactivity of siderophores produced by marine heterotrophic bacteria and cyanobacteria based on characteristic Fe(III) binding groups. *Limnol Oceanogr* **48**: 1069–1078.

Boyd PW, Ellwood MJ. 2010. The biogeochemical cycle of iron in the ocean. *Nature Geosciences* **3**: 675–682.

Boyd PW, Jickells T, Law CS, Blain S, Boyle EA, et al. 2007. Mesoscale iron enrichment experiments 1993–2005: Synthesis and future directions. *Science* **315**: 612–617.

Bricaud A, Stramski D. 1990. Spectral absorption-coefficients of living phytoplankton and nonalgal biogenous matter - A comparison between the Peru upwelling area and the Sargasso Sea. *Limnol Oceanogr* **35**: 562–582.

Buma AGJ, De Baar HJW, Nolting RF, Van Bennekom AJ. 1991. Metal enrichment experiments in the Weddell-Scotia Seas - Effects of iron and manganese on various plankton communities. *Limnol Oceanogr* **36**: 1865–1878.

Cisewski B, Strass VH, Losch M, Prandke H. 2008. Mixed layer analysis of a mesoscale eddy in the Antarctic Polar Front Zone. *J Geophys Res-Oceans* **113**: C05017. doi: 10.1029/2007JC004372.

De Baar HJW, Boyd PW, Coale KH, Landry MR, Tsuda A, et al. 2005. Synthesis of iron fertilization experiments: From the iron age in the age of enlightenment. *J Geophys Res-Oceans* **110**: C09S16.

De Jong J, Schoemann V, Mairq N, Mattielli N, Langhorne P, et al. 2013. Iron in land-fast sea ice of McMurdo Sound derived from sediment resuspension and wind-blown dust attributes to primary productivity in the Ross Sea, Antarctica. *Mar Chem* **157**: 24–40. doi: 10.1016/j.marchem.2013.07.001.

DiTullio GR, Garcia N, Riseman SF, Sedwick PN. 2007. Effects of iron concentration on pigment composition in *Phaeocystis antarctica* grown at low irradiance. *Biogeochemistry* **83**: 71–78.

Dutrieux P, De Rydt J, Jenkins A, Holland PR, Ha HK, et al. 2014. Strong sensitivity of Pine Island ice-shelf melting to climatic variability. *Science* **343**: 174–178. doi: 10.1126/science.1244341.

Falkowski PG, LaRoche J. 1991. Acclimation to spectral irradiance in algae. *J Phycol* **27**: 8–14.

Gerringa LJA, Alderkamp A-C, Laan P, Thuróczy C-E, De Baar HJW, et al. 2012. Iron from melting glaciers fuels the phytoplankton blooms in Amundsen Sea (Southern Ocean); iron biogechemistry. *Deep-Sea Res Pt II* **71–76**: 16–31.

Gorbunov MY, Kolber ZS, Falkowski PG. 1999. Measuring photosynthetic parameters in individual algal cells by Fast Repetition Rate fluorometry. *Photosynth Res* **2**: 141–153.

Greene RM, Geider RJ, Kolber ZS Falkowski PG. 1992. Iron-induced changes in light harvesting and photochemical energy-conversion processes in eukaryotic marine-algae. *Plant Physiol* **100**: 565–575.

Ha HK, Wåhlin AK, Kim TW, Lee JH, Lee HJ, et al. 2014. Circulation and modification of warm deep water on the central Amundsen shelf. *J Phys Oceanogr* **44**: 1493–1501.

Harazin KM, Lagerström M, Forsch KO, Severmann S, Sherrell RM. 2014. The metal-to-phosphorous ratio of natural phytoplankton assemblages in the Amundsen Sea Polynya and western Antarctic Peninsula, west Antarctica. 2014. *Ocean Sciences Meeting*. Honolulu, Hawaii.

Hassler CS, Schoemann V. 2009. Bioavailability of organically bound Fe to model phytoplankton of the Southern Ocean. *Biogeosciences* **6**: 2281–2296.

Hassler CS, Scoemann V, Nichols CM, Butler ECV, Boyd PW. 2011. Saccharides enhance iron availability to Southern Ocean phytoplankton. *P Natl Acad Sci USA* **108**: 1076–1081.

Hatta M, Measures CI, Selph KE, Zhou M, Hiscock WT. 2013. Iron fluxes from the shelf regions near the South Shetland Islands in the Drake Passage during the austral-winter 2006. *Deep-Sea Res Pt II* **90**: 89–101. doi: 10.1016/j.dsr2.2012.11.003.

Holm-Hansen O, Lorenzen C, Holmes RW, Strickland JDH. 1965. Fluorometric determination of chlorophyll in Chlorophyta, Chrysophyta, Phaeophyta, Pyrrophyta. *J Cons Perm Inter Explor Mer* **30**: 3–15.

Hopkinson BN, Seegers B, Hatta M, Measures CI, Mitchell BG, et al. 2013. Planktonic C:Fe ratios and carrying capacity in the southern Drake Passage. *Deep-Sea Res Pt II* **90**: 102–111.

Jacobs SS, Giulivi C, Dutrieux P, Rignot E, Nitsche F, et al. 2013. Getz ice shelf melting response to changes in ocean forcing. *J Geophys Res* **118**: 1–17. doi:10.1002/jgrc.20298.

Jacobs SS, Hellmer HH, Jenkins A. 1996. Antarctic ice sheet melting in the Southeast Pacific. *Geophys Res Lett* **23**: 957–960.

Jacobs SS, Jenkins A, Giulivi CF, Dutrieux P. 2011. Stronger ocean circulation and increasing melting under Pine Island Glacier ice shelf. *Nature Geosciences* **4**: 519–523.

Jacobs SS, Jenkins A, Hellmer HH, Giulivi C, Nitsche F, et al. 2012. The Amundsen Sea and the Antarctic Ice Sheet. *Oceanography* **25**: 154–163. doi: 10.5670/oceanog.2012.90.

Jenkins A, Dutrieux P, Jacobs SS, McPhail SD, Perrett JR, et al. 2010. Observations beneath Pine Island Glacier in West Antarctica and implications for its retreat. *Nature Geoscience* **3**: 468–472.

Johnson Z, Barber RT. 2003. The low-light reduction in the quantum yield of photosynthesis: potential errors and biases when calculating the maximum quantum yield. *Photosynth Res* **75**: 85–95.

Kishino M, Takahashi M, Okami N, Ichimura S. 1985. Estimation of the spectral absorption-coefficients of phytoplankton in the sea. *B Mar Sci* **37**: 634–642.

Klunder M, Laan P, Middag R, De Baar HJW. 2011. Dissolved iron in the Southern Ocean (Atlantic sector). *Deep-Sea Res Pt II* **58**: 2678–2694.

Kolber ZS, Prášil O, Falkowski PG. 1988. Measurements of variable chlorophyll fluorescence using fast repetition rate techniques: defining methodology and experimental protocols. *Biochim Biophys Acta* **1367**: 88–106.

Lannuzel D, Schoemann V, De Jong J, Pasquer B, Van der Merwe P, et al. 2010. Distribution of dissolved iron in Antarctic sea ice: Spatial, seasonal, and inter-annual variability. *J Geophys Res-Biogeosciences* **115**: G03022. doi: 10.1029/2009JG001031.

Lee SH, Kim BK, Yun MS, Joo H, Yang EJ, et al. 2012. Spatial distribution of phytoplankton productivity in the Amundsen Sea, Antarctica. *Polar Biol* **35**: 1721–1733.

Lewis MR, Smith JC. 1983. A small volume, short-incubation-time method for measurement of photosynthesis as a function of incident irradiance. *Mar Ecol-Prog Ser* **13**: 99–102.

Long MC, Thomas LN, Dunbar RB. 2012. Control of phytoplankton bloom inception in the Ross Sea, Antarctica, by Ekman restratification. *Global Biogeochem Cy* **26**: GB1006. doi: 10.1029/2010GB003982.

Lubbers GW, Gieskes WWC, DelCastilho P, Salomons W, Bril J. 1990. Manganese accumulation in the high pH micro-environment of *Phaeocystis*-sp (haptophyceae) colonies from the North-Sea. *Mar Ecol-Prog Ser* **59**: 285–293.

MacIntyre HL, Kana TM, Anning T, Geider RJ. 2002. Photoacclimation of photosynthesis irradiance response curves and photosynthetic pigments in microalgae and cyanobacteria. *J Phycol* **38**: 17–38.

Mackey MD, Mackey DJ, Higgins HW, Wright SW. 1996. CHEMTAX - A program for estimating class abundances from chemical markers: Application to HPLC measurements of phytoplankton. *Mar Ecol-Prog Ser* **144**: 265–283.

Maldonado MT, Boyd PW, Harrison PJ, Price NM. 1999. Co-limitation of phytoplankton growth by light and Fe during winter in the NE subarctic Pacific Ocean. *Deep-Sea Res Pt II* **46**: 2475–2485. doi: 10.1016/S0967-0645(99)00072-7.

Mankoff KD, Jacobs SS, Tulaczyk SM, Stammerjohn SE. 2012. The role of Pine Island Glacier ice shelf basal channels in deep-water upwelling, polynyas and ocean circulation in Pine Island Bay, Antarctica. *Ann Glaciol* **53**: 123–128, doi:10.3189/2012AoG60A062.

Marsay CM, Sedwick PN, Dinniman MS, Barrett PM, Mack SL, et al. 2014. Estimating the benthic efflux of dissolved iron on the Ross Sea continental shelf. *Geophys Res Lett* **41**: 7576–7583. doi: 10.1002/2014GL061684.

Mauzerall D, Greenbaum NL. 1989. The absolute size of a photosynthetic unit. *Biochim Biophys Acta* **974**: 119–140.

Mills MM, Alderkamp A-C, Thuróczy C-E, Van Dijken GL, De Baar HJW, et al. 2012. Phytoplankton biomass and pigment responses to Fe amendments in the Pine Island and Amundsen polynyas. *Deep-Sea Res Pt II* **71–76**: 61–76.

Mills MM, Kropuenske LR, Van Dijken GL, Alderkamp A-C, Berg GM, et al. 2010. Photophysiology in two major Southern Ocean phytoplankton taxa: photosynthesis and growth of *Phaeocystis antarctica* (Prymnesiophyceae) and *Fragilariopsis cylindrus* (Bacillariophyceae) under simulated mixed layer irradiance. *J Phycol* **46**: 1114–1127.

Mitchell BG, Brody EA, Holm-Hansen O, McClain C, Bishop J. 1991. Light limitation of phytoplankton biomass and macronutrient utilization in the Southern Ocean. *Limnol Oceanogr* **36**: 1662–1677.

Mitchell BG, Kiefer DA. 1988. Chlorophyll-alpha specific absorption and fluorescence excitation-spectra for light-limited phytoplankton. *Deep-Sea Res Pt I* **35**: 639–663.

Moore CM, Suggett DJ, Hickmann AE, Kim Y-N, Tweddle JF, et al. 2006. Phytoplankton photoacclimation and photo-adaptation in response to environmental gradients in a shelf sea. *Limnol Oceanogr* **51**: 936–949.

Mu L, Stammerjohn SS, Lowry KE, Yager PY. 2014. Spatial variability of surface $pCO_2$ and air-sea $CO_2$ flux in the Amundsen Sea Polynya, Antarctica. *Elem Sci Anth* **2**: 000036. doi: 10.12952/journal.elementa.000036.

Planquette H, Sherrell RM. 2012. Sampling for particulate trace element determination using water sampling bottles: methodology and comparison to in situ pumps. *Limnol Oceanogr Methods* **10**: 367–388.

Platt T, Gallegos CL, Harrison WG. 1980. Photoinhibition of photosynthesis in natural assemblages of marine-phytoplankton. *J Mar Res* **38**: 687–701.

Pritchard HD, Arthern RJ, Vaughan DG, Edwards LA. 2009. Extensive dynamic thinning on the margins of the Greenland and Antarctic ice sheets. *Nature* **461**: 971–975.

Raiswell R. 2011. Iceberg-hosted nanoparticulate Fe in the Southern Ocean: Mineralogy, origin, dissolution kinetics and source of bioavailable Fe. *Deep-Sea Res Pt II* **58**: 1364–1375.

Raiswell R, Benning LG, Tranter M, Tulaczyk S. 2008. Bioavailable iron in the Southern Ocean: the significance of the iceberg conveyor belt. *Geochem Trans* **9**: 7. doi: 10.1186/1467-4866-9-7.

Raiswell R, Tranter M, Benning LG, Siegert M, De'ath R, et al. 2006. Contributions from glacially derived sediment to the global iron (oxyhydr)oxide cycle: Implications for iron delivery to the oceans. *Geochim Cosmochim Ac* **70**: 2765–2780.

Randall-Goodwin E, Meredith PM, Sherrell RM, Yager PL, Alderkamp AC, et al. 2015. Meltwater distributions in the Amundsen Sea Polynya, Antarctica. *Elem Sci Anth:* under review for the ASPIRE Special Feature.

Raven JA. 1990. Predictions of Mn and Fe use efficiencies of phototrophic growth as a function of light availability for growth and of C assimilation pathway. *New Phytol* **116**: 1–18.

Rignot E, Jacobs SS, Mouginot J, Scheuchi B. 2013. Ice-shelf melting around Antarctica, *Science* **341**: 266–270.

Rijkenberg MJA, Gerringa LJA, Carolus VE, Velzeboer I, De Baar HJW. 2006. Enhancement and inhibition of iron photoreduction by individual ligands in open ocean seawater. *Geochim Cosmochim Acta* **70**: 2790–2805.

Rijkenberg MJA, Gerringa LJA, Timmermans KR, Fischer AC, Kroon KJ, et al. 2008. Marine diatoms enhance the reactive iron pool by modification of iron-binding ligands and stimulation of iron-photoreduction. *Mar Chem* **109**: 29–44.

Riley GA. 1957. Phytoplankton of the North Central Sargasso Sea, 1950–52. *Limnol Oceanogr* **2**: 252–270.

Sarmiento JL, Slater R, Barber R, Bopp L, Doney SC, et al. 2004. Response of ocean ecosystems to climate warming. *Global Biogeochem Cy* **18**: GB3003. doi: 10.1029/2003GB002134.

Schoemann V, Wollast R, Chou L, Lancelot C. 2001. Effects of photosynthesis on the accumulation of Mn and Fe by *Phaeocystis* colonies. *Limnol Oceanogr* **46**: 1065–1076.

Schofield O, Miles T, Alderkamp A-C, Lee SH, Haskins C, et al. 2015. *In situ* phytoplankton distributions in the Amundsen Sea Polynya measured by autonomous gliders. *Elem Sci Anth:* under review for the ASPIRE Special Feature.

Sedwick PN, DiTullio GR. 1997. Regulation of algal blooms in Antarctic shelf waters by the release of iron from melting sea ice. *Geophys Res Lett* **24**: 2515–2518.

Sedwick PN, DiTullio GR, Mackey DJ. 2000. Iron and manganese in the Ross Sea, Antarctica: Seasonal iron limitation in Antarctic shelf waters. *J Geophys Res-Oceans* **105**: 11321–11336.

Sedwick PN, Marsay CM, Sohst BM, Aguilar-Islas AM, Lohan MC, et al. 2011. Early season depletion of dissolved iron in the Ross Sea polynya: Implications for iron dynamics on the Antarctic continental shelf. *J Geophys Res-Oceans* **116**: C12019. doi: 10.1029/2010JC006553.

Shaw TJ, Raiswell R, Hexel CR, Vu HP, Moore WS, et al. 2011. Input, composition, and potential impact of terrigenous material from free-drifting icebergs in the Weddell Sea. *Deep-Sea Res* **58**:1376–1383.

Sherrell RM, Lagerström M, Forsch K, Stammerjohn SE, Yager PL. 2015. Dynamics of dissolved iron and other bioactive trace metals (Mn, Ni, Cu, Zn) in the Amundsen Sea Polynya, Antarctica. *Elem Sci Anth:* under review for the ASPIRE Special Feature.

Smith WO, Barber DG, eds. 2007. *Polynyas: Windows to the World.* Amsterdam: Elsevier Science. (Elsevier Oceanography Series. Vol. 74). 474 pp.

Smith WO, Tozzi S, Long MC, Sedwick PN, Peloquin JA, et al. 2013. Spatial and temporal variations in variable fluorescence in the Ross Sea (Antarctica): Oceanographic correlates and bloom dynamics. *Deep-Sea Res Pt I* 79: 141–155.

St-Laurent P, Klinck JM, Dinniman MS. 2013. On the role of coastal troughs in the circulation of warm Circumpolar Deep Water on Antarctic shelves. *J Phys Oceanogr* 43: 51–64.

Strzepek RF, Harrison PJ. 2004. Photosynthetic architecture differs in coastal and oceanic diatoms. *Nature* 431: 689–692.

Strzepek RF, Hunter KA, Frew RD, Harrison PJ, Boyd PW. 2012. Iron-light interactions differ in Southern Ocean phytoplankton. *Limnol Oceanogr* 57: 1182–1200. doi: 10.4319/lo.2012.57.4.1182.

Strzepek RF, Price NM. 2000. Influence of irradiance and temperature on the iron content of the marine diatom *Thalassiosira weissflogii* (Bacillariophyceae). *Mar Ecol-Prog Ser* 206: 107–117. doi: 10.3354/meps206107.

Suggett DJ, MacIntyre HL, Geider RJ. 2004. Evaluation of biophysical and optimal determinations of light absorption by photosystem II in phytoplankton. *Limnol Oceanogr Methods* 2: 316–332.

Tagliabue A, Arrigo KR. 2005. Iron in the Ross Sea: 1. Impact on $CO_2$ fluxes via variation in phytoplankton functional group and non-Redfield stoichiometry. *J Geophys Res-Oceans* 110: C03009. doi: 10.1029/2004JC002531.

Thoma M, Jenkins A, Holland D, Jacobs SS. 2008. Modelling Circumpolar Deep Water intrusions on the Amundsen Sea continental shelf, Antarctica. *Geophys Res Lett* 35: 1–6. doi:10.1029/2008GL034939.

Thuróczy C-E, Alderkamp A-C, Laan P, Gerringa LJA, De Baar HJW, et al. 2012. Key role of organic complexation of iron in sustaining the phytoplankton blooms in the Pine Island and Amundsen Polynyas (Southern Ocean). *Deep-Sea Res Pt II* 71–76: 49–60.

Twining BS, Baines SB. 2013. The trace metal composition of marine phytoplankton. *Ann Rev Mar Sci* 5: 191–215.

Twining BS, Baines SB, Fisher NS, Landry MR. 2004. Cellular iron contents of plankton during the Southern Ocean Iron Experiment (SOFeX). *Deep-Sea Res Pt I* 51: 1827–1850.

Van Hilst CM, Smith WO. 2002. Photosynthesis/irradiance relationships in the Ross Sea, Antarctica, and their control by phytoplankton assemblage composition and environmental factors. *Mar Ecol-Prog Ser* 226: 1–12.

Van Leeuwe MA, Stefels J. 2007. Photosynthetic responses in *Phaeocystis antarctica* towards varying light and iron conditions. *Biogeochemistry* 83: 61–70.

Vassiliev IR, Kolber Z, Wyman KD, Mauzerall D, Shukla VK, et al. 1995. Effects of iron limitation on photosystem-II composition and light utilization in *Dunaliella tertiolecta*. *Plant Physiol* 109: 963–972.

Vernet M, Sines K, Chakos D, Ekern L. 2011. Impacts on phytoplankton dynamics by free-drifting icebergs in the NW Weddell Sea. *Deep-Sea Res Pt II* 58: 1422–1435.

Visser F, Gerringa LJA, Van der Gaast SJ, De Baar HJW, Timmermans KR. 2003. The role of reactivity and iron content of aerosol dust on growth rates of two Antarctic diatom species. *J Phycol* 39: 1085–1094.

Wåhlin AK, Kalén O, Arneborg L, Björk G, Carvajal GK, et al. 2013. Variability of warm deep water inflow in a submarine trough on the Amundsen Sea shelf. *J Phys Oceanogr* 43: 2054–2070. doi: 10.1175/JPO-D-12-0157.1.

Webb WL, Newton M, Starr D. 1974. Carbon dioxide exchange of *Alnus rubra*. A mathematical model. *Oecologia* 17: 281–291.

Williams CM, Dupont A, Post AF, Riemann L, Dinasquet J, et al. 2015. Pelagic microbial heterotrophy in response to a highly productive bloom of the marine haptophyte *Phaeocystis antarctica* in the Amundsen Sea Polynya. *Elem Sci Anth:* under review for the ASPIRE Special Feature.

Wright SW, Jeffrey S, Mantoura R, Lewellyn C, Bjornland C, et al. 1991. An improved HPLC method for the analysis of chlorophylls and carotenoids from marine phytoplankton. *Mar Ecol-Prog Ser* 77: 183–196.

Wright SW, Thomas DP, Marchant HJ, Higgins HW, Mackey MD, et al. 1996. Analysis of phytoplankton of the Australian sector of the Southern Ocean: Comparisons of microscopy and size frequency data with interpretations of pigment HPLC data using the 'CHEMTAX' matrix factorisation program. *Mar Ecol-Prog Ser* 144: 285–298.

Wright SW, Van den Enden RL, Pearce I, Davidson AT, Scott FJ, et al. 2010. Phytoplankton community structure and stocks in the Southern Ocean (30–80° E) determined by CHEMTAX analysis of HPLC pigment signatures. *Deep-Sea Res Pt I* 57: 758–778.

Yager PL, Sherrell RM, Alderkamp AC, Ingall ED, Ducklow HW. 2014. Net community production and export in the Amundsen Sea Polynya (western Antarctica); with comparisons to Arctic polynyas and a link to climate sensitivity. 2014. *Ocean Sciences meeting.* Honolulu, Hawaii.

Yager PL, Sherrell RM, Stammerjohn SE, Alderkamp A-C, Schofield O, et al. 2012. ASPIRE: The Amundsen Sea Polynya International Research Expedition. *Oceanography* 25: 40–53. doi: 10.5670/oceanog.2012.73.

Zapata M, Jeffrey SW, Wright SW, Rodriguez F, Garrido JL, et al. 2004. Photosynthetic pigments in 37 species (65 strains) of Haptophyta: implications for oceanography and chemotaxonomy. *Mar Ecol-Prog Ser* 270: 83–102.

## Contributions

- Substantial contributions to conception and design: ACA, GLD, PLY, KRA
- Acquisition of data: ACA, GLD, KEL, TLC, ML, RMS, CH, ER, SES
- Analysis and interpretation of data: ACA, GLD, KEL, TLC, RMS, OS, SES, PLY, KRA
- Drafting the article or revising it critically for important intellectual content: all Authors
- Final approval of the version to be published: all Authors

Acknowledgments

We thank the captain, crew, technicians, and cruise participants of the NBP 10-05 ASPIRE cruise for their help and Jennifer Vreeland for HPLC analysis. Loes Gerringa and two anonymous reviewers are acknowledged for helpful comments on an earlier version. ASPIRE was part of "Oden Southern Ocean" (SWDARP 2010/11) organized by the Swedish Polar Research Secretariat and National Science Foundation Office of Polar Programs.

Funding information

This research was supported by the National Science Foundation Office of Polar Programs, Antarctic Organisms (ANT-0944727 to KRA, ANT-0839069 to PY, ANT-0838995 to RMS and OS, and ANT-0838975 to SS).

Competing interests

Authors declare that there are no competing interests.

# 6

# *In situ* measurements of thermal diffusivity in sediments of the methane-rich zone of Cascadia Margin, NE Pacific Ocean

Kira Homola [1,2]* • H. Paul Johnson [1] • Casey Hearn [1,2]

[1]School of Oceanography, University of Washington, Seattle, Washington, United States
[2]Graduate School of Oceanography, University of Rhode Island, Narragansett, Rhode Island, United States
*khomola@my.uri.edu

## Abstract

Thermal diffusivity (TD) is a measure of the temperature response of a material to external thermal forcing. In this study, TD values for marine sediments were determined *in situ* at two locations on the Cascadia Margin using an instrumented sediment probe deployed by a remotely operated vehicle. TD measurements in this area of the NE Pacific Ocean are important for characterizing the upslope edge of the methane hydrate stability zone, which is the climate-sensitive boundary of a global-scale carbon reservoir. The probe was deployed on the Cascadia Margin at water depths of 552 and 1049 m for a total of 6 days at each site. The instrumented probe consisted of four thermistors aligned vertically, one sensor exposed to the bottom water and one each at 5, 10, and 15 cm within the sediment. Results from each deployment were analyzed using a thermal conduction model applying a range of TD values to obtain the best fit with the experimental data. TD values corresponding to the lowest standard deviations from the numerical model runs were selected as the best approximations. Overall TDs of Cascadia Margin sediments of 4.33 and $1.15 \times 10^{-7}$ $m^2\,s^{-1}$ were calculated for the two deployments. These values, the first of their kind to be determined from *in situ* measurements on a methane hydrate-rich continental margin, are expected to be useful in the development of models of bottom-water temperature increases and their implications on a global scale.

**Domain Editor-in-Chief**
Jody W. Deming University of Washington

**Associate Editor**
Laurenz Thomsen, Jacobs University Bremen, Germany

**Knowledge Domains**
Earth and Environmental Science
Ocean Science

## 1. Introduction

Thermal diffusivity (TD) is defined as the ratio of the ability of a solid to conduct thermal energy to its capacity for storing thermal energy (Incropera et al., 2013); this ratio best describes how the solid responds to time-varying external thermal forcing. The TD of marine sediments is very poorly characterized due to the challenge of acquiring quantitative *in situ* values in an extreme environment. Previous measurements have used laboratory fabricated sediments or sediments recovered from coring (Hurwitz et al., 2012); methods which do not reproduce the *in situ* environment due to either mechanical disturbance or the ephemeral nature of pore water hydrate (liquid-gas) phases. Furthermore, a time-series of measurements on the order of days or longer is required to accurately compute TD for penetration depths of adequate length to be representative of the sediment column. Acquisition of such data is possible, however, with the use of a remotely operated vehicle (ROV) or similar submersible. To our knowledge, *in situ* thermal diffusivity measurements have not been obtained previously from any portion of the Cascadia Margin (Canadian, Washington, Oregon, northern California), although they have been made in other environments and in other sediment types (Jackson and Richardson, 2000; Wheatcroft et al., 2007). Near shore measurements of sediment TD are also uncommon, although the acquisition procedures are far less demanding, as described by Thomson (2010) for the collection of sediment TD values in the very shallow coastal waters of Washington State.

The goal of this experiment was to determine the range of TD values for marine sediments in 500 to 1100 m of water, which includes the upslope boundary of the methane hydrate stability zone on the Cascadia Margin in the NE Pacific. The Washington portion of the Cascadia North American margin is a focus area for two

US National Science Foundation programs (GeoPRISM and EarthScope). The accretionary wedge on this margin has been described at considerable length in the literature (e.g. Schmalzle et al., 2014; McCrory et al., 2014 and references within), as have the sediment types and physical properties for the alternating layers of pelagic clays and sandy silt turbidites on the Washington margin (see Davis and Hyndman, 1989; Atwater et al., 2014 and references within). Heat flow on the Washington margin varies from 120 mW m$^{-2}$ at the deformation front to less than 50 mW m$^{-2}$ near the edge of the shallow shelf (Johnson et al., 2013).

Characterizing the methane hydrate stability zone at mid-latitude continental margins has global significance, as methane hydrate deposits are the most climate-sensitive reservoirs of carbon on Earth. Globally, an estimated 99% of the hydrate reservoir is held in continental margin sediments at depths below 500 m (Collett et al., 2009; Ruppel, 2011). Slight increases in near-bottom water temperatures can potentially destabilize large quantities of hydrate, releasing bubbles of methane gas with 30 times the greenhouse potency of $CO_2$ (Denman et al., 2007; Riedel et al., 2010). One consequence of global warming over the past 100 years is the impact of rising temperatures of the near-bottom seawater that is in physical contact with the large deposits of methane hydrate present on continental slopes. Because the stability of methane hydrate within these massive carbon reservoirs depends on both temperature and pressure, warming seawater will cause dissociation of the hydrate phase and release of $CH_4$ gas on a time scale of only decades (Boswell and Collett, 2011; Ruppel, 2011; Phrampus and Hornbach, 2012). The time scale over which these thermal disturbances propagate into the sediment-hosted methane hydrate deposits depends directly on the thermal diffusivity of the uppermost seafloor. Sediment thermal diffusivity is a physical property that is only poorly known from laboratory studies; very few measurements have been made *in situ*.

Previous studies of contemporary methane hydrate decomposition have focused on high-latitude Arctic regions (Berndt et al., 2014; Westbrook et al., 2009), although more recent studies have shown that similar hydrate dissociation is occurring at mid-latitudes, with large volumes of methane currently being released on the Cascadia portion of the North American continent (Hautala et al., 2014). This release of oxidized carbon from hydrate decomposition can lower the pH of surrounding seawater, threatening local marine organisms sensitive to acidification. Key to understanding the dynamic upslope limit of methane hydrate stability is the rate of heat transfer from warming bottom water into the overlying sediments; accurate values of thermal diffusivity are required to advance this field of inquiry.

## 2. Methods

TD measurements were acquired during the R/V ATLANTIS cruise AT26-04, where the primary research goal was to study heat flow and fluid flux on the Washington continental margin (Johnson et al., 2013). During this cruise, an instrumented fiberglass probe was inserted into the seafloor for approximately 6 days at each of two sites on the continental slope; one deployment was at 552 m depth and the other at 1049 m. Site depths were selected to include the upper limit of methane hydrate stability and an additional region well below this depth to provide a control for future hydrate stabilities. Prior to deployment, high-resolution swath bathymetry and water column data were inspected to select sites with uniform low-slope bathymetry and no methane bubbles detected in the water-column data. In addition, a visual inspection of the selected deployment locations were performed via the high-resolution camera of the ROV Jason II; the probe was deployed only in flat, heavily sedimented regions with no visible evidence of hard surface material exposures (carbonate or hydrate) or fluid discharge.

The TD probe was approximately one meter in length and instrumented with four temperature sensors; it was inserted using the manipulator of the ROV Jason II. The probe was designed to hold four Onset TidbiT v2 temperature loggers at three sub-surface spacings: 5 cm, 10 cm, and 15 cm. To reduce vertical resistance during insertion and provide support during deployment, a $5.1 \times 5.1$ cm fiberglass L-channel with low thermal conductivity served as a rigid backbone for the probe. Mounting configuration ensured that no two consecutive sensors would lie directly above another to minimize insertion disturbance to the sediment. A schematic of the probe is shown in Figure 1.

The solid epoxy-encased TidbiT Version 2 sensors had a rated absolute accuracy of ± 0.21 °C, resolution of 0.02 °C, and a factory-rated depth limit of only 305 m. As deployment depths would exceed this limit, a pre-cruise pressure test was necessary to identify sensors that might fail. All four sensors were pressure-tested to a simulated depth of 700 m in the Pressure Test Vessel of the School of Oceanography at the University of Washington. All sensors passed the pressure test and were certified for use in the field.

All temperature sensors were calibrated prior to the cruise, as required for marine temperature sensors (Johnson et al., 2010). This laboratory calibration procedure included measurement of the offset of each thermistor with respect to a primary sensor (an Antares thermistor) by long-term soaking in a stirred ice-bath in a sediment core cold room (4°C). This calibration was done after the sensors had been placed in the high pressure vessel to accurately determine their sensitivity when deployed at water depths deeper than their factory certification. These offset values are reported in Table S1. For the calculation of thermal diffusivity, the need for highly accurate relative temperatures is much greater than the absolute temperatures

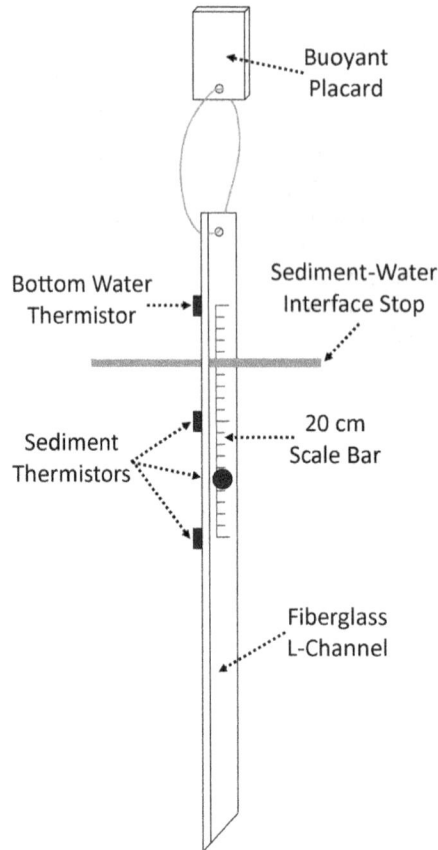

Figure 1

Line drawing of the assembled probe.

The sensors and their brackets are configured with 5 cm spacing. The gray bar, 30 cm long, is used to mark sediment-water interface during insertion. The scale bar on the probe is 20 cm long with 1 cm intervals.

measured, as it is the difference between sensor temperatures over the different measurement depths that determines the thermal diffusivity.

## 3. Data

The first probe deployment, at a water depth of 552 m, began on August 4, 2013 and lasted 147 h; the second deployment, at a depth of 1049 m, began on August 15 and lasted 140 h (Figure 2). The specifics of both deployments are listed in Table 1. All four thermistors logged temperature at 1 minute intervals from their respective positions on the probe. This configuration ensured that one sensor remained 5 cm above the sediment-water interface (gray bar in Figure 1), that the first buried sensor was 5 cm below the interface, and the last two sensors were at consecutive 5 cm depth intervals deeper within the sediment. Upon recovery of the probe, the calibration offsets were applied to the data acquired from each individual sensor.

The maximum observed temperature variation for a single sensor during the first deployment was ± 0.4°C, more than 25 times the resolution of the sensor. This bottom water temperature variation was sufficient to cause a fluctuation of ± 0.15°C at 15 cm, the maximum depth sampled. This amplitude variation is 7.5 times the resolution of the calibrated sensor, resulting in a dataset with statistically significant values for calculating TD at this site. A series of oscillations in bottom water temperature with a clear tidal period of approximately 24 h were observed in these data. There was also a systematic decrease in bottom water temperature observed over the full duration of the deployment (Figure 3). This observed long-term trend could be due to seasonal variability or other long time scale phenomena not fully captured in the six day window.

Table 1. Specifics of TD probe deployments

| Deployment | Latitude | Longitude | Depth (m) | Duration[a] (h) |
|---|---|---|---|---|
| 1 | 46.84954 | -124.9574 | 552 | 147 |
| 2 | 46.78218 | -125.2637 | 1049 | 140 |

[a]Duration indicates total hours the probe was deployed at each location.

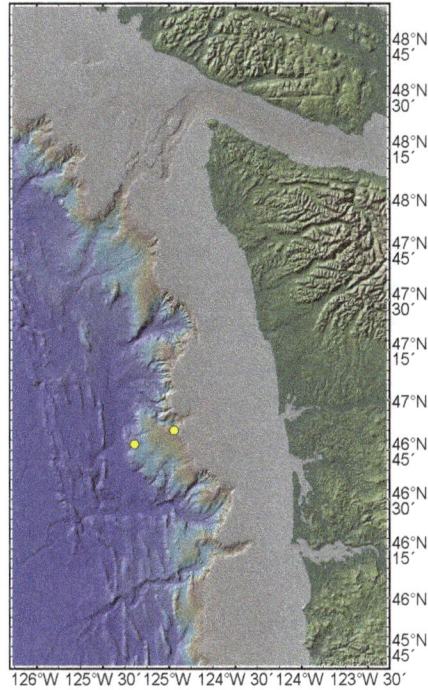

**Figure 2**

**Locations (yellow circles) of probe deployments on the Cascadia Margin.**

The eastern location (Deployment 1) was at a depth of 552 m; the western location (Deployment 2), at a depth of 1049 m.

The maximum temperature range for the diurnal variation of ± 0.25°C observed during the second, deeper deployment was similar to the first dataset. This 0.15°C loss of variation between sites only reduced the ratio between sensor resolution and magnitude of water temperature variation from a factor of 25 to 20. Although smaller in amplitude, the frequency of the bottom water temperature oscillations at the deeper site was unexpectedly higher than at the shallower site, with an average period of 7 h (Figure 4). The smaller amplitude temperature variation at the second site, combined with the much higher frequency of temperature oscillations, resulted in a considerably attenuated signal at greater depth in the sediment, with the 15 cm-deep sensor only observing two temperature steps through the entire deployment.

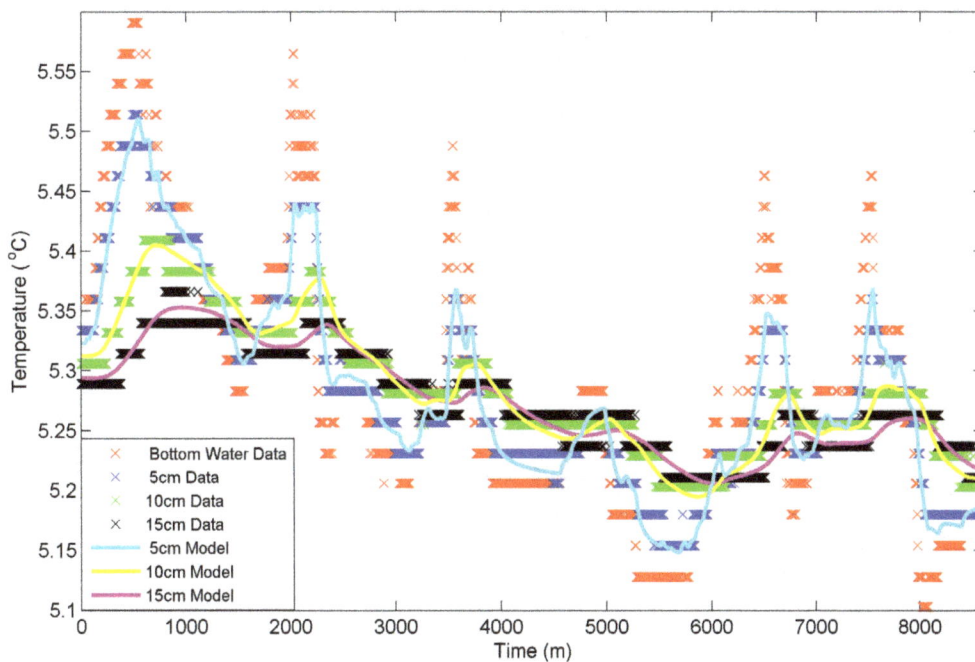

**Figure 3**

**Temperature versus time for Deployment 1 at a depth of 552 m.**

The measured temperatures, with offsets applied (x markers), and the best fits of each model set (solid lines) are shown for each depth. The bottom water data set has no model fit, as it provides the forcing function for the other models.

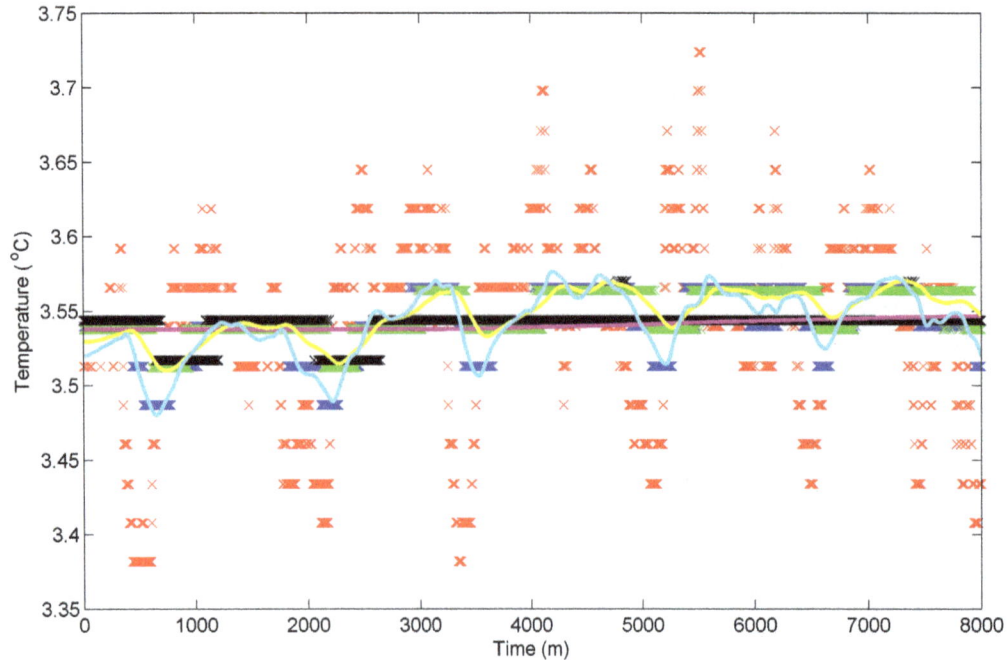

**Figure 4**
Temperature versus time for Deployment 2 at a depth of 1049 m.

See Figure 3 for legend and explanation of symbols.

## 4. Results and discussion

To estimate temperatures within the sediment column at depth $z$ and time $t$ due to bottom water thermal forcing, a range of TD values was modeled in Matlab Version R2011a using the differences between temperature at a given time from the bottom water sensor and the sensor at the depth of interest; see Text S1 for details of method. The program was custom-written and is also included in the Text S1. The analysis was run with 100 input TD values; the standard deviation (SD) between the model and true temperature values at each depth was calculated for each of the input TDs. These SDs were then plotted for the two deployments (Figure 5).

For each depth, the TD value corresponding to the lowest standard deviation between the modeled and observed temperature variations was chosen as the best approximation for the true TD. This process was repeated for the data from the second deployment, resulting in a total of six TD values; two for each sensor depth within the sediment column (Table 2). The model fit, calculated using the chosen TD value, is plotted for each sediment depth on Figures 3 and 4. The discretization of input TD values introduced a maximum error of 3.26% for each final TD value, as the nature of an input matrix cannot resolve an exact output value.

In order to compare the overall variation in thermal diffusivity between the two deployment depths, a second analysis was performed. This computation followed the same procedure as for the previous analysis, but the datasets from all three sensor depths at each site were analyzed using the same TD value. For each artificial TD value produced by the thermal model, the standard deviations were summed for all three depth intervals. Thus, for each artificial TD value, a measure of its fit across the full dataset was obtained. The TD value with the best fit (lowest summed deviation) for the full dataset was then chosen as the best representative value for the depth-integrated profile for each site. The purpose of this final technique was to reduce the noise for a series of calculations in what is assumed to be a 15-cm thick sediment layer of homogenous thermal diffusivity. The results yielded a TD of $4.33 \times 10^{-7}$ m² s⁻¹ for the first shallower location and $1.15 \times 10^{-7}$ m² s⁻¹ for the second, deeper location.

Table 2. TD values that correspond to the lowest standard deviation for each model run

| Deployment | 1 | 2 |
|---|---|---|
| Depth (cm) | Thermal diffusivity (x $10^{-7}$ m² s⁻¹) | |
| 5 | 7.05 | 1.00 |
| 10 | 2.66 | 1.63 |
| 15 | 2.54 | 0.215 |
| Overall[a] | 4.33 | 1.15 |

[a]The last row (overall depth) shows the results from the overall analysis (see text).

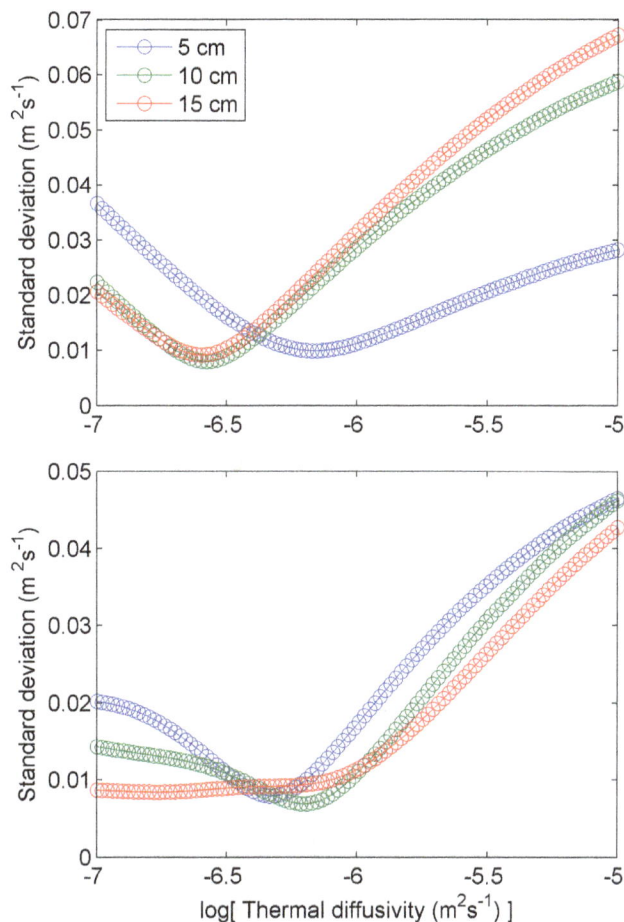

Figure 5

Standard deviation of measured minus modeled values versus (log) TD used in the model calculations.

Colors are used to indicate sensor depth in the sediments. Top panel shows the results from the first d eployment ( water d epth o f 554 m); bottom panel shows the results from the second deployment (water depth of 1049 m).

The magnitude of the temperature change recorded by the sensors has a direct effect on the accuracy of our final result. Thus, for the model runs where the corresponding dataset had greater variability, the *range* of calculated standard deviation values is likewise greater, making possible the selection of a minimum value with a high degree of confidence (see Figure 5). However, for the runs where the variability of the corresponding dataset was low (specifically the 15-cm depth of the second deployment), there was almost no difference in the standard deviation values, producing an unreliable TD estimate. For the two deployments conducted during this project, only the 15-cm sensor from the second deployment fell into this unreliable category, as all other sensors recorded significant variability and produced distinct standard deviation curves. Here, our significance criterion was a variation in temperature at least three times the resolution of the sensor over a temperature oscillation cycle. For our experiment, this translates to a change in temperature of at least 0.06°C in 14 h or less. The accuracy of our results were also affected by the insertion of the probe, as differences between the true and model sensor depths have the potential to introduce errors. A full sensitivity analysis was performed to address this and is included in the Text S1.

Turcotte and Schubert (1982) described a simple method for estimating penetration depth (also called diffusion length) of external thermal forcing:

$$L = \sqrt{\alpha * \Delta t} \tag{1}$$

Here L is the diffusion length, $\alpha$ is the overall thermall diffusivity found at each deployment locations, and $\Delta t$ is the time interval of interest. Calculations using overall diffusivity values from each deployment were performed for six different time intervals of interest (listed in Table 3). Penetration depths of 0.10, 3.69, and 11.68 m for bottom water forcing periods of $\Delta t$ = 7 h, 1 y, and 10 y using the overall diffusivity value were calculated from the first deployment (552 m water depth). The estimates for the second deployment (1049 m water depth) found diffusion lengths of 0.05, 1.90, and 6.02 m for time intervals of 7 h, 1 y, and 10 y. Though the diffusion length solutions to this equation are low in accuracy and do not account for expected variations in physical properties with depth due to compaction, they provide a reasonable first-order approximation based on actual *in situ* thermal diffusivity measurements and highlight the potential for deeper penetration of warming bottom waters at the upper boundary of this methane hydrate zone.

Table 3. Diffusion lengths for time intervals of interest[a]

| Deployment | 1 | 2 |
|---|---|---|
| Time interval | Diffusion length (m) | |
| 7 h | 0.10 | 0.05 |
| 24 h | 0.19 | 0.10 |
| 1 y | 3.69 | 1.90 |
| 10 y | 11.68 | 6.02 |
| 40 y | 23.37 | 12.04 |
| 100 y | 36.95 | 19.04 |

[a]Lengths were calculated using the method outlined in Turcotte and Schubert (1982) with the overall TD values from Deployments 1 and 2 (Table 2).

Recent studies (Pohlman et al., 2009; Phrampus and Hornbach, 2012; Brothers et al., 2014; Berndt et al., 2014; Hautala et al., 2014) have shown that increasing seawater temperatures at mid-water depths on continental margins globally are releasing methane hydrate-derived carbon at flux rates that can provide positive feedback to the present level of global warming. In some cases, hydrate-induced slope failures on the continental margin slopes can produce tsunamis capable of inundating coastal communities (Rao et al., 2002; Lopez et al., 2010). The Cascadia Margin in the NE Pacific has both over-steepened slopes and methane-rich sediments containing abundant solid hydrate that is vulnerable to warming-induced slope failure (Riedel et al., 2002; Booth-Rea et al., 2008; Torres et al., 2009). There is considerable societal need to examine the tsunami hazards associated with the present and on-going phase dissociation associated with ocean warming: our current experimental values of *in situ* thermal diffusivity can contribute to the boundary conditions of these models.

## 5. Conclusions

This study aimed to provide estimates of the thermal diffusivity of marine sediments on the Cascadia Margin through *in-situ* measurements acquired using a simple, yet highly functional device. The resulting, new *in situ* TD data for sediments located on a methane-rich continental margin span the relevant ocean depth interval where gas hydrates are known to be dissociating (Torres et al., 2009; Johnson et al., 2013). Eight TD values were obtained: three depth-sptecific values for each of two deployment locations, and two depth-independent values that characterize the surface sediments of the two deployment sites. The first deployment, at a water depth of 552 m, had an overall TD of $4.33 \times 10^{-7}$ m$^2$ s$^{-1}$, while the second deployment, at a water depth of 1049 m, had an overall TD value of $1.15 \times 10^{-7}$ m$^2$ s$^{-1}$. Using a simple analytical model for heat transfer, these diffusivity values were used to predict 1/e thermal penetration depths. These penetration depths can be improved using more complex, realistic numerical models, but are the first estimates based on data obtained *in situ* from a hydrate and methane-rich continental margin. These new data can be used to improve the ability to model and predict how rising seawater temperatures in the future may impact the relatively unstable reservoir of methane–derived carbon, which has the potential to provide a strong positive feedback to the already substantial anthropogenic greenhouse gas emissions.

## References

Atwater BF, Carson B, Griggs GB, Johnson HP, Salmi MS. 2014. Rethinking turbidite paleoseismology along the Cascadia subduction zone. *Geology* **42**(9): 827–830.

Berndt C, Feseker T, Treude T, Krastel S, Liebetrau V, et al. 2014. Temporal constraints on hydrate-controlled methane seepage off Svalbard. *Science* **343**: 284–287.

Booth-Rea G, Klaeschen D, Grevemeyer I, Reston T. 2008. Heterogeneous deformation in the Cascadia convergent margin and its relation to thermal gradient (Washington, NW USA). *Tectonics* **27**(4).

Boswell R, Collett TS. 2011. Current perspectives on gas hydrate resources. *Energy Environ Sci* **4**: 1206–1215.

Brothers DS, Ruppel C, Kluesner JW, Brink US, Chaytor JD, et al. 2014. Seabed fluid expulsion along the upper slope and outer shelf of the US Atlantic continental margin. *Geophys Res Lett* **41**(1): 96–101.

Collett TS, Johnson AH, Knapp CC, Boswell R. 2009. Natural gas hydrates—Energy resource potential and associated geologic hazards. *AAPG Mem* **89**: 146–219.

Davis EE, Hyndman RD. 1989. Accretion and recent deformation of sediments along the northern Cascadia subduction zone. *Geol Soc Am Bull* **101**: 1465–1480.

Denman KL, Brasseur G, Chidthaisong A, Ciais P, Cox PM, et al. 2007. Couplings between changes in the climate system and biogeochemistry, in Solomon S, Qin D, Manning M, Chen Z, Marquis M, et al. eds., *Climate Change 2007, The Physical Science Basis: Contribution of Working Group I to the Fourth Assessment Report of the IPCC*. New York: Cambridge University Press: pp. 499–587.

Hautala SL, Solomon EA, Johnson HP, Harris RN, Miller UK. 2014. Dissociation of Cascadia margin gas hydrates in response to contemporary ocean warming. *Geophys Res Lett*. doi: 10.1002/2014GL061606.

Hurwitz S, Harris R, Werner CA, Murphy F. 2012. Heat flow in vapor dominated areas of Yellowstone Plateau Volcanic Field: Implications for thermal budget of the Yellowstone Caldera. *J Geophys Res* 117: B10207. doi: 10.1029/2012JB009463.

Incropera FP, DeWitt DP, Bergman TL, Lavine AS. 2013. *Foundations of Heat Transfer 6th ed*. Singapore: John Wiley & Sons Singapore Pte. Ltd.

Jackson DR, Richardson MD. 2000. Seasonal temperature gradients within a sandy seafloor: implications for acoustic propagation and scattering. Stennis Space Center, MS: Naval Research Lab Marine Geosciences Div. Accession ADA393656.

Johnson HP, Solomon EA, Harris RN, Salmi MS, Berg RD. 2013. Heat flow and fluid flux in Cascadia's seismogenic zone. *Eos Trans AGU* 94(48): 457–458. doi: 10.1002/2013EO480001.

Johnson HP, Tivey MA, Bjorklund TA, Salmi MS. 2010. Hydrothermal circulation within the Endeavour Segment, Juan de Fuca Ridge. *Geochem Geophys Geosyst* 11: Q05002. doi: 10.1029/2009GC002957.

Lopez C, Spence GD, Hyndman RD, Kelley D. 2010. Frontal ridge slope failure at the northern Cascadia Margin-normal fault and gas hydrate control. *Geology* 38(11): 967–970.

McCrory PA, Hyndman RD, Blair JL. 2014. Relationship between the Cascadia fore-arc mantle wedge, nonvolcanic tremor, and the downdip limit of seismogenic rupture. *Geochem Geophys Geosyst* 15(4): 1071–1095.

Phrampus BJ, Hornbach MJ. 2012. Recent changes to the Gulf Stream causing widespread gas hydrate destabilization. *Nature* 490(7421): 527–530.

Pohlman JW, Kaneko M, Heuer VB, Coffin RB, Whiticar M. 2009. Methane sources and production in the northern Cascadia margin gas hydrate system. *Earth Planet Sci Lett* 287: 504–512.

Rao YH, Subrahmanyam C, Rastogi A, Deka B. 2002. Slope failures along the western continental margin of India: A consequence of gas-hydrate dissociation, rapid sedimentation rate, and seismic activity. *Geo-Mar Lett* 22(3): 162–169.

Riedel M, Spence GD, Chapman NR, Hyndman RD. 2002. Seismic investigations of a vent field associated with gas hydrates, offshore Vancouver Island. *J Geophys Res Solid Earth (1978–2012)* 107(B9): EPM-5.

Riedel M, Willoughby EC, Chopra S. 2010. Geophysical characterization of gas hydrates. *Soc Explor Geophys* 14. doi: 10.1190/1.9781560802197.

Ruppel CD. 2011. Methane hydrates and contemporary climate change. *Nature Edu Know* 3(10): 29.

Schmalzle GM, McCaffrey R, Creager KC. 2014. Central Cascadia subduction zone creep. *Geochem Geophys Geosyst* 15(4): 1515–1532.

Thomson J. 2010. Observations of thermal diffusivity and a relation to the porosity of tidal flat sediments. *J Geophys Res* 115(C05016). doi: 10.1029/2009JC005968.

Torres ME, Embley RW, Merle SG, Tréhu A M, Collier RW, et al. 2009. Methane sources feeding cold seeps on the shelf and upper continental slope off central Oregon, USA. *Geochem Geophys Geosyst* 10(11).

Turcotte DL, Schubert G. 1982. *Geodynamics 2nd ed*. Cambridge, UK: Cambridge University Press.

Westbrook GK, Thatcher KE, Rohling EJ, Piotrowski AM, Pälike H, et al. 2009. Escape of methane gas from the seabed along the West Spitsbergen continental margin. *Geophys Res Lett* 36(15): doi: 10.1029/2009GL03191.

Wheatcroft RA, Stevens AW, Johnson RV. 2007. In situ time-series measurements of subseafloor sediment properties. *IEEE J Oceanic Eng* 32(4): 862–871.

## Author contributions

- Contributed to conception and design: KLH, HPJ
- Contributed to acquisition of data: KLH, HPJ, CKH
- Contributed to analysis and interpretation of data: KLH, HPJ
- Drafted and/or revised the article: KLH, HPJ, CKH
- Approved the submitted version for publication: KLH, HPJ, CKH

## Acknowledgments

The crew of the R/V *Atlantis* and operating personnel for the ROV *Jason II* were essential to the success of this study. Special thanks are given to Tor Bjorklund for assistance with design and construction of the temperature probe, and to Robert Harris, OSU, for his assistance and advice during the acquisition and processing phases of this project.

## Funding information

Support for this project was provided by NSF Grant 1339635 to H. P. Johnson & E. A. Solomon.

## Competing interests

The authors have no competing interests.

## Supplementary material

- Text S1. Detailed model analysis.
- Table S1. Calibration offsets for temperature sensors.

## Data accessibility statement

Prior to publication, all data used in this study, including site location and thermistor logs, will be posted on the GeoPRISM data archives and the National Geophysical Data Center database, as part of the NSF Open Access policy for all data from this cruise.

# 7

# Meso- and macro-zooplankton community structure of the Amundsen Sea Polynya, Antarctica (Summer 2010–2011)

Stephanie E. Wilson[1] • Rasmus Swalethorp[2,3] • Sanne Kjellerup[2,3,4] • Megan A. Wolverton[5] • Hugh W. Ducklow[6] • Patricia L. Yager[7]

[1]Bangor University, School of Ocean Sciences, Menai Bridge, United Kingdom
[2]Department of Biology and Environmental Sciences, University of Gothenburg, Göteborg, Sweden
[3]DTU Aqua, Technical University of Denmark, Charlottenlund, Denmark
[4]Greenland Climate and Research Center, Greenland Institute of Natural Resources, Nuuk, Greenland
[5]Arizona State University, Tempe, Arizona, United States
[6]Lamont-Doherty Earth Observatory, Columbia University, Palisades, New York, United States
[7]University of Georgia, Athens, Georgia, United States

**Domain Editor-in-Chief**
Jody W. Deming, University of Washington

**Knowledge Domain**
Ocean Science

## Abstract

The Amundsen Sea Polynya (ASP) has, on average, the highest productivity per unit area in Antarctic waters. To investigate community structure and the role that zooplankton may play in utilizing this productivity, animals were collected at six stations inside and outside the ASP using paired "day-night" tows with a 1 $m^2$ MOCNESS. Stations were selected according to productivity based on satellite imagery, distance from the ice edge, and depth of the water column. Depths sampled were stratified from the surface to ~ 50–100 m above the seafloor. Macrozooplankton were also collected at four stations located in different parts of the ASP using a 2 $m^2$ Metro Net for krill surface trawls (0–120 m). The most abundant groups of zooplankton were copepods, ostracods, and euphausiids. Zooplankton biovolume (0.001 to 1.22 ml $m^{-3}$) and abundance (0.21 to 97.5 individuals $m^{-3}$) varied throughout all depth levels, with a midsurface maximum trend at ~ 60–100 m. A segregation of increasing zooplankton trophic position with depth was observed in the MOCNESS tows. In general, zooplankton abundance was low above the mixed layer depth, a result attributed to a thick layer of the unpalatable colonial haptophyte, *Phaeocystis antarctica*. Abundances of the ice krill, *Euphausia crystallarophias*, however, were highest near the edge of the ice sheet within the ASP and larvae:adult ratios correlated with temperature above a depth of 60 m. Total zooplankton abundance correlated positively with chlorophyll *a* above 150 m, but negative correlations observed for biovolume vs. the proportion of *P. antarctica* in the phytoplankton estimated from pigment ratios (19'hexanoyloxyfucoxanthin:fucoxanthin) again pointed to avoidance of *P. antarctica*. Quantifying zooplankton community structure, abundance, and biovolume (biomass) in this highly productive polynya helps shed light on how carbon may be transferred to higher trophic levels and to depth in a region undergoing rapid warming.

## Introduction

The structure of planktonic communities can play a defining role in the export, retention, and eventual fate of surface-derived organic carbon (Legendre and Rassoulzadegan, 1996; Boyd and Newton, 1999; Wilson et al., 2008). Variations in climate have been shown to affect planktonic communities in regions worldwide and will likely have an effect on export fluxes to the deep sea (Lavaniegos and Ohman, 2007; Richardson, 2008; Steinberg et al., 2012). In Antarctic waters, for instance, the observed shift in abundance from krill to salps as the dominant zooplankton species as a result of decreasing winter sea ice in the Western Antarctic Peninsula region may cause a significant change in carbon export (Atkinson et al., 2004; Gleiber et al., 2012).

Polynyas, seasonally occurring areas of open water surrounded by sea ice, are important sources of primary productivity along the continental shelf of Antarctica (Arrigo and Van Dijken, 2003; Arrigo and Alderkamp, 2012). Antarctic polynyas are hypothesized to be unusually productive because the additional iron input from melting sea ice can help fuel phytoplankton blooms (Raiswell et al., 2006; Alderkamp et al., 2012; Arrigo et al., 2012). Usually these areas are the earliest to be productive in a season and attract many higher trophic level Antarctic species (Arrigo and Van Dijken, 2003). The numbers of polynyas are expected to increase with global climate change, an increase that, with additional sea ice melting, will impact primary productivity, food web structures, and carbon export (Ainley et al., 2005; Stammerjohn et al., 2008; Alderkamp et al., 2012).

Few studies directly consider the impact of Antarctic polynyas on zooplankton, although there are more such studies for the Arctic (e.g., Ashjian et al., 1997; Tagliabue and Arrigo, 2003; Deibel and Daly, 2007; Lee et al., 2012). Factors influencing zooplankton abundance and biomass in polynyas include advection from local waters, diatom productivity, and the presence or absence of the colonial haptophyte, *Phaocystis antarctica* (Ashjian et al., 1997; Tagliabue and Arrigo, 2003; Deibel and Daly, 2007). Of particular interest is the Amundsen Sea Polynya, which has the highest phytoplankton biomass and rates of primary productivity per unit area in Antarctic waters (Arrigo and Van Dyjken, 2003; Alderkamp et al., 2012). *P. antarctica* dominates the phytoplankton assemblage in the ASP, with diatoms more prevalent closer to the ice edges (Arrigo and McClain, 1994; Lee et al., 2012). From Ross Sea studies, there is little evidence that *P. antarctica* is grazed frequently by zooplankton (Caron et al., 2000; Tagliabue and Arrigo, 2003).

There is little information available about the zooplankton biomass and community structure within the ASP itself, especially below 200 m. Lee et al. (2013) sampled Amundsen Sea zooplankton above 200 m in the Austral summer of 2010–2011 and observed high abundances of copepods outside the ASP and of the euphausiid *E. crystallarophias*, the major ASP grazer, inside the ASP. Further investigating depth-specific trends in zooplankton distribution from the surface to the seafloor is of interest as a prerequisite for understanding the fate of ASP primary production and potential implications of warming seas.

The objective of the Amundsen Sea Polynya International Research Expedition (ASPIRE) was to investigate factors as to why the Amundsen Sea Polynya has such high rates of primary productivity per unit area compared to other Antarctic polynyas (Yager et al., 2012). The ASPIRE team also desired to understand the fate of this productivity. The aims of this study were to describe and quantify depth-specific zooplankton community structure and biovolume within the ASP as part of ASPIRE during the austral summer of 2010–2011. The results of this study were compared to other biological-physical parameters; e.g., chlorophyll *a*, pigments ratios indicative of diatoms vs. *Phaeocystis* measured via high performance liquid chromatography (HPLC) and mixed layer depth (MLD), as measured during the cruise and presented in this and other papers (e.g., Yager et al., 2014) of this Special Feature. The main objectives of this study were to determine which factors are structuring the zooplankton community and begin to assess how this community may utilize the unusually high primary productivity of the polynya and the potential effects on particulate organic carbon (POC) flux out of the system.

## Methods

Zooplankton sampling and processing were carried out onboard the RV *Nathaniel B. Palmer* NBP 10–05 as part of the multidisciplinary Amundsen Sea Polynya International Research Expedition (ASPIRE), and in collaboration with the Swedish Antarctic Research Programme (SWEDARP), between 19 December 2010 and 8 January 2011 (Yager et al., 2012). For a more in-depth description of the study area, see other papers in this special feature.

Twelve oblique plankton tows using the Multiple Opening and Closing Net and Environmental Sensing System (MOCNESS) with a 1 m² opening (333 μm mesh) were executed within and around the ASP region at noon and midnight as "day/night" pairs (green labels, Figure 1). Day/night pairs were utilized at each station, as the intensity of the diel periodicity of zooplankton in the ASP is unknown. Zooplankton in the ASP had not been sampled prior to this study. The wire-up speed was 15 m min⁻¹ and ship speed was 2 knots. The tow locations and depth intervals were selected according to bathymetry, depth of the water column, productivity, and distance from the ice edge. Three shallow "MOC" tow locations were conducted on the shelf below the ASP (ASPIRE "long daily" stations 13, 25 and 35, respectively), two within a deep trough region on the shelf (Yager et al., 2012), where potential warmer and nutrient-rich Circumpolar Deep Water (CDW) may be intruding (ASPIRE stations 50 and 57), and one on the shelf slope outside the polynya (ASPIRE station 68), where CDW may be entering the shelf into the trough (e.g., Wahlin et al., 2010; Randall-Goodwin et al., 2014). Station 57 was also located near a large iceberg (Randall-Goodwin et al., 2014). Depths sampled were stratified at eight intervals from the surface to approximately 50–100 m above the seafloor.

In addition, four 2 m² Metro net (700 μm mesh) krill surface trawls (0–120 m) were also taken throughout the ASP (blue labels, Figure 1), with two tow stations selected due to close proximity to the ice margin (ASPIRE stations 57 and 66) and two selected in open water regions (ASPIRE stations 29 and 35). The 2 m² 700 um mesh Metro net was used in addition to the MOCNESS to focus on collecting larger krill

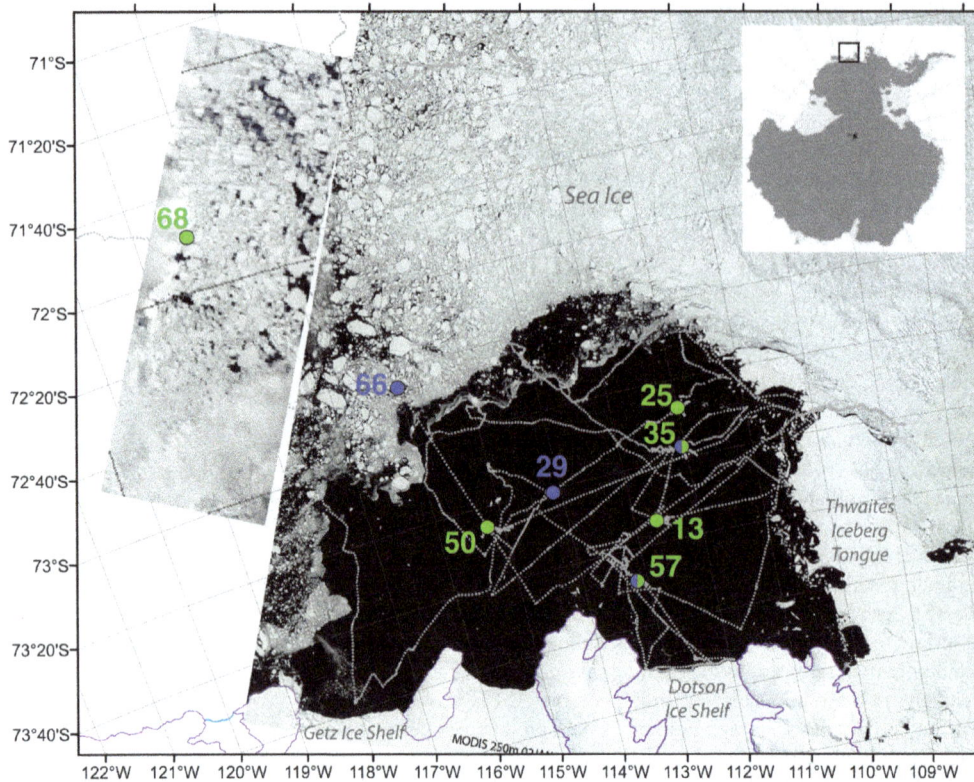

Figure 1

**Sampling Area.**

Zooplankton stations in the ASP region. Green labels represent all MOCNESS (MOC) sampling stations. Blue labels represent all Metro (MET) sampling stations. White dotted lines indicate cruise tracks. MOC tows correspond to ASPIRE stations 13, 25, 35, 50, 57, and 68, respectively. MET tows correspond to ASPIRE stations 29, 35, 57 and 66, respectively.

species that can avoid smaller nets and also to have results comparable to other studies in the region (e.g., Ross et al., 1988; Ross et al., 2008). Due to potential damage of the nets by brash ice accumulation around the edges of the polynya, no net sampling was initiated at the ice edge itself.

Immediately following collection, zooplankton > 5 mm were live-processed and quantified onboard the *NB Palmer* for species abundance and biovolume (as a proxy for biomass) measured as displacement volume. The large, > 5 mm fraction of zooplankton were counted and identified onboard. The biovolumes were measured individually for each species or group identified; total biovolume was measured for each tow. In cases where there was a visible (thick green mats) presence of the colonial haptophyte *Phaeocystis antarctica* in the cod-end buckets, all of the individual zooplankton species biovolumes were summed for that depth to measure a total *Phaeocystis*-free biovolume (generally in the top three or four depth intervals). The remaining individuals in the samples were split using a Folsom plankton splitter as follows: 1/2 of each tow was split and preserved in alcohol and archived, 1/4 was preserved in 10% buffered formalin for further taxonomic identification of the smaller individuals, and 1/4 was frozen in -80°C for separate analyses. Krill species collected from the Metro tows were also enumerated onboard, 50–100 were randomly selected and measured for total length, and total biovolume was measured. All tows were counted and preserved in entirety except MET 66, which was split 1/15. Laboratory analysis consisted of zooplankton taxonomic identification with size-fractionated aliquots of 5 mm–500 μm and 500–333 μm samples from the MOCNESS tows.

Depth-integrated zooplankton biovolume and abundance (ml m$^{-2}$ and ind m$^{-2}$, respectively) were also calculated to compare with independent sampling results, including chlorophyll *a* and pigment ratios associated with diatoms and *Phaeocystis* (Alderkamp et al., 2014). Chlorophyll *a* and pigments were normalized to the same depths as the MOCNESS tows; trapezoidal integration was utilized to calculate depth-integrated concentrations (mg m$^{-2}$). These results will help to determine if primary productivity or phytoplankton community structure play a role in zooplankton distribution and biomass within the ASP. Weighted mean depths (WMD) were calculated for taxa found in several depth intervals on every day-night sampling event using the following equation:

$$WMD = \frac{\sum n_s d_s}{\sum n_s},$$

where *d* is the mean depth of the sampled depth interval *s* and *n* is the abundance (ind m$^{-3}$) (Bollens and Frost, 1989).

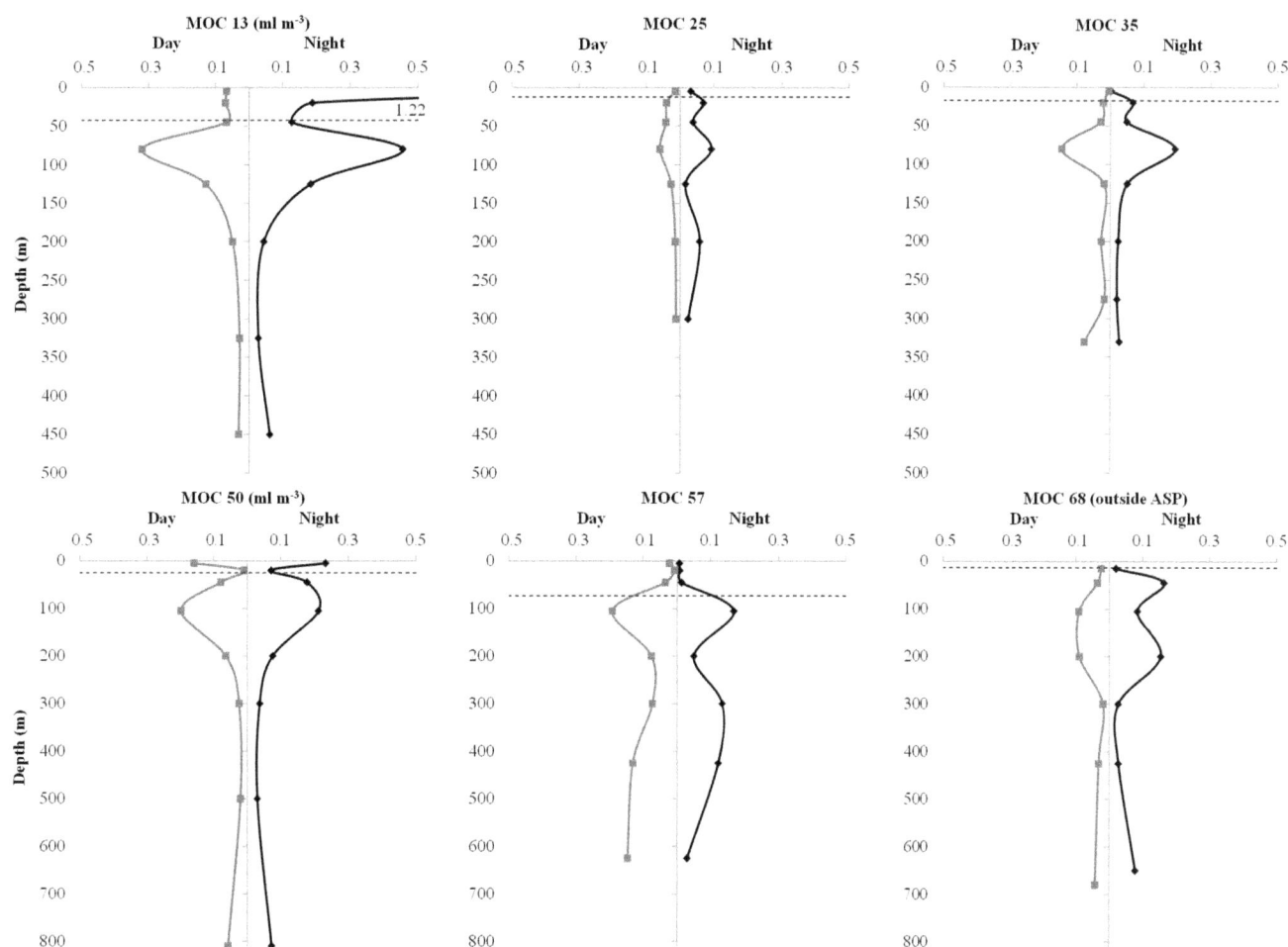

Figure 2

Mesozooplankton biovolumes (ml m⁻³) of MOCNESS (MOC) tows within and near the ASP.

Dashed line indicates mixed layer depth. Note different y axis.

Statistical testing was conducted using the software packages Minitab 16 and Primer 6. Biomass and abundance data for each station and depth were assumed nonparametric (Kolmogorov-Smirnov test for normal distribution); therefore, Mann-Whitney U tests were used to compare data between stations. When pooled across all depths and stations, and for individual species, a Mann-Whitney U test was also used to compare day to night values. The Pearson's R correlation coefficient was calculated for pooled data to compare abundance and biomass with salinity, chlorophyll $a$, and HPLC pigments. A Bray-Curtis similarity index was plotted with 2D Multidimensional Scaling (MDS) to compare grouped species abundances (ind m⁻³) between stations and depths. A further ANOSIM test on 4th root-transformed abundances was used to determine significance.

## Results

### Mesozooplankton distribution and community structure - MOCNESS

Zooplankton MOCNESS biomass in the form of biovolume density ranged from 0.001 ml m⁻³ at 10–30 m depth at MOC 57 to 1.22 ml m⁻³ at 0–10 m depth at MOC 13 (Figure 2). Abundance values ranged from 0.21 ind m⁻³ at 0–10 m depth at MOC 25 to 97.5 ind m⁻³ at 60–100 m depth at MOC 13, (Figure 3). Biovolume and abundance were significantly correlated (Pearson correlation, $R_p$ = 0.27, p = 0.01). Low zooplankton values were measured above 60 m (biovolume) and 30 m (abundance), with the exception of the value for 0–10 m at MOC 13 which was high due to the biovolumes of several individuals of the ice krill, *Euphausia crystallarophias*, at this depth. *E. crystallarophias* was the species with the largest total biovolume within the ASP (maximum MOCNESS biovolume of 0.85 ml m⁻³). The subsurface maximum biovolume and abundance of zooplankton was generally between 30 or 60 and 150 m depth and below the mixed layer depth (MLD; Alderkamp et al., 2014). This subsurface maximum was also immediately below the depth where thick aggregates of the haptophyte, *Phaeocystis antarctica*, were observed in the tows along with the zooplankton (Lee et al., 2012; Alderkamp et al., 2014).

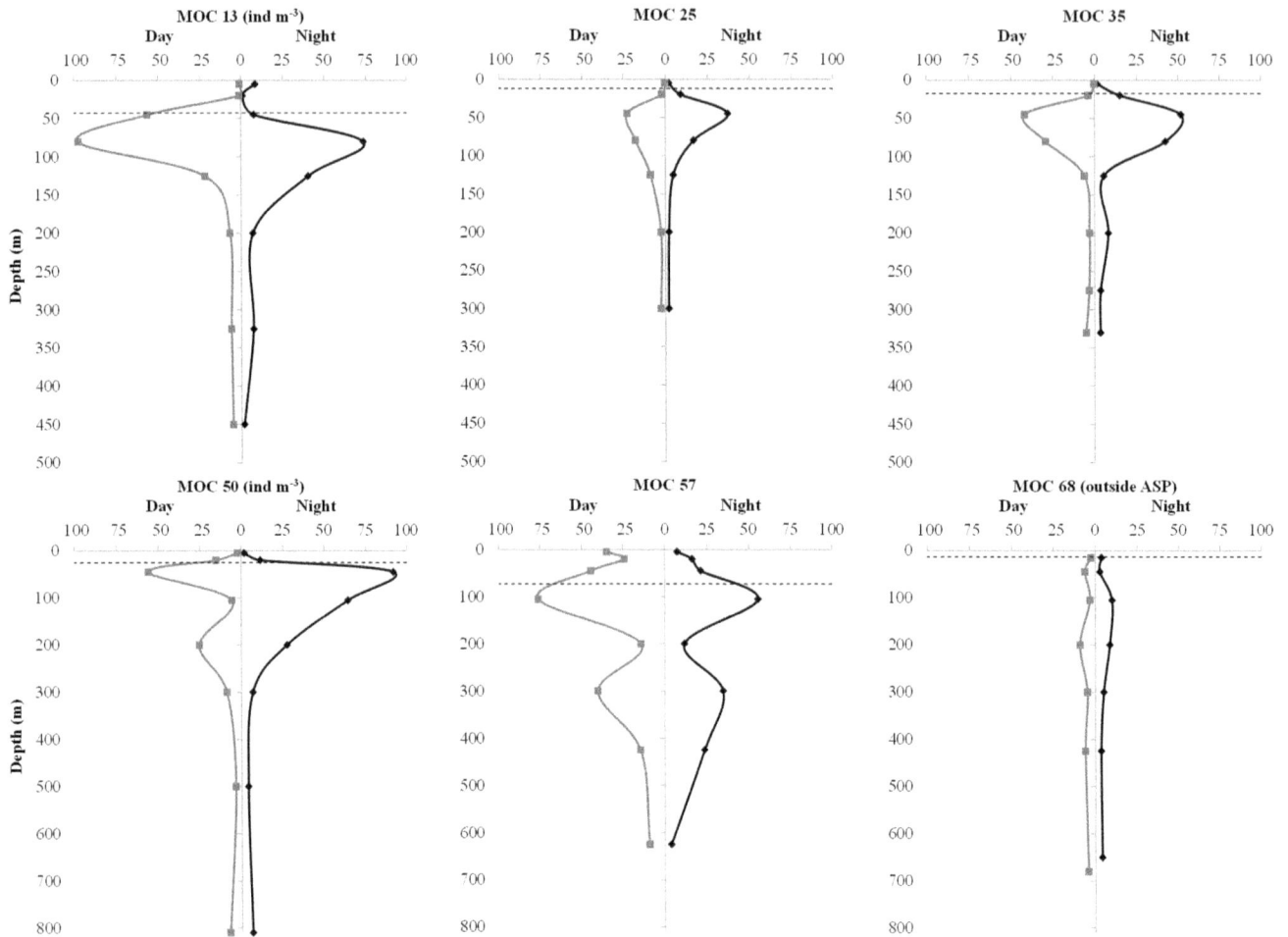

Figure 3

Mesozooplankton abundances (ind m$^{-3}$) of MOCNESS (MOC) tows within and near the ASP.

Dashed line indicates mixed layer depth. Note different y axis.

Despite 24 h of sunlight, an apparent (not statistically significant) diel periodicity of varying intensity was observed in zooplankton density. Mann-Whitney U tests of pooled MOCNESS abundance and biovolume densities showed insignificant differences between night and day samples (Mann-Whitney, biovolume: p = 0.14; abundance: p = 0.83). Although not significant, at most stations zooplankton abundance and biovolume were slightly higher during the night. The results of weighted mean depth analyses (WMD) show that zooplankton were distributed shallower in the water column (Table 1). Upon further investigations of the individual species observed in all tows at several depths, WMD calculation results for day and night of individual species were not significantly different (Mann-Whitney, p > 0.05), though at five of the six MOC stations *E. crystallarophias* (15 ± 24 m, night-day WDM difference ± SD), amphipods of the genus *Orchomene* (35 ± 70 m), and the calanoid copepods *Metridia gerlachei* (57 ± 59 m) and *Paraeuchaeta antarctica* (25 ± 38 m) were distributed at shallower depths (data not shown).

Crustacean zooplankton comprised the bulk of the zooplankton samples with > 90% of the total biomass and abundance. The majority of these zooplankton were calanoid copepods (maximum density of 93 ind m$^{-3}$ at 60–100 m depth at MOC 13), followed by euphausiids and ostracods (Figure 4). Euphausiids dominated in the shallower nets for most stations; copepods, the middle nets; and ostracods, the deeper nets. Proportionally, ostracods were the dominant zooplankton species at the shallower stations: at MOC 13–35 below 250 m and at MOC 68 at 350–500 m (Figure 4). Chaetognaths were observed in all tows and all stations but were present predominately below 100 m. Results of the Bray-Curtis MDS on the MOCNESS species abundance data show similarities between depths (above and below 60 m), but no pattern was apparent between stations (Figure S1). Results of the ANOSIM test show these depth similarities to be significantly different (Global R = 0.44, P = 0.001).

Table 1. Zooplankton depth-integrated cumulative abundance, biovolume, and weighted mean depth (WMD) at day and night from each MOCNESS station

| Station | Zooplankton abundance | | | | Zooplankton biovolume | | | |
|---|---|---|---|---|---|---|---|---|
| | Day (ind m$^{-2}$) | Night (ind m$^{-2}$) | Day WMD (m) | Night WMD (m) | Day (ind m$^{-2}$) | Night (ind m$^{-2}$) | Day WMD (m) | Night WMD (m) |
| MOC 13 | 8816 | 7262 | 105 | 52 | 35.7 | 62.2 | 96 | 108 |
| MOC 25 | 2396 | 2609 | 83 | 95 | 8.5 | 15.6 | 86 | 65 |
| MOC 35 | 3571 | 5047 | 151 | 103 | 16.1 | 18.2 | 89 | 79 |
| MOC 50 | 9244 | 15585 | 167 | 146 | 55.2 | 72.1 | 151 | 122 |
| MOC 57 | 18854 | 15121 | 303 | 261 | 89.6 | 60.0 | 145 | 185 |
| MOC 68 | 4315 | 4186 | 231 | 213 | 40.3 | 59.0 | 256 | 229 |
| Mean ± SE* | 7866 ± 2480 | 8301 ± 2313 | 173 ± 33 | 145 ± 32 | 40.9 ± 11.9 | 47.8 ± 10 | 137 ± 26 | 131 ± 26 |

*SE = standard error of the mean

The most abundant of the euphausiids was *E. crystallarophias* (maximum MOCNESS density of 3.1 ind m$^{-3}$ at 0–10 m depth at MOC 13). Other less common species of krill observed in the polynya were *Thysanoessa macrura*, *E. tricantha*, and *E. frigida*. There were few, if any, salps or *E. superba* observed within the polynya; however, two *E. superba* gravid adult females were found outside the polynya along the shelf edge at MOC 68 at depth intervals 300–500 and 500–800 m. The majority of the calanoid copepods were largely *Calanoides acutus* (maximum density of 62.3 ind m$^{-3}$ at 60–100 m depth at MOC 13) followed by *M. gerlachei*, and *Paraeuchaeta antarctica* (Figure 5). Copepods smaller than 500 µm (e.g., *Oithona* sp.) also comprised a significant proportion of the copepod abundance, especially above 150 m (Figure 4). The majority of *C. acutus* and *M. gerlachei* were distributed between 30 and 150 m at most stations, although *M. gerlachei* also displayed shallow diel vertical migration behavior. *P. antarctica* was the largest in size of the copepods and generally remained below 100 m, with their numbers increasing with depth.

Tow MOC 68, outside the polynya, did not show any significant difference with stations within the polynya in terms of biomass (Figure 3), including depth-integrated cumulative biomass and abundance (Table 1), due to the higher abundances and biovolumes of gelatinous zooplankton within the nets as well as the greater depths involved (Mann-Whitney, p > 0.05). However, total abundance densities at MOC 68 were significantly less than within the ASP (Mann-Whitney, p = 0.046; Figure 2). MOC 68 also showed a difference in species distribution from the stations within the polynya (Figure 4). Fewer euphausiids were observed at MOC 68 in proportion to other species along with higher proportions of appendicularians

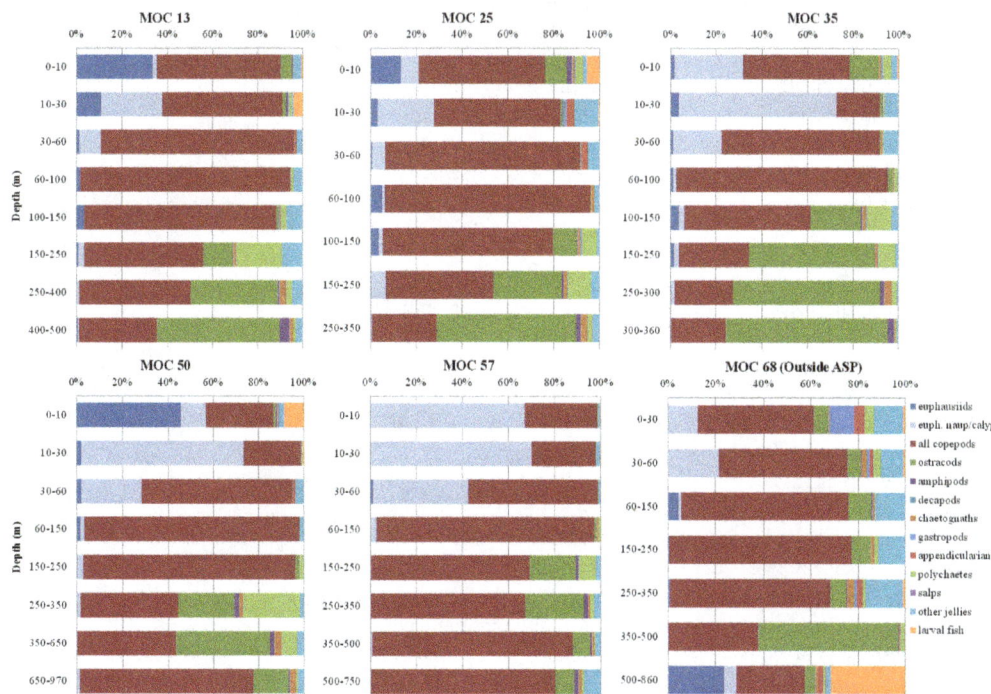

**Figure 4**

Mesozooplankton functional group numerical proportional abundance.

Abundance is mean of day and night tows for each depth interval. Note different y axis.

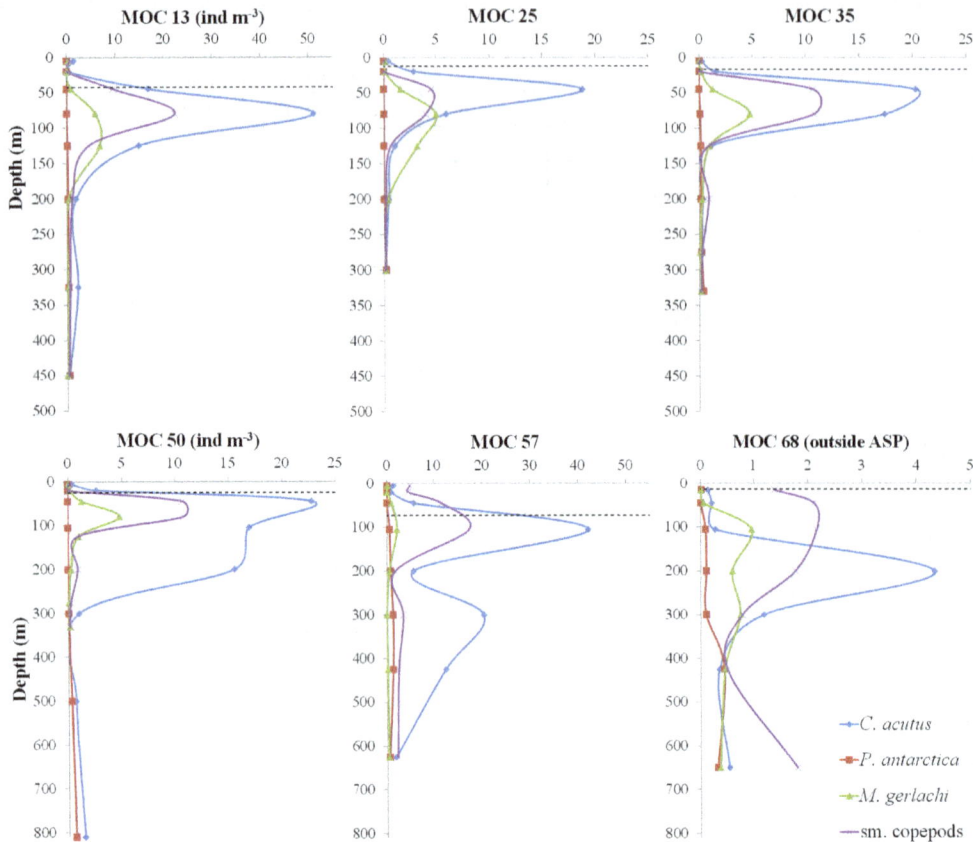

Figure 5

Calanoid    copepod    species
density.

Abundance is mean of day
and night tows for each depth
interval. Dashed line indicates
mixed layer depth. Note different
scales on axis.

and gelatinous zooplankton (Mann-Whitney: euphausiids, p = 0.07; appendicularians, p = 0.06; gelatinous, p = 0.001). Appendicularians were observed at most depths at MOC 68 although rarely inside the polynya.

### Euphausiid distribution and community structure – Metro net

Net avoidance can be an issue with larger euphausiids (Wiebe et al., 1982). In this case euphausiid species densities were comparable in range to MOCNESS euphausiid densities with the exception of MET 66, where we encountered a super swarm of *Euphausia crystallarophias*. The highest biovolume of *E. crystallarophias* was observed at MET 66, near the ice margin, where we collected a swarm of approximately 25,000 adults and juveniles (1.8 ml m⁻³, 11.1 ind m⁻³). In comparison, the highest biovolume and abundance of *E. crystal-larophias* in the MOCNESS tows was observed at MOC 13 with 0.85 ml m⁻³ and 3.13 ind m⁻³, respectively. MET 29–57 had much lower abundance and biovolume of all species compared to MET 66 (Figure 6A). The dominant krill species collected in the Metro net tows were *E. crystallarophias* and *Thysanoessa macrura* with few, if any, *E. frigida* or *E. superba* (Figure 6B).

### Chlorophyll, pigments, and other parameters

Log-transforming the data did not change the homoscedasticity of the data set. Zooplankton abundance at all stations and depths was negatively correlated with salinity with high variability (Pearson's R, $R_p$ = –0.26, p = 0.01; Figure 7A), although biovolume was not. Temperature, however, did not correlate with zooplankton abundance or biovolume. A correlation was observed between depth-integrated zooplankton abundance and integrated chlorophyll *a* in MOCNESS samples collected above 150 m (Pearson's R, $R_p$ = 0.32, p = 0.02; Figure 7B), but not with biovolume. With the ratio of 19'hexanoyloxyfucoxanthin to fucoxanthin (19-hex: fuco) measured from HPLC analysis (Alderkamp et al., 2014), indicative of the prevalence of *Phaeocystis*, the potentially less palatable phytoplankton than diatoms (Smith et al., 2010), a negative correlation was observed with both depth-integrated zooplankton abundance (marginal significance) and biovolume at polynya stations above 150 m (Pearson's R, abundance $R_p$ = –0.26, p = 0.07; biovolume Rp = –0.35, p = 0.01; Figure 7C). Euphausiid nauplii abundance and the calypotopis:adult ratio increased with increasing surface temperature (top 60 m of the water column where nauplii and calyptopis were located; Figure 8), peaking at the warm MOC 57 which also had the highest levels of chl *a* (916 mg chl *a* m⁻²).

A

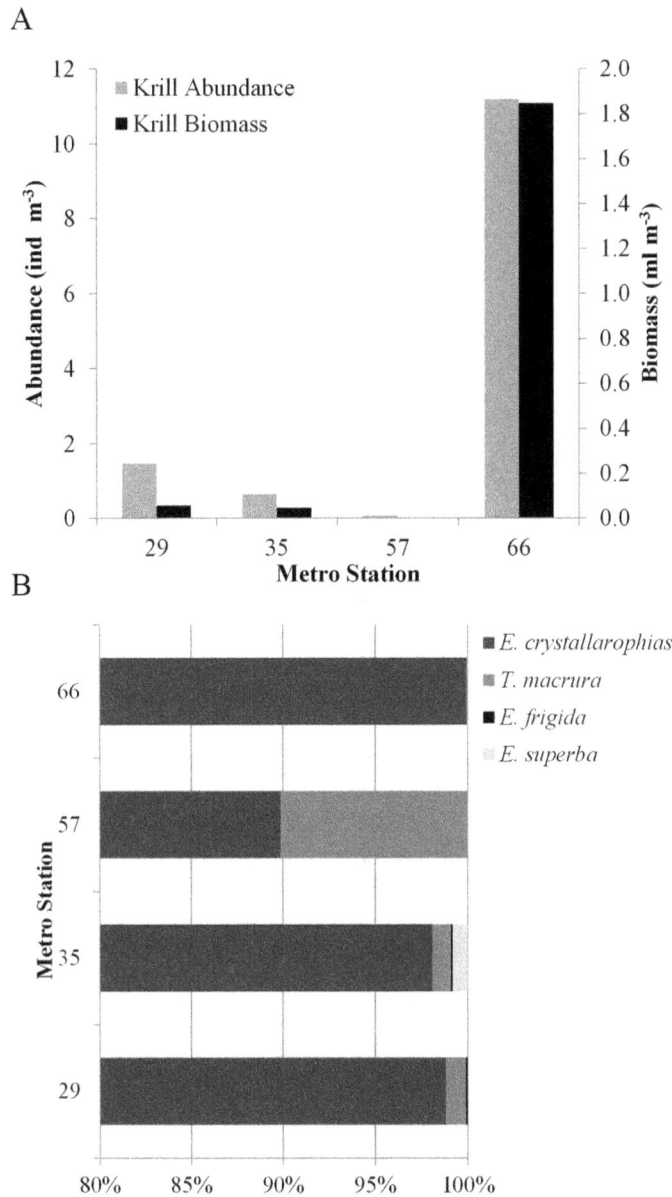

Figure 6
**Euphausiid distribution determined from Metro net.**

A) Abundance (ind m⁻³) and biomass (ml m⁻³) for euphausiids. B) Euphausiid proportional abundance (% of numerical density)

B

## Discussion

### Zooplankton abundance and biovolumes in the ASP

The results of high resolution (finer depth scale) sampling using the MOCNESS demonstrate that there was a distinct zooplankton biovolume and abundance depth distribution trend within this region during the sampling period. Within the polynya, a subsurface maximum of zooplankton was often observed at approximately 60–100 m, below the mixed layer depth (MLD; Figure 2 and 3). Directly outside of the polynya along the shelf slope at MOC68, where abundance was significantly lower, this depth-distribution trend was not nearly as well resolved. Further north, zooplankton sampling associated with the Swedish Antarctic Research Programme (SWEDARP) in Marguerite Bay during the same cruise (P. Moksness, *unpublished data*), showed that the highest biovolume was at the surface (0–10 m), nearly double the biovolume of the highest ASP values, and that it decreased step-wise with depth.

In other coastal regions such as the lower Antarctic Peninsula, the overall Metro net zooplankton biovolume was comparable. Zooplankton abundance and biovolume in January 2011 at the Charcot Island process station of the Palmer Long-term Ecological Research (LTER) study (the furthest south and furthest inshore of the LTER sampling grid) were slightly lower than the comparable Metro net tows during ASPIRE (data are available at the LTER DataZoo, http://oceaninformatics.ucsd.edu/datazoo/data/pallter/datasets, dataset number 199). *Euphausia crystallarophias* was also the dominant krill species at both locations

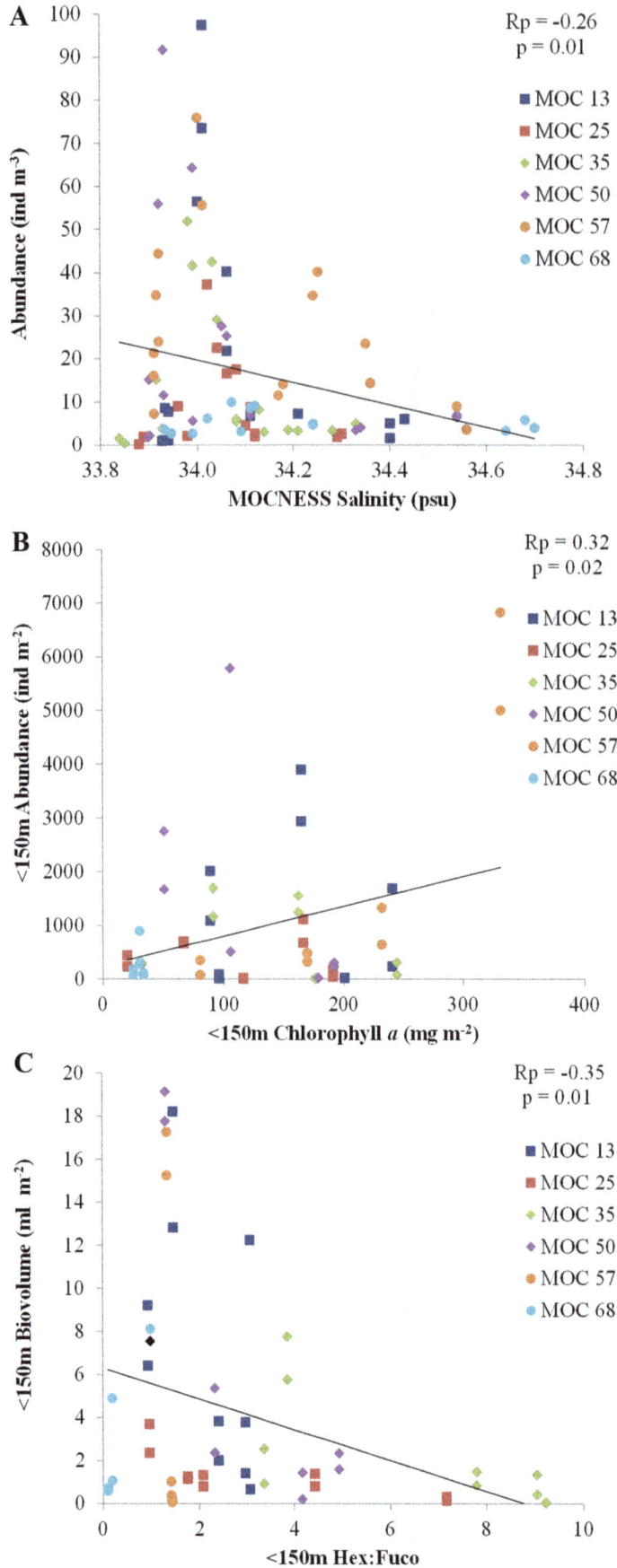

Figure 7

**Zooplankton distribution correlated to environmental parameters.**

A) Zooplankton abundance correlated to salinity (all stations and depths). B) Depth-integrated abundance correlated to depth-integrated chlorophyll *a* (all stations and depths above 150 m). C) Zooplankton depth-integrated biovolume correlated to hex:fuco (19'-hex :fucoxanthin) a ratio which compares prevalence of *Phaeocystis* to diatoms (all stations and depths above 150 m).

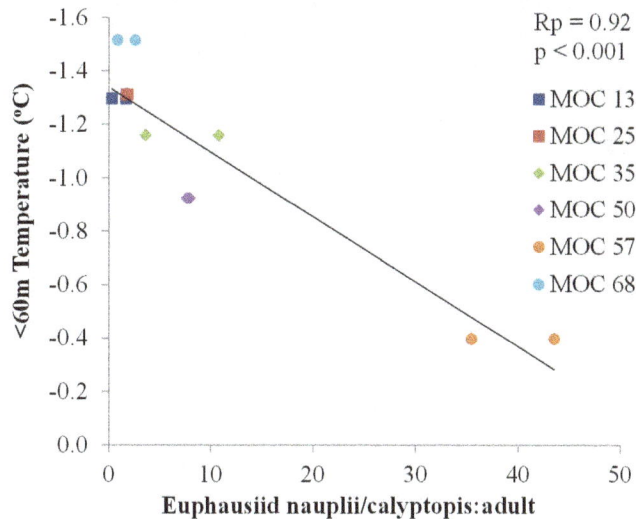

Figure 8

Larval to adult euphausiid ratio relative to water temperature.

Ratio of euphausiid larval stages to adults (integrated cumulative abundance from all stations) correlated to average temperature in the top 60 m of the water column.

(Charcot maximum, 2.65 ind m$^{-3}$; ASPIRE maximum, 10.24 ind m$^{-3}$); however, more *E. superba* were present at Charcot than ASP (Charcot maximum, 0.0211 ind m$^{-3}$; ASPIRE maximum, 0.0041 ind m$^{-3}$). Pakhomov and Perissinotto (1996) presented abundance data that led to their conclusion that polynyas provide favorable conditions for spawning and growth of *E. crystallarophias*, which may account for the higher maximum values observed within the ASP.

Within the nearby Ross Sea, where primary productivity is both high and dominated by *Phaeocystis antarctica*, zooplankton biomass was also low compared to other areas within the Southern Ocean such as the Western Antarctic Peninsula (Tagliabue and Arrigo, 2003; Ducklow et al., 2006). The low zooplankton biomass in the Ross Sea, Terra Nova Bay and potentially the ASP, may result from an inability to match the high growth rates of the phytoplankton blooms in the early spring (Tagliabue and Arrigo, 2003). The increase in phytoplankton biomass is due to the presence of *P. antarctica*, which is not necessarily grazed by zooplankton at the same rate due to its colony size and potential chemical deterrents (Bautista et al., 1992; Ducklow et al., 2006).

With high variability, there was a relationship between zooplankton abundance and salinity but not temperature. Similar results were also observed earlier by Lee et al. (2013) in the upper 200 m of the polynya. The relationship to salinity but not temperature may be due to mixing processes in the polynya that include melting glacier and sea ice (Randall-Goodwin et al., 2014). The positive correlation between the ratio of euphausiid nauplii and calyptopis to adults may indicate, however, that temperature was affecting euphausiid reproduction rates at the time of sampling. Euphausiid development rate increases with temperature (Ross et al., 1988; Pinchuk and Hopcroft, 2006) which may also reduce oocyte maturation time and affect when spawning is initiated during the productive season, as has been observed in the North Atlantic copepod *Calanus finmarchicus* (e.g., Niehoff, 2007). Furthermore, the highest chl *a* concentration found at MOC 57 is also likely to have shortened oocyte maturation time and maximized nutrient assimilation fueling the reproduction (Schmidt et al., 2012).

## Vertical and spatial distribution of species

We found some differences in community composition within the ASP compared to the outside station. Most notably were the low abundances of gelatinous zooplankton and rarity of appendicularians and salps within the ASP. Salps and appendicularians produce large, dense fecal pellets that can quickly transport small-particle biomass to the seafloor (Anderson, 1998; Phillips et al., 2009); their absence would have an effect on zooplankton-derived POC flux out of the system (Pakhomov et al., 2002; Ducklow et al., 2006). Salps are more common offshore, away from ice regions, and may have thermo-physiological limitations at higher latitudes (Pakhomov et al., 2002; Ward et al., 2004; Ross et al., 2008), which could explain their rarity in the ASP. Appendicularians are less well studied in Antarctic waters but are important producers of sinking carbon flux in Arctic polynyas (Deibel et al., 2007). Lindsey and Williams (2010) observed the highest abundances of larvaceans away from the shelf break and that numbers correlated with latitude in the southwest Indian Ocean region of East Antarctica between 30 and 80° East. Low abundances of appendicularians and salps, both filter feeders, in the ASP may also be due to the presence of *Phaeocystis antarctica*, which are able to alter colony size as a defense strategy against grazers (Tang 2003; Tang et al., 2008) and potentially clog feeding filters (Harbison et al., 1986; Acuña et al., 1989; Kawaguchi et al., 2004).

Along with biovolume and abundance, we also found that species differed in their vertical distribution (Figures 4 and 5). For example, large carnivorous copepods (e.g., *Paraeuchaeta antarctica*) and chaetognaths were observed deeper than many of the more herbivorous/omnivorous species (e.g., *Calanoides acutus* and *Metridia gerlachei*). Vertical position in the water column in the ASP may be a trade-off between light levels, feeding, competition, and predator avoidance (Fleddum et al., 2001; Coyle and Pinchuk, 2005; Marrari et al., 2011; Rabindranath et al., 2011) resulting in the observed segregation of increasing trophic position with depth.

Diel periodicity was small and night-day biovolume and abundance ratios tended to be larger at night (though not statistically significant; Table 1). Diel vertical migration (DVM) behavior in polar regions generally changes seasonally, with complete cessation during early summer (e.g., Cisewski et al., 2010). With some exceptions, the observed variations in general vertical distribution are not likely due to DVM given the time of year; however, specific organisms *M. gerlachei, Paraeuchaeta antarctica, E. crystallarophias*, and *Orchomene* sp. displayed weak DVM behavior. Although the sun was above the horizon around-the-clock during the ASPIRE cruise, irradiation levels were substantially lower at night, a difference that may be detectable by polar zooplankton (Berge et al., 2010). Engaging in minimal dial vertical migration may at least reduce the risk of predation while maintaining the ability to feed on the high productivity of the seasonal bloom (e.g., Dale and Kaartvedt, 2000; Berge et al., 2009; Cisewski et al., 2010).

Station MOC 57, in close proximity to a drifting iceberg, had some of the highest proportions of diatoms to *Phaeocystis* compared to other stations within the ASP (Figure 7C, Alderkamp et al., 2014), a pattern also reflected in zooplankton community structure, abundance and biovolume, and in euphausiid larvae ratio data. These results support earlier conclusions that proximity to an iceberg affects zooplankton distribution and abundance (e.g., Smith et al., 2007). Two studies investigating Weddell Sea iceberg-macrozooplankton interactions showed higher zooplankton biomass within close proximity of two large icebergs potentially due to aggregation via turbulent flows (Sherlock et al., 2011) and enhanced food availability (Kaufmann et al., 2011; Smith et al., 2007). Our results showed a slightly higher proportion of euphausiid nauplii and calyptopis at MOC 57, compared to the other stations, as well as slightly elevated abundance levels of *C. acutus* (at 30–60 m) and a second mid-depth maximum abundance (at 250–350 m). Although not statistically significant in this case (Kruskal–Wallis ANOVA, $p > 0.05$), higher proportions of euphausiid nauplii and calyptopis near icebergs may be due to ice being an essential stage in their developmental life history (Brinton and Townsend, 1991; Pakhomov and Perissinotto, 1996). The higher availability of diatoms would also increase the amount of energy allocated for reproduction (Schmidt et al., 2012).

The strongest biological correlations measured in this study were for zooplankton abundance with chl *a* and for zooplankton biovolume and abundance with the ratio of 19-'hex to fucoxanthin. These results suggest that, despite a positive relationship between overall abundance and chl *a*, a negative relationship exists with *Phaeocystis antarctica*. Negative correlations with *Phaeocystis* spp. suggest selective feeding, which Bautista et al. (1992) speculate "might indirectly contribute to the development of *Phaeocystis* spp. blooms because of the reduced grazing pressure and decrease in copepod abundances." Our current hypothesis for low surface-water biomass is that zooplankton are generally avoiding areas of high *P. antarctica* due to limited palatability and other defense mechanisms of these phytoplankton. We thus assume that, during the period sampled, phytodetritus and bacterial remineralization rather than zooplankton grazing and fecal pellet production are driving carbon flux in the ASP (Ducklow et al. 2006; Smetacek et al., 2004; Ducklow et al., 2014 et al., 2014; Yager et al., 2014). However, sediment trap data for this area do show a contribution by zooplankton to POC flux (as evidenced by the presence of fecal pellets in trap samples) as the bloom progressed (Ducklow et al., 2014). With the region around the ASP changing dramatically due to increased warming (Stammerjohn et al., 2014), we suggest that evaluating effects of these changes on the structure and biomass of the zooplankton on an expanded seasonal basis will be of further importance to determining carbon flux to both higher trophic levels and the benthos (Yager et al., 2012; Ducklow et al., 2014; Yager et al., 2014).

# References

Acuña JL, Deibel D, Bochdansky AB, Hatfield E. 1999. In situ ingestion rates of appendicularian tunicates in the Northeast Water Polynya (NE Greenland). *Mar Ecol Prog Ser* **186**: 149–160.

Ainley DG, Clarke ED, Arrigo K, Fraser WR, Kato A, et al. 2005. Decadal-scale changes in the climate and biota of the Pacific sector of the Southern Ocean, 1950s to the 1990s. *Antarct Sci* **17**: 171–182.

Alderkamp AC, Mills MM, van Dijken GL, Laan P, Thuroczy C, et al. 2012. Iron from melting glaciers fuels phytoplankton blooms in Amundsen Sea (Southern Ocean): Phytoplankton characteristics and productivity. *Deep-Sea Res Pt II* **71–76**: 32–48.

Alderkamp AC, van Dijken GL, Lowry KE, Connelly TL, Lagerström M. et al. 2014. Fe availability drives phytoplankton photosynthesis rates in the Amundsen Sea Polynya, Antarctica. *Elem Sci Anth:* Under review for the ASPIRE Special Feature.

Anderson V. 1998. Salp and pyrosomid blooms and their importance in biogeochemical cycles, in Bone Q, ed., *The Biology of Pelagic Tunicates*. Oxford: Oxford University Press. pp 125–137.

Arrigo KR, Alderkamp AC. 2012. Shedding dynamic light on Fe limitation (DynaLiFe) Introduction. *Deep-Sea Res Pt II* **71–76**: 1–4.

Arrigo KR, McClain CR. 1994. Spring phytoplankton production in the Western Ross Sea. *Science* **266**(5183): 261–263.

Arrigo KR, van Dijken GL. 2003. Phytoplankton dynamics within 37 Antarctic coastal polynya systems. *J Geophys Res* **108**: 3271.

Arrigo KR, Lowry KE, van Dijken GL. 2012. Annual changes in sea ice and phytoplankton in Polynyas of the Amundsen Sea, Antarctica. *Deep-Sea Res II* **71–76**: 5–15.

Ashjian C, Smith S, Bignami F, Hopkins T, Lane P. 1997. Distribution of zooplankton in the Northeast Water Polynya during summer 1992. *J Mar Sys* **10**(1–4): 279–298.

Atkinson A, Siegel V, Pakhomov E, Rothery P. 2004. Long-term decline in krill stock and increase in salps within the Southern Ocean. *Nature* **432**: 100–103.

Atkinson A, Schmidt K, Fielding S, Kawaguchi S, Geissler PA. 2012. Variable food absorption by Antarctic krill: Relationships between diet, egestion rate and the composition and sinking rates of their fecal pellets. *Deep-Sea Res Pt II* **59**: 147–158. doi: 10.1016/j.dsr2.2011.06.008

Bautista B, Harris RP, Tranter PRG, Harbour D. 1992. *In situ* copepod feeding and grazing rates during a spring bloom dominated by *Phaeocystis* sp. in the English Channel. *J Plankton Res* **14**(5): 691–703.

Berge J, Cottier F, Last K, Varpe O, Leu E, et al. 2009. Diel vertical migration of Arctic zooplankton during the polar night. *Biol Lett* **5**(1): 69–72.

Bollens SM, Frost BW. 1989. Predator-induced diet vertical migration in a planktonic copepod. *J Plankton Res* **11**: 1047–1065.

Boyd PW, Newton PP. 1999. Does planktonic community structure determine downward particulate organic carbon flux in different oceanic provinces? *Deep-Sea Res Pt I* **46**: 63–91.

Brinton E, Townsend AW. 1991. Development rates and habitat shifts in the Antarctic neritic euphausiid *Euphausia crystallarophias*, 1986–1987. *Deep-Sea Res* **38**(8–9): 1195–1121.

Caron DA, Dennett MR, Lonsdale DJ, Moran DM, Shalapyonok L. 2000. Microzooplankton herbivory in the Ross Sea, Antarctica. *Deep-Sea Res Pt II* **47**(15–16): 3249–3272.

Cisewski B, Strass VH, Rhein M, Krägefsky S. 2010. Seasonal variation of diel vertical migration of zooplankton from ADCP backscatter time series data in the Lazarev Sea, Antarctica. *Deep-Sea Res Pt I* **57**(1): 78–94.

Coyle KO, Pinchuk AI. 2005. Seasonal cross-shelf distribution of major zooplankton taxa on the northern Gulf of Alaska shelf relative to water mass properties, species depth preferences and vertical migration behaviour. *Deep Sea Res Pt II* **52**(1–2): 217–245.

Dale T, Kaartvedt S. 2000. Diel patterns in stage-specific vertical migration of *Calanus finmarchicus* in habitats with midnight sun. *ICES J Mar Sci* **57**: 1800–1818.

Deibel D, Daly KL. 2007. Zooplankton processes in Arctic and Antarctic polynyas, in Smith WO, Barber DG, eds., *Polynyas: Windows to the World*. (Elsevier Oceanography Series): pp. 271–322. doi: 10.1016/S0422-9894(06)74009-0

Deibel D, Saunders PA, Acuna J-L, Bochdansky AB, Shiga NR, et al. 2007. The role of appendicularian tunicates in the biogenic carbon cycle of three Arctic polynyas, in Gorsky G, Youngbluth MJ, Deibel D, eds., *Response of Marine Ecosystems to Global Change: Ecological Impact of Appendicularians*. Contemporary Publishing International: pp. 327–356.

Ducklow HW, Erickson M, Lee SH, Lowry K, Post A, et al. 2014. Particle flux over the continental shelf in the Amundsen Sea Polynya and Western Antarctic Peninsula. *Elem Sci Anth:* Under review for the ASPIRE Special Feature.

Ducklow HW, Fraser W, Karl DM, Quetin LB, Ross R, et al. 2006. Water-column processes in the West Antarctic Peninsula and the Ross Sea: Interannual variations and foodweb structure. *Deep-Sea Res Pt II* **53**: 834–852.

Fleddum A, Kaartvedt S, Ellertsen B. 2001. Distribution and feeding of the carnivorous copepod *Paraeuchaeta norvegica* in habitats of shallow prey assemblages and midnight sun. *Mar Biol* **139**(4): 719–726.

Gannon JE, Gannon SA. 1975. Observations on the narcotization of crustacean zooplankton. *Crustaceana* **28**(2): 220–224.

Gleiber MR, Steinberg DK, Ducklow HW. 2012. Time series of vertical flux of zooplankton fecal pellets on the continental shelf of the western Antarctic Peninsula. *Mar Ecol Prog Ser* **471**: 23–36.

Harbison GR, McAlister VL, Gilmer RW. 1986. The response of the salp, *Pegea confoederata*, to high levels of particulate material: Starvation in the midst of plenty. *Limnol Oceanogr* **31**(2): 371–382.

Kaufmann RS, Robison BH, Reisenbichler KR, Sherlock RS, Osborn K. 2011. Composition and structure of macro-zooplankton and micronekton communities in the vicinity of free-drifting Antarctic icebergs. *Deep-Sea Res Pt II* **58**(11–12): 1469–1484.

Kawaguchi S, Siegel V, Litvinov FF, Loeb VJ, Watkins JL. 2004. Salp distribution and size composition in the Atlantic sector of the Southern Ocean. *Deep-Sea Res Pt II* **51**(12–13): 1351–1367.

Lavaniegos BE, Ohman MD. 2007. Coherence of long-term variations of zooplankton in two sectors of the California Current System. *Prog Oceanogr* **75**(1): 42–69.

Lee DB, Choi KH, Ha HK, Yang EJ, Lee SH, et al. 2013. Mesozooplankton distribution patterns and grazing impacts of copepods and *Euphausia crystallorophias* in the Amundsen Sea, West Antarctica, during austral summer. *Polar Biol* **36**(8): 1215–1230.

Lee SH, Kim BK, Yun MS, Joo HT, Yang EJ, et al. 2012. Spatial distribution of phytoplankton productivity in the Amundsen Sea, Antarctica. *Polar Biol* **35**: 1721–1733.

Legendre L, Rassoulzadegan F. 1996. Food-web mediated export of biogenic carbon in oceans: hydrodynamic control. *Mar Ecol-Prog Ser* **145**: 179–193.

Lindsay MCM, Williams GD. 2010. Distribution and abundance of larvaceans in the Southern Ocean between 30 and 80°E, *Deep-Sea Res Pt II* **57**(9–10): 905–915.

Marrari M, Daly KL, Timoninc A, Semenovac T. 2011. The zooplankton of Marguerite Bay, western Antarctic Peninsula—Part II: Vertical distributions and habitat partitioning. *Deep-Sea Res Pt II* **58**(13–16): 1614–1629.

Niehoff B. 2007. Life history strategies in zooplankton communities: The significance of female gonad morphology and maturation types for the reproductive biology of marine calanoid copepods. *Prog Oceanogr* **74**: 1–47. doi: 10.1016/j.pocean.2006.05.005

Pakhomov EA, Froneman PW, Perissinotto R. 2002. Salp/krill interactions in the Southern Ocean: spatial segregation and implications for the carbon flux. *Deep-Sea Res Pt II* **49**: 1881–1907.

Pakhomov EA, Perissinotto R. 1996. Trophodynamics of the hyperiid amphipod *Themisto gaudichaudi* in the South Georgia region during late austral summer. *Mar Eco Prog Ser* **134**(1): 91–100.

Pinchuk AI, Hopcroft RR. 2006. Egg production and early development of *Thysanoessa inermis* and *Euphatisia pacifica* (Crustacea : Euphausiacea) in the northern Gulf of Alaska. *J Exp Mar Biol Ecol* **332**: 206–215. doi: 10.1016/j.jembe.2005.11.019

Rabindranath A, Daase M, Falk-Petersen S, Wold A, Wallace MI, et al. 2011. Seasonal and diel vertical migration of zooplankton in the High Arctic during the autumn midnight sun of 2008. *Mar Biodivers* **41**(3): 365–382.

Raiswell R, Tranter M, Benning LG, Siegert M, De'ath R, et al. 2006. Contributions from glacially derived sediment to the global iron (oxyhydr) oxide cycle: implications for iron delivery to the oceans. *Geochim Cosmochim Ac* **70**: 2765–2780.

Randall-Goodwin E, Meredith MP, Jenkins A, Sherrell RM, Abrahamsen EP et al. 2014. Water Mass Structure and Freshwater Distributions in the Amundsen Sea Polynya, Antarctica. *Elem Sci Anth:* Under review for the ASPIRE Special Feature.

Richardson AJ. 2008. In hot water: zooplankton and climate change. *ICES J Mar Sci* **65**(3): 279–295.

Ross RM, Quetin LB, Kirsch E. 1988. Effect of temperature on developmental times and survival of early larval stages of *Euphausia superba* dana. *J Exp Mar Biol Ecol* **121**: 55–71. doi: 10.1016/0022–0981(88)90023–8

Ross RM, Quetin LB, Martinson DG, Iannuzzi RA, Stammerjohn SE et al. 2008. Palmer LTER: Patterns of distribution of five dominant zooplankton species in the epipelagic zone west of the Antarctic Peninsula, 1993–2004. *Deep-Sea Res Pt II* **55**(18–19): 2086–2105.

Schmidt K, Atkinson A, Venables HJ, Pond DW. 2012. Early spawning of Antarctic krill in the Scotia Sea is fuelled by "superfluous" feeding on non-ice associated phytoplankton blooms. *Deep-Sea Res Pt II* **59**: 159–172. doi: 10.1016/j.dsr2.2011.05.002

Sherlock RE, Reisenbichler KR, Bush SL, Osborn KJ, Robison BH. 2011. Boundary layer zooplankton around free-drifting Antarctic icebergs. *Deep-Sea Res Pt II* **58**(11–12): 1457–1468.

Smetacek V, Assmy P, Henjes J. 2004. The role of grazing in structuring Southern Ocean pelagic ecosystems and biogeochemical cycles. *Antarct Sci* **16**(4): 541–558.

Smith Jr KL, Robison BH, Helly JJ, Kaufmann RS, Ruhl HA, et al. 2007. Free-drifting icebergs: Hotspots of chemical and biological enrichment in the Weddell Sea. *Science* **317**: 478–482.

Stammerjohn SE, Maksym T, Massom RA, Lowry KE, Arrigo KR, et al. 2014. Seasonal sea ice changes in the Amundsen Sea, Antarctica. *Elem Sci Anth:* Under review for the ASPIRE Special Feature.

Stammerjohn SE, Martinson DG, Smith RC, Yuan X, Rind D. 2008. Trends in Antarctic annual sea ice retreat and advance and their relation to El Niño–Southern Oscillation and Southern Annular Mode variability. *J Geophy Res* **113**: C03S90.

Steinberg DK, Lomas MW, Cope JS. 2012. Long-term increase in mesozooplankton biomass in the Sargasso Sea: Linkage to climate and implications for food web dynamics and biogeochemical cycling. *Global Biogeochem Cy* **26**: GB1004. doi: 10.1029/2010GB004026

Tagliabue A, Arrigo KR. 2003. Anomalously low zooplankton abundance in the Ross Sea: An alternative explanation. *Limnol Oceanogr* **48**(2): 686–699.

Tang KW. 2003. Grazing and colony size development in *Phaeocystis globosa* (Prymnesiophyceae): the role of a chemical signal. *J Plankton Res* **25**: 831–42.

Tang KW, Smith Jr WO, Elliott DT, Shields AR. 2008. Colony size of *Phaeocystis antarctica* (Prymnesiophyceae) as influenced by zooplankton grazers. *J Phycol* **44**: 1372–1378.

Wahlin AK, Yuan X, Bjork G, Nohr C. 2010. Inflow of warm circumpolar deep water in the central Amundsen Shelf. *J Phys Oceanogr* **40**(6): 1427–1434.

Walker DP, Brandon MA, Jenkins A, Allen JT, Dowdeswell JA, et al. 2007. Oceanic heat transport onto the Amundsen Sea shelf through a submarine glacial trough. *Geophys Res Lett* **34**(2): L02602.

Ward P, Grant S, Brandon M, Siegel V, Sushin V, et al. 2004. Mesozooplankton community structure in the Scotia Sea during the CCAMLR 2000 Survey: January–February 2000. *Deep-Sea Res Pt II* **51**(12–13): 1351–1367.

Wiebe PH, Boyd SH, Davis BM, Cox JL. 1982. Avoidance of towed nets by the euphausiid Nematoscelis megalops. *Fish Bull* **80**: 75–91.

Wilson SE, Steinberg DK, Buesseler KO. 2008. Changes in fecal pellet characteristics with depth as indicators of zooplankton repackaging of particles in the mesopelagic zone of the subtropical and subarctic North Pacific Ocean. *Deep-Sea Res Pt II* **55**(14–15): 1636–1647.

Yager PL, Sherrell RM, Stammerjohn SE, Alderkamp AC, Schofield O, et al. 2012. ASPIRE: The Amundsen Sea Polynya International Research Expedition. *Oceanography* **25**(3): 40–53.

Yager PL, Sherrell RM, Stammerjohn SE, Ducklow HW, Schofield OME, et al. 2014. A carbon budget for the Amundsen Sea Polynya, Antarctica: Estimating net community production and export in a highly productive polar ecosystem. *Elem Sci Anth:* Under review for the ASPIRE Special Feature.

Contributions
- Contributed to conception and design: SW, PY, RS, HD
- Contributed to acquisition of data: SW, RS, SK, PY, HD
- Contributed to analysis and interpretation of data: SW, MW, RS
- Drafted and/or revised the article: SW, RS
- Approved the submitted version for publication: SW, RS, HD, PY

Acknowledgments

The authors wish to thank the captain, technicians, engineers, participating scientists, and crew of the RV *Nathaniel B Palmer* NBP 10–05. Also Raytheon Polar Services, the support teams at Punta Arenas and McMurdo Station, P.-O. Moksness, J. Havenhand, S. Neuer, D. Steinberg, J. Cope, L. Gimenez, and K. Ruck. *Oden Southern Ocean* (SWEDARP 2010/11) was organized by the Swedish Polar Research Secretariat and National Science Foundation Office of Polar Programs.

Funding information

This project was funded by the National Science Foundation Office of Polar Programs, Antarctic Organisms and Eco-systems Award ANT-08–39012 to HD and ANT-08–39069 to PY. Funding was also provided by the Swedish Research Council (824-2008–6429 to P.-O. Moksness and J. Havenhand) for RS and SK to participate in the cruise.

Competing interests

None of the authors declare a competing interest for the work reported in this publication.

Data accessibility statement

All data will be publically available from BCO-DMO: http://www.bco-dmo.org/project/2132.

Supplemental material

- Figure S1. Multidimensional scaling (MDS) plot using a Bray-Curtis dissimilarity matrix on species abundance and depth.
  Numbers in plot indicate ASPIRE station, colors indicate depth of tow, and closed vs. open circles indicate time of tow (night or day).

# 8

# Temporal and spatial patterns of Cl⁻ and Na⁺ concentrations and Cl/Na ratios in salted urban watersheds

David T. Long[1,2]* • Thomas C. Voice[2,1] • Ao Chen[2] • Fangli Xing[2] • Shu-Guang Li[2]

[1]Department of Geological Sciences, Michigan State University, East Lansing, Michigan, United States
[2]Department of Civil and Environmental Engineering, Michigan State University, East Lansing, Michigan, United States

*long@msu.edu

**Domain Editor-in-Chief**
Joel D. Blum, University of Michigan

**Knowledge Domains**
Atmospheric Science
Earth & Environmental Science

## Abstract

The study of sodium and chloride in the environment has a long history with a particular focus on road salting in urban areas. In many studies, spatial scales are limited (e.g., city) and temporal measurements are coarse (e.g., monthly), with the result that our understanding of the hydrogeochemical dynamics is constrained. Through a unique set of spatial and temporal measurements from the State of Michigan we a) examine the spatial distribution of chloride across a broad geographic area, b) explore the temporal behavior of chloride and sodium over hydrologic events capturing snowmelt and rain through salting seasons, c) evaluate the use of chloride/sodium ratios as a tool for linking sources to concentrations, and d) develop a conceptual framework for processes responsible for their environmental concentrations. Results show 1) the short-term and local impact of urban areas on chloride concentrations is clearly delineated, 2) concentration and ratio variations over the hydrographs differ during salting and post-salting periods, 3) chloride/sodium ratios do not clearly indicate a halite source and can be very high (>5) and this is interpreted to be due to the different environmental behaviors of the two ions, and 4) during salting periods, chloride and sodium are quickly removed from the landscape during first flush and diluted as event water begins to dominate, but in post salting periods, only chloride is diluted. We also find evidence for upwelling of brine in some locations. These two solutes are easily measured indicators of human influences on water quality and it is recommended that they routine be included in water quality assessments. However, we suggest more research is necessary to better understand their cycling on shorter time scales and then how this knowledge can be used to inform our understanding of other chemical cycles in the environment.

## Introduction

Sodium (Na⁺) and chloride (Cl⁻) ions occupy a unique role in water quality studies because natural levels are frequently overwhelmed by halite that is intentionally released to the environment through anthropogenic uses (D'Itri, 1992). It is estimated that 50 million metric tons of halite are consumed each year in the U.S., with 30% being used for roadway icing (USGS, 2012). Five states (New York, Ohio, Michigan, Illinois, and Wisconsin) account for 75% of the halite applied, with the major portion of this amount being spread in urban areas (Sleeper, 2013). Chloride levels have been shown to have increased significantly in over 80 percent of urban streams in a U.S. Geological Survey study (USGS, 2014). Single-event application rates typically range from 200 to 400 lb/lane-mi, which can translate to annual applications of 50 tons per mile (Field et al., 1974). Food processing, agricultural, and industrial uses comprise 15, 5 and 4% of the halite used, respectively, with an unknown fraction of this being released directly to the environment in wastes from these operations. Water treatment, primarily home water softeners, constitute about 1% of the total, with nearly all of this being released on-site as regenerant solution. Domestic wastewater, regardless of whether it is treated on-site or collected and treated centrally, represents an indirect source, as most dietary salt is excreted. Considering all of these uses and release pathways, it can be said that with the exception of the small amount of salt that is incorporated into durable products and materials (for example, as chlorinated hydrocarbons), nearly all of the annual consumption of halite is released to the environment in either its crystalline or dissolved forms, and nearly all of the crystalline material dissolves as a result of the very high solubility of halite.

There are, of course, natural sources of $Na^+$ and $Cl^-$. In the mid-continent region of North America, these include atmospheric deposition, and interactions between water and soil, rocks, brines and salt deposits, while additional contributions from salt spray and seawater can be important in coastal regions. In the absence of brines, salt deposits and ocean influences, there are no major sources for $Na^+$ and $Cl^-$. Sodium in natural waters originates from the dissolution of feldspars or cation exchange reactions. The source for $Cl^-$ is as an impurity in minerals (e.g., biotite) or rocks (e.g., limestone). Thus, natural concentrations are typically below about 15 mg/L for each ion and are highly variable (Hem, 1970).

There is a large body of literature dating back to the 1970s and continuing to the present, reporting levels of $Na^+$ and $Cl^-$ in surface and ground waters and discussing sources and transport processes. From these studies it has been established that road salt is the primary source of significant salinization in many locales (Demers et al., 1990; Gardner and Royer, 2010; Kelly et al., 2010, MacLeod et al., 2011; Dailey et al., 2014). Long-term studies suggest increases over several decades across broad regions including in remote locations with relatively low salt levels (Godwin et al., 2003; Kaushal et al., 2005; Kelly et al., 2008; Medalie, 2012; Perera et al., 2013). Techniques including spatial statistics using geographic information systems, multi-variate analysis, and mass-balance modeling have been used to link these broader trends to land-use in general, and specifically to roadways and urban areas (Mattson and Godfrey, 1994; Fitzpatrick et al., 2007; Novotny et al., 2009; Peters et al., 2009; Halstead et al., 2014).

Attempts to identify specific indicators for salt sources have proved useful, but not definitive. Ratios of $Cl^-$ to other halogens (iodide, fluoride, and bromide) are the most widely reported, but there are concerns about whether signals are locally specific, that they often overlap, and that there are measurement difficulties at low salinity levels (Howard and Beck, 1993; Davis et al., 1998; Panno et al., 2006; Dailey et al., 2014). Chloride isotope ratios have also been used (Dailey et al., 2014). Other dissolved constituents have been considered, but in most studies, the extent to which these indicate source, as opposed to other environmental processes, is unclear.

Early work suggested that $Na^+$ to $Cl^-$ ratios might be used for source identification, but there now seems to be general agreement that using these ratios is complicated by differences in how these two ions interact with environmental solids (Neal and Kirchner, 2000). Chloride is usually considered non-reactive, or conservative, in the environment, as most soils have only limited anion exchange capacity at relevant soil pH values (Foth, 1999). However, cation exchange is significant for most solids, and although highly variable, sodium interactions result in changes in the ratios during transport that must be considered (Shanley, 1994). There is now clear evidence from soil column experiments, environmental data over several seasons, mathematical modeling, and a field "experiment" involving high-level salt contamination, that $Na^+$ is sequestered by soils and sediments by cation exchange as dissolved concentrations increase, for example during spring snowmelt, resulting in initial Na/Cl ratios less than one. With continued inputs of salt as it is released from the landscape, the cation exchange capacity can be approached or exceeded, such that little or no $Na^+$ removal takes place and the ratio decreases, approaching a value of one. As the salt is flushed out of the system, concentrations decrease, $Na^+$ is released from the exchange sites, and ratios increase, often to levels well above one. Finally, as the $Na^+$ sequestered on the solids is depleted, soils approach equilibrium with the non-salted water, and ratios return to one (Werner and diPretoro, 2006; Sun et al., 2012, 2014).

The scenario outlined above can be applied conceptually to an idealized one-dimensional system along a flow path, but it is complicated in most real systems by multiple flow paths that can change occur over seasons, or even over hydrologic events. For example, water reaching a stream at any point in time represents a mixture of water that has moved through these different flow paths and can have different solute concentrations and ratios. Thus, changes in pathways and concentration histories along pathways both affect the observations, and unless it can be argued that a particular pathway dominates at a particular time, it is difficult to separate the two effects. Because of this complexity, the competing roles of mixing of different water sources and $Na^+$ exchange have not been fully investigated for short-term hydrologic events. Cherkauer (1975) compared $Na^+$ and $Cl^-$ levels and ratios in an urban and a rural stream in Wisconsin for an autumn rainfall of 2.2 cm. Although the study only discusses the effects of dilution of the streams by the presumably low salinity precipitation, evidence of exchange processes can be found in the differing responses of the two watersheds. Both streams show molar ratios close to one prior to the precipitation event, but the ratio increases for the urban stream in a manner consistent with release of residual sequestered $Na^+$, while the rural stream shows only a decrease attributed to dilution. In addition, two sequential peaks were found for ratio increases in the urban stream suggesting two primary flow paths. Similar findings were reported by Ostendorf (2013) and Cooper et al. (2014) in observing one or more first-flushes of conductivity, accompanied by either more specific measures (but not of $Na^+$ and $Cl^-$ across the event) or modeling supporting an interpretation that road salt was the likely source and that exchange mechanisms were operative. Reisch and Toran (2014) took a similar approach for a snowmelt event, using conductivity data to suggest changes in flow paths over the course of the event, and how these relate to temperature and precipitation.

Hysteresis loop graphs are an approach that has been used to explore decoupling of variables that are mechanistically related but do not fully correlate in complex hydrologic systems. Andermann et al. (2012) investigated relationships between precipitation and discharge over annual cycles to quantify the effects of transient storage, while Williams et al. (1989) compared sediment concentration to discharge ratios on the rising and falling limbs of the hydrograph to explore the different patterns of sediment transport during events. Evans and Davies (1998) used the approach for dissolved solutes, linking observed concentration vs. discharge patterns to those predicted by a model based on three water sources whose contributions change over the event, and an assumption that the solutes are conservative. Aubert et al. (2013) employed annual hysteresis loops to investigate how several non-conservative water quality parameters (nitrate, sulfate, dissolved organic carbon, dissolved inorganic carbon) varied as a function of climatic conditions. Sun et al., (2014) used the technique of Aubert et al. (2013) to study annual changes in average monthly Na/Cl ratios. The data spanned a period from 1994 to 2012, but were averaged for periods (e.g., 1944 to 1960, 1961 to 1975). They report a change in hysteresis loop from a counter clock-wise pattern in the early data to a mix of counter clock-wide and clock-wide patterns in the later data. They attributed this change to increases in road salting and in impervious surfaces. To the best of our knowledge, the approach has not been used to investigate the roles of flow path and $Na^+$ exchange for deicing salt in the environment over short (i.e., hydrograph) time scales, but we suggest that it may have potential.

The general hypothesis driving this study, and much of the body of past work, is that the use of halite is the dominant process adding contaminant $Na^+$ and $Cl^-$ to most environments and that urban areas are "hot spots" because of road salting. We have attempted to be more specific and consider that both the spatial and the temporal distribution of these ions are affected by the time and location of sources, the dynamic behavior of transport processes, and $Na^+$ exchange reactions. Therefore, the goals of this paper are to a) examine the spatial distribution of $Na^+$ and $Cl^-$ across a broad geographic area, b) explore the temporal behavior of $Na^+$ and $Cl^-$ over the primary hydrologic events that release these materials, spring precipitation and snowmelt, c) evaluate the use of Cl/Na molar ratios as a tool for linking sources to observed concentrations, and d) to develop a conceptual framework for of processes responsible for $Na^+$ and $Cl^-$ levels found in the environment.

## Materials and methods

### Spatial patterns data

Two unique data sets were explored. One set resulted from data mining the Michigan Department of Environmental Quality (MDEQ) Waterchem and Wellogic databases (MDEQ, 2014). Concentrations for $Na^+$ and $Cl^-$ in drinking water wells were extracted from this database for the whole State of Michigan using the Michigan Ground Water Management Tool developed for MDEQ by Michigan State University (MSU, 2014). This tool allowed not only for the extraction of concentration values from the databases, but also the spatial visualization of the data. The MDEQ-MSU database comprises over 550,000 samples, but does not differentiate the aquifers from which the water is being drawn from. The second database arises from the U.S. Geological Survey's Regional Aquifer-System Analysis (RASA) program (Mandle, 1986; Dannemiller and Baltusis, 1990) with additions from Long et al. (1988). The RASA-MSU database comprises 700 samples from the three main aquifers in the Michigan Basin. These are the bedrock Marshall and Saginaw Formations (Mississippian and Pennsylvanian ages, respectively) (Fig. 1) and the overlying Glacial Drift (Bauer et al., 1996, Meissner et al., 1996, Wahrer et al, 1996). It should be noted that the dominate deicer used in the State of Michigan is halite (NaCl) (Rustem et al., 1993).

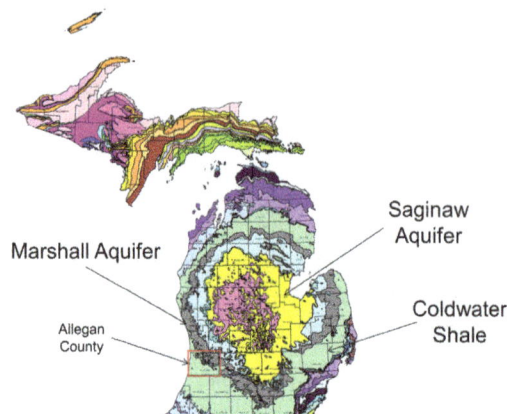

**Figure 1**

Bedrock geology of the State of Michigan showing the three main aquifers in the Michigan Basin.

Map modified from http://w3.salemstate.edu/~lhanson/gls210/gls210_struct.htm. Last assessed December 7, 2014.

**Figure 2**
Map of the Red Cedar River, Michigan showing sampling site in East Lansing, Michigan.

Map modified from http://en.wikipedia.org/wiki/Red_Cedar_River_(Michigan). Last assessed December 7, 2014.

## Temporal patterns data

The temporal aspects of the study involved the Red Cedar River, Michigan, U.S.A. (Fig. 2) which has a watershed area of approximately 461 mi$^2$ (1194 km$^2$) and a lengths of 51.1 mi (82.2 km). The watershed is a glaciated landscape characterized predominantly as medium-textured glacial till with abundant eskers. Two-thirds of the watershed is characterized as agriculture, wetlands, and grasslands. At its confluence with the Grand River, the land use is highly urbanized and includes the cities of Lansing and East Lansing, and the campus of Michigan State University (MSU). The population of East Lansing is approximately 100,000 during the time of the sampling, as MSU was in session. Upstream communities of Williamston, Okemos, Webberville and Fowlerville contribute another 30,000 inhabitants. Landuse/land cover is diverse consisting grasslands, wetlands, forests, agricultural, and developed residential, commercial, and industrial land. In portions of the watershed there is high potential for the influence of septic systems on river chemistry. However, much of the land is engineered to drain water from the soil by way of drain tiles, buried conduits that collect and convey groundwater from the soil to the river. Urban areas are largely drained by storm sewers. Spring snowmelt and summer rains are normally the dominant hydrologic events for this river. When precipitation occurs, much of it runs off impervious surfaces and goes directly into storm drains. Waters drained by both drain tiles and storm sewers are not treated and run through pipes to outfalls along the river. A series of maps showing data relevant to the watershed can be found at http://redcedarriver.weebly.com/map-gallery.html.

The data reported here originate from a collection of research and training activities over several decades including independent undergraduate and graduate student projects and student laboratory exercises in an environmental geochemistry class. Data sets from 1994 and 2013 have been selected because they were specifically designed to study chemical behavior during first flush. However, the two projects had different foci and therefore only measurements of river discharge, dissolved Cl$^-$, and meteorological data are similar. Regardless of the task, all samples were collected using the same methodology in which river water was collected in precleaned and river rinsed polyethylene bottles, filtered through 0.45 μm Millipore filter, and stored at 4 °C. The sampling site was on the campus of MSU.

Discharge measurements were obtained from the USGS gaging station (Hydrologic Unit 04050004) located along the river on the MSU campus. This station has been collecting 15 minute discharge data since 1902. By using this station the chemical data reflect the upper 355 mi$^2$ of the Red Cedar watershed. The precipitation data were collected from Michigan State University Enviro-weather (Formerly Michigan Automated Weather Network (MAWN), which has a station on the campus of MSU).

Measurements for total dissolved solids were done at the time of collection using Myron L 4PII conductivity/TDS/resistivity meter. Na$^+$ concentrations were measured immediately after collection using flame atomic emission spectroscopy. Because of its concentration stability in the samples, Cl$^-$ concentrations were measured in batches following collection.

In this paper we use the ratio Cl/Na rather than Na/Cl, which is frequently used. Our rational is that we found very high Cl/Na ratios that are not typically appreciated using Na/Cl ratios. Considering halite as the

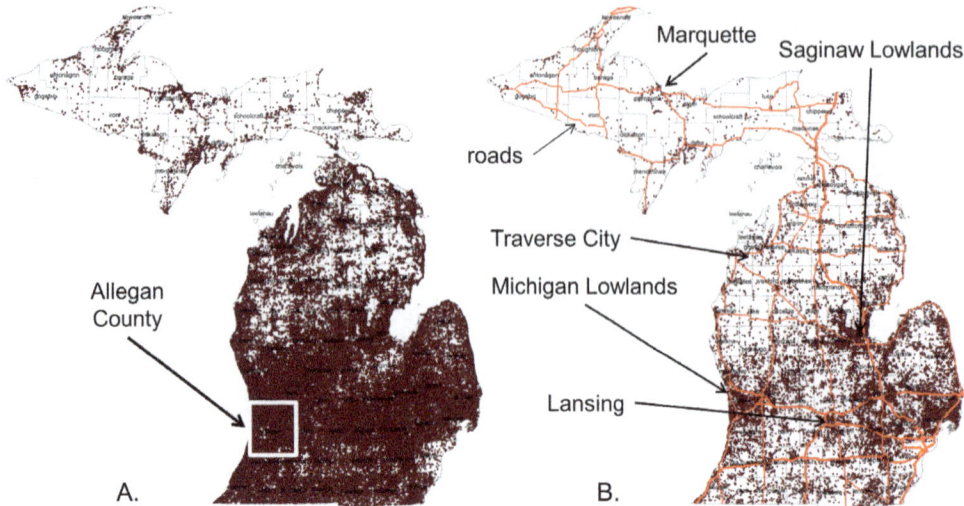

Figure 3

Chloride concentrations in drinking waters wells of Michigan (data from Michigan Department of Environmental Quality).

A. concentrations > 10 mg/L and B. concentrations > 100 mg/L.

dominant contaminate source, these high ratios serve to emphasize the different control on the environmental behaviors of Cl⁻ and Na⁺.

## Results and discussion

### Spatial patterns of Cl, Na, and Cl/Na

As mentioned, dissolved concentrations of Cl⁻ in near surface waters can come from various sources, but in most environments, there are no natural mineral/rock sources for this ion. Thus, natural concentrations should be relatively low, a condition that would define natural near-surface waters in most of the State of Michigan. In these aquifers, 10 mg/L has been estimated as the natural concentration of Cl⁻ (e.g., Wharer et al., 1996). Here the dissolution of limestone, which is prevalent in the bedrock and glacial drift might be considered one of the dominant sources. This number is not fixed, of course, and can vary across the state. It has been shown that in a similar setting in a watershed in north eastern Illinois that a clear link to the influence of halite near surface water chemistry could not be established until dissolved Cl⁻ and Na⁺ concentrations were over 50 mg/L might be consider elevated (Long and Saleem, 1974). We considered for illustration purposes, 100 mg/L as representing Cl⁻ concentrations that are clearly elevated from natural concentrations. The prevalence of Cl⁻ in the near surface waters of the State of Michigan's is illustrated in Figure 3 (a, b) which shows the distribution for concentrations above 10 mg/L and 100 mg/L, respectively from the MDEQ-MSU data set.

The population density in Michigan is highest in the southern half of the Lower Peninsula so it must be considered that interpretation of this distribution data is biased by the number of wells in a particular area. However, the spatial pattern of Cl⁻ with concentrations above 100mg/L suggests two clear patterns. The first is the cluster of high concentrations in the near-surface waters of the Saginaw and Michigan lowland areas. Population density cannot account for these high concentrations as these areas are highly agricultural and Long et al. (1988) and Hoaglund et al. (2004) have shown that the high concentrations of Cl⁻ in the Saginaw Lowlands are due to the upwelling of brine. Since the geohydrologic situation in the Michigan Lowlands is similar to that of the Saginaw Lowlands (Westjohn and Weaver, 1996; Westjohn et al., 1994), it has been hypothesized that a similar process is causing elevated Cl⁻ concentrations there (Bauer et al., 1996; Fitzpatrick et al., 2007). The second pattern is the higher concentrations of Cl⁻ in near-surface waters in areas such as southeastern Michigan, the Greater Lansing area, Traverse City, Marquette and along highways (e.g., I-94). The like cause of the second pattern is road salt, particularly in urban environments. It is clear from this figure that the impact of anthropogenic activities associated with the use of halite is widespread.

Theoretically, waters highly influenced by halite would have molar Cl/Na ratio near 1, but reported ratios are highly variable, and this has not proven to be a definitive indicator of source.

This is illustrated in the range of Cl/Na ratios in over 12,000 groundwater samples collected from drinking water wells in Allegan County, Michigan, shown in Figure 4, which is a sub set of the MDEQ-MSU database. Thus, these data might reflect the range of values to be expected in most near-surface salted environments. This region is characterized by a mix of agricultural, forested, low-density urban and suburban land use. Sources for Cl⁻ and Na⁺ included road salting in urban areas and county and state roads, septic systems and other animal wastes, spray irrigation and possible brine upwelling as a portion of the county is in the Michigan Lowlands and overlies the Marshall Formation contains Na-Cl brines at depth (Bauer et al., 1996).

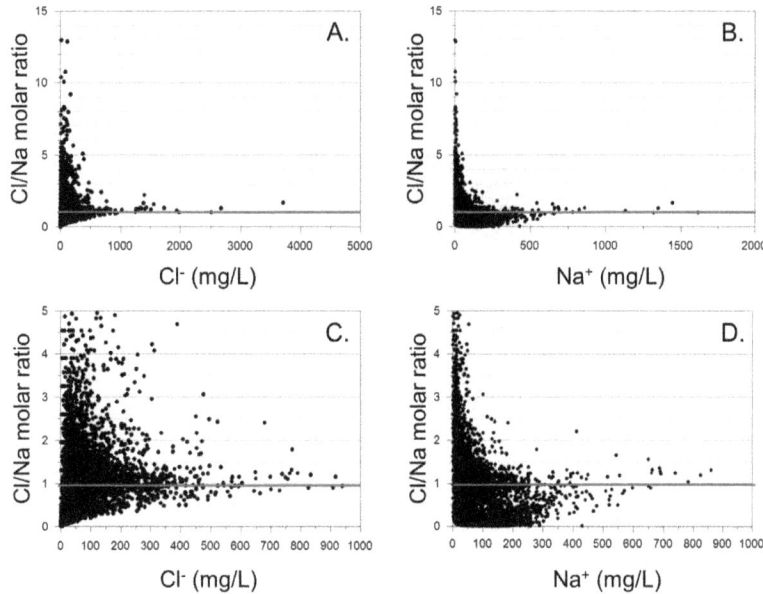

Figure 4

Cl/Na molar ratios versus Cl⁻ and Na⁺ concentrations (mg/L) in drinking waters wells of Allegan County Michigan (data from Michigan Department of Environmental Quality).

A. Cl⁻ with Cl/Na ratios < 15, B. Na⁺ with Cl/Na ratios < 15, Cl⁻ with Cl/Na ratios < 5, and D. Na⁺ with Cl/Na ratios < 5.

The highest Cl/Na ratios (approaching 15) are constrained to low Cl⁻ and Na⁺ concentrations (Fig. 4a, b). At Cl⁻ and Na⁺, concentrations greater than about 600 mg/L, ratios tend to cluster around or slightly above one. Ratios below one for the Cl⁻ plots show a clustering bounded by what might appear to be a mixing curve between solutions containing low Cl/Na ratios and low Cl⁻ concentrations with solutions containing higher ratios (above 1) and high Cl⁻ concentrations. Such a trend is not as clear for Na⁺, however, there appears to be dominant cluster "peaking" around 1 at concentrations below 200 mg/L In addition, the Na⁺ plot shows a second dominant clustering bounded by a Cl/Na ratio of 0.5 and Na⁺ concentrations < 300 mg/L.

Cl⁻ and Na⁺ are not correlated to total dissolved solids (TDS) at concentrations below 500 mg/L (Fig. 5a, b). Fewer samples are shown here because only a portion of the Allegan County data reported TDS values. At concentrations above 500 mg/L, there appears to be a rough correlation, but the extent of the scatter suggests that other ions contribute to TDS and that these are variable in a manner unrelated to Cl⁻ and Na⁺. Ratios of Cl/Na show no trend with respect to TDS at any concentration increment below TDS values of 1500 mg/L (Fig 5c). The lack of correlation with TDS is typical for groundwater chemistry in the Lower Peninsula of Michigan (Fig. 5d).

These data show that there is no simple relationship between Cl⁻, Na⁺, Cl/Na ratios, and TDS. Cl/Na ratios below 1 might be expected in uncontaminated near-surface waters as there are no natural rock sources

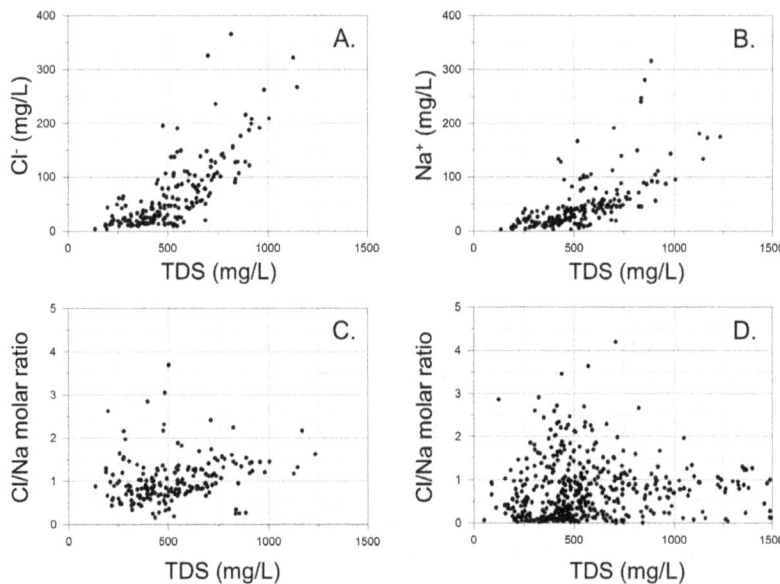

Figure 5

Graphs showing relationships of Cl⁻ and Na concentrations and Cl/Na molar ratios to TDS.

A. Cl⁻ for Allegan County, B. Na⁺ for Allegan County, C. Cl/Na for Allegan County, and D. Cl/Na for the three main aquifers in the Michigan Basin. Data are from MDEQ-MSU and RASA-MSU data bases, respectively. (See text for references)

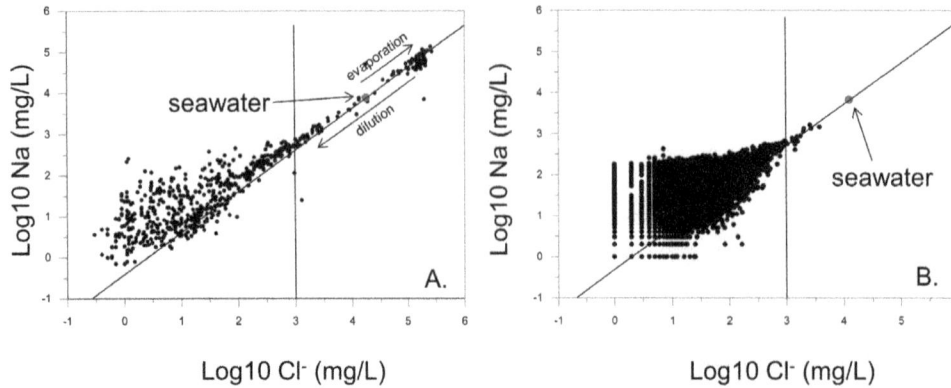

Figure 6
Graphs showing relationships of
Log10 Cl⁻ versus Log10 Na⁺ for
A. the three main aquifers in the
Michigan Basin and B. Allegan
County groundwater.

Seawater is shown for reference.
Data are from RASA-MSU
and MDEQ-MSU data bases,
respectively. (See text for
references)

for Cl⁻ in this area, while dissolved Na⁺ can be derived from weathering of rock and exchange reactions. Weathering processes are most likely limited to low Na⁺ concentrations, whereas exchange reactions involving release of halite-derived Na⁺ could produce much higher levels. Hypotheses for ratios above 1 include application of alternative deicers and selective retardation of Na⁺ during transport. Applications of $CaCl_2$ and KCl for deicing could account for some high Cl/Na ratios but these deicers are not used widely so their effect on water chemistry would likely be minimal compared to the large amount of halite used on landscapes.

Upwelling of brine or its application for deicing might also explain the high Na⁺ and Cl⁻ concentrations observed in some samples. Brine in the lower formations of the Michigan Basin was created from the evaporation of ancient seawater (e.g., Wilson and Long, 1993a, 1993b). A common technique to examine for the influence of brine with near-surface water is to plot an indicator chemical that reflects the degree of evaporation against solute concentrations (Long et al., 2009). The concentration changes of the solutes in seawater during evaporation are well known. A comparison is then made between the groundwater date and expected trends. Typically, Br⁻ is used as the indicator of the degree of evaporation as its behavior is conservative through much of the evaporation sequence. When Br⁻ data are not available, Cl⁻ is often used. When sea water evaporates, Cl⁻ and Na⁺ concentrations increase conservatively along the trajectory shown on Figure 6a (McCaffrey et al., 1987). When the brines mix with the near-surface freshwater, they are diluted down this trajectory. The groundwater from the three main aquifers in the Michigan Basin (Fig 6a) characterize the expected trajectories when near surface waters interact with brine from the evaporation of seawater (e.g., Wilson and Long, 1993a). At concentrations below 1000 mg/L, there is much scatter in the data with Cl/Na ratios varying from high to low. A larger proportion of the data plot above (Na⁺ rich) the potential mixing line than below (Cl⁻ rich). The Na⁺ concentration appear to be converging to a concentration of about 300 mg/L at low Cl⁻ concentrations, while there is no clear trend for Cl⁻. Above Cl⁻ concentrations of 1,000 mg/L the data tightly cluster along the evaporation trajectory line. The data at higher Cl⁻ concentrations on Figure 6a that plot off of the evaporation trajectory have been concentrated to a point that Cl⁻ and Na⁺ are no longer conservative as halite begins to precipitate out (McCaffrey et al., 1987). These solutions are not influencing Cl⁻ and Na⁺ concentrations in near-surface waters.

The trends in Cl⁻ and Na⁺ concentrations for the Allegan County data are similar (Fig. 6b). The MDEQ-MSU data set at low concentrations involved only one significant figure causing the data patterns at low concentrations. Similar to Michigan Basin groundwater, the majority of the data plot above the potential mixing with Na⁺ converging near 300 mg/L. A difference from the Michigan Basin groundwater trends is that there are significantly more data that plot below the potential mixing line (Cl⁻ rich) with a possible trend to 10 mg/L at low Na⁺ concentrations. The preponderance of data plot below a Cl⁻ concentration of 1000 mg/L, but because of the scatter in Cl/Na ratios, a specific source is not apparent. However, it is clear that there is an influence of solutions with high concentrations of Na⁺ and Cl⁻ and the trend towards seawater indicates the potential for near-surface water chemistry to be influenced by brines.

## Temporal patterns of Cl⁻, Na⁺, and Cl/Na: 1994 data set

The 1994 data (Fig 7) were collected during the first 100 days of the year. The collection period was characterized as a strong El Niño and included four storm events. Only Cl⁻ concentrations were measured. The low-flow or base-flow discharge appears to be about 100 ft³/sec (cfs) for the period. For the first 28 days, the discharge is relatively constant even though there are some small precipitation events, as these were snowfalls that did not immediately affect stream discharge. The first major high discharge event occurred around the 28th day. The delay from precipitation to the discharge peak is approximately 5 days, which is typical for the Red Cedar River at this location. There are four major peak discharge events during this time period. It can be noted that discharge does not return to the base flow value of 100 cfs, rather it increases over the course of the four events.

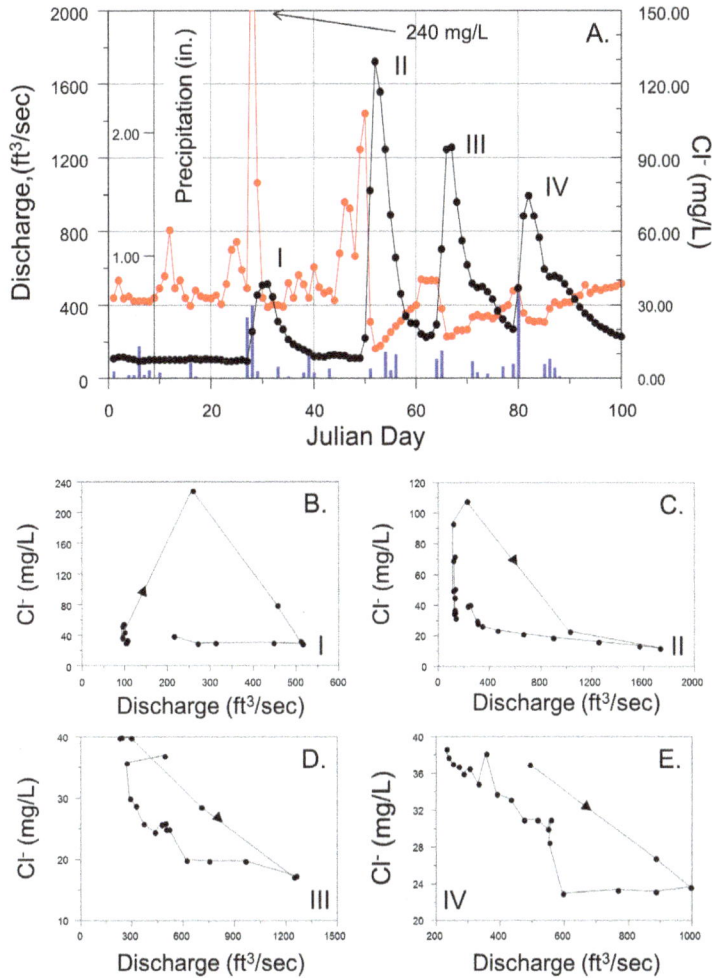

Figure 7

**Red Cedar hydrograph for the first 100 days of 1994.**

A. changes in discharge, and Cl⁻ concentrations, and B. C. D, and E. are Cl⁻ discharge hysteresis diagrams for events, I, II, III, IV respectively shown in A.

Chloride concentrations at low-flow discharge are about 34 mg/L. While the discharge is constant for the first 28 days, there are two peaks in the Cl⁻ concentration that indicate additional inputs. The peaks around days 10 and 24 are from minor snowmelts, although not enough melting occurred in either case to measurably change discharge. The precipitation event on Day 27 produced a first-flush of Cl⁻, peaking at 240 mg/L on Day 28, well before peak discharge on Day 32. It can be noted that the Cl⁻ peak is extremely sharp, suggesting that the precipitation immediately liberated a pool of Cl⁻ from the landscape as a pulse that reached the stream quickly. The return in Cl⁻ to levels at or below those observed at base-flow suggests that this mass was likely carried by surface runoff.

A similar temporal pattern is seen for a snowmelt event around day 45, with Cl⁻ concentrations preceding the discharge peak by several days. It can also be seen, however, that the concentration peaks at a much lower level, about 100 mg/L, and the peak is much broader. This seems to suggest that the additional Cl⁻ being liberated is either less accessible to the water that transports it, or the transport pathway is less direct. This second event produces a much larger change in discharge, which may explain why the Cl⁻ levels drop to less than half the base-flow levels, and then recover linearly on the falling limb of the hydrograph. This observation is consistent with the effects of dilution by the precipitation and snowmelt that follows the first flush of Cl⁻. The first flushes of Cl⁻ associated with events 3 and 4 again show a similar temporal pattern but the magnitude of the concentration peak is diminished. This suggests that after the first two events, there is little remaining road salt on the landscape. Decreases in Cl⁻ concentrations at peak discharges still occur however. It also appears that the amount of the decrease in Cl⁻ concentrations is related to the magnitude of the discharge peak and the trajectory of recovery related to shape of the falling limb curve. For example for event 2, the falling limb is rapid and relatively linear. The recovery of the Cl concentration increases linearly. As the shapes of the falling limb become nonlinear (e.g., events 3 and 4) the recovery trajectory of Cl is also nonlinear but still inversely reflects the trajectory of the falling limb.

The relationship between discharge and Cl⁻ for the Red Cedar River in the 1994 data might be considered typical for rivers (e.g., Cooper et al., 2014). These patterns would include a delay in peak discharge after the start of an event (snowmelt, rain), and evidence of first flush of chemicals. However, not typically reported

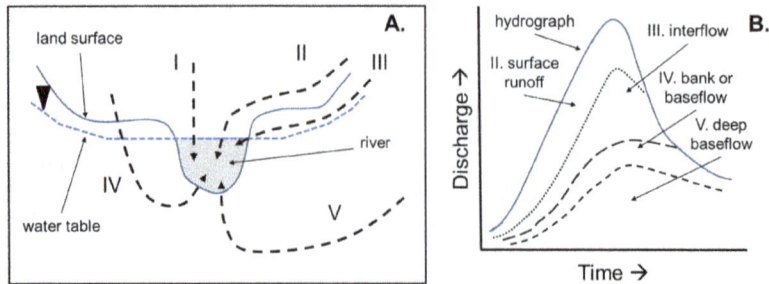

Figure 8

Conceptual framework for dissolved chemicals such as Cl⁻ and Na⁺ to streams.

A. pathways of water of water to streams where I. is direct precipitation input, II. is surface runoff, III is interflow, IV. is bank flow or baseflow, and V deep baseflow; and B. relative contributions of the various water pathways in a to the total stream hydrograph.

in the literature is the actual change in Cl⁻ concentrations over a hydrograph and the concentration decrease following peak discharge. The patterns observed in the 1994 data are similar to data sets for the Red Cedar River collected for the same time period every year from 1999 to 2014.

These patterns can be related to changes in the pathways of water to the river and the relative contributions of water from these pathways to the stream hydrograph (Fig. 8a and b). We will ignore the effect of evaporation because it would not play a role during the course of a hydrologic event. Pathway I, direct precipitation input is not considered to be a major contributor to the mass of water or of chemical input into the river. Pathway II, overland flow or surface runoff caused by the precipitation event, is directly responsible for the concentration increase (i.e., first flush) of Cl⁻ in the river. This water can profoundly change the chemistry of the river, but not necessarily the mass of water and discharge. Increasing discharge is slightly delayed and is the result of Pathway II and the increasing importance of pathway III (interflow). At the discharge peak, Cl⁻ concentrations are low due to dilution and then increase as pathways IV (base or bankflow) and eventually V (deep base flow) become more dominant during the falling limb.

The hysteresis loop plots also shown in Figure 7 (b, c, d, e) provide a different way of looking at the changes across the four events. The plot for event I represents simple first-flush behavior, where we see a clockwise progression with concentration, increasing and peaking with the first increases in flow, decreasing at flow continues to increase until base-flow levels are reached, followed by a horizontal return toward the starting point as discharge returns to pre-event levels. The plot for event II also clearly shows first-flush behavior, but in addition, the effect of dilution is present with the concentrations at high flow reaching levels below those found pre-event, and concentration increasing as discharge subsides. In the plots for events III and IV, we see little evidence of first flush (note the different scale), but dilution behavior remains clear.

### Temporal patterns of Cl, Na, and Cl/Na: 2013 data sets

Two series of data are available for 2013. Both data series are characterized by a period of a very weak La Niña. The first chemical data series (Fig. 9a) covers Julian days 7 to 89 (January 7 to March 29) with four larger events and three relatively smaller events observed. Values for Cl⁻ and TDS are available for this data set. The second data series (Fig. 9b) covers a later time period of Julian days 97 to 163. (April 7 to June 11). There are two large events and three small events during this time period. The largest event exceeded the bankfull discharge level. Cl⁻ and Na⁺ values and Cl/Na ratios are available for this data series. Low-flow discharge values are similar to those of 1994. Low-flow Cl⁻ concentrations (~70 mg/L) are higher than in 1994. To better explore the patterns Cl⁻, Na⁺, and Cl/Na, values for discharge, Cl⁻, Na⁺, TDS were normalized to their highest value during the measurement period. Normalization allows for better characterization of the relationships and rates of change in the relationships over time.

In the early data series (Fig. 9a), Cl⁻ and TDS concentration changes do not always correlate. For the first hydrologic event, Cl⁻ shows evidence of first flush while TDS does not. If fact, TDS decreases slightly as Cl⁻ concentrations rise. Preceding the second event, which is the highest discharge for the period, both Cl⁻ and TDS show first-flush behavior, as well as reaching their highest values for the time period. Both Cl⁻ and TDS concentrations decrease similarly in magnitude to low values at peak discharge and then increase, however the trajectories of their recoveries differ. Total dissolved solids concentrations recover more quickly and the Cl⁻ trajectory is similar to that observed in the 1994 data. For both Cl⁻ and TDS, first flush clearly precedes events 3 and 6 and again their trajectories for recovery are not the same. The magnitude of the concentration increase for event 6 is second highest for these parameters with that for Cl⁻ being greater than that for TDS. Although concentration changes for event 6 are relatively high for the period, the discharge peak for 6 is also one of the lowest. Concentration decreases at peak discharge are not observed. Event 7, the second largest event during this period, starts at Julian day 69 with a precipitation event. Cl⁻ concentrations increase slightly before the precipitation event as result of snow melt. There is no change in TDS concentration. Both concentrations greatly decrease in a similar magnitude at peak discharge, but the slopes for recovery differ.

For the second data series (Fig. 9b), Cl⁻ and Na⁺ concentration changes do not always correlate. Both Cl⁻ and Na⁺ exhibit first-flush behaviors during hydrologic event 1.. Their slopes of increase, decrease, and recovery are similar. The pattern in their recovery trajectories are similar to those observed for Cl⁻ in 1994

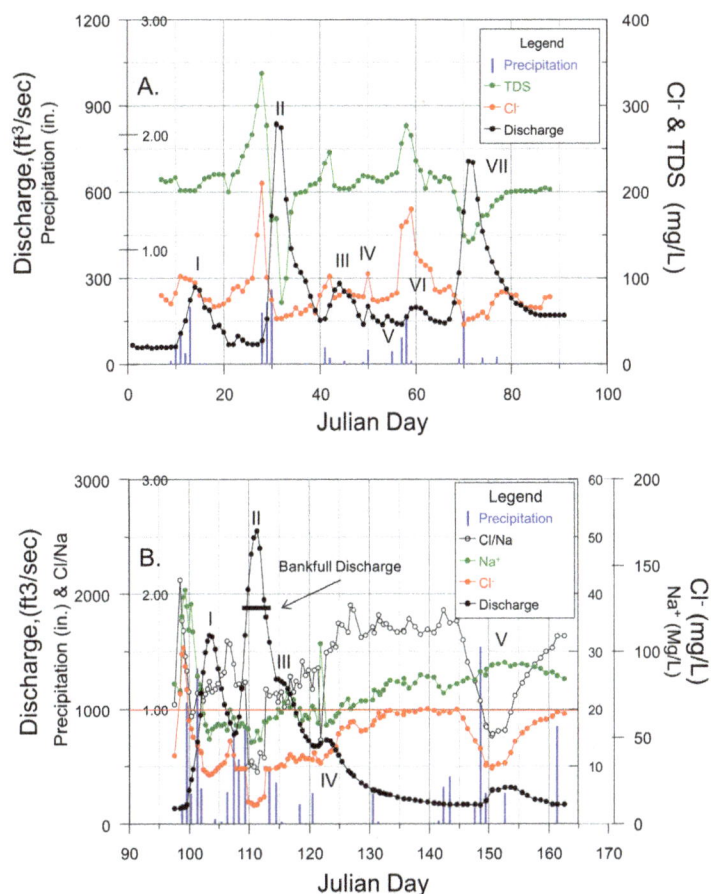

Figure 9

Red Cedar hydrograph for selected periods of 2013.

A. Julian days 7 to 89 (January 7 and March 29, respectively) showing changes in discharge, and Cl⁻ and TDS concentrations and B. Julian days 97 to 163 (April 7 and June 11, respectively) showing discharge and Cl⁻, Na⁺, concentrations and Cl/Na molar ratios. Discharge events are noted with Roman Numerals.

and the earlier measurement period in 2013. Prior to event 2, the highest discharge event in this period, both Cl⁻ and Na⁺ show first flush curves with concentration peaks that are significantly less than in event 1. Both of their concentrations decrease at peak discharge but the decrease for Na⁺ is much less than that for Cl⁻. The shapes of decrease and recovery trajectories below the peak discharge differ from those observed in the 1994 and early 2013 data. After this event, changes in Cl⁻ and Na⁺ concentrations generally parallel one another until event 5, where Cl⁻ concentrations substantially decrease and recover while those of Na⁺ slightly increase and recover. Event 5 produced only a modest increase in discharge, and appeared in late May, long after the deicing season.

As might be expected, the Cl/Na ratios vary because of the differential changes in Cl⁻ and Na⁺ concentrations. During the first flush for event 1, the ratio increases and then declines before the peak concentrations. Low ratios (near 1) occur before the concentration lows for Cl⁻ and Na⁺ and then increase before a rapid rise during first flush for event 2. Ratios fall below 1 at peak discharge and then increase to relatively steady ratios around 1.6. Ratios decrease at peak discharge for event 5. It appears that event 1 releases much of the Cl⁻ and Na⁺ on the landscape through pathway II, while we see significant dilution of Cl⁻ with the very high discharge associated with event II. Na⁺ shows very little dilution behavior, perhaps because pathway III has become more important, and the release of previously sequestered Na⁺ maintains dissolved concentration levels. This is seen even more dramatically in event 4, perhaps because pathway IV is of greater importance and the pool of Na⁺ available for release has increased. It is interesting to note the Na⁺ levels appear to start decreasing at about Day 150, possibly suggesting that the supply of sequestered Na⁺ is starting to by depleted.

Despite the visual complexity of the integrated plots, the interpretation of the Cl⁻ concentration hysteresis loops (Fig. 10) is essentially the same as offered for the 1994 data. We can clearly see the first flush events, the dilution that occurs at peak discharge, and the recovery to pre-event conditions. A hysteresis plot of the Cl/Na ratios (Fig. 11) for the second 2103 data set is also informative. The highest ratios occur for the first couple of first-flush events, and we only see ratios less than 1 under the highest discharge conditions. This indicates that over most of this period (the first half of the year), Na⁺ is being sequestered. We can speculate from the release that occurs under diluting conditions and the literature on annual cycles that events in the second half of the year are responsible for most of the Na⁺ release that eventually occurs.

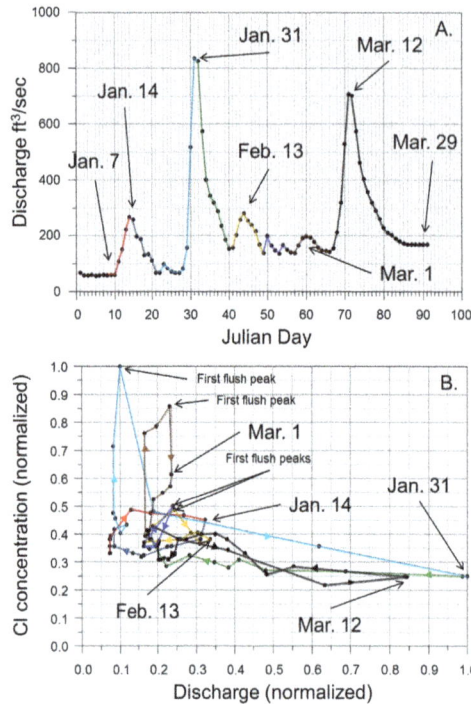

Figure 10

Patterns of change in the Red Cedar for 2013 Julian days 7 to 89 (January 7 to March 29).

A. River hydrograph for the period showing events and B. Hysteresis diagram (Cl⁻ versus discharge) for the period. Cl⁻ concentrations and discharge values are normalized to their highest value for the period. Discharge events are noted with Roman Numerals.

## Summary and conclusions

The results of the spatial analysis of Cl⁻ and Na⁺ across the state of Michigan show that elevated Cl⁻ and Na⁺ are widespread. The higher Cl⁻ concentrations in urban areas and along interstates are consistent with halite as the dominant source. In addition, there is evidence for the influence of upwelling brine in some locations. Ratios of Cl/Na do not approach a diagnostic value (e.g. ~1) until Cl⁻ concentrations reach 1000 mg/L. Below 1000 mg/L the ratios are highly variable above and below 1. Very high Cl/Na ratios (> 5) are not observed at Cl⁻ and Na⁺ concentrations above about 100 mg/L. Exchange reactions of Na⁺ are more dominant at low Na⁺ concentrations. Because of the dynamic nature of exchange processes, and the likelihood of mixed water sources/pathways, Cl/Na ratios are not all that useful for spatially distributed data with no temporal control.

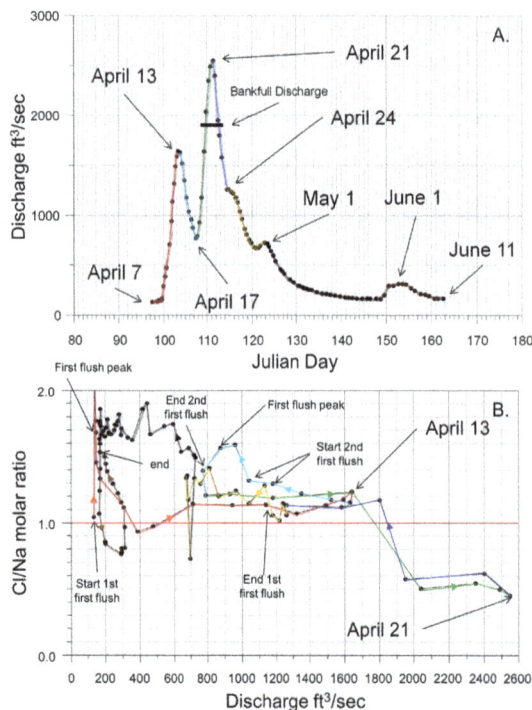

Figure 11

Patterns of change in the Red Cedar for 2013 Julian days 97 to 163. (April 7 to June 11).

A. River hydrograph for the period showing events and B. Hysteresis diagram (Cl⁻ versus discharge) for the period. Cl⁻ concentrations and discharge values are normalized to their highest value for the period. Discharge events are noted with Roman Numerals.

One might speculate however, that in this setting, any ratio above 1 at concentrations below 1000 mg/L can be attributed to the use of halite. Chloride and sodium concentrations are not related to TDS at TDS values < 600 mg/L, patterns in relationships differ for Cl⁻ and Na⁺ at TDS concentrations > 600 mg/L and there is no relationship of Cl/Na to TDS at TDS values < 1500 mg/L.

The results of the temporal analysis show that most of the release of Na⁺ and Cl⁻ from the application of road salt occurs in a first-flush when overland flow is dominant. Chloride, sodium, and total dissolved solids decrease to low values at peak discharge as interflow and bank flow becomes more dominant. Chloride concentrations recover from dilution along a trajectory that is inversely related to the trajectory of falling limb as base flow and bank flow become more dominant as "event" water is flushed out. Sodium concentration changes track that of chloride during first flush, decrease, and recovery phases. However, the magnitude of concentration changes is dampened compared to Cl⁻ as a result of exchange reactions. As the spring season progresses (in the absence of road salting), the concentration changes of Na⁺ and Cl⁻ can be decoupled as the pathways for Na⁺ and Cl⁻ transport shift from being direct (e.g., overland flow) to increasingly less direct. Sodium concentrations become more controlled by water-rock interactions, with sequestration occurring during most of the spring season. Chloride behavior remains conservative: its concentration is influenced by the legacy of road salt in the system and it is more sensitive to changes in discharge than is sodium. Therefore, during storm events Na⁺ is sourced from water-rock reactions (e.g., desorption) while Cl⁻, with no direct contaminant source or rock source, is diluted. The differential behavior of Cl⁻ and Na⁺ result in changes in Cl/Na ratios over a hydrograph. The pattern of the change in Cl/Na ratios is primarily driven by changes in Cl⁻ concentrations. Similar to what was learned in the spatial analysis, Cl⁻ concentrations does not always track TDS values, suggesting that one must use caution in using TDS or specific conductance values to infer Cl⁻ cycling.

In terms of our goals, these results show that 1) the impact of urban areas on Cl⁻ concentrations in the environment can be clearly delineated, 2) concentration and ratio changes over the hydrographs differ during salting and post-salting periods with Cl⁻ concentrations the dominant factor causing ratio changes, 3) Cl/Na ratios do not clearly indicate a halite source (e.g., ratio of 1) and can be very high because the different environmental behaviors of the two solutes whose concentration trends are often decoupled, and 4) during salting periods, Cl⁻ and Na⁺ are quickly flushed from the landscape during first flush and diluted as event water begins to dominate, while in post salting periods, only Cl⁻ is diluted. Additional findings provide evidence for 1) the influence of upwelling brine on Cl⁻ and Na⁺ concentrations and 2) possible concentrations limits on Cl⁻ and Na⁺ in the environment. These limits might be on the order of 300 mg/L for Na⁺ at low Cl⁻ concentrations and 10 mg/L for Cl⁻ at low Na⁺ concentrations.

Dissolved Cl⁻ and Na⁺ are two of the most easily measured chemical indicators of human influence on the environment. As demonstrated here, Cl⁻ concentrations in near-surface waters definitely show the fingerprint of human activities on the landscape. We conclude, however, from our results and those of others, that the study of the general hypothesis that halite is the dominant process adding contaminant Na⁺ and Cl⁻ to most environments and that urban areas are "hot spots" because of road salting, has been fully tested. However, we suggest that more measurements are necessary to better understand the Cl⁻ and Na⁺ patterns observed here including the unique Cl/Na ratios in terms of their different environmental behaviors, dynamic nature of pathways and the combination of sources supplying each pathway. Finally, we suggest that this knowledge might be used to inform our understanding of the cycling of other chemicals in the environment.

# References

Andermann C, Longuevergne L, Bonnet S, Crave A, Davy P, et al. 2012. Impact of transient groundwater storage on the discharge of Himalayan rivers. *Nat Geosci* **5**: 127–132.

Aubert AH, Gascuel-Odoux C, Merot P. 2013. Annual hysteresis of water quality: A method to analyze the effect of intra- and inter-annual climatic conditions. *J Hydrol* **478**: 29–39.

Bauer P, Long D, Lee R. 1996. Selected Geochemical Characteristics of Ground Water from the Marshall Aquifer, Michigan Basin. U.S. Geological Survey Water Resources Investigations 94-4220.

Cherkauer DS. 1975. Urbanization impact on water quality during a flood in small watersheds. *Water Resour Bull* **11**: 987–998.

Cooper CA, Mayer PM, Faulkner BR. 2014. Effects of road salts on groundwater and surface water dynamics of Na⁺ and Cl⁻ in an urban restored stream. *Biogeochemistry* **121**: 149–166.

Dailey KR, Welch KA, Lyons WB. 2014. Evaluating the influence of road salt on water quality of Ohio rivers over time. *Appl Geochem* **47**: 25–35.

Dannemiller GT, Baltusis MA. 1990. Physical and chemical data for groundwater in the Michigan Basin 1986–89. U.S. Geological Survey Open-File Report 90-368. http://pubs.usgs.gov/of/1990/0368/report.pdf.

Davis SN, Whittemore DO, Fabryka-Martin J. 1998. Uses of chloride/bromide ratios in studies of potable water. *Ground Water* **36**: 338–350.

Demers CL, Sage J, Richard W. 1990. Effects of Road Deicing Salt on Cl⁻ Levels in Four Adirondack Streams. *Water Air Soil Poll* **49**: 367–373.

D'Itri FM. 1992. *Chemical Deicier in the Environment*. Chelsea, Michigan: Lewis Publishers. 587 p.

Evans C, Davies TD. 1998. Causes of concentration/discharge hysteresis and its potential as a tool for analysis of episode hydrochemistry. *Water Resour Res* **34**: 129–137.

Field R, Struzeski EJ, Masters HE, Tafuri AN. 1974. Water Pollution And Associated Effects From Street Salting, in Jewell WJ, Swan R, eds., *Water Pollution Control in Low Density Areas. Proceedings of a Rural Environmental Engineering Conference. Warren, Vermont, U.S.A. Sept. 26–28, 1973.* Hanover, NH: University Press of New England, U.S.A. ISBN 0-87451-105-4. pp. 317–340. 498p.

Fitzpatrick ML, Long DT, Pijanowski BC. 2007. Exploring the effects of urban and agricultural land use on surface water chemistry, across a regional watershed, using multivariate statistics. *Appl Geochem* **22**: 1825–1840.

Foth HD. 1999. *Fundamentals of Soil Science.* John Wiley and Sons. 384 pp.

Gardner KM, Royer TV. 2010. Effect of Road Salt Application on Seasonal Cl⁻ Concentrations and Toxicity in South-Central Indiana Streams. *J Environ Qual* **39**: 1036–1042.

Godwin KS, Hafner SD, Buff MF. 2003. Long-term trends in Na⁺ and Cl⁻ in the Mohawk River, New York: the effect of fifty years of road-salt application. *Environ Pollut* **124**: 273–281.

Halstead JA, Kliman S, Berheide CW, Chaucer A, Cock-Esteb A. 2014. Urban stream syndrome in a small, lightly developed watershed: a statistical analysis of water chemistry parameters, land use patterns, and natural sources. *Environ Monit Assess* **186**: 3391–3414.

Hem JD. 1970. Study and Interpretation of the Chemical Characteristics of Natural Water, 2nd Ed. Geological Survey Water-Supply Paper 1473.

Hoaglund JR, Kolak JJ, Long DT, Larson GJ. 2004. Analysis of modern and Pleistocene hydrologic exchange between Saginaw Bay (Lake Huron) and the Saginaw Lowlands area. *Geol Soc Am Bull* **116**: 3–15.

Howard KWF, Beck PJ. 1993. Hydrochemical implications of groundwater contamination by road deicing chemicals. *J Contam Hydrol* **12**: 245–268.

Kaushal SS, Groffman PM, Likens GE, Belt KT, Stack WP, et al.2005. Increased salinization of fresh water in the northeastern United States. *P Natl Acad Sci USA* **102**: 13517–13520.

Kelly VR, Lovett GM, Weathers KC, Findlay SEG, Strayer DL, et al. 2008. Long-term Na⁺ Cl⁻ retention in a rural watershed: Legacy effects of road salt on stream water concentration. *Environ Sci Technol* **42**: 410–415.

Kelly WR, Panno SV, Hackley KC, Hwang H-H, Martinsek AT, et al. 2010. Using Cl⁻ and other ions to trace sewage and road salt in the Illinois Waterway. *Appl Geochem* **25**: 661–673.

Long DT, Lyons WB, Hines ME. 2009. Influence of hydrogeology, microbiology and landscape history on the geochemistry of acid hypersaline waters, NW Victoria. *Appl Geochem* **24**: 285–296.

Long DT, Saleem ZA. 1974. Hydrogeochemistry of carbonate groundwaters of an urban area. *Water Resour Res* **10**: 1229–1238.

Long DT, Wilson TP, Takacs MJ, Rezabek DH. 1988. Stable-isotope geochemistry of saline near surface ground water – east central Michigan Basin. *Geol Soc Am Bull* **100**: 1568–1577.

MacLeod A, Sibert R, Snyder C, Koretsky CM. 2011. Eutrophication and salinization of urban and rural kettle lakes in Kalamazoo and Barry Counties, Michigan, USA. *Appl Geochem* **26**: S214–S217.

Mandle RJ. 1986. Plan of study for the regional aquifer systems analysis of the Michigan Basin. U.S. Geological Survey Open-File Report 86–494.

Mattson MD, Godfrey PJ. 1994. Identification of road salt contamination using multiple regression and GIS. *Environ Manage* **18**: 767–773.

McCaffrey MA, Lazar B, Holland HD. 1987. The evaporation path of seawater and the coprecipitation of Br⁻ and K⁺ with halite. *J Sediment Petrol* **57**: 928–937.

MDEQ. 2014. Wellogic. Michigan Department of Environmental Quality [Online]. http://www.michigan.gov/deq/0,4561,7-135-6132_6828-16124--,00.html. Accessed December 12, 2014.

Medalie L. 2012. Temporal and Spatial Trends of Cl⁻ and Na⁺ in Groundwater in New Hampshire, 1960–2011. U.S. Geological Survey Open-File Report 2012-1236.

Meissner B, Long D, Lee R. 1996. Characteristics of Ground Water from the Grand River-Saginaw Aquifer, Michigan Basin. U.S. Geological Survey Water Resources Investigations 93-4220.

MSU. 2014. Michigan Ground Water Management Tool. East Lansing, Michigan: Department of Environmental Engineering, Michigan State University.

Neal C, Kirchner JW. 2000. Na⁺ and Cl⁻ levels in rainfall, mist, streamwater and groundwater at the Plynlimon catchments, mid-Wales: inferences on hydrological and chemical controls. *Hydrol Earth Syst Sc* **4**: 295–310.

Novotny EV, Sander AR, Mohseni O, Stefan HG. 2009. Cl⁻ ion transport and mass balance in a metropolitan area using road salt. *Water Resour Res* **45**.

Ostendorf DW. 2013. Hydrograph and Cl⁻ pollutograph analysis of Hobbs Brook reservoir subbasin in eastern Massachusetts. *J Hydrol* **503**: 123–134.

Panno SV, Hackley KC, Hwang HH, Greenberg SE, Krapac IG, et al. 2006. Source identification of Na⁺ and Cl⁻ in natural waters: Preliminary results. *Ground Water* **44**: 176–187.

Perera N, Gharabaghi B, Howard K. 2013. Groundwater Cl⁻ response in the Highland Creek watershed due to road salt application: A re-assessment after 20 years. *J Hydrol* **479**: 159–168.

Peters NE. 2009. Effects of urbanization on stream water quality in the city of Atlanta, Georgia, USA. *Hydrol Process* **23**: 2860–2878.

Reisch CE, Toran L. 2014. Characterizing snowmelt anomalies in hydrochemographs of a karst spring, Cumberland Valley, Pennsylvania (USA): evidence for multiple recharge pathways. *Environ Earth Sci* **72**: 47–58.

Rustem WR, Long DT, Cooper WE, Armoudlian A. 1993. The Use of Selected Deicing Materials on Michigan Roads: Environmental and Economic Impacts [Online]. Prepared by Public Sector Consultants for the Michigan Department of Transportation. http://cdm16110.contentdm.oclc.org/cdm/ref/collection/p9006coll4/id/97524. Accessed December 12, 2014. 135 p.

Shanley JB. 1994. Effects of ion exchange on stream solute fluxes in a basin receiving highway deicing salts. *J Environ Qual* **23**: 977–986.

Sleeper F. 2013. Winter Deicer Maintenance Practices on Impervious Surfaces: Impacts on the Environment [Online]. University of Minnesota Water Resources Center. http://wrc.umn.edu/prod/groups/cfans/@pub/@cfans/@wrc/documents/asset/cfans_asset_467144.pdf. Accessed December 12, 2014.

Sun H, Alexander J, Gove B, Pezzi E, Chakowski N, et al. 2014. Mineralogical and anthropogenic controls of stream water chemistry in salted watersheds. *Appl Geochem* **48**: 141–154.

Sun H, Huffine M, Husch J, Sinpatanasakul L. 2012. Na/Cl molar ratio changes during a salting cycle and its application to the estimation of Na+ retention in salted watersheds. *J Contam Hydrol* **136**: 96–105.

USGS. 2012. Minerals Commodities Summaries 2102 – Salt [Online]. http://minerals.usgs.gov/minerals/pubs/commodity/salt/mcs-2014-salt.pdf. Accessed December 12, 2014.

USGS. 2014. Urban Stream Contamination Increasing Rapidly Due to Road Salt [Online]. http://www.usgs.gov/newsroom/article.asp?ID=4076#.VSvjPUbTBhE. Accessed April 12, 2015.

Wahrer M, Long D, Lee R. 1996. Selected Geochemical Characteristics of Ground Water from the Grand River-Saginaw Aquifer, Michigan Basin. U.S. Geological Survey Water Resources Investigations 94-4017.

Werner E, diPretoro RS. 2006. Rise and fall of road salt contamination of water-supply springs. *Environ Geol* **51**: 537–543.

Westjohn D, Weaver T. 1996. Hydrogeologic framework of Pennsylvanian and late Mississippian rocks in the central Lower Peninsula of Michigan. U.S. Geological Survey Water-Resources Investigations Report 94-4107.

Westjohn D, Weaver T, Zacharias K. 1994. Hydrogeology of Pleistocene glacial deposits and Jurassic "red beds" in the central Lower Peninsula of Michigan. U.S. Geological Survey Water-Resources Investigations Report 93-4152.

Williams GP. 1989. Sediment concentration versus water discharge during single hydrologic events in rivers. *J Hydrol* **111**: 89–106.

Wilson TP, Long DT. 1993a. Geochemistry and isotope chemistry of Ca-Na-Cl brines in Silurian strata, Michigan Basin, U.S.A. *Appl Geochem* **8**: 507–524.

Wilson TP, Long DT. 1993b. Geochemistry and isotope chemistry of Michigan Basin brines – Devonian Formations. *Appl Geochem* **8**: 81–100.

## Contributions

- Contributed to conception and design: DTL, TCV, AC, FX
- Contributed to acquisition of data: DTL, TCV, AC, FX, S-G L
- Contributed to analysis and interpretation of data: DTL, TCV, AC, FX
- Drafted and/or revised the article: DTL, TCV
- Approved the submitted version for publication: DTL, TCV, AC, FX, S-G L

## Acknowledgments

We would like to thank Tina Beals, Eunsang Lee, Wu Huiyun, Amira Oun, and the 2013 MSU Environmental Geochemistry class for assistance in sample collection and analysis. We very much appreciate the insightful discussions with Dr. Irene Xagoraraki about this research.

## Funding information

Portions of the work were funded by the U.S.G.S. Michigan RASA project, Michigan State University, MSU Water Initiative.

## Competing interests

The authors have no competing interests.

## Data accessibility statement

The data for the project can be obtained from:
U.S.G.S. RASA data: Dannemiller and Baltusis, 1990.
MDEQ-MSU: MDEQ, 2014; MSU, 2014.
Red Cedar River event data: These data are part of ongoing student thesis research and are not currently available.

# 9

# The influence of light and water mass on bacterial population dynamics in the Amundsen Sea Polynya

Inga Richert [1,2] • Julie Dinasquet [3,4] • Ramiro Logares [5] • Lasse Riemann [3] • Patricia L. Yager [6] • Annelie Wendeberg [2] • Stefan Bertilsson [1*]

[1]Department of Ecology and Genetics, Limnology and Science for Life Laboratory, Uppsala University, Uppsala, Sweden
[2]Department of Environmental Microbiology, Helmholtz Centre for Environmental Research – UFZ, Microbial Ecosystem Services Group, Leipzig, Germany
[3]Marine Biological Section, University of Copenhagen, Helsingør, Denmark
[4]Marine Biology Research Division, Scripps Institution of Oceanography, UCSD, San Diego, California, United States
[5]Institute of Marine Sciences, CSIC, Barcelona, Spain
[6]University of Georgia, Department of Marine Science, Athens, Georgia, United States

*stebe@ebc.uu.se

**Domain Editor-in-Chief**
Jody W. Deming, University of Washington

**Knowledge Domains**
Ocean Science
Ecology

## Abstract

Despite being perpetually cold, seasonally ice-covered and dark, the coastal Southern Ocean is highly productive and harbors a diverse microbiota. During the austral summer, ice-free coastal patches (or polynyas) form, exposing pelagic organisms to sunlight, triggering intense phytoplankton blooms. This strong seasonality is likely to influence bacterioplankton community composition (BCC). For the most part, we do not fully understand the environmental drivers controlling high-latitude BCC and the biogeochemical cycles they mediate. In this study, the Amundsen Sea Polynya was used as a model system to investigate important environmental factors that shape the coastal Southern Ocean microbiota. Population dynamics in terms of occurrence and activity of abundant taxa was studied in both environmental samples and microcosm experiments by using 454 pyrosequencing of 16S rRNA genes. We found that the BCC in the photic epipelagic zone had low richness, with dominant bacterial populations being related to taxa known to benefit from high organic carbon and nutrient loads (copiotrophs). In contrast, the BCC in deeper mesopelagic water masses had higher richness, featuring taxa known to benefit from low organic carbon and nutrient loads (oligotrophs). Incubation experiments indicated that direct impacts of light and competition for organic nutrients are two important factors shaping BCC in the Amundsen Sea Polynya.

## Introduction

Despite the cold conditions, heterotrophic bacterioplankton communities are thriving in the Southern Ocean (SO). These cold-adapted microbes mediate the transformation and remineralization of organic and inorganic nutrients and contribute significantly to elemental cycles and to the marine carbon pump. Numerous cold-water studies have measured rates of bacterial activity similar to those measured in temperate oceanic regions (e.g., Cota et al., 1990; Granéli et al., 2004; Williams et al., 2014). Hence, polar marine bacterioplankton have a central role in these ecosystems, yet it remains challenging to elucidate how the combination of habitat-specific drivers affects the growth, distribution and eventually functional role of individual populations in these cold oceanic regions.

Spatial partitioning caused by hydrographical separation and contrasting light conditions (photic/aphotic) likely select for bacterioplankton populations carrying specific traits with regards to energy and nutrient acquisition, as well as predation resistance and many other metabolic functions (Pernthaler, 2005;

Violle et al., 2007; Comte and del Giorgio, 2011). In surface waters across the SO, seasonally recurring but patchy summer areas of open water also promote biological activity because of the increased surface water irradiance. Extremely productive phytoplankton blooms, fueled by a combination of this higher radiation, enhanced stratification from sea ice meltwater, and nutrient supply from bottom water interaction with the continental ice sheet (Sherrell et al., 2014), are typical for these polar waters and affect the entire food web (Ducklow et al., 2001; Arrigo and Dijken, 2003; Alderkamp et al., 2012). Transient and patchy inputs of phytoplankton-derived organic substrates from such blooms drastically change the bottom-up factors controlling heterotrophic bacterioplankton (Billen et al., 1990) and likely also promote shifts in their abundance and community structure.

Light can directly influence the growth of several bacterial groups that are able to sustain, or at least supplement, their energy demands by harvesting photons (Bryant and Frigaard, 2006). Besides the canonical oxygenic photosynthesis characteristic for cyanobacteria and eukaryotic phytoplankton, some bacterial phototrophs are capable of anaerobic anoxygenic photosynthesis (e.g., *Rhodobacteraceae*), or proteorhodopsin-mediated energy harvesting (e.g., *Polaribacter* [Cottrell and Kirchman, 2009; Koh et al., 2010], SAR11 [Giovannoni et al., 2005] and some *Gammaproteobacteria* [Stingl et al., 2007]). However, direct exposure to high solar radiation is also known to inhibit the growth of some heterotrophic populations (Doudney and Young, 1962; Okubo and Nakayama, 1967; Cabiscol et al., 2010). Thus, the dramatic increase in light intensity during austral summer is likely to cause a rapid shift in surface-water community structure by favoring taxa that harvest light at the expense of those that are susceptible to photoinhibition.

Hydrographic separation can partition and isolate bacterioplankton communities, as observed in the North Atlantic (Agogué et al., 2011) and in polar oceans (Galand et al., 2010; Alonso-Sáez et al., 2011; Hamdan et al., 2013), but also in more global surveys of marine bacterioplankton (Ghiglione et al., 2012). Hydrographic separation was recently recognized to be a common feature separating the bacterioplankton communities in the Southern Ocean (Wilkins et al., 2012). The Antarctic shelf region that encompasses the Amundsen Sea Polynya (ASP; the open-water area of this study), can be separated into three water masses during summer: at depth is the warmer but more saline modified Circumpolar Deep Water (mCDW), overlain by the colder and less saline Winter Water (WW) and by the warmer and less saline Antarctic Surface Water (AASW), which is influenced by freshwater from melting sea ice and present only temporarily during austral summer in close connection to the ice retreat (Randall-Goodwin et al., 2014).

In the present study, the remote ASP was used as a model-system to consider how light and water mass interactively influence the composition of local bacterioplankton communities. We carried out experiments during the Amundsen Sea International Research Expedition (ASPIRE, Yager et al., 2012) with the broader aim of assessing how individual bacterial taxa respond to light, by experimentally simulating two contrasting light regimes experienced by bacterioplankton in this region during the course of the year. Communities emerging in experimental incubations were also compared to the communities residing in the different water masses that characterize the ASP. Two complementary datasets of ASP bacterioplankton community composition were collected by use of 454 pyrosequencing of amplified 16S rRNA gene fragments. Field observations across the different water masses were combined with a factorial experiment where bacterioplankton communities from the different water masses were exposed to darkness and to photic-zone levels of irradiation. Focus was on the distribution and responses of a subset of dominant bacterial groups present in the ASP. These previously studied groups with inferred metabolic traits were highlighted to identify and illustrate individual and contrasting population-level responses to light regime and water mass. The underlying hypothesis was that differing light conditions and other distinctions between water masses select for particular metabolic traits, resulting in the emergence of microbial populations specifically adapted to these local conditions.

## Methods

### Sampling

Sampling was conducted during the austral summer (November 2010 to January 2011) from the icebreaker Nathaniel B. Palmer. Samples from 15 stations were obtained to include samples from the three major water masses of the Amundsen Sea Polynya and its margins (71–75°S, 110–120°W; Figure 1) during the summer season. Seawater was collected in 12 L Niskin bottles attached to a 24-bottle SBE 32 rosette; coupled to the rosette was a system of sensors reading depth-resolved profiles of temperature [° C], conductivity [S $m^{-1}$], oxygen [mg $L^{-1}$], photosynthetically active radiation (PAR) [$\mu$mol photons $s^{-1}$ $m^{-2}$] and fluorescence [mg $m^{-3}$ chl-a] for each cast (SBE 911, Sea-Bird Electronics, Bellevue, Washington, USA). Water samples for incubation experiments were processed immediately at 2°C in a temperature-controlled room.

### Bacterial community analysis

Bacteria from sampled seawater were collected by filtration onto 0.2 $\mu$m membranes in Sterivex filter-cartridges (Millipore, Solna, Sweden) using peristaltic pumps and acid-washed silicone tubing. For each of 2–5 depths

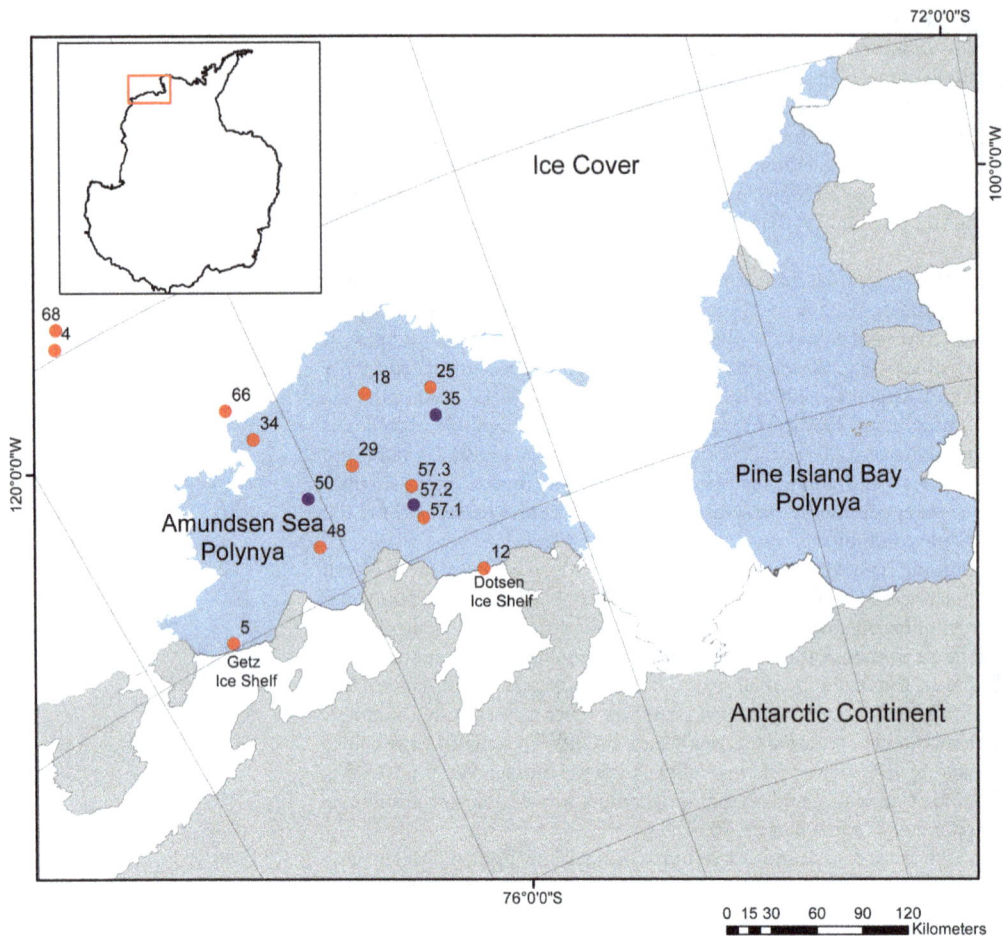

**Figure 1**

**Map of the Amundsen Sea Polynya and stations sampled.**

Map illustrating the extent of seasonal ice retreat in the Amundsen Sea at the time of sampling. Stations sampled to examine *in situ* bacterial distributions are marked by solid circles (red and purple). The seawater inocula and media for the shipboard incubations were obtained from stations marked in purple (st35, st50, and st57.2).

per station, a volume of approximately 5 L of water was filtered. The filters were subsequently covered with a sucrose lysis buffer (20% sucrose, 50 mM EDTA, 50 mM Tris HCL, pH 8) and stored at -80°C.

## Incubation experiments

To assess bacterioplankton responses to light and dark conditions in the absence of larger predators and eukaryotic phytoplankton, 0.2 μm filtered seawater in 1 L acid washed polycarbonate bottles was inoculated with 5% [v/v] 0.6 μm filtered seawater from the same depth using a vacuum pump to a total volume of 1 L. The experimental design by station included two factors (water mass source of inoculum: epipelagic and mesopelagic) with two treatments (dark and light) for each inoculum. Triplicate incubations were conducted under the dark and light conditions using water from each of three stations: 35 (73°27′95″S, 112°10′41″W), 50 (73° 41′60″S, 115°25′03″W) and 57.2 (73°70′73″S, 113°26′5″W) (Figure 1). Each experiment included one inoculum from the light-exposed AASW, and one from the mesopelagic zone from either WW or mCDW. Each 0.2-μm filtered seawater medium came from the same depth and station as its inoculum. The light source imitated light levels at approximately 20–50 m below the surface (Philips TLD-18W/18 blue, 1.5–1.99 ⋅ $10^{-2}$ μmol photons s$^{-1}$ m$^{-2}$) in a PAR range of 400–500 nm (according to manufacturer), excluding the short-wavelength UV. After a 7-day incubation at near in situ temperature (0.5°C), bacteria from the full sample volume were collected by vacuum filtration onto 0.2-μm, 47-mm Supor filters (Pall, Lund, Sweden) and stored at -80°C in sucrose lysis buffer (20% sucrose, 50 mM EDTA, 50 mM TrisHCl, pH = 8). Darkness was achieved by covering the bottles with black and lightproof foil.

## Bacterial abundance

For measuring *in situ* bacterial abundance and bacterial abundance in the incubation experiments, 1.5 ml water was sub-sampled at time zero and three subsequent occasions during the incubation and fixed in 1% EM grade glutaraldehyde (Sigma Aldrich), flash-frozen in liquid nitrogen, and stored at -80°C. Bacterial abundance was determined with a FASCanto II flow cytometer (Becton Dickinson, USA) (Gasol and del

Giorgio, 2000) after staining the fixed cells with SYBR green (Invitrogen). The flow rate was calibrated with fluorescent beads. Net growth was determined for each incubation by the change in bacterial abundance after 7 days, relative to the control (time zero), and reported as the mean percentage for the triplicate experiments.

## Molecular analysis

The DNA was extracted using a phenol-chloroform extraction approach as previously described (Riemann et al., 2000). Prior to extraction, microorganisms were enzymatically digested for 30 min with lysozyme at 37°C followed by an overnight digestion with Proteinase K (both 20 mg ml$^{-1}$, Sigma Aldrich) at 55°C (Boström et al., 2004). The 16S rRNA genes were amplified using the bacterial primers Bakt_341F (CCTACGGGNGGCWGCAG) and Bakt_805R (GACTACHVGGGTATCTAATCC) with 454-Lib-L adapters and sample-specific barcodes on the reverse primer (Herlemann et al., 2011). Each set consisted of up to 72 samples with individual barcodes pooled for sequencing (Table 1). Triplicate PCR reactions for each sample were carried out with 10 to 70 ng extracted environmental DNA as template. Each 20-µl reaction also contained Phusion Hot Start high-fidelity DNA polymerase (Thermo Scientific). Amplification was carried out by initial denaturation at 98°C for 30 seconds followed by 25 cycles of an initial 98°C denaturation for 30 seconds, subsequent annealing at 50°C for 30 seconds and 30-second extension at 72°C. These 25 cycles were followed by a final 7-min extension at 72°C. Triplicate reactions for each sample were pooled and PCR products were purified using the Agencourt AMPure XP kit according to manufacturer instructions (Beckman Coulter) and quantified with a Picogreen quantification essay (Invitrogen). Equimolar amounts of amplicon from each sample were pooled and sequenced by 454 pyrosequencing using Titanium chemistry at the SNP/SEQ SciLifeLab platform hosted by Uppsala University (Sweden).

Table 1. Individual barcode sequences used in the two 16S rRNA amplicon batches

| Primer[a] ID | Barcode | Batch 1 | Batch 2 |
|---|---|---|---|
| | | Station_depth (m)_ Treatment | |
| 4 | TATCGCA | 57/cast71_30 | |
| 5 | TACTAGC | 4_638 | |
| 6 | TACTCTC | 4_2 | |
| 7 | TACTCGA | 4_25 | |
| 8 | TACTGAC | 4_70 | |
| 9 | TACTGCA | 4_560 | |
| 10 | TACGTCA | 5_2 | |
| 11 | TACGAGT | 5_25 | 35_120_dark |
| 12 | TACGCTA | 18_20 | 35_120_dark |
| 13 | TAGTCAC | 5_750 | 35_120 dark |
| 14 | TAGACTC | 5_1227 | 35_12_dark |
| 15 | TAGACGA | 12_22 | 35_12_dark |
| 16 | TAGAGAC | 12_80 | 35_12_dark |
| 17 | TAGAGCA | 12_240 | 35_120_light |
| 18 | TAGCTCA | 12_600 | 35_120_light |
| 19 | TAGCACT | 12_900 | 35_120_light |
| 20 | TAGCAGA | | 35_12_light |
| 21 | TAGCGTA | 5_70 | 35_12_light |
| 22 | TCTACTC | 18_50 | 35_12_light |
| 23 | TCTCTCA | 18_350 | 50_120_dark |
| 24 | TCTCATC | 18_422 | 50_120_dark |
| 25 | TCTCACT | 25_2 | 50_120_dark |
| 26 | TCTCAGA | 25_18 | 50_10_dark |
| 27 | TCTGAGT | 25_80 | 50_10_dark |
| 28 | TCATAGC | 25_350 | 50_10_dark |
| 29 | TCATCTC | 25_400 | 50_120_light |
| 30 | TCATCGA | 12_180 | 50_120_light |
| 31 | TCATGAC | 29_2 | 50_120_light |

| Primer[a] ID | Barcode | Batch 1 | Batch 2 |
|---|---|---|---|
| | | Station_depth (m)_ Treatment | |
| 32 | TCATGCA | 29_20 | 50_10_light |
| 33 | TCACTAC | 29_80 | 50_10_light |
| 34 | TCACTCT | 29_655 | 50_10_light |
| 35 | TCACTGA | 29_733 | 57/cast72_680_dark |
| 36 | TCACACA | 34_2 | 57/cast72_680_dark |
| 37 | TCACAGT | 34_10 | 57/cast72_680 dark |
| 38 | TCACGTA | 34_50 | 57/cast72_10_dark |
| 39 | TCACGAT | 34_360 | 57/cast72_10_dark |
| 40 | TCAGTCA | 34_672 | 57/cast72_10_dark |
| 41 | TCAGATC | 35_2 | 57/cast72_680_light |
| 42 | TCAGAGA | 35_12 | 57/cast72_680_light |
| 43 | TCAGCTA | 35_120 | 57/cast72_680_light |
| 44 | TCGTAGA | 35_360 | 57/cast72_10_light |
| 45 | TCGTGTA | 35_420 | 57/cast72_10_light |
| 46 | TCGATCA | | 57/cast72_10_light |
| 47 | TCGACTA | 48_2 | |
| 46 | TCGATCA | 48_25 | |
| 48 | TCGCATA | 48_120 | |
| 49 | TGTACGA | 48_500 | |
| 50 | TGTAGCA | 48_983 | |
| 51 | TGTCACA | 50_5 | |
| 52 | TGTCGTA | 50_10 | |
| 53 | TGTGTCA | 50_170 | |
| 54 | TGTGCTA | 50_320 | |
| 55 | TGATCAC | 50_1031 | |
| 56 | TGACTCA | 57/cast71_4 | |
| 57 | TGACACT | 57/cast71_140 | |
| 58 | TGAGTAC | 57/cast71_500 | |
| 59 | TGAGTCT | 57/cast71_735 | |
| 61 | TGAGCAT | 57/cast72_10 | |
| 62 | TGCTAGA | 57/cast72_150 | |
| 63 | TGCTGTA | 57/cast72_300 | |
| 64 | TGCATCA | 57/cast72_772 | |
| 65 | TGCACTA | 57/cast72_680 | |
| 66 | TGCGATA | 57/cast84_5 | |
| 67 | ATACTGC | 57/cast84_10 | |
| 68 | ATACGCT | 57/cast84_150 | |
| 69 | ATAGCGT | 57/cast84_300 | |
| 70 | ATCTCAC | 57/cast84_625 | |
| 71 | ATCATGC | 66_2 | |
| 72 | ATCACTC | 66_10 | |
| 73 | ATCACGT | 66_100 | |
| 74 | ATCAGAC | 66_636 | |
| 75 | ATCAGCT | 68_25 | |
| 76 | ATCGTGT | 68_500 | |
| 79 | ATGTCGT | 68_820 | |

[a] Reverse Primer 805R with Titanium Adapter A

In order to obtain a list of observed operational taxonomic units (OTUs) suitable for statistical analysis, low-quality sequences were removed from the dataset, and noise was reduced using AmpliconNoise v1.24 (Quince et al., 2011) with default parameters. AmpliconNoise implements algorithms that remove PCR single-base and 454-pyrosequencing errors, as well as the chimera removal tool Perseus. Reads that did not carry the exact primer sequence were removed. With a length-cutoff of 425 base pairs (bp), the remaining reads were processed using the Quantitative Insights Into Microbial Ecology software (QIIME v1.3, Caporaso et al., 2010). Sequences were clustered into OTUs at 99% pairwise identity using Uclust (Edgar, 2010). Taxonomic assignments of representative sequences from each OTU were obtained according to the SILVA111 database (Quast et al., 2013) by using the RDP classifier implemented in QIIME (Wang et al., 2007). After excluding non-bacterial taxa and singletons, altogether 488,028 reads classified into 452 OTUs, which were kept for further analysis.

## Statistical analysis

The curated OTU table generated after our quality control was used to calculate alpha diversity (Simpson diversity, Shannon diversity, observed OTUs, Chao1 richness) with QIIME. Reads obtained for each OTU were transformed into relative abundances across samples. For alpha diversity, 2,000 reads were subsampled by rarefaction in order to reduce any possibility of biases due to uneven sequencing efforts across samples. Samples with less reads were excluded.

For analyzing the *in situ* populations, first the degree of group separation was tested statistically by Analysis of Similarity (ANOSIM) in RStudio. In a following step SIMPER (Similarity Percentage) analysis was conducted using Primer6 (Clarke, 1993; Clarke and Warwick, 2001), based on Bray Curtis dissimilarity, to identify the OTUs responsible for significant differences in community composition across the different water masses. The individual samples were eventually grouped by their water mass of origin based on their temperature and salinity signatures (Randall-Goodwin et al., 2014).

The experimental incubations were evaluated with regards to the responses in alpha diversity and net-growth of the total community, and to population-level responses for specifically targeted populations. It was hypothesized that: (i) light will favor phototrophic or photoheterotrophic bacterial populations and (ii) the origin of the water sample and inoculum will generate differences in the emergent communities because of a water-mass-specific species pool and contrasting local nutrients and organic resources available in the filtered seawater media.

For statistical analysis, experimental data were checked for normality and homogeneity of variance before applying parametric tests. Response in terms of net-growth (increase in bacterial abundance relative to initial bacterial concentration), Simpson diversity, Shannon diversity and Chao1 richness across the four treatments was assessed with a two-way ANOVA testing for the factors light and epipelagic/mesopelagic water mass. All of these analyses were carried out in RStudio. The Simpson diversity estimator and Shannon diversity indices for alpha diversity applied here were calculated based on the proportional abundance of taxa; in comparison, the Simpson diversity is more sensitive to changes in dominant taxa than the rare ones. The estimator applied for capturing richness was Chao1, accounting for rare taxa that might have been missed due to the sequencing approach (see Hill et al., 2003, for an overview of alpha diversity metrics).

## Population dynamics analysis

To assess the response of individual taxonomically defined populations rather than the combined and rather complex communities, a subset of abundant bacterial populations (defined by OTUs classified at the genus level) was selected for further analysis. This subset was chosen to illustrate population dynamics under different environmental conditions using examples of populations with contrasting metabolic features. The examples represent key organisms within the bacterioplankton community of the Southern Ocean, the metabolic features and ecology of which have been highlighted and discussed in several recent publications (Koh et al., 2010; Williams et al., 2012; Tripp, 2013). The selected genera were identified by SIMPER to contribute substantially to the observed differences in community composition between the water masses. They were comprised of pairs of closely related bacterial taxa selected from each of four broader taxonomic groups: the *Alphaproteobacteria* (*Roseobacter* and SAR 11), the *Flavobacteriales* of the *Bacteriodetes* (*Ulvibacter* and *Polaribacter*), and the *Oceanospirillales* (*Balneatrix* and SAR86) and *Alteromonadales* (*Colwellia* and SAR92) within the *Gammaproteobacteria* (according to the Silva111 reference database). We did not test for significance here as the low relative abundance might cause a bias.

# Results

## In situ bacterial community composition of the water masses

Characteristics of the samples collected for the *in situ* survey of bacterioplankton communities, which allowed water mass origin to be confirmed, are summarized in Table 2. Bacterial communities differed

**Table 2.** Characteristics of samples collected to survey *in situ* bacterial populations[a]

| Station[a] | Latitude (°S) | Longitude (°W) | Depth (m) | Water mass | Salinity | Temperature (°C) | Total bacteria (number x 10[5] ml[-1]) | Simpson Diversity | Shannon Diversity | Chao1 richness | Observed taxa |
|---|---|---|---|---|---|---|---|---|---|---|---|
| 4 | 71°95′29″ | 118°47′26 | 70 | WW | 34.03 | −1.80 | 2.18 | 0.90 | 4.13 | 79 | 70 |
| 4 | 71°95′29″ | 118°47′26 | 638 | mCDW | 34.66 | 0.81 | 0.66 | 0.90 | 4.30 | 126 | 98 |
| 5 | 73°96′66″ | 118°03′48″ | 2 | WW | 33.99 | −1.41 | 4.70 | 0.81 | 3.19 | 79 | 57 |
| 5 | 73°96′66″ | 118°03′48″ | 25 | WW | 33.99 | −1.69 | 4.83 | 0.82 | 3.19 | 63 | 51 |
| 5 | 73°96′66″ | 118°03′48″ | 750 | mCDW | 34.53 | 0.59 | 0.88 | 0.90 | 4.41 | 115 | 95 |
| 5 | 73°96′66″ | 118°03′48″ | 1227 | mCDW | 34.58 | 0.79 | 0.96 | 0.89 | 4.27 | 133 | 99 |
| 12 | 74°21′85″ | 112°33′62″ | 240 | WW | 34.03 | −1.26 | 1.27 | 0.93 | 4.73 | 110 | 92 |
| 12 | 74°21′85″ | 112°33′62″ | 600 | mCDW | 34.41 | 0.27 | 1.18 | 0.88 | 4.10 | 112 | 86 |
| 12 | 74°21′85″ | 112°33′62″ | 900 | mCDW | 34.55 | 0.60 | 1.09 | 0.87 | 4.18 | 115 | 90 |
| 18 | 73° | 113°30′23″ | 20 | WW | 33.95 | −1.26 | 7.97 | 0.73 | 2.56 | 73 | 47 |
| 18 | 73° | 113°30′23″ | 50 | WW | 34.02 | −1.72 | 6.45 | 0.84 | 3.44 | 74 | 56 |
| 18 | 73° | 113°30′23″ | 350 | WW | 34.31 | −0.30 | 1.25 | 0.89 | 4.35 | 117 | 94 |
| 18 | 73° | 113°30′23″ | 422 | mCDW | 34.52 | 0.43 | 1.44 | 0.81 | 3.79 | 109 | 87 |
| 25 | 73°12′01″ | 112°00′07″ | 2 | AASW | 33.83 | −0.61 | 4.22 | 0.72 | 2.68 | 66 | 48 |
| 25 | 73°12′01″ | 112°00′07″ | 18 | AASW | 33.89 | −1.13 | 6.00 | 0.76 | 2.77 | 54 | 43 |
| 25 | 73°12′01″ | 112°00′07″ | 350 | WW | 34.30 | −0.42 | 1.12 | 0.86 | 4.21 | 116 | 95 |
| 25 | 73°12′01″ | 112°00′07″ | 400 | mCDW | 34.44 | 0.22 | 1.30 | 0.84 | 4.06 | 106 | 90 |
| 29 | 73°35′04″ | 114°12′68″ | 2 | AASW | 33.91 | −0.90 | 4.99 | 0.56 | 1.82 | 54 | 36 |
| 29 | 73°35′04″ | 114°12′68″ | 20 | AASW | 33.91 | −0.92 | 5.36 | 0.65 | 2.14 | 56 | 39 |
| 29 | 73°35′04″ | 114°12′68″ | 80 | WW | 34.03 | −1.62 | 2.87 | 0.83 | 3.27 | 70 | 54 |
| 29 | 73°35′04″ | 114°12′68″ | 655 | mCDW | 34.52 | 0.48 | 1.33 | 0.87 | 4.15 | 122 | 97 |
| 29 | 73°35′04″ | 114°12′68″ | 733 | mCDW | 34.62 | 0.75 | 1.59 | 0.85 | 3.94 | 106 | 87 |
| 34 | 72°96′35″ | 115°75′96″ | 2 | AASW | 33.74 | −1.07 | 4.56 | 0.74 | 2.69 | 66 | 49 |
| 34 | 72°96′35″ | 115°75′96″ | 10 | AASW | 33.76 | −1.25 | 4.56 | 0.70 | 2.60 | 62 | 50 |
| 34 | 72°96′35″ | 115°75′96″ | 50 | WW | 34.00 | −1.72 | 4.98 | 0.88 | 3.71 | 73 | 52 |
| 35 | 73°27′95″ | 112°10′41″ | 2 | AASW | 33.83 | −0.17 | 4.53 | 0.73 | 2.60 | 74 | 42 |
| 35 | 73°27′95″ | 112°10′41″ | 12[b] | AASW | 33.83 | −0.20 | 5.04 | 0.67 | 2.31 | 47 | 40 |
| 35 | 73°27′95″ | 112°10′41″ | 420 | mCDW | 34.45 | 0.25 | 1.57 | 0.90 | 4.41 | 105 | 89 |
| 48 | 73°70′13″ | 115°44′99″ | 2 | AASW | 33.89 | −0.18 | 2.54 | 0.70 | 2.22 | 36 | 29 |
| 48 | 73°70′13″ | 115°44′99″ | 120 | WW | 34.04 | −1.74 | 1.71 | 0.87 | 3.54 | 56 | 51 |
| 48 | 73°70′13″ | 115°44′99″ | 500 | mCDW | 34.35 | −0.26 | 1.14 | 0.92 | 4.73 | 115 | 96 |
| 48 | 73°70′13″ | 115°44′99″ | 983 | mCDW | 34.57 | 0.67 | 1.18 | 0.85 | 4.13 | 122 | 96 |
| 50 | 73°41′61″ | 115°25′03″ | 5 | AASW | 33.87 | −0.33 | 5.11 | 0.63 | 2.19 | 58 | 37 |
| 50 | 73°41′61″ | 115°25′03″ | 10[b] | AASW | 33.88 | −0.37 | 4.49 | 0.65 | 2.17 | 48 | 33 |
| 50 | 73°41′61″ | 115°25′03″ | 320 | WW | 34.13 | −1.74 | 1.62 | 0.94 | 4.95 | 110 | 89 |
| 57.1 | 73°80′17 | 113°16′5″ | 4 | AASW | 33.93 | −0.50 | 4.35 | 0.57 | 1.88 | 38 | 30 |
| 57.1 | 73°80′17 | 113°16′5″ | 30 | AASW | 33.93 | −0.53 | 4.34 | 0.51 | 1.78 | 40 | 33 |
| 57.1 | 73°80′17 | 113°16′5″ | 140 | WW | 33.98 | −1.38 | 2.99 | 0.84 | 3.61 | 105 | 75 |
| 57.2 | 73°70′73″ | 113°26′53″ | 10[b] | AASW | 33.93 | −0.50 | 3.80 | 0.69 | 2.62 | 67 | 49 |
| 57.2 | 73°70′73″ | 113°26′53″ | 772 | mCDW | 34.59 | 0.71 | missing | 0.87 | 4.14 | 111 | 91 |
| 57.3 | 73°60′22″ | 113°14′88″ | 5 | AASW | 33.92 | −0.26 | 4.52 | 0.70 | 2.40 | 42 | 33 |
| 57.3 | 73°60′22″ | 113°14′88″ | 10 | AASW | 33.92 | −0.28 | 3.99 | 0.71 | 2.40 | 45 | 36 |
| 57.3 | 73°60′22″ | 113°14′88″ | 150 | WW | 34.05 | −1.17 | 3.09 | 0.88 | 3.73 | 62 | 57 |
| 66 | 72°74′09 | 116°01′99″ | 2 | AASW | 33.60 | −1.17 | 4.15 | 0.80 | 2.96 | 67 | 43 |
| 66 | 72°74′09 | 116°01′99″ | 10 | AASW | 33.69 | −1.28 | 4.35 | 0.82 | 3.06 | 68 | 43 |
| 66 | 72°74′09 | 116°01′99″ | 100 | WW | 34.07 | −1.61 | 2.96 | 0.86 | 3.69 | 80 | 63 |

[a] Only samples with read counts > 2000 were included

[b] These samples, along with three others not listed here,  were used for inocula and media in the incubation experiments

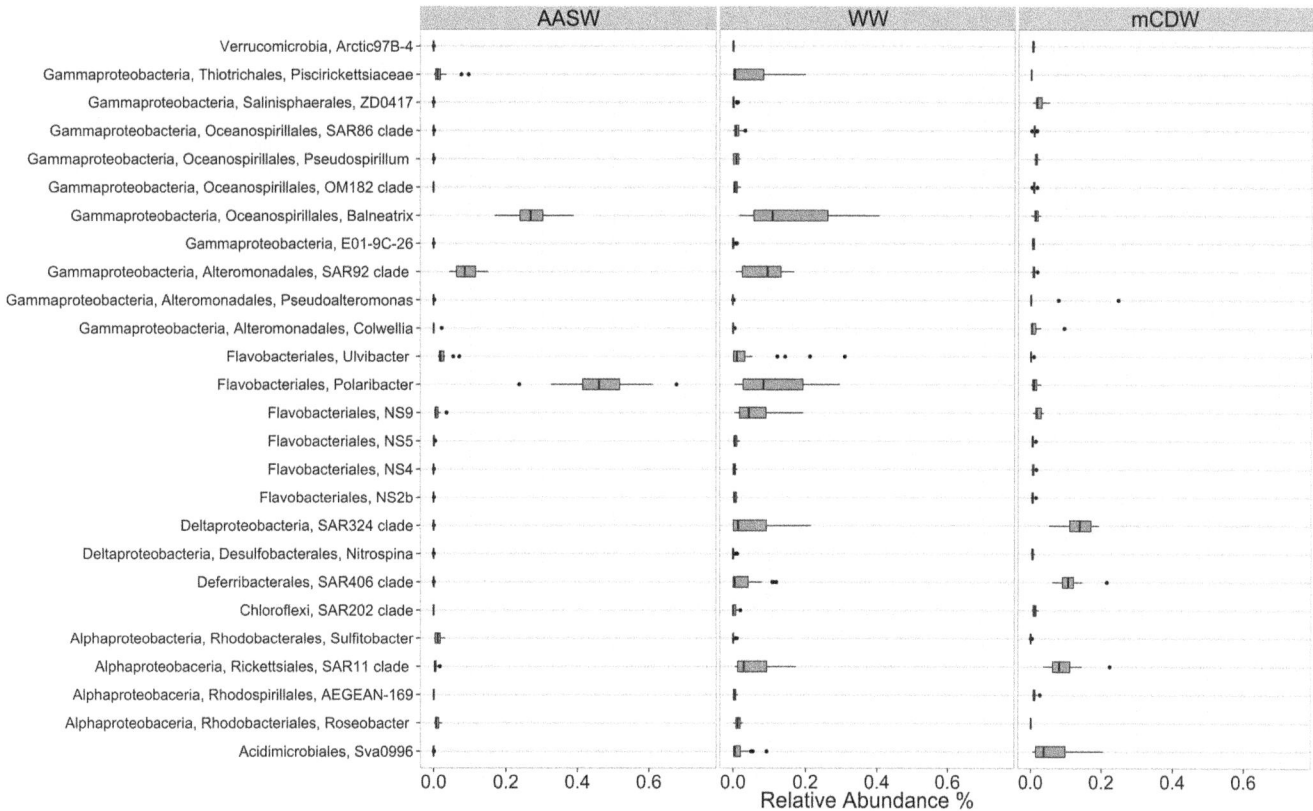

significantly between the three water masses (AASW, WW, and mCDW) according to the ANOSIM analysis, even though the R values pointed to different degrees of separation, with R = 0.28 for AASW and WW, R = 0.41 for WW and mCDW, and R = 0.87 for AASW and mCDW (p = 0.001 for all three cases). According to SIMPER, about 30 OTUs alone explained 90% of the dissimilarity in bacterial community composition between each of the three water masses (Figure 2). AASW was characterized by low species richness caused by the dominance of *Flavobacteria/Polaribacter* and *Oceanospirillaceae/Balneatrix* clades with a similarity of 74% within the group. The observed dissimilarity in community composition between WW and mCDW was mainly due to shifts in the relative abundances of broadly distributed taxa, but mCDW hosted additional abundant taxa that were rare in both AASW and WW. For example, *Deltaproteobacteria* clade SAR324, *Deferribacterales* clade SAR 406 and *Acidomicrobiales* clade Sva0996 were prevalent bacterial community members in mCDW. SIMPER analysis further identified that the bacterioplankton community in mCDW was more homogenous in composition than in WW with a within-group similarity of 65% and 53%, respectively. Alpha diversity differed between the water masses, in that mCDW and WW harbored more diverse bacterioplankton communities than AASW (Table 3). Total bacterial abundance was highest in AASW, averaging 4.4 ($\pm$ 0.6) $\times$ $10^5$ bacteria ml$^{-1}$, whereas in WW and mCDW, abundance averaged 2.8 ($\pm$ 1.8) $\times$ $10^5$ and 1.2 ($\pm$ 0.3) $\times$ $10^5$ bacteria ml$^{-1}$, respectively (Figure 3).

**Figure 2**

**Relative abundance of bacterial taxa across the water masses.**

Bacterial genus-level taxa identified by Similarity Percentage (SIMPER) analysis to explain 90% of the Bray-Curtis dissimilarities between pairs of sampled water masses. Multiple OTUs affiliated with the same taxon were combined for the analysis and visualization. For each taxon, boxplots illustrate the 25–75 percentile values (box), median (bar inside box), standard deviation (whiskers), and outliers (solid circles).

## Experimental treatments

At the time of sampling, bacterial inocula originating from the epipelagic zone at 2–10 m had experienced PAR in the range of 5.6–105 µmol photons s$^{-1}$ m$^{-2}$, while no PAR reached the mesopelagic zone or the bacteria used as inocula from this zone. Experimental incubations with inocula from the mesopelagic water

Table 3. Alpha diversity[a] in the three water masses of the summer Amundsen Sea Polynya

| Water mass | Simpson diversity | Shannon diversity | Observed OTUs | Chao1 richness |
|---|---|---|---|---|
| AASW (n = 18) | 0.68 ± 0.07 | 2.4 ± 0.37 | 40 ± 7 | 55 ± 12 |
| WW (n = 15) | 0.86 ± 0.05 | 3.75 ± 0.62 | 67 ± 18 | 86 ± 21 |
| mCDW (n = 13) | 0.87 ± 0.03 | 4.2 ± 0.23 | 92 ± 5 | 115 ± 8 |

[a] Alpha diversity metrics, with standard deviations, were calculated based on rarefaction, with 2000 reads per sample included

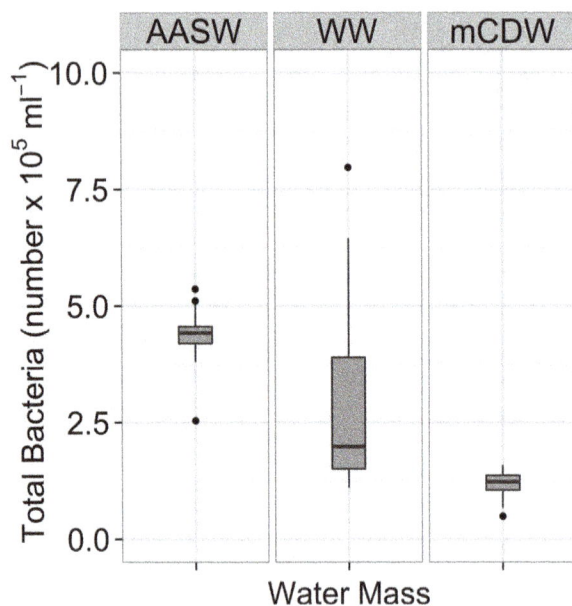

Figure 3
Total bacterial abundance in the water masses.

Bacterial cell counts derived from flow cytometry assays. The boxplots give the 25–75 percentile values (box), median (bar inside box), standard deviation (whiskers), and outliers (solid circles).

Table 4. Characteristics[a] of bacterial communities in experimental incubations after 7 days

| Sample origin[b] | | | Dark treatment | | | | | Light treatment | | | | |
|---|---|---|---|---|---|---|---|---|---|---|---|---|
| Station | Depth (m) | Water mass | Simpson diversity | Shannon diversity | Chao1 richness | Observed OTUs | % net growth[c] | Simpson diversity | Shannon diversity | Chao1 richness | Observed OTUs | % net growth[c] |
| 35 | 12 | AASW | 0.50 ± 0.12 | 1.47 ± 0.5 | 28 ± 11 | 22 ± 8 | 110 ± 92.3 | 0.3 | 2.31 ± 1.7 | 30.5 | 84 ± 8 | 17 ± 23.3 |
| 35 | 120 | WW | 0.90 ± 0.01 | 4.2 ± 0.02 | 82 ± 8 | 68 ± 3 | 13.7 ± 8.1 | 0.86 ± 0.02 | 3.9 ± 0.14 | 81 ± 3 | 65 ± 2 | 0.7 ± 1.2 |
| 50 | 10 | AASW | 0.40 ± 0.05 | 1.2 ± 0.13 | 22 ± 3 | 16 ± 2 | 152 ± 186 | 0.07 ± 0.02 | 0.31 ± 0.1 | 19 ± 10 | 12 ± 5 | 135 ± 117 |
| 50 | 120 | WW | 0.87 ± 0.01 | 3.7 ± 0.08 | 84 ± 6 | 65 ± 1 | 95.7 ± 135 | 0.41 ± 0.29 | 1.7 ± 1.23 | 56 ± 33 | 41 ± 23 | 40.5 ± 65.1 |
| 57.2 | 10 | AASW | 0.13 ± 0.09 | 0.52 ± 0.3 | 27 ± 3 | 19 ± 3 | 48.2 ± 25 | 0.14 ± 0.13 | 0.54 ± 0.4 | 30 ± 7 | 19 ± 3 | 3.6 ± 4.2 |
| 57.2 | 680 | mCDW | 0.85 ± 0.03 | 3.58 ± 0.2 | 88 ± 7 | 67 ± 5 | 50.6 ± 84.4 | 0.91[d] | 4.39 | 87 | 82 | 0 |

[a] Alpha diversity metrics, with standard deviations, were calculated based on rarefaction, with 2000 reads per sample included; n = 3 except where indicated by missing standard deviation

[b] Seawater inocula and incubation media were taken from three different sites and two depths at each site to cover all three water masses in the ASP

[c] Net growth was calculated based on the difference in total bacterial abundance between time zero and 7 days

[d] Sequencing data retrieved only for one sample at this location

masses generated the highest alpha diversity compared to those with inocula from the photic zone (AASW), while highest net growth was measured in dark treatments from AASW (Table 4). The two-factorial ANOVA of Simpson diversity as response variable for light and water mass demonstrated that water mass had a significant influence on this diversity index at all stations (Station 35: F = 27.5, p = 0.002; Station 50: F = 16.8, p = 0.0046; Station 57.2 F = 139, p < 0.001). Light was a significant factor only for Stations 50 (F = 12.9, p = 0.008) and 57.2 (F = 7.3, p = 0.035). When testing for Shannon diversity, the ANOVA revealed the same significant treatment effects as for Simpson diversity. Changes in species richness, represented by the Chao1 metric, were significantly affected by water mass at station 50 (F = 18.6, p = 0.0035) and 57.2 (F = 241, p < 0.001) and by light at station 57.2 (F = 12.9, p = 0.011). Despite considerable variation in growth among the incubations, there was a general trend across all three independent experiments that communities from the photic surface waters exhibited the highest net growth but also the greatest light inhibition (each experimental incubation; Tables 5, 6, 7).

## Dynamics of a subset of populations in experimental treatments

Among the *Alphaproteobacteria* (Figure 4A), the SAR11 clade was a major community component in mCDW at 8.9 ± 4.4% and in WW at 5.2 ± 4.8%, but was insignificant in AASW (0.5 ± 0.5%). After the 7-day experimental incubations, SAR11 averaged 2.2 ± 2.5% and consistently represented an abundant community member in the incubated samples from WW and mCDW, but was rare in those from AASW, regardless of

Table 5. Results of experimental incubations with samples from station 35 (73°27′95″ S, 112°10′41″W)

| Depth (m) | Water mass | Treatment | Total bacteria (number × 10^5 ml^-1) | | | | Simpson diversity | Shannon diversity | Chao1 richness | Observed taxa |
|---|---|---|---|---|---|---|---|---|---|---|
| | | | 0 h | 67 h | 134 h | 169 h | | | | |
| 120 | WW | dark | 0.09 | 0.05 | 0.09 | 0.10 | 0.90 | 4.16 | 73 | 65 |
| 120 | WW | dark | 0.08 | 0.06 | 0.09 | 0.09 | 0.90 | 4.19 | 86 | 67 |
| 120 | WW | dark | 0.09 | 0.09 | 0.09 | 0.10 | 0.91 | 4.21 | 88 | 72 |
| 12 | AASW | dark | 0.16 | 0.09 | 0.23 | 0.83 | missing | missing | missing | missing |
| 12 | AASW | dark | 0.24 | 0.28 | 0.41 | 1.23 | 0.58 | 1.82 | 36 | 28 |
| 12 | AASW | dark | 0.32 | 0.12 | 0.15 | 0.47 | 0.41 | 1.12 | 20 | 16 |
| 120 | WW | light | 0.07 | 0.06 | 0.05 | 0.06 | 0.84 | 3.79 | 81 | 64 |
| 120 | WW | light | 0.08 | 0.07 | 0.07 | 0.07 | 0.85 | 3.80 | 83 | 68 |
| 120 | WW | light | 0.08 | 0.08 | 0.09 | 0.08 | 0.89 | 4.03 | 77 | 64 |
| 12 | AASW | light | 0.57 | 0.12 | 0.14 | 0.29 | missing | missing | missing | missing |
| 12 | AASW | light | 0.35 | 0.10 | 0.14 | 0.27 | 0.66 | 3.49 | 182 | 143 |
| 12 | AASW | light | 0.15 | 0.13 | 0.28 | 0.12 | 0.30 | 1.13 | 31 | 25 |

Table 6. Results of experimental incubations with samples from station 50 (73°41′60″ S, 115.25′03″ W)

| Depth (m) | Water mass | Treatment | Total bacteria (number × 10^5 ml^-1) | | | | Simpson diversity | Shannon diversity | Chao1 richness | Observed taxa |
|---|---|---|---|---|---|---|---|---|---|---|
| | | | 0 h | 46 h | 98 h | 171 h | | | | |
| 120 | WW | dark | 0.06 | 0.06 | 0.41 | 0.70 | 0.86 | 3.72 | 80 | 64 |
| 120 | WW | dark | 0.02 | 0.03 | 0.08 | 0.03 | 0.87 | 3.83 | 88 | 65 |
| 10 | AASW | dark | 0.34 | 0.00 | 1.43 | 1.17 | 0.34 | 1.08 | 18 | 15 |
| 10 | AASW | dark | 0.11 | 0.23 | 0.74 | 1.43 | 0.45 | 1.34 | 24 | 18 |
| 10 | AASW | dark | 0.07 | 0.22 | 0.78 | 4.30 | 0.42 | 1.26 | 22 | 16 |
| 120 | WW | light | 0.03 | 0.03 | 0.03 | 0.04 | 0.67 | 2.86 | 74 | 59 |
| 120 | WW | light | 0.03 | 0.03 | 0.03 | 0.03 | 0.47 | 1.84 | 76 | 49 |
| 120 | WW | light | 0.03 | 0.04 | 0.03 | 0.27 | 0.10 | 0.42 | 18 | 15 |
| 10 | AASW | light | 0.06 | 0.13 | 0.43 | 1.44 | 0.05 | 0.21 | 8 | 7 |
| 10 | AASW | light | 0.05 | 1.02 | 0.41 | 0.92 | 0.07 | 0.33 | 27 | 16 |
| 10 | AASW | light | 0.05 | 0.13 | 0.00 | 1.59 | 0.09 | 0.39 | 21 | 13 |

Table 7. Results of experimental incubations with samples from station 57.2 (73°70′73″ S, 113.26′53″ W)

| Depth (m) | Water mass | Treatment | Total bacteria (number × 10^5 ml^-1) | | | Simpson diversity | Shannon diversity | Chao1 richness | Observed taxa |
|---|---|---|---|---|---|---|---|---|---|
| | | | 0 h | 56 h | 173 h | | | | |
| 680 | mCDW | dark | 0.03 | 0.04 | 0.05 | 0.88 | 3.76 | 95 | 71 |
| 680 | mCDW | dark | 0.05 | 0.05 | 0.37 | 0.87 | 3.63 | 83 | 61 |
| 680 | mCDW | dark | 0.05 | 0.04 | 0.05 | 0.82 | 3.36 | 85 | 69 |
| 10 | AASW | dark | 0.63 | 0.25 | 1.02 | 0.24 | 0.86 | 25 | 19 |
| 10 | AASW | dark | 0.32 | 0.22 | 0.87 | 0.06 | 0.27 | 26 | 16 |
| 10 | AASW | dark | 0.50 | 0.15 | 0.29 | 0.09 | 0.45 | 31 | 22 |
| 680 | mCDW | light[a] | 0.03 | 0.03 | 0.03 | missing | missing | missing | missing |
| 680 | mCDW | light | 0.03 | 0.05 | 0.02 | 0.91 | 4.39 | 87 | 82 |
| 10 | AASW | light | 0.39 | 0.24 | 0.27 | 0.07 | 0.36 | 38 | 22 |
| 10 | AASW | light | 0.79 | 0.28 | 0.39 | 0.28 | 1.01 | 25 | 20 |
| 10 | AASW | light | 0.36 | 0.23 | 0.22 | 0.05 | 0.26 | 27 | 16 |

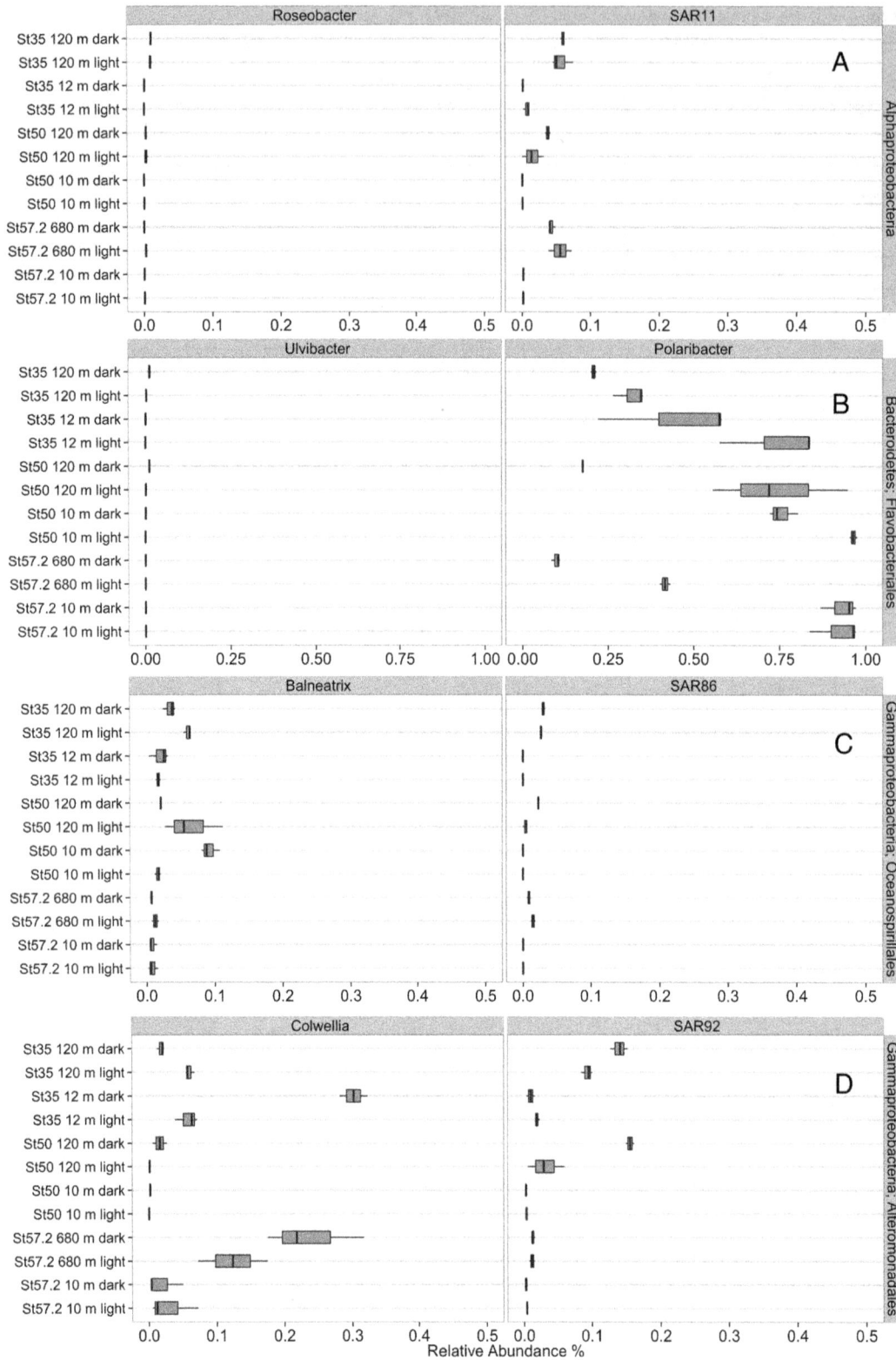

## Figure 4

**Distribution patterns of targeted bacterioplankton populations in incubation experiments.**

Observed differences in the relative abundance of targeted populations at the end of the experimental incubations. Differences are plotted for each station (st35, st50, and st57.2) by water mass (where 10 and 12 m indicate AASW, 120 m indicates WW, and 680 m indicates mCDW) and by treatment (dark, light). Corresponding alpha diversity and net growth are given in Table 4. Targeted populations are (A) *Roseobacter* and SAR11 (*Alphaproteobacteria*), (B) *Ulvibacter* and *Polaribacter* (*Flavobacteriales* of the *Bacteriodetes*), (C) *Balneatrix* and SAR86 (*Oceanospirillales* of the *Gammaproteobacteria*), and (D) *Colwellia* and SAR92 (*Alteromonadales* of the *Gammaproteobacteria*). For each treatment, boxplots illustrate the 25–75 percentile values (box), median (bar inside box), and standard deviation (whiskers). No outliers were detected.

light conditions. Similar to SAR11, members of the *Roseobacter* clade were scarce following treatments of water originating from AASW, despite the fact that *Roseobacter* was originally abundant in AASW at 1.0 ± 0.8% and in WW at 1.3 ± 0.8%. They were not significant in mCDW (0.05 ± 0.05%) and were reduced to 0.3 ± 0.4% in the incubated samples.

Within the *Flavobacteriales* (Figure 4B), the highly abundant *Polaribacter* contributed to the bacterioplankton communities in AASW at 45.7 ± 11.2%, in WW at 10.6 ± 9.4% and in mCDW at 1.2 ± 0.8%. *Polaribacter* reached a maximum average contribution of 57.7 ± 32% in the incubated samples, whereas the related *Ulvibacter* was much less abundant in each of the water masses and the incubated samples, with an average relative abundance of 2.5 ± 1.6% in AASW, 4 ± 7.5% in WW, 0.15 ± 0.2% in mCDW and 0.3 ± 0.4% in the incubated samples. *Ulvibacter* was nevertheless identified by SIMPER as one of the taxa contributing to the major part of the *in situ* community dissimilarity between water masses, but was clearly outcompeted in the incubations. In contrast *Polaribacter* was a dominant community member in all treatments, though it clearly preferred light-exposed treatments from WW and AASW.

Two representative clades from the *Oceanospirillales* (Figure 4C) within the *Gammaproteobacteria*, *Balneatrix* and SAR86, were selected for further scrutiny as they displayed contrasting abundance patterns. Of the total reads *Balneatrix* contributed 27.6 ± 6.25% in AASW, 15.6 ± 12.1% in WW, and 1.5 ± 0.7% in mCDW. It occurred in highest abundance in AASW and WW (similar to *Polaribacter*) but dropped following incubation to 3 ± 3%. SAR86 was present in all incubations, but at low relative abundances, averaging 0.8 ± 1.1%, similar to those observed in all of the water masses: 0.02 ± 0.03% in AASW, 1.1 ± 0.8% in WW, and 0.98 ± 0.41% in mCDW.

Within the *Alteromonadales* (Figure 4D), two ubiquitous marine clades, *Colwellia* and SAR92, displayed contrasting abundance patterns across the water masses. The contribution of *Colwellia* to the total communities was on average 0.2 ± 0.56% in AASW, 0.1 ± 0.1% in WW, and 1.2 ± 2.1% in mCDW. In the incubated samples, relative *Colwellia* abundance increased the most in samples from mCDW (8.5±14.8%) where its relative *in situ* abundance was also the greatest. In contrast, the average relative abundance of SAR92 was greater in both AASW at 8.9 ± 3.3% and WW at 8.1 ± 5.2% and lower in mCDW at 0.8 ± 0.4%, averaging 3.7 ± 5.2% following the incubations.

## Discussion

Here we combined a spatial community survey of *in situ* bacterioplankton communities with controlled and replicated experimental incubations to illustrate how the environmental heterogeneity of the Amundsen Sea affects bacterioplankton population dynamics and community structure. Our results demonstrated that a relatively small number of taxa are responsible for the majority of observed dissimilarities in bacterial community composition between the three summer water masses of the Amundsen Sea Polynya, with observed differences originating mainly from shifts in the relative abundance of these taxa. The greatest differences appear between the seasonal AASW and the two mesopelagic WW and mCDW water masses. Our results agree with results from previous surveys based on clone libraries (Gentile et al., 2006) or pyrosequencing of 16S rRNA genes (Delmont et al., 2014; Kim et al., 2014; Wilkins et al., 2012), but go further by considering separation by water mass in the mesopelagic zone. Normally, a variety of factors in combination shape marine bacterial community structure and create spatiotemporal distribution patterns (biogeography). In AASW for instance, factors known to control the population dynamics of bacterioplankton include light (this study), the availability of organic compounds (Sipler and Connelly, 2014), inorganic nutrients (Alderkamp et al., 2014; Sherrell et al., 2014) and complex interactions with eukaryotic phytoplankton blooms consisting mostly of *Phaeocystis antarctica* or diatoms (Delmont et al., 2014). Even if interactions between these factors ultimately shape bacterioplankton communities, individually they can cause different kinds of physiological responses in individual bacterial groups. Light, for instance, can induce a shift from heterotrophy to photoautotrophy, while eukaryotic phytoplankton may control resource availability by exudation of organic matter. While the present study focuses on bacterial population dynamics shaped by light and hydrographic separation, the overall mechanism of environmental gradients determining microbial biogeography in this region at the level of combined communities will be presented and discussed elsewhere. Future applications of metagenomic and metatranscriptomic techniques may also shed light on these issues.

### *Diversity in experimental treatments*

In the incubation experiments, the origin of the seawater, serving as both inoculum and medium, explained most of the differences in the composition of emerging bacterial communities, while light exposure caused a modest inhibition of growth in all but the deepest water mass. Community richness after 7 days of experimental incubation was reduced compared to the bacterioplankton communities residing in the respective original water mass. This result is not surprising, as the preparation of dilution-extinction series for environmental samples is known to reduce diversity (Garland and Lehman, 1999).

Furthermore, particle-associated bacteria as well as larger grazers were removed from the seawater medium by pre-filtration, which will alter the overall diversity and activity in the experiments (Delmont et al., 2014; Williams et al., 2014). Grazing as a top-down regulating factor can, for instance, prevent copiotrophic populations from becoming dominant under resource-deplete conditions (Hahn and Höfle, 2001). Nevertheless, the community composition after incubation generally reflected the community composition in the original water masses, suggesting that grazing is of minor importance in controlling the bacterioplankton communities in this study. We cannot exclude that a longer incubation time may have allowed for additional shifts in community composition in response to the experimental manipulations; however, we chose a relatively short incubation period to minimize the risk of confinement effects (Massana et al., 2001). Mechanical disruption caused by pre-filtration of the seawater inoculum might also have affected the BCC or media composition. The changes observed in each of the three independent experiments were all in the same direction, with growth inhibited by light and greater richness observed in communities emerging from the mesopelagic water masses than from the surface water mass. This consistency of response implies a true treatment effect induced by light and not an indirect effect from active (or inactive) phytoplankton. The estimated growth during incubation was similar to that estimated previously for related polar ocean habitats (Rivkin et al., 1996).

## Population dynamics across water masses

The *Flavobacteriales* genus *Polaribacter* contributed most to the emerging communities in the incubation experiments. This genus was dominant both *in situ* and after incubation for AASW and WW while present in much lower abundance in mCDW. Corresponding with its highest relative abundance in samples from the photic AASW, *Polaribacter* also appeared dominant in all incubations exposed to light, despite the removal of phytoplankton as a source of dissolved organic matter (DOM). Enhanced solar radiation has the capacity to hamper bacterial growth in surface waters (Sommaruga et al., 1997; Alonso-Sáez et al., 2006), a general effect also observed in our treatments. However, *Polaribacter* appears to benefit from light, either by direct phototrophic energy metabolism or by being more photoresistant than their competitors. The *Flavobacteria* contains bacteria that function as both free-living and particle-attached populations, is highly abundant in marine and freshwater environments, and has been linked specifically to the uptake and decomposition of polymeric DOM (Kirchman, 2002; 2008). Some *Flavobacteria* exhibit maximal growth rates that are nearly 2-fold higher than those of other abundant bacterial groups, resulting in pronounced growth during phytoplankton blooms (Zeder et al., 2009). Their abundant presence in the ASP region has been reported before (Kim et al., 2014; Delmont et al., 2014), as well as their increased activity during episodes of high primary production (Grzymski et al., 2012), presumably due to their extensive metabolic capabilities and photoheterotrophy (Williams et al., 2012). These findings are all consistent with the extraordinarily high productivity in the surface waters (AASW) of the Amundsen Sea Polynya (Yager et al., 2012; Williams et al., 2014).

The high abundance of *Polaribacter* in AASW and in light-exposed incubations could represent a seasonal population bottleneck, defined as a reduction of the gene pool of a community due to superior growth of one specific population in a community. Such bottlenecks typically happen during cultivation or in artificial laboratory conditions (Nei et al., 1975; Koskiniemi et al., 2012). However, in the present study the copiotroph *Polaribacter* appears to expand in an analogous way in the natural ecosystem, where it is as dominant in epipelagic waters as in the experimental incubations. Accordingly, this type of population bottleneck may reduce diversity in natural ecosystems, an explanation that is consistent with the lower estimated richness in AASW where *Polaribacter* contributed 50–70% of the total community.

Similar to *Polaribacter*, members of the genus *Balneatrix* (*Oceanospirillales*) are known to be present at high concentrations in photic AASW, often associated with intense phytoplankton blooms (Nikrad et al., 2013). The response of *Balneatrix* in the experimental incubations, however, differed from that of *Polaribacter*. *Balneatrix* was less dominant in the experiments independently of light conditions than in the upper water masses. An explanation for this response in the experiments could be, on the one hand, the lack of phototrophic energy metabolism in these heterotrophic bacteria and, on the other hand, the uncoupling from primary production and associated release of autochthonous organic matter in the experiments, as phytoplankton had been removed from the incubation medium.

Several other bacterial taxa remained in low abundance across all water masses and in the experimental incubations. Those populations are likely to have an oligotrophic lifestyle without the capacity for rapid proliferation. Members of the chemoheterotrophic *Ulvibacter*, for instance, belonging to the family *Flavobacteriaceae*, exhibited opposite abundance patterns to the closely related *Polaribacter*. *Ulvibacter*, previously isolated from the Southern Ocean (Choi et al., 2007) did not increase in abundance during our experimental incubations. Nevertheless, it was sometimes abundant, though highly variable in WW, possibly related to changes in local resources in this water mass. Isolates of *Ulvibacter* have gliding motility as a typical trait (Choi et al., 2007), indicating a particle-associated lifestyle. Hence, one possible explanation for the low representation of this group in the incubations, and variable abundance *in situ*, is that many of these bacteria were particle-associated and thus removed during pre-filtration of the seawater inoculum.

In general, the incubated samples, which originated from three different stations, responded similarly to light and by water mass source, yet the emerging communities varied in Simpson diversity and richness among stations, indicating that there was a site-specific response. For instance, the abundance of *Colwellia* after incubation varied by station. *Colwellia* is considered to be a genus of obligate psychrophiles involved in the production of extracellular polymeric substances required for biofilm formation, as well as enzymes for the breakdown of such high-molecular-weight organic compounds (Huston et al., 2004; Methé et al., 2005); *Colwellia* species are also known from sea ice communities (Bowman et al., 1998). This group of bacteria may be responsible for much of the important breakdown of export attributed to particle-associated, mesopelagic bacteria in the ASP (Ducklow et al., 2014; Williams et al., 2014), where *Colwellia* has been found associated with particulate matter (Delmont et al., 2014); such particle association may have led to selective removal of this group in the experimental pre-filtration. Chemoheterotrophic *Colwellia* have also been identified as being active in dark bicarbonate uptake (Alonso-Sáez et al., 2010), an intriguing possible contribution to their emergence in the dark treatments.

### Response to light

The phototrophic *Roseobacter* lineage was a minor component both *in situ* and in the experimental incubations, even though it is one of the most abundant and well-studied clades of bacteria in the marine environment, where it can contribute up to 20% of bacterioplankton communities in some coastal regions (Buchan et al., 2005; Wagner-Döbler and Biebl, 2006). A number of metabolic traits have been identified in the *Roseobacter*, enabling members of this genus to interact closely with phytoplankton (Geng and Belas, 2010), including chemotaxis for motility towards free nutrients and fimbrial adhesins facilitating adhesion to phytoplankton. Coupled to the latter is a better opportunity to exploit DOM and organic sulfur compounds (e.g. dimethylsulfoniopropionate-DMSP) released from phytoplankton (Ruiz-González et al., 2012; Tortell et al., 2012). In the ASP, *Roseobacter* was rather scarce and, despite being known to share many metabolic and ecological traits with *Polaribacter*, was apparently not competitive in any of the incubation experiments. The ability of *Roseobacter* to adhere to particles and thus be removed during pre-filtration may have contributed to these results.

Light can have contrasting effects on different populations of bacteria. Light-inhibition was observed for the *Gammaproteobacteria* clade SAR92 in WW. This result seems to contradict earlier findings, where isolates of this group tested positive for the light-driven proton pump proteorhodopsin and were considered oligotrophic (Stingl et al., 2007). Our apparently contradictory result emphasizes that the presence of individual genes, as identified in isolates, does not always correlate with functional responses *in situ*. Isolates related to the abundant SAR92 clade found in the ASP have also been characterized as members of the oligotrophic marine *Gammaproteobacteria* (OMG) (Cho and Giovannoni, 2004), indicating a predominantly oligotrophic life strategy in the pre-bloom WW. Within clade SAR86 (*Gammaproteobacteria*), some subgroups carry proteorhodopsin and display enhanced growth when exposed to light (Schwalbach et al., 2005). Nevertheless, this clade did not respond to light treatments. This result is not surprising, given that SAR86 in the ASP was mainly abundant in the dark water masses and in experimental incubations originating from dark waters, and thus not likely to rely on phototrophic energy metabolism.

The ubiquitous *Alphaproteobacteria* clade SAR11 from different samples responded in contrasting ways to the experimental incubations, but rarely responded to light. SAR11 is known to prefer low molecular weight and labile organic compounds and also to carry proteorhodopsin (Giovannoni et al., 2005; Tripp, 2013). Interestingly, we detected a much lower abundance of SAR11 compared to Kim et al. (2014) or Delmont et al. (2014), which might be related to different sampling conditions or the application of different primer pairs, though an explanation for this difference is not obvious.

## Conclusion

This study presents an example of how environmental heterogeneity affects bacterial community composition in the ASP. We show in particular how hydrographical separation accounts for shifts in the abundant fraction of taxa, which we attribute to differences in bacterioplankton population dynamics between the water masses driven in part by light availability. The BCC in the photic AASW was characterized by low richness, favoring dominant bacterial populations related to taxa known to expand under bacterioplankton bloom events. In contrast, the BCC in the mesopelagic water masses had higher richness, featuring taxa known to benefit from oligotrophic conditions. Grazing appeared of minor importance, whereas (in hindsight) our method of pre-filtration may have removed an important fraction of particle-associated bacteria. Incubation experiments indicated the importance of light as a regulating factor of the BCC: light in general inhibits community growth, but some populations, particularly *Polaribacter* in the ASP, profited directly from light conditions and increased in abundance. In parallel with the strong influence of phytoplankton and their exudates, our experimental treatments help to explain how changing light conditions in surface waters, with the seasonally variable ice conditions, may influence the dynamics of BCC between photic surface and dark mesopelagic waters.

# References

Agogué H, Lamy D, Neal PR, Sogin ML, Herndl GJ. 2011. Water mass-specificity of bacterial communities in the North Atlantic revealed by massively parallel sequencing. *Mol Ecol* **20**(2): 258–274. doi: 10.1111/j.1365-294X.2010.04932.x.

Alderkamp AC, Dijken GL, Lowry KE, Connelly TL, Lagerström M, et al. 2014. Fe availability drives phytoplankton photosynthesis rates in the Amundsen Sea Polynya, Antarctica. *Elem Sci Anth*: under review for the ASPIRE Special Feature.

Alderkamp AC, Mills MM, van Dijken GL, Laan P, Thuróczy CE, et al. 2012. Iron from melting glaciers fuels phytoplankton blooms in the Amundsen Sea (Southern Ocean): Phytoplankton characteristics and productivity. *Deep-Sea Res Pt II* **71–76**: 32–48. doi: 10.1016/j.dsr2.2012.03.005.

Alonso-Sáez L, Andersson A, Heinrich F, Bertilsson S. 2011. High archaeal diversity in Antarctic circumpolar deep waters. *Environ Microbiol Rep* **3**(6): 689–697. doi: 10.1111/j.1758-2229.2011.00282.x.

Alonso-Sáez L, Galand PE, Casamayor EO, Pedrós-Alió C, Bertilsson S. 2010. High bicarbonate assimilation in the dark by Arctic bacteria. *ISME J* **4**(12): 1581–1590. doi: 10.1038/ismej.2010.69.

Alonso-Sáez L, Gasol JM, Lefort T, Hofer J, Sommaruga R. 2006. Effect of natural sunlight on bacterial activity and differential sensitivity of natural bacterioplankton groups in northwestern mediterranean coastal waters. *Appl Environ Microbiol* **72**(9): 5806–5813.

Arrigo K, van Dijken G. 2003. Phytoplankton dynamics within 37 Antarctic coastal polynya systems. *J Geophys Res* **108**(C8). doi: 10.1029/2002JC001739.

Billen G, Servais P, Becquevort S. 1990. Dynamics of bacterioplankton in oligotrophic and eutrophic aquatic environments: bottom-up or top-down control? *Hydrobiologia* **207**(1): 37–42. doi: 10.1007/BF00041438.

Boström K, Simu K, Hagström Å, Riemann L. 2004. Optimization of DNA extraction for quantitative marine bacterioplankton community analysis. *Limnol Oceanogr* **2**: 365–373.

Bowman JP, Gosink JJ, McCammon SA, Lewis TE, Nichols DS, et al. 1998. *Colwellia demingiae* sp. nov., *Colwellia hornerae* sp. nov., *Colwellia rossensis* sp. nov. and *Colwellia psychrotropica* sp. nov.: psychrophilic Antarctic species with the ability to synthesize docosahexaenoic acid (22:ω63). *Int J Syst Bacteriol* **48**(4): 1171–1180. doi: 10.1099/00207713-48-4-1171.

Bryant DA, Frigaard NU. 2006. Prokaryotic photosynthesis and phototrophy illuminated. *Trends Microbiol* **14**(11): 488–496. doi: 10.1016/j.tim.2006.09.001.

Buchan A, Gonzalez JM, Moran MA. 2005. Overview of the marine *Roseobacter* lineage. *Appl Environ Microbiol* **71**(10): 5665–5677. doi: 10.1128/AEM.71.10.5665-5677.2005.

Cabiscol E, Tamarit J, Ros J. 2010. Oxidative stress in bacteria and protein damage by reactive oxygen species. *Int Microbiol* **3**(1): 3–8.

Caporaso J, Kuczynski J, Stombaugh J, Bittinger K, Bushman F, et al. 2010. QIIME allows analysis of high-throughput community sequencing data. *Nat Methods* **7**(5): 335–336. doi: 10.1038/nmeth.f.303.

Cho J-C, Giovannoni SJ. 2004. Cultivation and growth characteristics of a diverse group of oligotrophic marine Gammaproteobacteria. *Appl Environ Microbiol* **70**(1): 432–440. doi: 10.1128/AEM.70.1.432-440.2004.

Choi T-H, Lee HK, Lee K, Cho JC. 2007. *Ulvibacter antarcticus* sp. nov., isolated from Antarctic coastal seawater. *Int J Syst Evol Microbiol* **57**(12): 2922–2925. doi: 10.1099/ijs.0.65265-0.

Clarke K. 1993. Non-parametric multivariate analyses of changes in community structure. *Aust J Ecol* **18**(1): 117–143. doi: 10.1111/j.1442-9993.1993.tb00438.x.

Clarke KR, Warwick RM. 2001. *Change in marine communities: An approach to statistical analysis and interpretation.* 2nd ed. Plymouth, UK: Plymouth Marine Laboratory.

Comte J, del Giorgio P. 2011. Composition influences the pathway but not the outcome of the metabolic response of bacterioplankton to resource shifts. *PLoS ONE* **6**(9). doi: 10.1371/journal.pone.0025266.

Cota G, Kottmeier S, Robinson D, Smith Jr W, Sullivan C. 1990. Bacterioplankton in the marginal ice zone of the Weddell Sea: biomass, production and metabolic activities during austral autumn. *Deep-Sea Res* **37**(7): 1145–1167. doi: 10.1016/0198-0149(90)90056-2.

Cottrell MT, Kirchman DL. 2009. Photoheterotrophic microbes in the Arctic Ocean in summer and winter. *Appl Environ Microbiol* **75**(15): 4958–4966. doi: 10.1128/AEM.00117-09.

Delmont TO, Hammar KM, Ducklow HW, Yager PL, Post AF. 2014. *Phaeocystis antarctica* blooms strongly influence bacterial community structures in the Amundsen Sea polynya. *Front Microbiol* **5**: 646. doi: 10.3389/fmicb.2014.00646.

Doudney CO, Young CS. 1962. Ultraviolet light induced mutation and deoxyribonucleic acid replication in Bacteria. *Genetics* **47**: 1125–1138.

Ducklow H, Carlson C, Church M, Kirchman D, Smith D, et al. 2001. The seasonal development of the bacterioplankton bloom in the Ross Sea, Antarctica, 1994–1997. *Deep-Sea Res Pt II* **48**(19–20): 4199–4221. doi: 10.1016/S0967-0645(01)00086-8.

Ducklow HW, Erickson M, Lee SH, Lowry KE, Post A, et al. 2014. Particle flux over the continental shelf in the Amundsen Sea Polynya and Western Antarctic Peninsula. *Elem Sci Anth*: under review for the ASPIRE Special Feature.

Edgar R. 2010. Search and clustering orders of magnitude faster than BLAST. *Bioinformatics* **26**(19): 2460–2461. doi: 10.1093/bioinformatics/btq461.

Galand PE, Potvin M, Casamayor EO, Lovejoy C. 2010. Hydrography shapes bacterial biogeography of the deep Arctic Ocean. *ISME J* **4**: 564–576. doi: 10.1038/ismej.2009.134.

Garland JL, Lehman RM. 1999. Dilution/extinction of community phenotypic characters to estimate relative structural diversity in mixed communities. *FEMS Microbiol Ecol* **30**(4): 333–343. doi: 10.1111/j.1574-6941.1999.tb00661.x.

Gasol JM, del Giorgio PA. 2000. Using flow cytometry for counting natural planktonic Bacteria and understanding the structure of planktonic bacterial communities. *Sci Mar* **64**(2): 197–224.

Geng H, Belas R. 2010. Molecular mechanisms underlying Roseobacter–phytoplankton symbioses. *Curr Opin Biotechnol, Energy biotechnology – Environmental biotechnology* **21**(3): 332–338. doi: 10.1016/j.copbio.2010.03.013.

Gentile G, Giuliano L, D'Auria G, Smedile F, Azzaro M, et al. 2006. Study of bacterial communities in Antarctic coastal waters by a combination of 16S rRNA and 16S rDNA sequencing. *Environ Microbiol* **8**(12): 2150–2161. doi: 10.1111/j.1462-2920.2006.01097.x.

Ghiglione JF, Galand PE, Pommier T, Pedrós-Alió C, Maas EW, et al. 2012. Pole-to-pole biogeography of surface and deep marine bacterial communities. *Proc Natl Acad Sci USA* **109**(43):17633–17638. doi: 10.1073/pnas.1208160109.

Giovannoni SJ, Bibbs L, Cho JC, Stapels MD, Desiderio R, et al. 2005. Proteorhodopsin in the ubiquitous marine bacterium SAR11. *Nature* **438**(7064): 82–85. doi: 10.1038/nature04032.

Granéli W, Carlsson P, Bertilsson S. 2004. Bacterial abundance, production and organic carbon limitation in the Southern Ocean (39–62°S, 4–14°E) during the austral summer 1997/1998. *Deep-Sea Res Pt II* **51**(22–24): 2569–2582. doi: 10.1016/j.dsr2.2001.01.003.

Grzymski JJ, Riesenfeld CS, Williams TJ, Dussaq AM, Ducklow H, et al. 2012. A metagenomic assessment of winter and summer bacterioplankton from Antarctica Peninsula coastal surface waters. *ISME J* **6**:1901–1915. doi: 10.1038/ismej.2012.31.

Hahn MW, Höfle MG. 2001. Grazing of protozoa and its effect on populations of aquatic bacteria. *FEMS Microb Ecol* **35**(2): 113–121. doi: 10.1111/j.1574-6941.2001.tb00794.x.

Hamdan LJ, Coffin RB, Sikaroodi M, Greinert J, Treude. 2013. Ocean currents shape the microbiome of Arctic marine sediments. *ISME J* **7**(4): 685–696. doi: 10.1038/ismej.2012.143.

Herlemann D, Labrenz M, Jürgens K, Bertilsson S, Waniek, et al. 2011. Transitions in bacterial communities along the 2000 km salinity gradient of the Baltic Sea. *ISME J* **5**(10): 1571–1579. doi: 10.1038/ismej.2011.41.

Hill TCJ, Walsh KA, Harris JA, Moffett BF. 2003. Using ecological diversity measures with bacterial communities. *FEMS Microbiol Ecol* **43**(1): 1–11. doi: 10.1111/j.1574-6941.2003.tb01040.x.

Huston AL, Methé B, Deming JW. 2004. Purification, characterization, and sequencing of an extracellular cold-active aminopeptidase produced by marine psychrophile *Colwellia psychrerythraea* strain 34H. *Appl Environ Microbiol* **70**(6): 3321–3328. doi: 10.1128/AEM.70.6.3321-3328.2004.

Kim JG, Park SJ, Quan ZX, Jung MY, Cha IT, et al. 2014. Unveiling abundance and distribution of planktonic Bacteria and Archaea in a polynya in Amundsen Sea, Antarctica. *Environ Microbiol* **16**(6): 1566–1578. doi: 10.1111/1462-2920.12287.

Kirchman DL. 2002. The ecology of Cytophaga–Flavobacteria in aquatic environments. *FEMS Microbiol Ecol* **39**(2): 91–100. doi: 10.1111/j.1574-6941.2002.tb00910.x.

Kirchman DL. 2008. New light on an important microbe in the ocean. *PNAS* **105**(25): 8487–8488. doi: 10.1073/pnas.0804196105.

Koh EY, Atamna-Ismaeel N, Martin A, Cowie ROM, Beja O, et al. 2010. Proteorhodopsin-bearing bacteria in Antarctic sea ice. *Appl Environ Microbiol* **76**(17): 5918–5925. doi: 10.1128/AEM.00562-10.

Koskiniemi S, Sun S, Berg OG, Andersson DI. 2012. Selection-driven gene loss in bacteria. *PLoS Genet* **8**(6). doi: 10.1371/journal.pgen.1002787.

Massana R, Pedrós-Alío C, Casamayor EO, Gasol JM. 2001. Changes in marine bacterioplankton phylogenetic composition during incubations designed to measure biogeochemically significant parameters. *Limnol Oceanogr* **46**(5): 1181–1188. doi: 10.4319/lo.2001.46.5.1181.

Methé BA, Nelson KE, Deming JW, Momen B, Melamud E, et al. 2005. The psychrophilic lifestyle as revealed by the genome sequence of *Colwellia psychrerythraea* 34H through genomic and proteomic analyses. *Proc Natl Acad Sci USA* **102**(31): 10913–10918. doi: 10.1073/pnas.0504766102.

Nei M, Maruyama T, Chakraborty R. 1975. The bottleneck effect and genetic variability in populations. *Evolution* **29**(1): 1–10.

Nikrad MP, Cottrell MT, Kirchman DL. 2013. Growth activity of gammaproteobacterial subgroups in waters off the west Antarctic Peninsula in summer and fall. *Environ Microbiol* **16**(6): 1513–1523. doi: 10.1111/1462-2920.12258.

Okubo S, Nakayama H. 1967. DNA synthesis after ultraviolet light irradiation in UV-sensitive mutants of *Bacillus subtilis*. *Mutat Res-Fund Mol M* **4**(5): 533–541.

Pernthaler J. 2005. Predation on prokaryotes in the water column and its ecological implications. *Nat Rev Microbiol* **3**: 537–546.

Quast C, Pruesse E, Yilmaz P, Gerken J, Schweer T, et al. 2013. The SILVA ribosomal RNA gene database project: improved data processing and web-based tools. *Nucleic Acids Res* **41**: D590–D596. doi: 10.1093/nar/gks1219.

Quince C, Lanzen A, Davenport R, Turnbaugh P. 2011. Removing noise from pyrosequenced amplicons. *BMC Bioinformatics* **12**(1). doi: 10.1186/1471-2105-12-38.

Randall-Goodwin E, Meredith MP, Jenkins A, Sherrell RM, Abrahamsen EP, et al. 2014. Water mass structure and freshwater distributions in the Amundsen Sea Polynya, Antarctica. *Elem Sci Anth*: under review for the ASPIRE Special Feature.

Riemann L, Steward GF, Azam F. 2000. Dynamics of bacterial community composition and activity during a mesocosm Diatom bloom. *Appl Environ Microbiol* **66**(2): 578–87. doi:10.1128/AEM.66.2.578-587.2000.

Rivkin RB, Anderson MR, Lajzerowicz C. 1996. Microbial processes in cold oceans. I. Relationship between temperature and bacterial growth rate. *Aquat Microb Ecol* **10**: 243–254.

Ruiz-González C, Galí M, Gasol JM, Simó R. 2012. Sunlight effects on the DMSP-sulfur and leucine assimilation activities of polar heterotrophic bacterioplankton. *Biogeochemistry* **110**: 57–74. doi: 10.1007/s10533-012-9699-y.

Schwalbach MS, Brown M, Fuhrman JA. 2005. Impact of light on marine bacterioplankton community structure. *Aquat Microb Ecol* **39**: 235–245. doi: 10.3354/ame039235.

Sherrell R, Lagerström M, Forsch KM, Stammerjohn S, Yager PL. 2014. Dynamics of dissolved iron and other bioactive trace metals (Mn, Ni, Cu, Zn) in the Amundsen Sea Polynya, Antarctica. *Elem Sci Anth*: under review for the ASPIRE Special Feature.

Sipler RE, Connelly TL. 2014. Bioavailability of surface dissolved organic matter to Antarctic aphotic bacterial communities. *Elem Sci Anth*: under review for the ASPIRE Special Feature.

Sommaruga R, Obernosterer I, Herndl GJ, Psenner R. 1997. Inhibitory effect of solar radiation on thymidine and leucine incorporation by freshwater and marine bacterioplankton. *Appl Environ Microbiol* **63**(11): 4178–4184.

Stingl U, Desiderio RA, Cho JC, Vergin KL, Giovannoni SJ. 2007. The SAR92 clade: an abundant coastal clade of culturable marine bacteria possessing proteorhodopsin. *Appl Environ Microbiol* **73**(7): 2290–2296. doi: 10.1128/AEM.02559-06.

Tortell PD, Long MC, Payne CD, Alderkamp AC, Dutrieux P, et al. 2012. Spatial distribution of pCO$_2$, ΔO$_2$/Ar and dimethylsulfide (DMS) in polynya waters and the sea ice zone of the Amundsen Sea, Antarctica. *Deep-Sea Res Pt II: Topical Studies in Oceanography, Shedding Dynamic Light on Fe limitation (DynaLiFe)* **71–76**: 77–93. doi: 10.1016/j.dsr2.2012.03.010.

Tripp HJ. 2013. The unique metabolism of SAR11 aquatic bacteria. *J Microbiol* **51**(2): 147–153. doi: 10.1007/s12275-013-2671-2.

Violle C, Navas ML, Vile D, Kazakou E, Fortunel C, et al. 2007. Let the concept of trait be functional! *Oikos* **116**(5): 882–892. doi: 10.1111/j.0030-1299.2007.15559.x.

Wagner-Döbler I, Biebl H. 2006. Environmental biology of the marine roseobacter lineage. *Annu Rev Microbiol* **60**: 255–280. doi: 10.1146/annurev.micro.60.080805.142115.

Wang Q, Garrity G, Tiedje J, Cole J. 2007. Naïve Bayesian classifier for rapid assignment of rRNA sequences into the new bacterial taxonomy. *Appl Environ Microbiol* **73**(16): 5261–5267. doi: 10.1128/AEM.00062-07.

Wilkins D, Lauro FM, Williams TJ, Demaere MZ, Brown MV, et al. 2012. Biogeographic partitioning of Southern Ocean microorganisms revealed by metagenomics. *Environ Microbiol* **15**(5): 1318–1333. doi: 10.1111/1462-2920.12035.

Williams CM, Dupont AM, Loevenich J, Post AF, Dinasquet J, et al. 2014. Pelagic microbial heterotrophy in response to a highly productive bloom of *Phaeocystis antarctica* in the Amundsen Sea Polynya, Antarctica. *Elem Sci Anth*: in production.

Williams TJ, Wilkins D, Long E, Evans F, Demaere MZ, et al. 2012. The role of planktonic Flavobacteria in processing algal organic matter in coastal East Antarctica revealed using metagenomics and metaproteomics. *Environ Microbiol* **15**(5): 1302–1317. doi: 10.1111/1462-2920.12017.

Yager P, Sherrell R, Stammerjohn S, Alderkamp A, Schofield O, et al. 2012. ASPIRE: The Amundsen Sea Polynya International Research Expedition. *Oceanogr* **25**: 40–53. doi: 10.5670/oceanog.2012.73.

Zeder M, Peter S, Shabarova T, Pernthaler J. 2009. A small population of planktonic *Flavobacteria* with disproportionally high growth during the spring phytoplankton bloom in a prealpine lake. *Environ Microbiol* **11**(10): 2676–2686. doi: 10.1111/j.1462-2920.2009.01994.x.

## Contributions

- Contributed to planning and experimental design: SB LR IR
- Contributed to acquisition of data: IR JD RL PY
- Contributed to analysis and interpretation of data: IR JD RL AW LR SB, PY
- IR wrote the paper with the help and inputs from all coauthors

## Acknowledgments

We thank the captain and crew of the RVIB Nathaniel B Palmer (NBP 10-05) and the ASPIRE team. SWEDARP 2010/11 was organized by the Swedish Polar Research Secretariat and the National Science Foundation Office of Polar Programs. Logistic support was provided from the Swedish Polar Research secretariat and Raytheon Polar Services. 454 pyrosequencing was handled by the SciLifeLab SNP/SEQ facility hosted by Uppsala University and bioinformatic analyses was supported by the UPPMAX Next Generation Sequencing Cluster (UPPNEX). We also acknowledge Alexander Eiler for assistance with handling of the raw pyrosequencing data, Shona Whatson, who created the GIS map and Jody W. Deming for valuable comments on the manuscript.

## Funding information

The research was funded by the Swedish Research Council (grants to SB) and by the US National Science Foundation Office of Polar Programs (ANT-0839069 to PY). Sequencing was made possible by an instrument grant from the K&A Wallenberg foundation.

## Competing interests

The authors have declared that no competing interest exists.

## Data accessibility statement

Accession No.: PRJEB4866 (European Nucleotide Archive).

# 10

# Recovery of a mining-damaged stream ecosystem

Christopher A. Mebane[1]* • Robert J. Eakins[2] • Brian G. Fraser[2] • William J. Adams[3]

[1]Idaho Water Science Center, U.S. Geological Survey, Boise, Idaho, United States
[2]EcoMetrix Incorporated, Mississauga, Ontario, Canada
[3]Rio Tinto, Lake Point, Utah, United States

*cmebane@usgs.gov

Domain Editor-in-Chief
Joel D. Blum, University of Michigan

Knowledge Domains
Earth and Environmental Science
Ecology

## Abstract

This paper presents a 30+ year record of changes in benthic macroinvertebrate communities and fish populations associated with improving water quality in mining-influenced streams. Panther Creek, a tributary to the Salmon River in central Idaho, USA suffered intensive damage from mining and milling operations at the Blackbird Mine that released copper (Cu), arsenic (As), and cobalt (Co) into tributaries. From the 1960s through the 1980s, no fish and few aquatic invertebrates could be found in 40 km of mine-affected reaches of Panther Creek downstream of the metals contaminated tributaries, Blackbird and Big Deer Creeks.

Efforts to restore water quality began in 1995, and by 2002 Cu levels had been reduced by about 90%, with incremental declines since. Rainbow Trout (*Oncorhynchus mykiss*) were early colonizers, quickly expanding their range as areas became habitable when Cu concentrations dropped below about 3X the U.S. Environmental Protection Agency's biotic ligand model (BLM) based chronic aquatic life criterion. Anadromous Chinook Salmon (*O. tshawytscha)* and steelhead (*O. mykiss*) have also reoccupied Panther Creek. Full recovery of salmonid populations occurred within about 12-years after the onset of restoration efforts and about 4-years after the Cu chronic criteria had mostly been met, with recovery interpreted as similarity in densities, biomass, year class strength, and condition factors between reference sites and mining-influenced sites. Shorthead Sculpin (*Cottus confusus*) were slower than salmonids to disperse and colonize. While benthic macroinvertebrate biomass has increased, species richness has plateaued at about 70 to 90% of reference despite the Cu criterion having been met for several years. Different invertebrate taxa had distinctly different recovery trajectories. Among the slowest taxa to recover were *Ephemerella, Cinygmula* and *Rhithrogena* mayflies, Enchytraeidae oligochaetes, and *Heterlimnius* aquatic beetles. Potential reasons for the failure of some invertebrate taxa to recover include competition, and high sensitivity to Co and Cu.

## 1. Introduction

The ecological impairment of lotic environments by metal mine contamination is a longstanding problem that has occurred in many areas (Woody et al., 2010; Byrne et al., 2012; Hogsden and Harding, 2012). In the USA, at least 156 hard-rock mining sites requiring restoration have been inventoried and could cost as much as $24 billion USD to address (Gustavson et al., 2007). Yet while river restoration has become a >$1 billion USD per year industry, its practice has been severely criticized for lacking scientific rigor and assessment (Bernhardt et al., 2005; Palmer et al., 2005; Lake et al., 2007; Palmer, 2009). While many of these criticisms are directed towards projects seeking to restore physical habitats, linking water quality restoration efforts to ecosystem responses has also been difficult (Harris, 2012). With some notable exceptions such as Adams et al. (2002) (Tennessee), Clements et al. (2010) (Colorado), and Murphy et al. (2014) (UK), the effectiveness of interventions designed to improve freshwater environments have been unclear because of projects that were based more on faith than science and with a lack of rigorous effectiveness monitoring (Bernhardt et al., 2005; Hilderbrand et al., 2005; Jähnig et al., 2011). A counterpart to this situation is that environmental assessment or cleanup projects in the USA tend to produce massive reports that may be data-rich, but difficult to access and in print formats that are difficult to extract data from. These factors may lead to a "data-rich and information-poor syndrome" (Ward et al., 1986; Gustavson et al., 2007).

The recovery of aquatic ecosystems can be controversial to even define (Ormerod, 2003). Interpreting recovery from metals pollution must draw upon the science of diverse practices including ecotoxicology, geochemistry, stream ecology, monitoring, and fisheries. The concept of recovery at least implicitly relies on underlying ecological concepts and assumptions relating to natural variability, disturbance, dispersal, and succession (Connell and Slatyer, 1977; Fisher, 1990; Palmer et al., 1997; Parker and Wiens, 2005; Lake et al., 2007). Many ecological questions arise when interpreting recovery in streams. For instance, after a long-term ("press") disturbance from metal contamination is relaxed, will the stream community reassemble itself similarly to nearby reference areas? With water quality restoration efforts driven by regulatory criteria, a key assumption is that numeric chemical criteria represent necessary and sufficient thresholds for ecological recovery. This leads to the question, as metals pollution declines, at what thresholds do organisms' internal limiting factors (physiological tolerance) give way to external limiting factors such as dispersal from colonist pools and biological tolerance or inhibition by early colonists of later arrivals? These concepts and questions give context to interpreting biological responses following water quality restoration efforts.

These factors are among the motivations for our present article. Over a 30+ year record, we examine changes in stream communities associated with declines in metals contamination from an inactive hard-rock mine in Idaho, USA. Our objectives include (1) assessing the effectiveness of water quality restoration efforts in reducing contamination in different stream media (water, sediment, periphyton, and macroinvertebrate tissues), (2) examining differing recovery trajectories for stream invertebrate and fish communities in response to improving water quality, (3) identifying apparent field thresholds for recovery of different taxa, and (4), considering whether the "recovering" stream ecosystems are "recovered."

## 1.1 Study area

The Blackbird Mine was a cobalt (Co) and copper (Cu) producer that operated from about 1948 to 1967. The mine is located on a high divide and flows south and north to the Blackbird Creek and Big Deer Creek drainages, respectively. To the south, mine drainage enters Blackbird Creek, a steep $2^{nd}$ order stream which flows for about 10 km before reaching Panther Creek, a $4^{th}$ order stream. To the north, the Blackbird Mine forms the headwaters of Bucktail Creek which flows to Panther Creek via the South Fork Big Deer Creek and Big Deer Creek (Figure 1, Figure S1). Blackbird and Big Deer Creeks each contribute about 12 to 13% of the streamflow in Panther Creek, as calculated immediately below their respective confluences (U.S. Geological Survey, 2012). Panther Creek is a tributary of the Salmon River, Idaho, USA and eventually drains to the Pacific Ocean about 1160 km downstream of the mouth of Panther Creek. Other ecologically important natural features include a series of waterfalls and cascades on Big Deer Creek situated about 1 km upstream of its mouth which prevents upstream passage of fish from Panther Creek, and several debris jams in the South Fork Big Deer Creek which impede upstream fish passage from Big Deer Creek.

The ore body consisted of the mineral cobaltite with equal portions of arsenic (As) and Co and lesser amounts of Cu. Ore was excavated from both open pit and underground operations and was processed into concentrate on-site. Mill effluents, which were highly enriched with Cu, Co, As, and iron (Fe) were run through a pipeline for about 5 km to a tailings dam and pond in the lower West Fork Blackbird Creek drainage where they were decanted. However, reagent spills, icing, pipeline breaks, and bypasses were frequent and the tailings frequently entered Blackbird Creek, and were transported downstream to Panther Creek. Acid mine drainage developed in both underground workings and waste rock dumps, which flowed to both the Big Deer Creek and Blackbird Creek drainages (Mebane, 1994; Gray and Eppinger, 2012).

Prior to mine development, Panther Creek supported abundant populations of anadromous Chinook Salmon (*Oncorhynchus tshawytscha*) and steelhead (*O. mykiss)*. By the mid-1950s, the salmon were in decline in Panther Creek as water quality problems worsened, and no spawning redds were observed after 1962 during annual aerial surveys of index reaches. Electrofishing surveys in 1967 and 1980 found no fish in Panther Creek downstream of Blackbird and Big Deer Creeks, and close to 100% mortalities occurred during short-term caged fish tests conducted at both locations in 1985. Unsuccessful reintroduction efforts of steelhead and Chinook Salmon were made in the 1970s and 1980s (Mebane, 1994). In September 2001, about 1053 adult hatchery-origin Chinook Salmon were released in Panther Creek (Smith et al., 2012).

Concerted efforts to restore water quality began in 1995. The efforts included a variety of measures to divert clean water around disturbed areas, and to intercept, collect, and treat contaminated water. Measures included relocation and containment of mine waste and sediment, construction of reservoirs, water and sediment control structures, and tunneling through the mountain to re-purpose the mine workings to capture and convey mine water to a water treatment plant. (USEPA, 2003, 2013). In addition to the water quality restoration, interim losses of ecosystem services (natural resource damages) were compensated for by off-site habitat improvement projects (Chapman and Julius, 2005).

Other than chemical water quality, the watershed was largely free from disturbances that constrain recovery. The watershed is lightly populated, with anthropogenic disturbances other than mining mostly limited to a

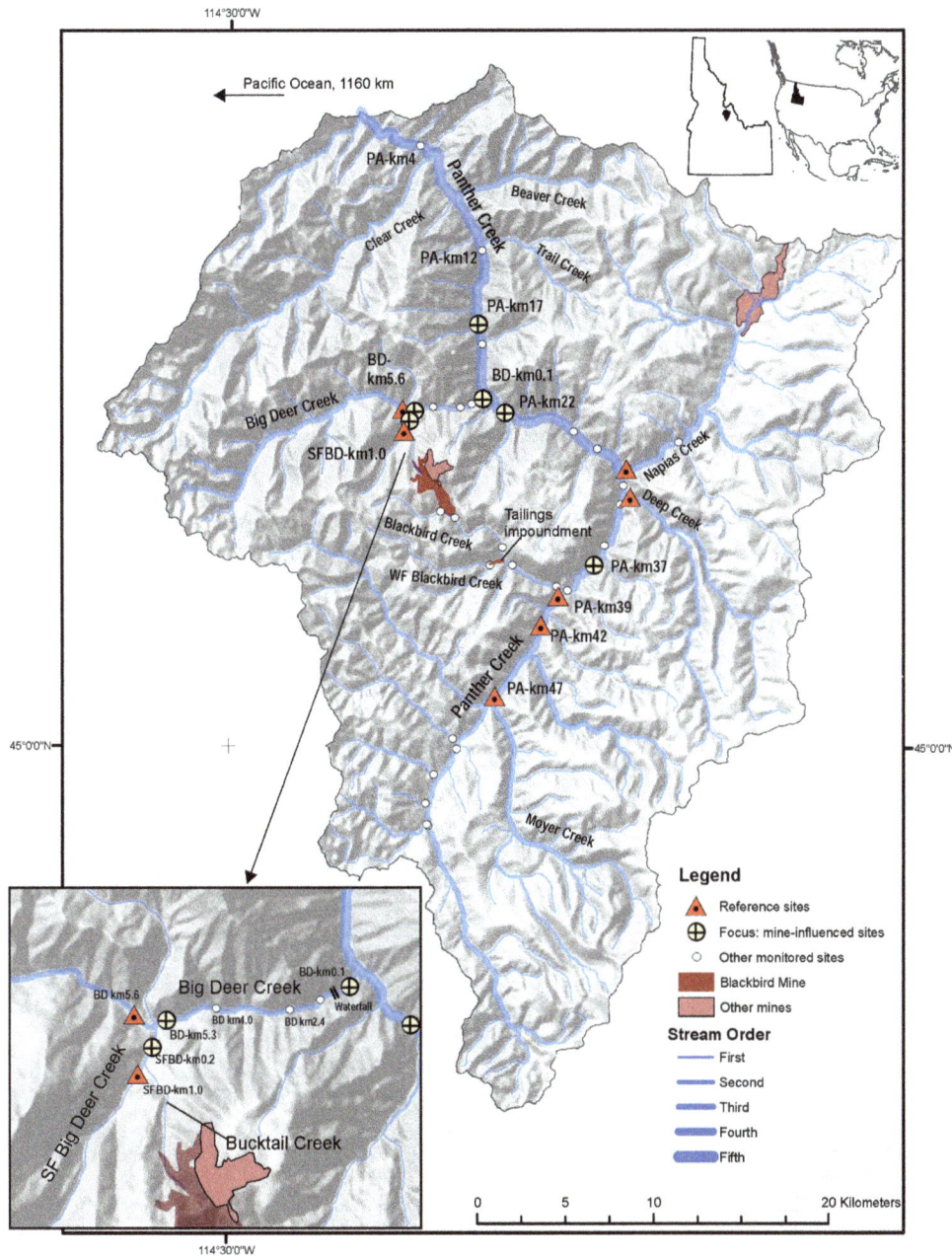

**Figure 1**

**Panther Creek study area, Idaho.**

"Focus: mine-influenced sites" are those sites on which we focus our analyses on here. The mining-affected areas are approximate and were traced from satellite imagery.

network of forest roads and seasonal cattle grazing in the upper watershed. Thus stream hydrology, channel morphology and riparian zones are largely intact and upstream reaches and tributaries provide water quality refugia and colonizing sources of invertebrates and fish. The four mining-affected tributaries (Big Deer Creek, South Fork Big Deer Creek, Blackbird Creek, and the West Fork Blackbird Creek), have their headwaters in near pristine, roadless watersheds (Figure 1). Two other mines are located within the Panther Creek watershed (Figure 1), although as of 2013 they had not noticeably affected the water quality in Panther Creek (Text S2).

In 2000, a large forest fire burned about 83 km² in the watershed (Eppinger et al., 2003). Fire intensity was high in the Big Deer Creek drainage, with almost the entire riparian forest canopy burning at both reference and mining-influenced assessment sites. A large debris flow from the Clear Creek drainage near stream km 4, temporarily dammed Panther Creek, realigned the channel, and filled pools. More information on the fire related disturbances is given in Text S2 and Figure S3.

Previous investigations include a synthesis of chemical and biological surveys through the early 1990s (Mebane, 1994), geochemical studies (Mok and Wai,1989; MacRae et al., 1999; Gray and Eppinger, 2012), toxicity testing of Cu and Co in laboratory waters intended to reflect stream water characteristics (Marr et al., 1996, 1998, 1999), avoidance and olfactory toxicity testing of Cu and Co to salmonids (Hansen et al., 1999), toxicity testing of Co in Panther Creek water (Pacific EcoRisk, 2005), toxicity testing of sediments (Mebane, 1994), and field surveys of macroinvertebrate communities and salmonid populations (LeJeune et al., 1995; Beltman et al., 1999), in addition to many unpublished reports.

## 2. Methods

Evaluating "recovery" requires defining of the concepts of recovery and restoration in the context of stream ecology and water pollution. We consider a "recovered" ecosystem to be unconstrained by chemical disturbance and similar in community composition to what would be expected in natural settings, absent mine-drainage pollution.

In short, our assessment of recovery followed a multi-year, before-after-control-impact (BACI) design that incorporates spatial and temporal variability, where samples were collected concurrently at both the "impact" and "control" (reference) sites in an annual time-series before and after the perturbation (Stewart-Oaten et al., 1992; Parker and Wiens, 2005). In a turn from usual BACI designs, in our study the "impacts" are the imposition of pollution controls, and the "before" conditions are the degraded conditions that persisted for the prior ~50 years. Because no single measure of metals exposure or biological response is likely adequate to indicate recovery (e.g., Niemi et al., 1993; Adams et al., 2002), we collected a suite of exposure and response endpoints. We evaluated chemical recovery of the streams by comparing chemical measures against numeric guidelines, and biological recovery using various macroinvertebrate and fish condition, population, and community level biological metrics (Table 1).

Biological metrics from mining-influenced streams were evaluated relative to concurrently sampled reference sites. Expressing biological metrics as a proportion of concurrently sampled reference sites has two purposes. First, because both reference and mining-influenced sites were affected by the 2000 wildfire, and because regional factors such as weather and hydrologic pattern similarly affect both reference and mining-influenced sites, natural temporal variability should be dampened. Second, prior to the 2002–2013 period, datasets were collected with different sampling and analysis methods. While different sampling and taxonomic efforts prevent direct data comparisons across studies, normalizing samples from mining-influenced sites to values collected from concurrent reference sites puts all data on the same scale. Normalizing data from mining-influenced sites against concurrent reference sites allowed us to extend our record across the older datasets, which has often been a limitation in "long-term" aquatic ecology studies (Jackson and Füreder, 2006).

Project-specific biological recovery goals for Panther Creek were to "restore and maintain water quality and aquatic biota conditions capable of supporting all life stages of resident and anadromous salmonids and other fishes." The recovery goals for Big Deer Creek and South Fork Big Creek were similar, but for Blackbird Creek these goals were considered unattainable for the foreseeable future. Instead, a more limited recovery goal for Blackbird Creek was that water quality could be improved "such that cleanup levels are not exceeded in Panther Creek and to support some aquatic life in Blackbird Creek" (USEPA, 2003).

**Table 1.** Conceptual framework for monitoring recovering stream communities subject to metals stress

| Endpoint | Expected response to reductions in metals stress |
|---|---|
| Benthic macroinvertebrate community | |
| Species richness | Increase. As concentrations decline, metals-sensitive taxa will expand their ranges and reoccupy habitats (Raddum and Fjellheim, 2003). Although simple and non-specific, species richness is often one of the most sensitive responses to many environmental stressors (Niemi et al., 1993) |
| Biomass | Variable. Under severe metals stress, species shifts to dominance by small-bodied taxa such as midges or aquatic mites would result in low biomass even though total abundance was high (Beltman et al., 1999; Hogsden and Harding,2012). |
| Mayfly and stonefly abundance | Increase. Most mayflies and some stoneflies are metals-sensitive (Clements et al., 2000; Hogsden and Harding, 2012). |
| Specific taxa | Variable. Within groups thought to be generally metals sensitive or resistant, specific taxa may differ in response (Courtney and Clements, 2002). Metals resistant taxa may decline with competition as metals-sensitive taxa recover. |
| Fish populations | |
| Species richness | Increase. In open-systems, large-bodied, more motile species such as salmonids will recolonize newly suitable habitats sooner than small-bodied fish with higher site fidelity (Gibbons et al., 1998; Milner et al., 2008). |
| Population size | Increase. In stressed systems, abundance may be limited by recruitment failure or indirectly from reduced prey base (Munkittrick and Dixon, 1989; Power, 1997; Milner et al., 2003). |
| Age structure | Shift toward younger. Natural fish populations are typically age-structured and numbers decrease with age in the total population. Selective mortality of sensitive fry may lead to recruitment failure in metal stressed populations, causing a shift to older fish (Schindler et al., 1985; Munkittrick and Dixon, 1989; Campbell et al., 2003; Milner et al., 2003). |
| Condition factor | Increase. Fish may have decreased growth or condition in metals-stressed systems from energy requirements of detoxification or food limitation/prey shifting. Condition factor and size are surrogates for energy reserves for overwinter survival and reproductive fitness (Munkittrick and Dixon, 1989; Farag et al., 1995; Cunjak et al., 1998; Campbell et al., 2003). |

We further explain our concepts for evaluating recovery in the context of the expected natural template for the affected streams, ecological assumptions, interpretations of chemical guidelines and project-specific goals in Text S2.

## 2.1 Sampling and analysis methods

A synthesis of data collected by different investigators over a 30+ period necessarily involves non-identical sampling and analyses, and details were sometimes sparse for older data. Our analyses center on the 2002–2013 time-series monitoring that repeatedly sampled the same locations at the same time of year (mid–September) using identical protocols, with continuity in field crews. Specific data sources and summaries are included in online supplemental datasets.

Site selection and sampling and analysis methods details are described in more detail in Text S2. In brief, all metals data in water are for "dissolved" fraction analyzed after 0.45 μm filtration. Water sampling effort varied over the years with effort biased toward the more variable spring runoff period. Since 2004, about 40–60 dissolved Co and Cu samples were collected per site per year, among other analytes (Table 2; Mebane et al., 2015). Metals in sediment were from the <2mm fraction from surficial samples, metals in periphyton (also referred to as biofilms or aufwuchs) were collected from rock scrapings, and metals in macroinvertebrate tissues were collected as a composite of species and sizes that were intended to represent the prevailing benthic community at each site. In 2012, additional collections targeted specific species. Hydrospychid caddisflies were targeted for Cu because they have been recommended as an indicator taxa for interpreting metals in tissues of aquatic insects (Rainbow et al., 2012). Hydrospychid caddisflies and two stonefly taxa at different trophic levels were also targeted for As speciation analyses in 2012 because of reports linking dietary exposure of inorganic As to reduced growth of Rainbow Trout (Erickson et al., 2010).

Benthic macroinvertebrate samples were all collected from riffle habitats using fixed area, substrate disturbance methods. In a survey undertaken in 1993 prior to water-quality restoration efforts, fish population counts were made by direct observation by snorkelers, but subsequent fish community surveys were conducted through electrofishing using blocknets and multiple-pass depletion. All fish were weighed and measured to enable calculation of fish condition, which is a measure of both individual and cohort (e.g., age- or size-group) wellness and is expected to decline under stress (Munkittrick and Dixon, 1989).

Methods, rationale, and study sites are described in more detail in Text S2.

## 2.2 Data analyses

Copper concentrations in water were evaluated across years in comparison to USEPA's (2007) biotic-ligand model (BLM) based aquatic life criteria. In this model, predicted toxicity is primarily a function of pH and dissolved organic carbon (DOC). Alkalinity, calcium (Ca), sodium (Na), magnesium (Mg), sulfate, and chloride are also included, but are less influential than DOC and pH. The BLM-based Cu criteria values tend to be higher during spring than baseflow conditions because streams tend to have elevated DOC during snowmelt runoff. For time periods without sufficient BLM-data being available (DOC, pH, etc.), we made conservative seasonal estimates of the BLM-based criteria of 5.3 and 2.3 μg/L for spring and near baseflow conditions respectively. No national or state criteria for Co have been established in the USA. We show Co concentrations in relation to both a site-specific chronic Co target of 86 μg/L and Environment Canada's (2013) guideline of 2.5 μg/L Co for freshwater habitats. Additional information on numeric, chemical-specific guidelines and calculations is given in the Text S2, and in the online data sets (Mebane et al., 2015).

To standardize comparisons across biological data, downstream measurements were generally calculated as a proportion of the concurrent reference value, which makes the reference condition always equal to 1. Where more than one reference site was concurrently monitored (e.g., Panther Creek from 2002–2013), reference site data were pooled for a single reference condition value for that point in time. To illustrate whether different biological measurements from the mine-influenced locations were similar to those from concurrently measured upstream reference comparison sites, some definition of "similar" is needed. If a biological measurement from the mine-influenced, downstream location was within the 95% confidence interval (CI) of the mean of the reference condition values, the values were considered similar to reference conditions (Di Stefano et al., 2005). Data used to calculate the confidence intervals were restricted to the periods with most consistent methods, which were 2002–2013 and 2003–2013, for the fish and macroinvertebrate data, respectively. The CIs ranged from 4% of reference for Panther Creek macroinvertebrate taxa richness to 35% of reference for Big Deer Creek benthic biomass. However, we did not interpret similarity to reference rigidly, as no two stream sites are identical and with sufficient sampling effort, the biology from any two sites can be statistically demonstrated to indeed be from two different sites (e.g., Martínez-Abraín, 2007). Further, classic inferential statistical tests for differences between groups are better suited for manipulative experiments than observational field studies. Samples in field studies are not independent of one another but are expected to occur in spatially and temporally autocorrelated gradients. For example, sample results from biological communities one year will be influenced by the community present at the location the year

Table 2. Summary of dissolved copper and cobalt concentrations in the Panther Creek watershed, pre- and post-water quality restoration effort[a]

| Location | Dissolved Copper (μg/L) | | Dissolved Cobalt (μg/L) | |
|---|---|---|---|---|
| | Average (range), n | | Average (range), n | |
| | 1993–1994 | 2013 | 1993–1994 | 2013 |
| Panther Cr, PA-km39, above Blackbird Cr. (Reference) | 1.1 (0.4–3.6), 12[b] | 0.25 (<0.1–0.7), 48[c] | 0.1 (0.05–0.2), 12[b] | 0.85 (0.2–2.3), 48 |
| Blackbird Cr., BB-km0.1 | 380 (94–1525), 38 | 8.7 (2.–24), 16 | 845 (329–1880), 38 | 120 (50–251), 3 |
| Panther Cr, PA-km37, below Blackbird Cr. | 51 (12–140), 91 | 1.0 (<0.1–2.9), 68[c] | 89 (36–204), 19 | 22 (7.4–51), 65 |
| Panther Cr, PA-km22, above Big Deer Cr | 27 (7–73), 7 | 1.2 (<0.1–4.8) 50[c] | 32 (17–69), 7 | 12 (4.3–20), 50 |
| Big Deer Cr., BD-km5.6 (Reference) | 3.0 (0.9–4.7), 6[d] | 0.26 (<0.1–0.7) 47[c] | <0.1 (<0.1), 6 | 0.54 (0.1–1.2) 48 |
| SF Big Deer Cr., SFBD-km0.1 | 1611 (560–2740) 5 | 14 (7.6–17) 7 | 704 (480–912) 5 | 13 (5.7–15) 7 |
| Big Deer Cr., BD-km5.3 | 139 (77–234), 3 | 2.6 (0.4–24), 66 | 63 (3–80), 3 | 2.0 (0.6–5.3), 63 |
| Big Deer Cr., BD-km0.1 | 105 (40–199), 6 | 4.7 (3.0–10.8), 49 | 58 (38–84), 6 | 2.0 (1–3), 49 |
| Panther Cr, PA-km17, below Big Deer Cr. | 25 (12–59), 19 | 1.4 (<0.1–4.1) 58[c] | 31 (13–63), 31 | 10 (4.2–19), 55 |

[a] Tributaries are indented and italicized.

[b] Values from 1994 only because of suspected Cu contamination in 1993 values, 1993 mean Cu at PA-km39 was 2.6 μg/L We consider the up to 10-fold "declines" in Cu concentrations at reference sites from 1993 to 2013 to be artifacts of better contemporary sample collection and processing, and improved laboratory practices, rather than environmental changes. However, with the higher concentrations present in mining-influenced sites in the 1990s, these low part-per-billion contamination artifacts would not affect interpretations.

[c] Averages estimated using the Kaplan-Meier procedure for datasets with censored (nondetected) values (Helsel, 2005). Nondetection rates in 2013 for Cu were 35% at PA-km39, 9% at PA-km37, 2% at PA-km22, 14% at PA-km17, and 14% at BD-km5.6.

[d] Data available from 1993 only, tablenote "b" is relevant.
Data sources: (Mebane et al., 2015), where 1993-1994 data were obtained by RCG/HaglerBailley, Boulder, CO, (unpublished) and 2013 data were obtained by Golder Associates, Redmond, WA, (unpublished).

before, and by nearby upstream conditions. Thus in field studies of unplanned impacts, simple graphs and correlations may be more appropriate than inferential statistics. Impact of contaminants can be inferred if metals concentrations are correlated with biological metrics, and recovery can then be inferred as the disappearance of such a correlation through time (Wiens and Parker, 1995). Locally weighted (Loess) regression was used to smooth and visualize patterns in time series graphs (Cleveland and Devlin, 1988). We tested for bivariate and multivariate correlations between biological metrics, metals, and two potentially confounding factors, stream temperature, and streamflow. Details are given in Text S2.

# 3. Results

## 3.1 Metals in sediment, periphyton, and benthic macroinvertebrate tissues

At reference sites in both Panther Creek and Big Deer Creek, Cu, Co, and As were consistently low in sediment, periphyton, and macroinvertebrate tissues (Figure 2). In mining-influenced Panther Creek sites, Cu concentrations in sediment, periphyton, and tissue all declined by about an order of magnitude between 1993 and 2006, and remained relatively low and stable since. Panther Creek Co concentrations in sediment and tissue also declined over time, but were more variable and magnitudes of decline were less than with Cu. Cobalt in invertebrate tissues collected at PA-km17 downstream of Big Deer Creek were stable throughout the 1993–2013 period. Likewise, Co in periphyton in Panther Creek samples declined by less than a factor of two from 1993–2013. Arsenic concentrations in sediment, periphyton, and tissue were at least factor of 4 lower in 2013 than in 1993. Arsenic concentrations in Panther Creek sediment and periphyton increased in 2008 following a freshet in Blackbird Creek that overtopped banks and remobilized contaminated soils and sediment. Arsenic concentrations then steadily declined through 2013. However, arsenic in invertebrate tissues did not increase following the increases in sediment and periphyton arsenic, suggesting that the arsenic measured in sediment and periphyton had low bioavailability Since 2006, As concentrations in benthic invertebrate tissues have been near or <20 mg/kg dw (Figure 2).

In Big Deer Creek in 1993, sediments were exorbitantly high in Cu, at >11,000 mg/kg downstream of the confluence with South Fork Big Deer Creek and were still >6000 mg/kg in Big Deer Creek near its mouth. By 2013, Cu concentrations in sediments had declined 100-fold. No pre-restoration data for metals in periphyton and benthic macroinvertebrate tissues were collected in Big Deer Creek. From 2006 to 2013 Cu in periphyton declined by less than a factor of 2, and at site BD-km5.3, Cu in macroinvertebrate tissue

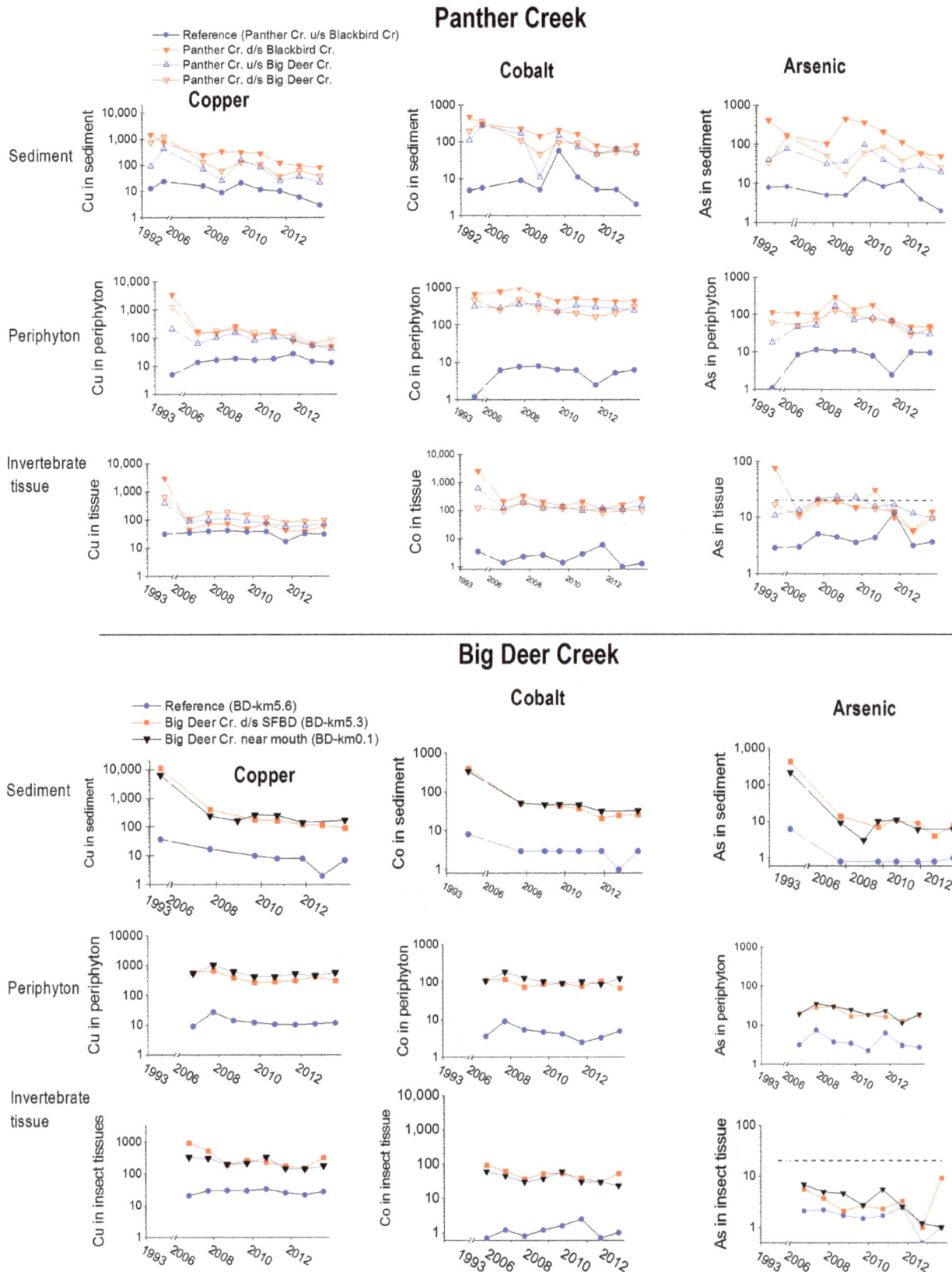

## Figure 2

Changes in arsenic, cobalt, and copper residues in sediment, periphyton, and aquatic macroinvertebrate tissue.

All in mg/kg dry weight. Note differing vertical scales. Dashed line for arsenic in tissue is an approximate threshold for dietary effects of inorganic arsenic to salmonids (see text). Surveys with more than one sample per location were averaged.

declined by a factor of 6. Co and As concentrations in periphyton and aquatic invertebrate tissues have been uniformly low since 2006 (Figure 2).

Despite the different patterns apparent in the individual time series graphs in Figure 2, from a broader view, when all matched samples were pooled across sites and years (n=47), concentrations of Cu, Co, and As in water, sediment, periphyton, and macroinvertebrate tissue residues were all strongly correlated with each other. The strongest correlations were between Co in water, periphyton, and tissue (Pearson's $r$ coefficient values of 0.93 to 0.95), followed by Cu in water, periphyton, and tissue ($r$ values of 0.89 to 0.93) (Text S2).

While the aquatic macroinvertebrate tissue samples in our study were pooled to reflect the general potential dietary exposure of fish to contaminants, in 2012, we also collected Hydropsychid caddisflies for tissue Cu analyses. In Big Deer Creek only *Arctopsyche* sp. were found, and in Panther Creek mostly *Hydropsyche* sp. were found. *Hydropsyche* and *Arctopsyche* have similar metal bioaccumulation patterns (Cain et al., 2004). Pooling both genera, Cu in Hydropsychidae caddisflies was highly correlated with, and slightly lower than Cu in the general community samples ($y = 1.45 \cdot x + 9.1$, $r^2 = 0.97$, P= 0.002, where $x$ is Cu in Hydropsychidae tissues, and $y$ is Cu in the community samples, as mg/kg dw). Both Cu in water and Cu in macroinvertebrate tissue were strongly correlated with benthic community metrics such as taxa richness or mayfly abundance (Text S2).

The three species individually targeted for As analyses did show differences: *Arctopsyche* sp., a net spinning caddisfly that feeds by collecting diatoms and animal matter, the stonefly *Pteronarcys californica* that feeds by shredding and consuming plant material, and the predatory stonefly *Hesperoperla pacifica*, all collected from station PA-km37 in September 2012. The range of As concentrations (n=3 for each organism) was about 1 to 1.5 mg/kg dw (20% organic As) for *Hesperoperla*, 3 to 7 mg/kg dw for *Arctopsyche* (50% organic As), and 7.5 to 11 mg/kg (80% organic As) for *Pteronarcys* (R.J. Erickson, U.S. Environmental Protection Agency, Duluth, MN, personal communication, 16Jul2013). The corresponding pooled community sample of 6 mg/kg dw was within the range of the *Arctopsyche* samples, slightly lower than the range for *Pteronarcys* range, and considerably higher than *Hesperoperla*. Since the predatory *Hesperoperla* stonefly had the lowest As residues, these results suggest bio-dilution through trophic transfer.

### 3.2 Metals in water

**Natural background concentrations**

In 2013 in Panther Creek upstream of Blackbird Creek, the average background Cu concentration was 0.25 (range <0.1 to 0.7) µg/L, and in Big Deer Creek, upstream of South Fork Big Deer Creek, the average Cu concentration was 0.2 (range <0.1 to 0.3) µg/L. Background Co concentrations were similar to Cu (Table 2). Background total As concentrations in Panther Creek were about 1 µg/L in the mid-1980s (Mok and Wai, 1989). However, because total As in Panther Creek downstream of Blackbird was usually only in the range of 2–6 µg/L (Mok and Wai, 1989), subsequent water sampling seldom included As.

**Panther Creek downstream of Blackbird Creek**

Prior to the onset of water quality restoration efforts, dissolved Cu concentrations ranged from 12 to 140 µg/L in Panther Creek downstream of Blackbird Creek, which exceeded the estimated Cu BLM-based chronic criterion (CCC) by greater than a factor of 10. In 2013, Cu concentrations ranged only from <0.1 to 2.9 µg/L, with a maximum Cu chronic criterion exceedance factor of 0.6 and 0.2 (Figure 3, Table 2). Cobalt declines have been proportionally less than Cu declines. From 1992 to 2013, the locally-weighted (Loess) average Co concentration declined by about a factor of 6.5, from 108 µg/L to 16 µg/L, whereas over the same period at this site, Loess smoothed Cu concentrations declined by about a factor of 25 (Figure 3). Because of the large number of Cu and Co samples (~750 each for PA-km37 alone), we emphasize the local Loess-smoothed values to simplify and visualize patterns. In most years at most sites, the maximum Cu and Co concentrations were about 3X and 2X greater, respectively (Figure 3; Table 1; Mebane et al. 2015).

**Panther Creek downstream of Big Deer Creek**

Patterns of decline with Cu and Co concentrations at this location were similar to those in Panther Creek downstream of Blackbird Creek, although pre-restoration, the absolute concentrations of both Cu and Co were lower in Panther Creek downstream of Big Deer Creek than downstream of Blackbird Creek. In 1993 and 1994, low flow Cu concentrations downstream of Big Deer Creek ranged from 12–15 µg/L, with high flow concentrations ranging from 25 to 97 µg/L. By 2004, nearly all Cu concentrations were below the BLM-based water quality chronic criterion, and as of 2013, Cu ranged only from 1.7 to 4.1 µg/L during high flows, and <0.1 to 1.1 during low flows.

**Big Deer Creek**

The only discrete source of metal loading to Big Deer Creek is via the South Fork of Big Deer Creek (Figure 1). In 1993–1994, dissolved Cu concentrations in Big Deer Creek immediately after mixing with the South Fork Big Deer Creek averaged 139 µg/L (range 71–234), compared with a 2013 average of 2.6 µg/L (range <0.1 to 24 µg/L, Table 2). The 2013 peak concentration of 24 µg/L occurred in response to a

## Panther Creek d/s of Blackbird Creek (PA-km37)

## Panther Creek d/s of Big Deer Creek (PA-km17)

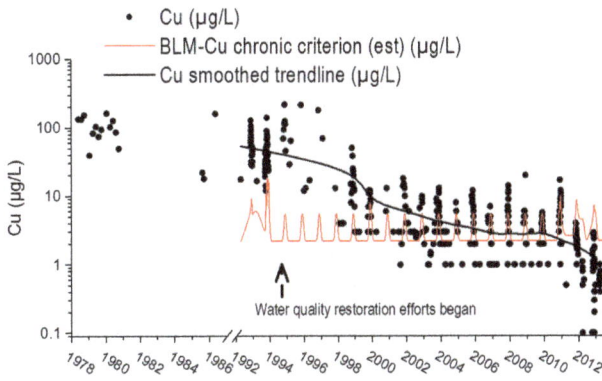

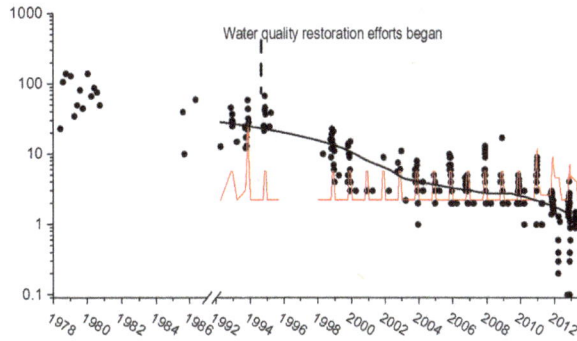

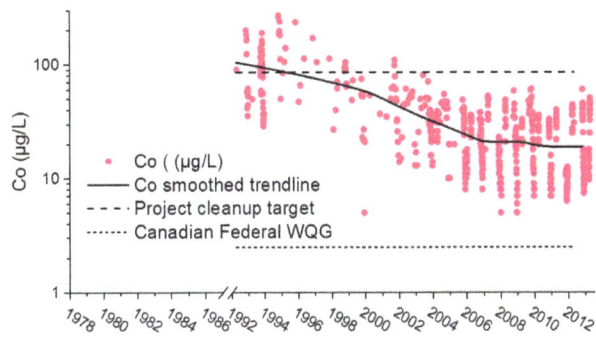

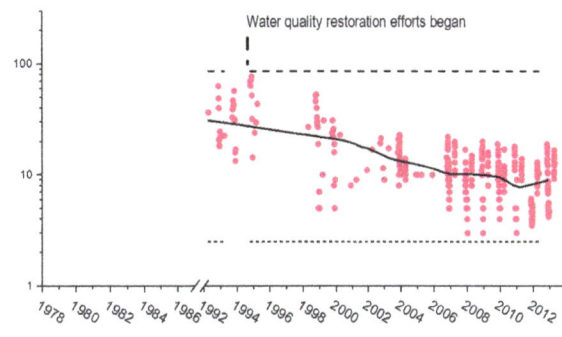

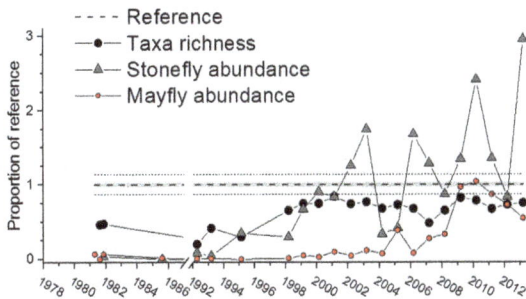

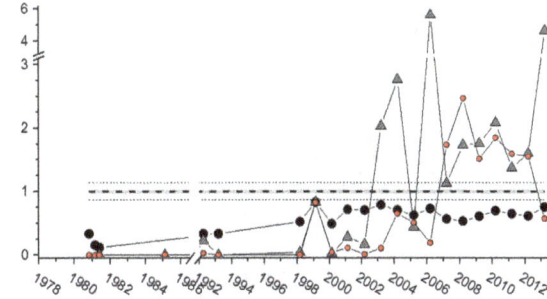

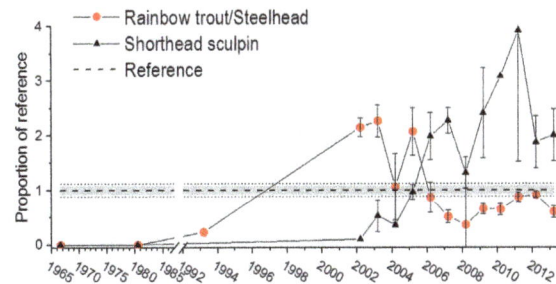

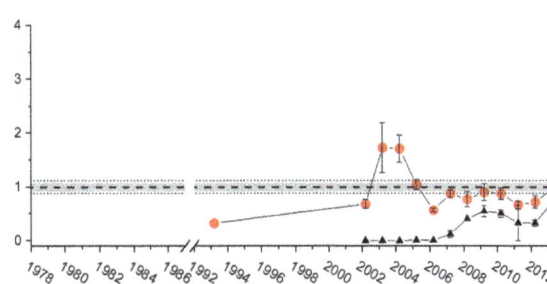

### Figure 3

**Declines in dissolved copper and cobalt and concurrent changes in aquatic communities in Panther Creek.**

Comparisons are relative to pooled upstream reference conditions; d/s- downstream. Shading and lines bracketing the reference condition line indicate the 95% confidence interval (CI) on the mean reference condition values. Shading around the reference lines in their respective plots indicates the narrower CIs for inverterate taxa richness and Shorthead Sculpin densities; dotted lines bracketing the reference lines indicate CIs for mayfly and stonefly abundance (identical), and Rainbow Trout.

thunderstorm on September 5, 2013, which was captured by an autosampler triggered by a stage increase in the headwaters source area (Bucktail Creek). Copper was elevated above background for about 12-hours, reaching 24 µg/L within 3-hours of the onset of the event. Eight-hours after the peak concentration, Cu had declined to 3 µg/L, similar to the base flow concentration. The same storm also caused an increase in stream stage on Blackbird Creek, triggering a downstream autosampler on Panther Creek, but no increase in Cu or Co was detected. This storm event was the only one detected since June 2008 when the stage triggered, telemetered autosamplers were installed.

In 1992–1994, Cu concentrations were lower at the mouth of Big Deer Creek (site BD-km0.1) than just downstream of the confluence of the South Fork (BD-km5.3) by a median factor of 0.67 (n=4 sample pairs). However as Cu loads from the Bucktail Creek drainage were reduced, this pattern switched, and Cu tended to increase in Big Deer Creek with distance away from the mine sources. As of 2013, Cu concentrations were higher at the mouth of Big Deer Creek by a median factor of 2.3 (n=43 sample pairs) than they were just downstream of the South Fork (site BD-km5.3). Cobalt declined in Big Deer Creek by about a factor of 30 over the same time frame. In the 1992–1994 sampling, Co at the Big Deer Creek sites averaged about 60 µg/L, and in 2013 averaged about 2 µg/L.

### Blackbird Creek

From the 1993–1994 period to 2013, average Cu concentrations at the mouth of Blackbird Creek dropped from 380 to 8.7 µg/L and average Co concentrations declined from 845 to 120 µg/L (Table 2).

## 3.3 Correlations between biological and physiochemical covariates

During the early restoration period of 1993–2002, Cu was strongly, negatively correlated with macroinvertebrate taxa richness, mayfly abundance, stonefly abundance, and overall macroinvertebrate biomass. These metrics were also negatively correlated with Co, although not as strongly as with Cu. Macroinvertebrate biomass was also positively correlated with summer stream flows (Text S2).

During the latter restoration period of 2003–2013, correlations between Cu and Co and biological metrics had declined or disappeared relative to the earlier period. Mayfly and stonefly abundances and biomass were no longer correlated with Cu or Co, nor were trout or sculpin densities correlated with Cu or Co. For instance, from the 1993–2002 period, Cu and mayfly abundance had an r value of -0.90 (n=44 samples) whereas from the 2003–2013 period, the r value was only -0.17 (n=46 samples). Only taxa richness remained strongly correlated with metals (r = -0.80 with both Cu and Co). Trout and sculpin densities were negatively correlated (r -0.49, Text S2).

In Big Deer Creek, Rainbow Trout densities were not strongly correlated with any environmental variable analyzed, with the highest r value of only 0.37 between trout density and summertime flows. Macroinvertebrate taxa richness was negatively correlated with both Cu and Co (r values of -0.60 and -0.59, respectively), but the other biological metrics analyzed were not consistently correlated with metals. Invertebrate biomass was negatively correlated with temperature (r -0.68) and also Co and Cu (-0.66 and -0.62, respectively). Stonefly abundance was negatively correlated with summertime temperatures (r -0.72) (Text S2).

## 3.4 Aquatic community changes over time

### Panther Creek downstream of Blackbird Creek (PA-km37)

Benthic macroinvertebrate taxa richness at this location increased from 20% of reference in the fall of 1992 to 75% of reference by 1999 (Figure 3). From 1999 through 2013, species richness had no consistent pattern over time, ranging from 60 to 90% of reference. Mayflies (Ephemeroptera) and stoneflies (Plecoptera) had been effectively extirpated from Panther Creek downstream of Blackbird and Big Deer Creeks prior to water quality restoration efforts. Stoneflies reappeared by 1998, and by 2002 reached the abundance of upstream reference sites. Subsequently, stonefly densities have been highly variable, but were usually abundant. Mayflies were sparse prior to 2008 and abundant since (Figure 3). The 2008 survey was about two years after the Loess smoothed Cu concentrations dropped below the chronic Cu criterion, and as of 2008 the low-flow Loess smoothed Co concentrations had dropped to about 20 µg/L.

No fish were found in Panther Creek downstream of Blackbird Creek in 1967 or 1980 (Figure 3). By 1993, Rainbow Trout had reappeared but were sparse, at 23% of reference. By 2002, Rainbow Trout abundance was 2X that of reference, at which time Cu concentrations during spring high-flow conditions had dropped to less than 2X the chronic criterion, and during low-flow concentrations were similar to the Cu chronic criterion. Shorthead sculpin (*Cottus confusus*) were detected in 2002, but were sparse. Sculpin abundance greatly increased after 2002, and trout abundances declined coincident with the sculpin increases (Figure 3).

## Panther Creek downstream of Big Deer Creek

Benthic taxa richness at site PA-km17 increased from 33% of reference in 1993 to 71% in 2001. From 2001 through 2013, taxa richness had no obvious further increases, ranging from 63 to 90% of reference. Stoneflies were absent or sparse until 2003, with an irruption to 6X reference in 2006. Mayflies were also absent or sparse until 2007, but have mostly been abundant since. The increase in mayfly abundance was coincident with Loess smoothed Cu concentrations dropping below the Cu chronic criterion and Loess smoothed Co concentrations dropping to about 10 µg/L. Rainbow Trout were more abundant than at reference sites during 2003 and 2004, but then declined and were mostly similar to reference from 2005–2013. Shorthead Sculpin recovery lagged Rainbow Trout recovery by at least 6 years, only becoming common at this site in 2008 (Figure 3).

## Big Deer and South Fork Big Deer Creeks

Recovery patterns of invertebrates and fish are contrasted in three locations with subtly differing Cu concentrations. Relative to Panther Creek, Co concentrations in Big Deer Creek are low (0.7 to 4 µg/L in 2013 at BD-km0.1. Figures 3 and 4). Prior to water quality restoration efforts, virtually all life had been extirpated in Big Deer Creek downstream of the mine-contaminated South Fork of Big Deer Creek.

Benthic invertebrate taxa richness in Big Deer Creek in 1993 was 18% of reference, when sampled at site BD-km5.3, located only 0.2 km downstream from the confluence of Cu and Co contaminated South Fork Big Deer Creek (Figure 4). Taxa richness steadily increased as Cu concentrations decreased, and by 2005 richness at BD-km5.3 was similar to reference, coincident with Loess smoothed Cu concentrations dropping to about 1.7X the CCC at BD-km5.3 (Figure 4). At the mouth of Big Deer Creek (BD-km0.1), 5.5km downstream of upstream sources of colonizing organisms, taxa richness was only 2% of reference in 1993. Taxa richness exceeded 50% of reference by 1999 as Cu declined to about 10X the CCC, and in 2012 coincident with Cu declining to about 1.3X the CCC during base flow, taxa richness approached 90% of reference. Taxa richness at sites BD-km0.1 and BD-km5.3 rose in parallel as Cu declined, yet richness was always lower at BD-km0.1, averaging 67% of that at BD-km5.3. Cu declines were roughly parallel between the two sites, but from 2002 to 2013, Cu concentrations at BD-km0.1 were typically about twice those of BD-km5.3. No invertebrate taxa were observed in South Fork Big Deer Creek prior to 2001. As Cu declined to about 8X the CCC in 2006, taxa richness in South Fork Big Deer Creek reached about 50% of reference, with little further increase by 2013 despite further Cu decreases to about 4X the CCC (Figure 4).

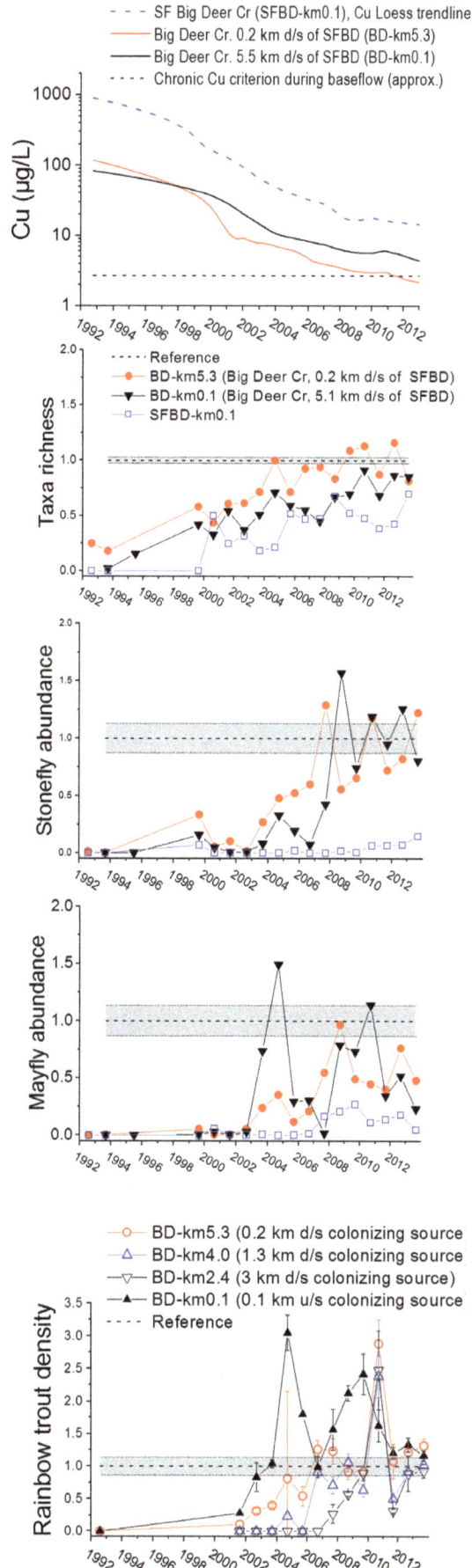

## Figure 4

**Differing recovery patterns at three locations with differing declines in dissolved copper.**

Fish and invertebrates recovered slower at the sites with progressively less reduction in copper in water. Invertebrate data from Big Deer Creek and SF Big Deer Creek (SFBD) are shown relative to matched collections from their respective upstream (u/s) reference sites. Rainbow trout abundance increased later with increasing distance downstream of both copper and colonizing sources (d/s -downstream). Shading indicates 95% confidence i nterval o n t he reference condition mean value.

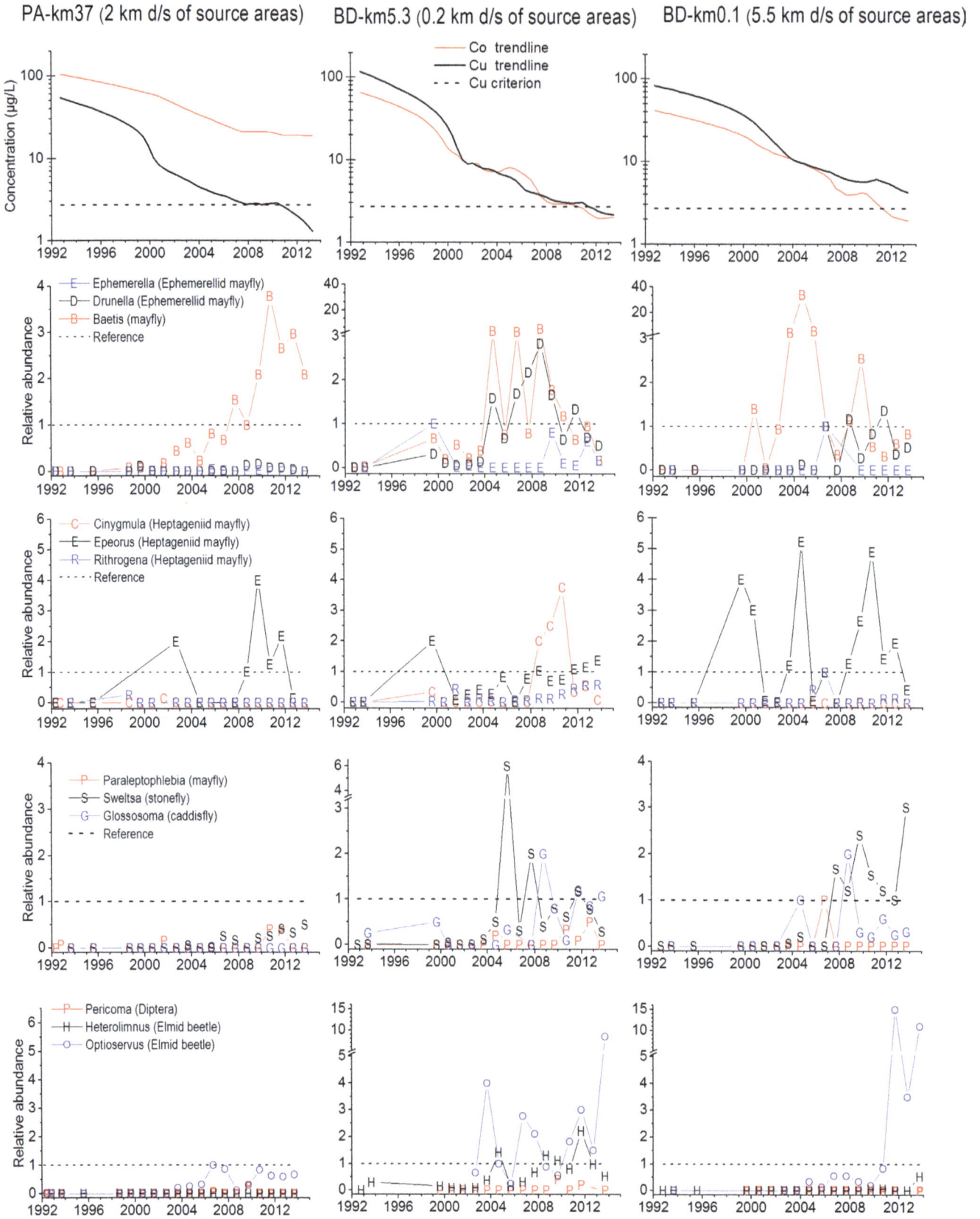

**Figure 5**

Differing taxa recovery trajectories with differing distance to colonists and differing Cu and Co declines.

Plotted as abundance relative to concurrent, upstream reference samples.

No stoneflies or mayflies were found in Big Deer Creek in 1993 (Figure 4). As smoothed Cu concentrations dropped to about 2X the CCC at sites BD-km5.3 and BD-km0.1 in 2003 and 2004 respectively, stonefly abundance began increasing at both sites. By 2007, when smoothed Cu concentrations had dropped to about 1.3 to 1.6X the CCC, stoneflies at these two sites were similar in abundance to upstream reference. Stoneflies remain rare in South Fork Big Deer Creek. Mayflies, taken as a group, reoccupied the Big Deer Creek sites at about the same times as the stoneflies, but were much more variable. An irruption of mayflies (mostly *Epeorus grandis*) occurred in 2004 in lower Big Deer Creek, followed by a collapse in 2007 (Figure 4). From 2008–2013, mayflies were always common, but still tended to be less abundant than reference, ranging from 45 to 113% of reference. Mayflies (mostly *Baetis tricaudatus*) have been detected in South Fork Big Deer Creek since 2008 (Figure 4), at which time Cu declined to about 5X CCC (~15 µg/L) during base flows.

Rainbow Trout had been extirpated from Big Deer Creek downstream of South Fork Big Deer Creek (Figure 4). By 2002, Rainbow Trout were present at BD-km5.3 and steadily increased in abundance until 2007. From 2007 through 2013, Rainbow Trout abundances at this site have usually been similar to upstream reference abundances. In 2002, when Rainbow Trout were present but sparse, Cu concentrations during low flows averaged 15 µg/L, about 4X higher than the CCC. About 3km downstream at station BD-km2.4, recovery of Rainbow Trout abundances to those similar with reference abundances lagged that at station BD-km5.3 by about 5 years. No fish were captured at site BD-km2.4 from 2002–2006, and when fish were first detected at this site in 2007, only age 2+ fish were observed. Young-of-year were captured the following year, indicating that reproduction was occurring in this formerly unsuitable and unoccupied habitat. By 2009, the year class strength and overall abundance was similar to reference. Fish in South Fork Big Deer Creek showed a similar temporal pattern, with first captures of pioneering adults in 2003, with increasing abundances of adult fish through 2008. From 2008–2012, fish were always present but did not generally increase in abundance, nor was there evidence of reproduction. The first evidence of successful reproduction was in 2013, with the capture of a single YOY Rainbow Trout. Smoothed Cu concentrations remained at about 3 to 4X the CCC as of 2013.

### Blackbird Creek

In 1993, Blackbird Creek just upstream of its mouth was virtually lifeless, with no more than 1 organism found per benthic sample. For fish, conditions in lower Blackbird Creek were acutely toxic in 1993, with 100% mortality of caged Rainbow Trout within 48-hours (Hansen et al. 1995). By 2002, conditions had improved to the point that Rainbow Trout, Chinook Salmon, and Bull Trout (*Salvelinus confluentus*) were found in Blackbird Creek near the mouth. As of 2013, fish occurrences downstream of mine sources remained limited to the lower 2 km of the creek. Further upstream, colonization by fish appears to be constrained by a combination of increasing substrate densities of oxyferrihydroxide floc and increasing Cu concentrations. Because of the limited recovery objectives for Blackbird Creek, it has received far less monitoring than the other streams, and the results are more qualitative. Aquatic community changes and factors limiting the recovery of Blackbird Creek are described in more detail in Text S2 and Figure S3.

### 3.5 Recovery trajectories of invertebrate taxa

As Cu and Co concentrations decreased, the recoveries of mayflies and stoneflies tended to lag the overall taxa increases (Figures 3 and 4). Recovery patterns among taxa had recurring patterns among the Panther Creek and Big Deer sites, and are illustrated for 12 genera at three sites which differ in the magnitude of Cu and Co declines, and in distance from colonizing sources (Figure 5).

*Baetis* was among the first mayfly to increase in abundance as Cu and Co concentrations declined. *Baetis* was similar in abundance to reference by 2002 in lower Big Deer Creek and by 2005 in Panther Creek. At Big Deer Creek at km 0.1, *Baetis* had a 3-year irruption from 2004–2006, reaching 34X reference before declining to abundances within a factor of 2 of reference. *Drunella* (Ephemerellidae) had recovered to reference densities at BD-km0.1 by 2008, but remained scarce at PA-km37. *Ephemerella* (Ephemerellidae) remain absent or scarce as of 2013. The recovery trajectories of three Heptageniid mayflies differed noticeably. *Epeorus* was the first mayfly to recover to reference abundances in lower Big Deer Creek, exceeding 4X of reference in 1999. Abundances have been highly variable since. For instance in 2007, no *Epeorus* (and few mayflies of any taxa) were found at BD-km0.1, but by 2008 had increased to 0.7X of reference, and then ranged from 0.4 to 4.9X through 2013. *Cinygmula* and *Rhithrogena* were seldom seen at PA-km37 or BD-km0.1 in any year surveyed, and the mayfly *Paraleptophlebia* (Paraleptophlebidae) was also scarce in most years. Only at site BD-km5.3 were the mayflies *Ephemerella*, *Cinygmula*, and *Rhithrogena* frequently collected, although their abundances were still less than those at the reference site only 0.3 km upstream (Figure 5).

The stonefly *Sweltsa* (Chloroperlidae) was absent or scarce in Big Deer Creek prior to 2007, but subsequently has remained abundant through 2013. This pattern is roughly similar to the overall pattern of recovery of stoneflies as a group at this location (Figure 4). In contrast, in Panther Creek stoneflies as a group were as abundant as at reference sites, but *Sweltsa* was seldom detected until 2006, and remained at 20–40% of reference abundance through 2013. The caddisfly *Glossosoma* (Trichoptera, Glossosomidae) was distinguished

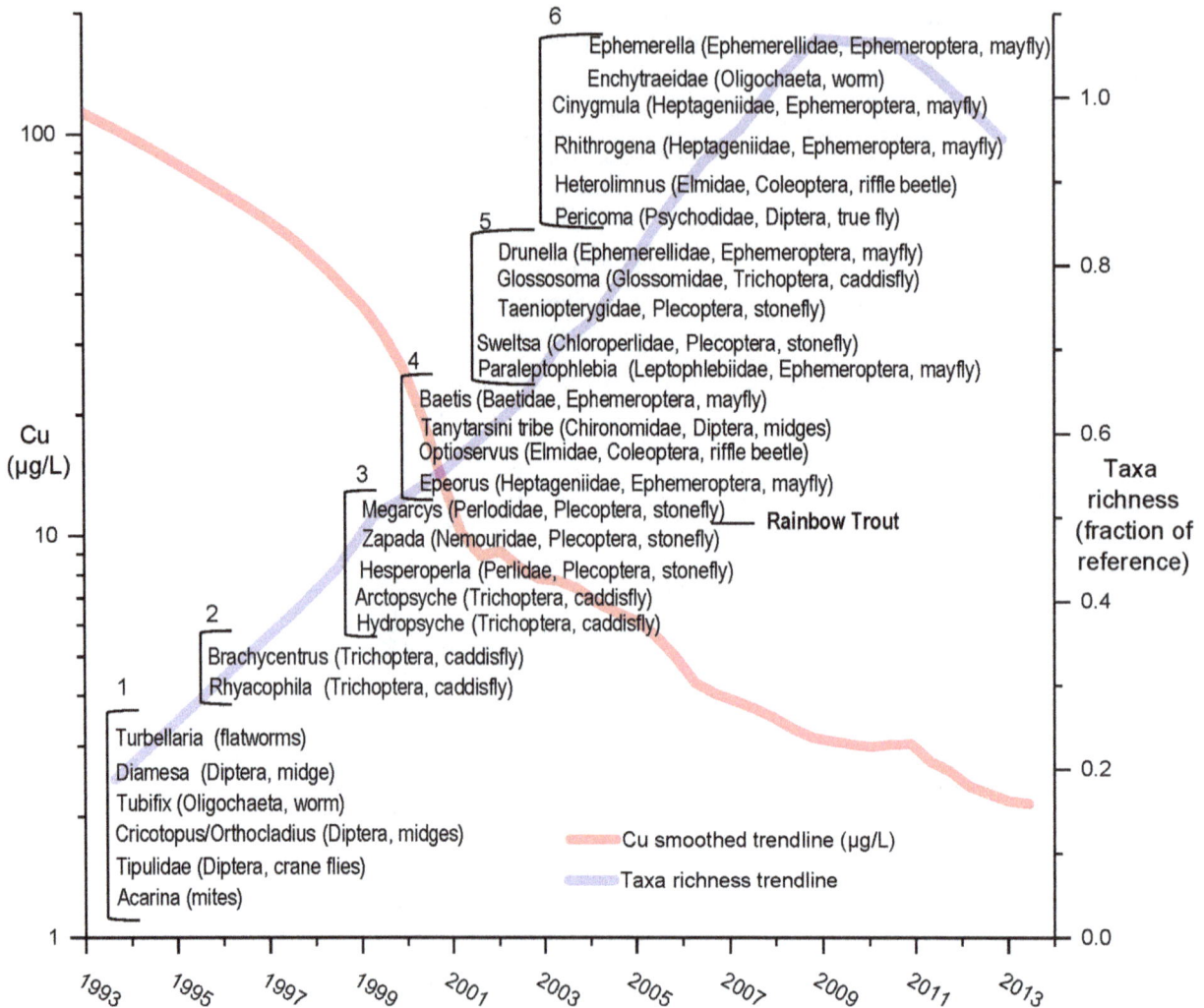

Figure 6

General order of appearance of commonly occurring taxa in study streams as copper concentrations declined.

Copper and taxa richness curves are from site BD-5.3km, but the progression was similar at other Big Deer sites. Brackets, numbered on their shoulders, group taxa with similar rank orders of appearance.

by lagging behind caddisflies as a group before recovering to abundances similar to reference at BD-km5.3. The true fly *Pericoma* (Diptera, Psychodidae) remains scarce. Two beetles in the Elmidae family, *Heterlimnius* and *Optioservus* showed distinctly different recovery patterns. Whereas *Optioservus* was reasonably common most years, *Heterlimnius* has remained mostly absent, except at BD-km5.3 (Figure 5).

Oligochaetes in the family Enchytraeidae were common in reference samples across stream sizes (i.e., upper Panther Creek, Deep Creek, upper South Fork Big Creek and upper Big Deer Creek). However, enchytraeidids were never abundant at mining influenced sites downstream of the reference sites, with no obvious temporal or spatial patterns among the downstream sites. Average Enchytraeidae densities from pooled Panther Creek reference sites from 2003–13 were 43 per m² (CI 23–66), compared to 0.8 per m² (CI 0.2–1.4) from the pooled, downstream, mining-influenced sites. Tributary sites showed similar contrasts between the reference and downstream sites over the same time period, with Enchytraeidae densities averaging 38 per m² (CI 14–61) versus 0.9 per m² (CI 0.2–1.6), respectively. Aquatic insects were numerically dominant in 240 of 242 benthic macroinvertebrate samples collected from all sites between 2003 to 2013, accounting for 91% of all taxa. Molluscs were rare at both reference and mining-influenced sites.

Across sites, taxa repopulated the streams in a recurring progression. Out of the total taxa identified during the 2003–2013 sampling, which ranged from 127 to 211 taxa per event, about 26 commonly occurring taxa could be grouped among the earliest, intermediate, and later colonizers. We illustrate the progression at Big Deer Creek at km5.3 because of its extreme changes in Cu concentrations, low Co concentrations, and close proximity to upstream colonizing sources. In the initial surveys with smoothed Cu concentrations exceeding 125 μg/L, only six taxa persisted (Figure 6, bracket 1). These included Turbellaria flatworms, Acarina mites, Tubificidae oligochaete worms, and a few Diptera in the Chironomidae and Tipulidae families. Only in the

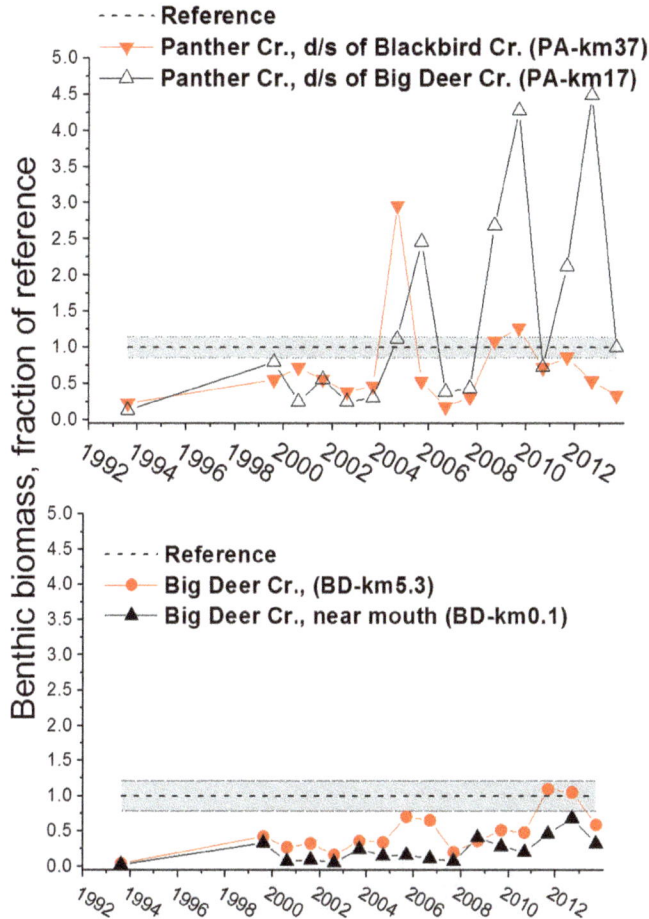

Figure 7

Changes in total biomass of benthic invertebrates in Panther Creek and in Big Deer Creek.

Shading indicates minimum detectable difference from reference.

extreme conditions of South Fork Big Deer Creek with >1000 µg/L Cu were these (and all) taxa absent. The caddisflies *Brachycentrus* and *Rhyacophila* appeared in Big Deer Creek by 1995, and by 1998 Hydropsychid caddisflies and the stoneflies *Hesperoperla*, *Zapada* and *Megarcys* appeared. *Baetis* and *Epeorus* mayflies and the Elmid beetle *Optioservus* appeared the following year as smoothed Cu concentrations dropped to 30 µg/L. As Cu dropped below 6 µg/L in 2005, the stonefly *Sweltsa*, the caddisfly *Glossosoma*, and the Elmid beetle *Heterolimnus* appeared. The final arrivals, coincident with smoothed Cu concentrations dropping below 3 µg/L, were the mayflies *Cinygmula* and *Ephemerella* (Figure 6).

The progression of recovery at other locations in Big Deer Creek was similar, although the taxa in bracket 6 of Figure 6 diminish with distance downstream and as of 2013 remained absent or uncommon at station BD-km0.1. In Panther Creek, general patterns were similar with some noticeable taxa differences. The mayfly *Epeorus* appeared in lower Panther Creek (PA-km17 and PA-km2.7) by 2001 but not at PA-km37 until 2010. While upstream colonization sources were closer at PA-km37, Co is highest at PA-km37, at about 20 µg/L in 2010. *Sweltsa* is the only taxa grouped in Figure 6, brackets 5 and 6 that has been commonly found in Panther Creek downstream of the reference sites.

## 3.6 Benthic community biomass

We used total community biomass as an indicator of whether benthic invertebrates provided a sufficient prey base for fish. Because high counts of small organisms that contribute little to biomass (such as Acarina, aquatic mites) can be misleading in simple counts of total organisms, we considered biomass of benthic invertebrates a better indicator of the available prey base for fish. Because the vast majority of benthic invertebrate taxa collected were aquatic insects that were classified as being vulnerable to salmonid or sculpin predation (Suttle et al., 2004), we considered total biomass to reflect the available prey base for fish.

In mining-influenced reaches of Big Deer Creek prior to 2011, benthic biomass was less than half that of Big Deer Creek upstream of the mining-influenced segment (Figure 7). Biomass was consistently lower at the mouth of Big Deer Creek (BD-km0.1) than at site BD-km5.3. These differences correspond both with increased downstream Cu concentrations and increased distances from upstream colonizing sources. In Panther Creek, while benthic biomass was low prior to about 2003, the biomass in Panther Creek appears to track expected regional natural upstream to downstream patterns for mid-order streams with biomass increasing

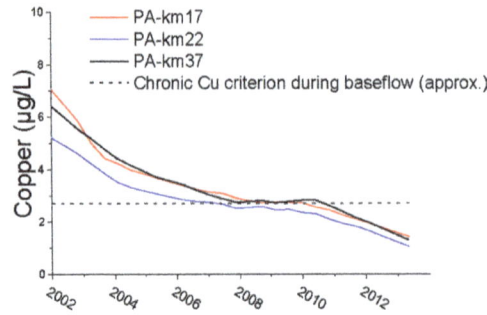

Distance from potential sources of colonizing fish

PA-km37   2 km downstream of colonizing source
PA-km12   8 km upstream of potential source
PA-km17   15 km upstream of potential source
PA-km22   20 km upstream and 12 km downstream of
potential sources

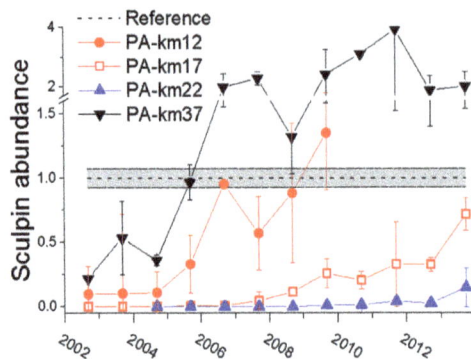

**Figure 8**
Recovery trajectories for Shorthead Sculpin relative to copper concentrations and distance from source populations.

At Panther Creek downstream of Blackbird Creek (PA-km37), it seems certain that the fish moved downstream from above Panther Creek. However, the source populations for the other sites are inferred from the downstream to upstream pattern in first detections and subsequent abundance increases.

as the streams increased in size and temperature (Vannote et al., 1980). Prior to 2003, biomass was contrary to this expected natural pattern, with downstream biomass lower than upstream biomass, suggesting a more limited prey base for fish, relative to reference. Occurrences of high biomass in Panther Creek downstream of Big Deer Creek was driven by increased downstream abundance in large stoneflies, *Pteronarcys* and *Hesperoperla*, and the large caddisfly *Hydropsyche* which all do well in warmer water (Ott and Maret 2003). A 2005 spike in biomass downstream of Blackbird Creek was driven by a "bloom" of the oligochaete worm *Nais bicuspidalis,* which was rare in the reference sites.

In 2013 benthic biomass declined relative to reference across different Panther Creek and Big Deer Creek sites. The reasons for this decline in Panther Creek are not obvious. Measured metals concentrations did not increase in Panther Creek in 2013. Panther Creek fish densities in 2013 were the highest measured, about 150% of the long-term average, but the 2013 increase in fish density was similar at the reference sites. Streamflows in 2013 were unremarkable. Stream temperatures in 2013 were warmer than average (e.g., maximum summer temperatures of 20.8°C in 2013 vs. 19.6°C average maximum from 2003–2013 at PA-km37). Warm stream temperatures could plausibly be a factor, as some previous dips in benthic biomass occurred in previous warm years (2006 and 2007) and some higher biomasses occurred in cool years (2009, 2010, 2012) although these patterns were not apparent at all sites and all years. In Big Deer Creek, the plausible factors for decreased benthic biomass differ from those in Panther Creek. A 10X pulse in Cu concentrations 4 days prior to sampling site BD-km5.3 could have initiated drift, leaving lower biomass when sampled. Fish densities were also higher at BD-km5.3 in 2013 (150% of the long-term average for the site) whereas densities at the reference site, BD-km5.6 were only 109% of the long-term average density. Temperature patterns were similar between the streams.

### 3.7 Shorthead Sculpin

For the lower Panther Creek sites PA-km12, PA-km17, and PA-km 22, sculpin appeared to expand their range from downstream to upstream, based upon the later increases at the more upstream sites (Figure 8). The distances from potential source populations are given from Clear Creek, in which Shorthead Sculpin were the most abundant species in a 1998 survey by the Idaho Department of Environmental Quality, whereas no sculpin were found in two other potential refuge habitats, Beaver and Trail Creeks, (http://mapcase.deq.

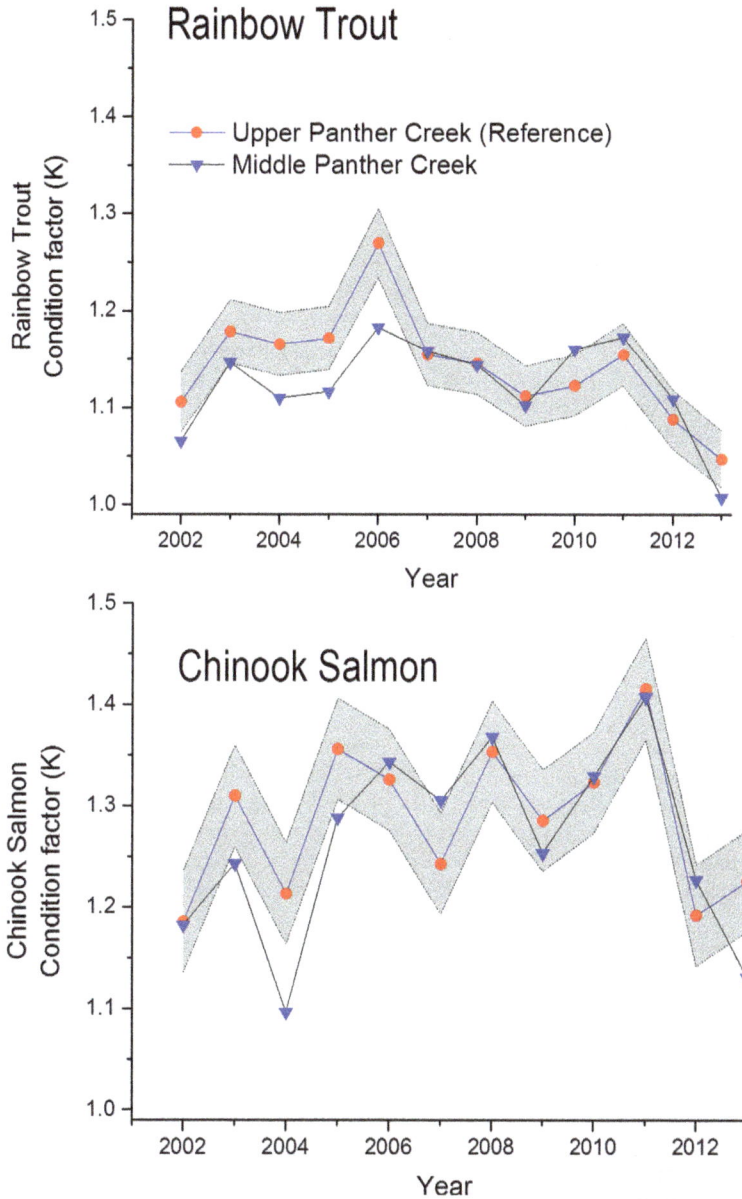

Figure 9

Rainbow Trout and Chinook Salmon body condition factor (*K*) in Panther Creek upstream and downstream of Blackbird Creek.

Prior to 2007, body condition was subtly, but consistently lower downstream of Blackbird Creek. ["Upper Panther Creek" pools data for all sites upstream of Blackbird Creek; "Middle Panther Creek" pools data for all sites in between Blackbird and Big Deer Creeks. Comparison was limited to these two adjacent reaches to minimize the influence of warmer temperatures in lower Panther Creek.]

idaho.gov/wq2010/, accessed December 2014). Sculpin were present but sparse at site PA-km12 in lower Panther Creek at the time of our first survey, and thus their actual colonizing source is unknown. While no fish were found in Panther Creek just downstream of Big Deer Creek in 1980, not until 2002 was a quantitative electrofishing survey conducted in multiple locations in Panther Creek. Presumably Shorthead Sculpin began reoccupying lower Panther Creek by at least the early 1990s, by which time Cu had attenuated to below critical limits for Rainbow Trout to become re-established in Panther Creek. While the overall declining Cu concentrations are similar at the three different locations monitored, the site with the lowest Cu and greatest distance from source populations was last to be reoccupied by sculpin (Figure 8). By 2005, once smoothed Cu concentrations had declined to less than about 1.3X the chronic criterion, Cu did not appear to exert any constraint on sculpin recovery. Water chemistry was not monitored at site PA-km12, but Cu concentrations were likely similar to those monitored near PA-km17 since no large tributaries enter Panther Creek between these two sites.

As of our first survey (2002), sculpin were also abundant at the most downstream site on Panther Creek (PA-km4), but then disappeared following a massive debris flow entering Panther Creek from Clear Creek (1 km upstream) in June 2003. Sculpin did not return to their pre-landslide abundances at this location until 2012 (Text S2).

### 3.8 Fish population characteristics

Prior to 2007, average Rainbow Trout condition factors tended to be lower in mining-influenced sites, relative to reference sites, whereas from 2007 on, average Rainbow Trout condition factors were similar in upper (reference) and middle (mine-affected) reaches (Figure 9). While differences were subtle, the patterns were reasonably consistent during the early years of the water quality restoration efforts, and vanished in the later years as Cu concentrations continued to drop. Chinook Salmon showed roughly similar patterns. The timing of decreasing Cu concentrations and increasing mayfly abundances both corresponded with the timing of the disappearance of apparent growth differences (Figures 3 and 9). Upstream of direct mine-influences, Rainbow Trout condition was negatively correlated with Rainbow Trout density (r = -0.74), and Chinook Salmon condition was most strongly correlated with average summertime streamflow (r = 0.63) (Text S2).

The age-class structure of trout in Panther Creek in 1993 was skewed toward older fish, relative to reference reaches. Young-of-year (YOY) trout in particular were less common in Panther Creek relative to reference reaches, and trout of all ages were less abundant in mine-influenced reaches of Panther Creek than in reference reaches (Figure 10). For the 1993 data, the comparisons are made with all trout because most of the reference reaches were located in the Middle Fork Salmon River basin where Cutthroat Trout were the dominant species, while in Panther Creek, Rainbow Trout made up more than 95% of the total trout. By 2003, fish in mine-influenced reaches were as abundant as fish in upstream reference reaches. Also by 2003, a balanced age class structure was present in mine-influenced reaches, similar to reference reaches. In 2013, the densities and age class structure of Rainbow Trout in Panther Creek were similar to those in 2003, and also similar to expectations for salmonids in streams subject to natural population controls, discussed earlier.

Other native fish including Bull Trout, Mountain Whitefish, Longnose Dace (*Rhinichthys cataractae*) and Cutthroat Trout were regularly observed within the study area, but their distributions were variable and appeared unrelated to water quality, described in Text S2.

### 3.9 Chinook Salmon recolonization

The Panther Creek Chinook Salmon population appears to be self-sustaining as of 2013. The relatively high density of fry observed in 2002 followed a release of about 1053 hatchery-origin adults the previous year, and the cyclical density peaks in 2006 and 2011 indicate first and second generation progeny from this release (Figure 11A). Because about 99% of Chinook Salmon in the Salmon River, Idaho have either a 4- or 5-year life cycle (Mebane and Arthaud, 2010; Kennedy et al., 2013), the presence of YOY in off-cycle years from the 2001 release indicate natural-origin reproduction of wild salmon independent of the 2001 hatchery release. The "off-cycle" years with only natural-origin Chinook Salmon YOY present were 2003–2005, 2008–2009, and 2013 (Figure 11B).

## 4. Discussion

### 4.1 Attenuation of metals contamination

Streambed sediments can represent a persistent reserve of exchangeable metals that can be remobilized, released to the water column, and cause delays in biological recovery following pollution source controls (Hamilton, 2012). While only a fraction of the metal present in sediments is bioavailable at any one given time due to binding to particulate organic carbon and iron and manganese oxyhydroxides and sediment burial (Costello and Burton, 2014), trace contaminants associated with sediments will remain in the river-floodplain system and be subject to remobilization until stored deposits become depleted (Hamilton, 2012; Moore and Langner, 2012).

Patterns with Big Deer Creek Cu concentrations are consistent with Hamilton's (2012) concept of streambed and alluvial sediments acting as a reserve of Cu that is slowly dissipated as it is conveyed downstream, released in trace amounts to the overlying water, and diluted with sediments from the upper, unaffected watershed. In Big Deer Creek prior to restoration efforts, Cu concentrations in both water and sediment were higher at the site nearest to the contaminated mine drainage (BD-km5.3) than at the farthest site (BD-km0.1). However, as source controls became effective and Cu concentrations in water dropped, this rank order was reversed, with consistently higher Cu concentrations in water and sediment at the more distant site from the mine source (Figure 2; Figure 4).

When examining changes in metals concentrations in different media through time-series, the graphs do show differences. For example, As concentrations in macroinvertebrate tissues did not track spikes in sediment or periphyton As concentrations in 2008. Copper in Panther Creek declined sharply during the first decade of recovery, whereas declines in Co and As concentrations have been slower and more variable (Figure 2). Yet when viewing the data broadly across all sites and years, Co and Cu were particularly strongly correlated within media, with Pearson r values ≥0.8 among water, sediment, periphyton, and invertebrate tissue (Text S2). The strength of these correlations suggests redundancy in the information provided by the exposure data from the different media, and possible monitoring efficiencies. For instance, with a correlation coefficient of

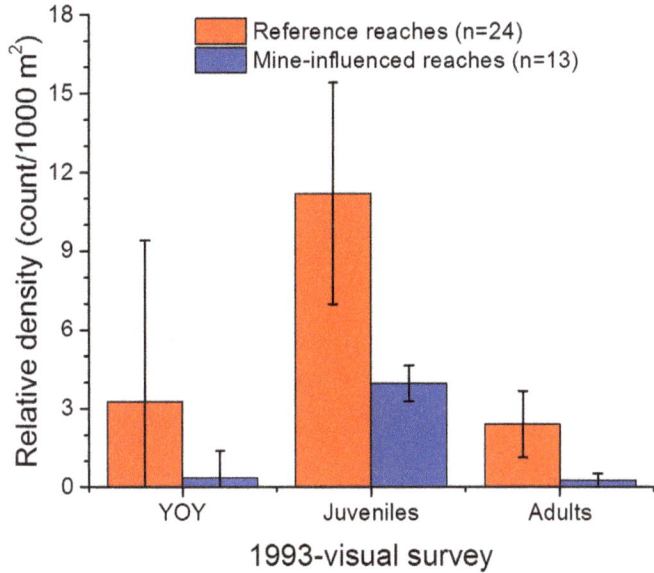

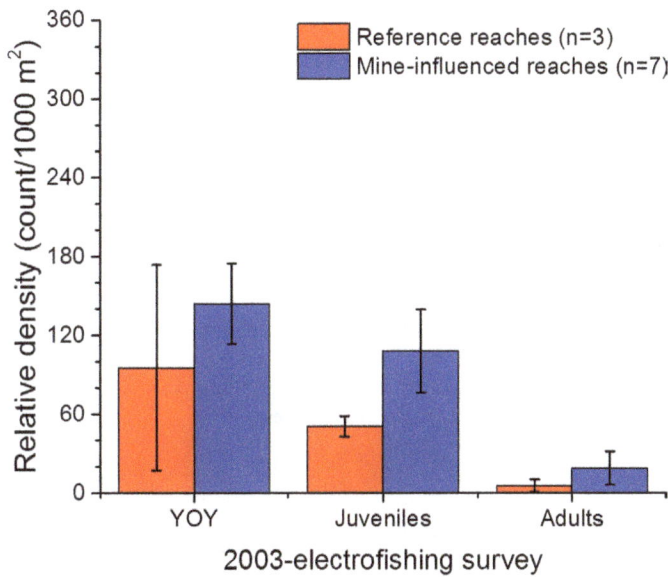

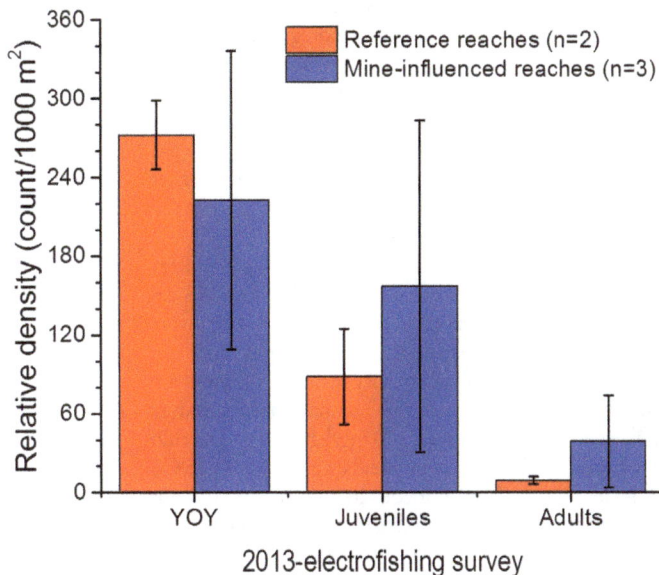

Figure 10

Mean relative densities of all trout by life stages in Panther Creek downstream of mine discharges, relative to reference reaches, at decade intervals.

Fish ≤75 mm in length were considered to be young-of-year (YOY), "Juveniles" are fish 76 to 179 mm in length and are probably 1 and 2 year old fish, and "Adults" are fish >180mm in length. Lengths in the 1993 snorkel survey were visually estimated, and fish lengths in the 2003 and 2013 surveys were measured fork lengths. Error bars show standard deviations. Contrasts between the 1993 visual, snorkel survey and the 2003 and 2013 electrofishing surveys need to be relative to the within-survey reference conditions because the different survey methods had markedly different detection efficiencies (note 20X difference in axis units).

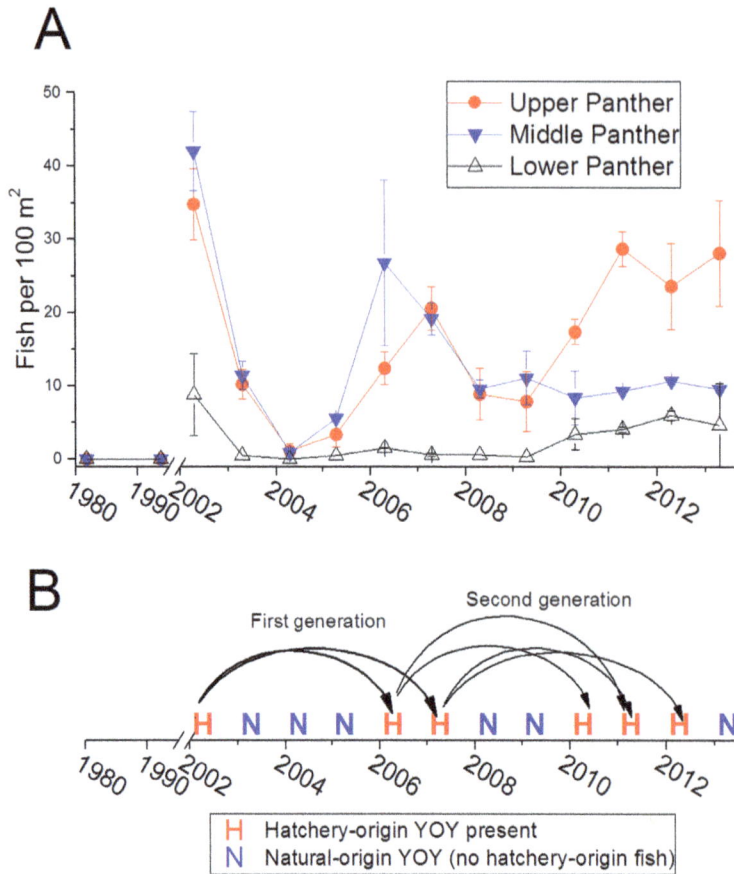

**Figure 11**

Chinook Salmon juvenile densities in Panther Creek, by water quality reach.

a. "Upper Panther" is upstream of all mine influence, average of 3 to 4 sites depending on year; "Middle Panther" is between Blackbird and Big Deer Creeks (average of 2 to 6 sites per year); and "Lower Panther" is downstream of Big Deer Creek (average of 2 to 4 sites).

b. Years in which the monitored densities of young-of-year (YOY) Chinook Salmon were likely influenced by hatchery-origin progeny from the 2001 release of hatchery adult fish, based on the 4- or 5-year life cycle of Chinook Salmon in the Salmon River, Idaho (see text).

0.95 between annual average Co in water and annual macroinvertebrate tissue sampling, monitoring one or the other might be sufficient to evaluate time trends. Strong but variable relations between metals concentrations in different media have been reported from other systems (Hornberger et al., 2009), suggesting that site-specific data would be needed before using concentrations in one media to predict another.

## 4.2 Benthic macroinvertebrates

The rank ordering of reoccurrences of common taxa as Cu declined formed a pattern reminiscent of species-sensitivity distributions from toxicity testing compilations (e.g., Brix et al., 2011). The patterns were also similar to recovery or sensitivity distributions developed from field studies of taxa recolonization streams recovering from acidification (Raddum and Fjellheim, 2003; Masters et al., 2007) and taxa occurrence in relation to freshwater ionic strength (Cormier et al., 2013). These similar distributions suggest ionoregulatory disruption as a common mode of toxicity, for Cu is known to block sodium transport. The colonization order as Cu declined in Big Deer Creek was also consistent with the susceptibility of insects to accumulate cadmium (Cd), which was suggested as being relevant to Cu iono-transport as well (Buchwalter et al., 2008). Taxa that were among the last to appear as Cu declined (*Ephemerella* and *Rhithrogena* mayflies, Figure 6) were the most susceptible to Cd accumulation. *Drunella* mayflies with intermediate Cd susceptibility appeared earlier than *Ephemerella* and *Rhithrogena* mayflies, and the caddisfly *Rhyacophila*, which had low susceptibility to Cd accumulation was an early colonist as Cu declined (Buchwalter et al., 2008; Figure 6).

The similarity between recovery patterns of benthic taxa in our study area expands and generalizes the observations of metals sensitive taxa from Colorado streams that were enriched with Cd, Cu, and Zn mixtures (Clements et al., 2000; Courtney and Clements, 2002; Clements et al., 2010). We are not aware of any previous reports of oligochaetes in the family Enchytraeidae appearing to be sensitive to metals in lotic systems. However, terrestrial enchytraeids (potworms) are a standard soil toxicity test organism and are sensitive to elevated Cu and other metals in moist soil. Severe declines in taxa richness have been noted by about 200 mg/kg dw Cu in soil, with low-effects thresholds occurring at less than 100 mg/kg dw Cu (Maraldo et al., 2006; Cedergreen et al., 2013). We were unable to evaluate whether enchytraeids were sensitive to Cu or Co in other stream studies, because most studies we reviewed on effects of metals in lotic systems did not report actual taxonomic data, and of those that did, oligochaetes were not identified beyond "Oligochaeta."

This in turn may reflect a prevalent but incorrect perception that aquatic oligochaetes are 'tolerant' to pollution, and thus not appropriately sensitive for use in ecological assessments (Chapman, 2001).

Two other exceptions to previous work were the mayfly *Epeorus* and the caddisfly *Rhyacophila*, both of which appeared to be sensitive to metals in Colorado field studies (Clements et al., 2000) but not here. *Rhyacophila* was one of the early colonizers in Big Deer Creek and was one of the most Cu tolerant species, while *Epeorus* appeared less sensitive to Cu than other mayflies. *Rhyacophila* was more prevalent in cooler Big Deer Creek, where it was common in both reference and high-Cu downstream areas, than in Panther Creek. Ott and Maret (2003) showed that in reference streams *Rhyacophila* and *Epeorus* were obligate coldwater taxa that declined with increasing stream temperatures. This suggests that the apparent sensitivity of *Rhyacophila* and *Epeorus* to metals in some surveys could actually have reflected an upstream/downstream temperature gradient, since reference streams are often located upstream of disturbed areas.

A conundrum in the literature on the effects of metals on stream insects has been the dichotomy between laboratory and field-based studies. Conventional short-term toxicity tests with field-collected stream insects have produced exorbitantly-high effect values relative to apparent effects from field studies, longer-term community microcosm tests, or to short-term tests with fish or cultured, small crustaceans such as amphipods or daphnids (Brinkman and Johnston, 2008; Brix et al., 2011; Mebane et al., 2012; Clements et al., 2013; Poteat and Buchwalter, 2014). For example, 137 µg/L Cu was required to reduce the survival of the mayfly *Rhithrogena hageni* by 50% in a 96-hour laboratory toxicity test (Brinkman and Johnston, 2008), yet recovery of *Rhithrogena* sp. in Big Deer Creek was apparently prevented by about 5 µg/L Cu or less (Figure 5, BD-km0.1). In our results, the insect-dominated benthic macroinvertebrate community was apparently very responsive to Cu and possibly Co at concentrations that did not produce discernable effects in fish populations. We think the present results support the views that (1) benthic macroinvertebrate communities may be susceptible to metals in the environment at concentrations far lower than effects concentrations produced from conventional short-term toxicity tests with aquatic insects, and (2) relative sensitivity rankings of aquatic insects to metals based on the latter may be misleading (e.g., von der Ohe and Liess, 2004; Malaj et al., 2012).

Tissue-residues have been advanced as a more meaningful measure of metals exposure and risk than metals in water because by definition, metals in tissues have already been taken up by the organism and reflect contributions from both dietary and waterborne exposures (Adams et al., 2011). Most of our macroinvertebrate tissue data were from community samples pooled from multiple species. The initial rationale for this community sample approach was to relate metals concentrations in benthic invertebrates to potential dietary toxicity in fish (Woodward et al., 1994; Beltman et al., 1999). While mixed taxa samples are less than ideal for trends monitoring, we have retained the original method over the years for continuity in the long-term monitoring record. In contrast, because of interspecies differences in metals bioaccumulation, others have related metals in tissues of targeted taxa such as the cosmopolitan, metals tolerant Hydropsychid caddisflies to predict alteration of sensitive stream benthos due to metals (Cain et al., 2004; Rainbow et al., 2012; Balistrieri et al., 2015). Consistent with Rainbow et al.'s (2012) prediction, Cu in Hydropsychidae was correlated with mayfly abundance ($r^2$ =0.63) and when limited to Big Deer Creek with its minimal Co concentrations, Hydropsychidae was very strongly correlated with mayfly abundance ($r^2$ =0.96).

Copper in both Hydropsychidae and the macroinvertebrate community tissue samples was strongly correlated with both Cu in water and periphyton, e.g., $r^2$ of 0.99 between Cu in periphyton and Hydropsychidae and $r^2$ of 0.97 for Cu in water and Hydropsychidae. Overall, the correlations between Cu and Co in water, periphyton, and tissue residues support the idea of using metals tissue residues as a predictor of effects to benthic communities. Metals in the diet of aquatic insects have been shown to be the predominant route of total tissue accumulation in controlled laboratory studies (Poteat and Buchwalter, 2014). Yet, the present results indicate that water chemistry also exerts a strong control on the levels of metals accumulated by periphyton and aquatic macroinvertebrates, which in turn has implications for toxicity.

## 4.3 Fish populations

### Sampling issues

While bias, representativeness, and comparability of data are key concerns in all components of long-term monitoring efforts, the explicit emphasis of recovery goals on fish populations and controversies in the literature give these issues focus with fish population data. Although we did not perform any capture efficiency experiments, Meyer and High (2011) evaluated capture efficiency and bias in population estimates from electrofishing surveys that used methods very similar to our 2002–2013 collections. On the average their methods overestimated capture efficiency and underestimated absolute population size by about 22 to 27% (Meyer and High, 2011).

However, because we are evaluating recovery from disturbance rather than stock assessment, we are not as interested in absolute abundances as we are in relative abundances between reference and exposure locations. For comparisons with concurrently (within a few days) sampled reference sites, this requires an assumption of similar detection efficiency at reference and exposure sites. Our study sites were selected to have similar habitat features (Text S2 and Figure S3), and we believe this assumption is reasonable. However, the pre-restoration fish surveys used visual snorkel count methods that were not directly comparable to our post-restoration methods. Densities at reference sites estimated from snorkel surveys were about 20X lower than those we obtained by electrofishing at reference sites, with YOY fish underrepresented (Figure 10). Thus it is again necessary to assume that the snorkel data are internally valid, that is, the fish detection efficiency in the snorkel survey was similar between reference and mine-influenced sites, and that when snorkel counts are standardized as fractions of the reference condition, the standardized fractions are comparable between the snorkel and electrofishing data. Because the factors that likely influence the ability of snorkelers to detect fish (such as water clarity, depth, substrate, observer spacing, and lighting) were generally similar between the reference and mining-influenced sites (LeJeune et al., 1995), we assume similar detection efficiency at reference and mining-influenced sites.

The use of upstream reference sites presumes that in flowing waters, upstream conditions are little affected by downstream conditions. This assumption is not always true, especially with anadromous or other migratory species which need to transit downstream conditions in order to complete their life cycles. Anadromous salmon and steelhead use olfactory cues to return to their natal streams to spawn (Quinn, 2005), and Cu can disrupt chemoreception at sublethal concentrations (Hansen et al., 1999; Meyer and Adams, 2010). Declines of returning adult Pacific salmon also led to decreased delivery of marine-derived nutrients to streams, and locations with no or few salmon carcasses may in turn have lower productivity than streams with more abundant carcasses (Ebel et al., 2014). Thus if anadromous fish migration were impeded by water quality conditions, upstream sites would make unsuitable reference sites. For this reason, LeJeune et al. (1995) used geomorphically similar reaches from regional wilderness streams that were uninfluenced by downstream water pollution as reference sites for mining-influenced Panther Creek reaches. Because expeditionary wilderness sampling was infeasible for an annual monitoring strategy, we assumed that with water quality improvements, anadromous fish would not be impeded from reaching upper Panther Creek. While this assumption was met (Figure 10), regional salmon populations remain influenced by dams and other factors that likely limit adult returns and subsequent marine-derived nutrient delivery (Ebel et al., 2014).

### Resident fish

Rainbow Trout rapidly occupied new habitats as they became marginally habitable. About 4 years elapsed from the first detections to when densities were similar to reference in Big Deer Creek. In Big Deer Creek, the trout population advanced downstream at about 0.5 km/year (Figure 4). Abundances of Rainbow Trout in Panther Creek appeared to initially overshoot reference densities, and then declined. This nonlinear recovery trajectory of bloom and decline of Rainbow Trout densities was similar to patterns observed in the recovery of an experimentally acidified lake. The first species to recover was White Sucker (*Catostomus commersonii*), and as the pH of the lake recovered to circumneutral levels, abundance greatly overshot pre-disturbance abundance but then declined as other species recovered (Mills et al., 2000). With Panther Creek Rainbow Trout, the decline following the initial rebound was coincident with increases in Shorthead Sculpin densities. In Panther Creek downstream of Blackbird Creek, trout outnumbered sculpin by 3 to 1 in 2002, but since 2006, sculpin have outnumbered trout by as much as 10 to 1. Because both sculpin and juvenile salmonids tend to preferentially feed on baetid mayflies, chironomids, and simuliids (Brocksen et al., 1968; Johnson et al., 1983; Boag, 1987; Riehle and Griffith, 1993), this suggests potential competition for food resources between Shorthead Sculpin and Rainbow Trout.

The age structure of Rainbow Trout in Panther Creek shifted toward younger fish contributing a greater proportion of the total number. In 1993, young-of-year were less abundant in mine-influenced sites in Panther Creek than in reference sites, yet by 2003 YOY were similarly abundant in mine-influenced and reference sites. In the quantitative electrofishing surveys, YOY were more abundant than older fish (Figure 10). While for space, we only showed plots at decadal intervals, YOY were the most abundant age group in all Panther Creek surveys from 2002–2013. This shift toward younger fish in the population age structure is consistent with expectations for recovering populations (Table 1).

Condition factors of Rainbow Trout in the upper Panther Creek reference sites and in the middle Panther Creek sites clearly rose and fell in synchrony (Figure 9). However prior to 2007, rainbow trout condition factors were subtly lower than those from the upstream reference reaches. Within the range of mean summer temperatures occurring in Panther Creek (about 10-15°C), we would expect higher growth in lower, warmer reaches (Railsback and Rose, 1999). Thus, the lower condition in Rainbow Trout collected in the mining-influenced middle reach of Panther Creek was counter to expected natural patterns. While co-occurrence alone does not indicate cause, the lower condition factors in the middle reaches of Panther Creek prior to 2007 and subsequent disappearance of condition factor differences as Cu declined and mayfly abundance increased is congruent with a scenario of metals stress and recovery (Figure 3; Figure 9; Table 1). While patterns were subtle, small differences in relative condition or size may be important to survival of juvenile salmonids in streams (Mebane and Arthaud, 2010). The subtleness of differences in fish condition in mine-influenced and reference streams may be inherent to the measure, as differences in condition factor reported even from areas with well documented metals stress were usually on the order of 10% or less (Schindler et al., 1985; Gauthier et al., 2009).

The interannual differences in Rainbow Trout condition upstream of the mine-influenced reaches, appeared to be density dependent, with a strong negative correlation between condition and Rainbow Trout density (r = -0.77, Text S2). Contrary to expectations that growth would be related to prey abundance, neither total invertebrate biomass or mayfly density were correlated with Rainbow Trout condition with r values of 0.00 and -0.11 respectively. Unlike Rainbow Trout, Chinook Salmon condition factor in Panther Creek upstream of mine-influences was not strongly correlated with either Chinook Salmon or Rainbow Trout density (r = 0.05 and -0.18 respectively, Text S2). This suggests that Chinook Salmon densities remain too low for density dependent growth limitation, and limited competition occurs between Rainbow Trout and Chinook Salmon relative to intraspecific Rainbow Trout competition.

Shorthead Sculpins have been slower to return than Rainbow Trout, presumably because of their limited home ranges and movements. In Panther Creek downstream of Blackbird Creek, about 2 km downstream from upstream source areas, Shorthead sculpin increases in abundance lagged Rainbow Trout by about 3-years. Yet at the sites downstream and upstream of Big Deer Creek, about 9 to 12 km from presumed source areas respectively, sculpin recovery lagged Rainbow Trout recovery by over 10 years (Figures 3 and 8). Shorthead Sculpin and Mottled Sculpin (*Cottus bairdii*) are closely related, and Mottled Sculpin and Rainbow Trout have overlapping sensitivities to prolonged Cu exposures (Besser et al., 2007). Thus differences in the intrinsic sensitivities to metals doubtfully explain the slower recovery of Shorthead Sculpin than Rainbow Trout. Sculpin usually have very restricted home ranges with typical annual movements on the order of 10 m or less, and their maximum measured annual movements have only been up to about 500 m, whereas stream-resident trout may move tens of kilometers or more per year (Brown and Downhower,1982; Schmetterling and Adams, 2004; Breen et al., 2009). Sculpin have been observed to recover more slowly than more motile fish elsewhere. Milner et al. (2008, 2011) reported salmonids colonized newly accessible stream habitat much faster than sculpin. In the recovery of an experimentally acidified lake, sculpin had not recolonized the lake 13 years after chemical recovery even though it was connected to a seed population by an inlet stream (Mills et al., 2000). Niemi et al. (1990) found that after relatively small stream habitat areas were decimated, such as by experimental pesticide pulses to small streams where the pollutant would be quickly flushed out and refugia was nearby, the median time for the first appearance of salmonids was 10X shorter than for cottids (0.17 years for salmonids vs. 2.0 years for cottids). However, for less severe disturbances that did not completely extirpate local populations, recovery times were similar between salmonids and cottids (Niemi et al., 1990), likely because recovery times were limited by maximum reproductive rates and resource availability, rather than fish movements and distance from refugia.

### Anadromous fish

Natural recolonization of Chinook Salmon into Panther Creek was occurring by 2002, as evidenced by observations of spawning adults during September of that year. As of 2013, the Chinook Salmon population was naturally reproducing and self-sustaining with abundances within the range of unpolluted streams in the region. In quantitative surveys using sodium cyanide to obtain a complete kill of fish within block netted sections of tributary streams to the mid-Columbia River, the median and $90^{th}$ percentile Chinook Salmon densities at 69 stream sites were 8 and 36 fish/100 $m^2$ (Mullan et al., 1992), whereas we measured 10-30 fish per 100 $m^2$ in the middle and upper reaches of Panther Creek in 2010–2013 (Figure 11). Higher densities of Chinook Salmon in Upper Panther Creek above Blackbird Creek than in lower reaches are likely related to proximity to hatching sites. About 80% of the suitable habitat for Chinook Salmon spawning in the Panther Creek watershed occurs upstream of Blackbird Creek, which is mostly due to lower channel gradients found in upper Panther Creek and in Moyer Creek (Reiser, 1986). Young-of-year salmon do not disperse widely from their hatching location, and their distribution in streams generally reflects redd distribution (Richards and Cernera, 1989).

The recolonization of Chinook Salmon in Panther Creek was complicated by a large release of hatchery-origin adult salmon in the fall of 2001. Because Chinook Salmon from the Salmon River have either a 4- or 5-year life cycle (Mebane and Arthaud, 2010; Kennedy et al., 2013), YOY found in Panther Creek in 2002, 2006–2007, and 2010–2012 can be attributed back to the 2001 transplants of South Fork Salmon River fish. Genetic analyses of Chinook Salmon tissue samples collected in 2010 and 2011 showed that 85 to 90% of the fish could be linked to hatchery origins and the remainder resembled fish from the nearby Middle Fork Salmon River and Upper Salmon River populations as well as a small component from an uncharacterized population (Smith et al., 2012). This implies that in 2010–2011, the descendants of the hatchery salmon swamped any natural origin salmon. However because of their rigid life cycle timing, adult Chinook Salmon observed in 2002, 2004, 2007–2008 and 2012, and YOY observed in 2003–2005, 2008–2009 and 2013 could not have resulted from the 2001 hatchery releases. The original source(s) of the naturally colonizing fish are uncertain. Possible source populations for the natural-origin fish in Panther Creek may include early recolonizers from Clear Creek. Clear Creek, located low in the Panther Creek watershed, may have served as a clean water refuge with limited exposure of migrants to elevated Cu concentrations. At least the surveyed lower 6 km of Clear Creek had suitable steelhead and Chinook Salmon spawning and rearing habitat (Reiser, 1986). During the 1980s and early 1990s when Cu concentrations in middle Panther Creek upstream of Clear Creek were still likely too high to support fish, repeated anadromous fish sightings were reported from lower Panther Creek near Clear Creek. (Reiser, 1986; Mebane, 1994).

Anadromous steelhead are difficult to assess because the pre-smolt juveniles are physically indistinguishable from resident Rainbow Trout. The historical steelhead population was presumably extirpated or at least greatly diminished along with Chinook Salmon in the 1960s, but began to return to the lower reaches of Panther Creek near Clear Creek by the mid-1980s (Mebane, 1994). Wild and hatchery steelhead populations in the Salmon River basin, including Panther Creek, have been evaluated through tagging out-migrants with PIT tags and genetic stock identification. As of 2011, the modeled abundance of steelhead spawners returning to Panther Creek was 485 adults (Copeland et al., 2013). Because of plasticity between the resident and anadromous forms of Rainbow Trout (Sloat et al., 2014) it is possible that a portion of the present Panther Creek steelhead population was contributed by straying adults from other drainages or from the existing resident Rainbow Trout that persisted in upstream or tributary refugia such as Clear or Beaver Creeks.

## 4.4 Factors affecting recovery

### Copper

Particularly in Big Deer Creek, declining Cu concentrations clearly corresponded with progressive range expansions and increasing abundances of invertebrates and fish. At the outset of our study, Big Deer Creek downstream of the South Fork Big Deer Creek was nearly lifeless and because of the waterfall in the lower reaches, colonization could only occur from upstream to downstream. As of 2007 and later, the biological communities in Big Deer Creek immediately after mixing with the South Fork Big Deer Creek (site BD-km5.3) were comprised of mostly similar taxa as those present upstream. Cobalt is low in Big Deer Creek relative to Panther Creek, averaging 2.0 µg/L at BD-km5.3 vs. 22 µg/L at PA-km37 in 2013 (Table 2), suggesting that colonization in Big Deer Creek was limited more by Cu and distance from colonizing sources than Co.

The timing of first appearances of different taxa at the sites close to upstream colonizing source areas suggests thresholds above which Cu appeared to prevent occupancy. With Rainbow Trout in Big Deer Creek and South Fork Big Deer Creek (BD-km5.3 and SFBD-km0.2), the pioneering adults first appeared when Cu dropped to below about 4X the chronic criterion during baseflow conditions (Figure 4). When Cu dropped below about 3X the baseflow chronic criterion, the presence of YOY fish indicated some reproduction and survival of early life stages was occurring. By the time (2006) that the Rainbow Trout population characteristics at BD-km5.3 were indistinguishable from reference, smoothed Cu concentrations had dropped to 1.2X the baseflow chronic criterion. Benthic macroinvertebrate taxa richness at BD-km5.3 first reached that of reference in 2004, at which time smoothed Cu had dropped to about 1.8X the baseflow chronic criterion. However, while few taxa present at the reference site (BD-km5.6) were absent from the samples at BD-km5.3, abundances of some mayfly taxa that were common upstream at BD-km5.6 remained consistently fewer at BD-km5.3. As of 2013, abundances of the common mayflies *Caudatella hystrix*, *Ephemerella*, *Cinygmula*, and *Rhithrogena* were 50 to 60% that of reference, at which time smoothed Cu concentrations were about 0.6X the chronic criterion. Reductions in total mayfly abundance on the order of 50% have been observed with 10-day aquatic insect microcosm exposures to Cu at only 0.8X the mean BLM-based chronic criterion (Clements et al., 2013), which suggests that Cu concentrations at BD-km5.3 averaging about 0.6X the Cu criterion could plausibly have caused reduced abundances of particularly Cu sensitive mayflies.

Shorthead Sculpin in Panther Creek downstream of Blackbird Creek were present in low numbers during our initial surveys in 2002 when smoothed Cu was at about 1.5X the chronic criterion. Sculpin densities greatly increased after 2004, which corresponded to smoothed Cu concentrations only 1.2X higher than the Cu chronic criteria (Figure 7). However, it is not possible to detangle whether this increase in sculpin densities was related to relaxed Cu stress, increased aquatic macroinvertebrate prey base, or simply reflects the exponential phase of a population growth curve in newly colonized habitats.

### Pulsed vs. stable Copper exposures

Copper concentrations fluctuate seasonally in the study area, which has implications for interpreting Cu concentrations in relation to apparent sensitivities with co-occurring aquatic organisms. Prior to 2012, Cu was elevated for about a week to a month during the spring snowmelt runoff period with concentrations 5X or more higher than the mostly stable concentrations occurring during the other ~11 months of the year. In 2012–13, average concentrations during the runoff were about 2X higher than baseflow concentrations (Figure 2). While concentrations are mostly stable outside the spring runoff season, isolated summer thunderstorms may cause brief, pulsed Cu exposures, such as the September 5, 2013 event with a several hour pulse of increased Cu that peaked at 10X pre-storm baseline. This situation raises the question, are effects attributable to Cu more likely from the higher, brief pulse exposures or from the much lower and longer press exposures to Cu during low flow periods? Because our biological sampling was annual, organisms were exposed to both the annual short-term pulse and to long-term, less variable and lower concentrations. These are difficult influences to separate even in controlled experiments (Johnston and Keough, 2002), and our data alone are insufficient to untangle these influences. However, three lines of reasoning suggest that the lower, long-term Cu exposures had the greater influence. First, the timing of the snowmelt-driven Cu pulses occurs at times that larger and less metals-sensitive life stages of aquatic insects would be expected to be present (Clark and Clements, 2006). The sensitivity of fish to metals toxicity is also size-dependent. Egg and alevin stages are particularly resistant to metals, and the most sensitive sizes appear to be YOY salmonids of about 6 to 16 weeks post-hatch, and newly emerged YOY sculpin (Chapman, 1978; Besser et al., 2007; Mebane et al., 2008, 2012). In nearby streams of similar elevation, Chinook Salmon, Rainbow Trout, and Shorthead Sculpin tend emerge from the gravels in June to August, after peak runoff (Bailey, 1952; Orcutt et al., 1968; Richards and Cernera, 1989). During the annual April to early June Cu pulses, less sensitive eggs and alevins would be present. Second, DOC mitigates Cu toxicity and because DOC also tended to increase during the spring runoff, the DOC-influenced Cu criteria tends to rise and fall in synchrony with the ambient Cu concentrations (Figure 3). Third, correlations between biological endpoints and average Cu concentrations tended to be stronger than with peak annual Cu concentrations (Text S2).

The September 2013 storm-pulse of Cu in Big Deer Creek appeared to influence the benthic community structure at BD-km5.3, sampled 4-days later. Taxa richness (presence of taxa) was subtly lower and some mayfly abundances declined in Big Deer Creek in 2013 relative to 2012. Most notable was *Baetis* with the lowest relative abundance since 1993 (Figure 5). Massive drift of mayflies and *Baetis* in particular has been reported within minutes of a chemical disturbance in streams (Ormerod et al., 1987), and drift appears to be an avoidance response of some aquatic insects to short-term, novel metals exposures (Clements, 2004). Because recovery times for benthic abundance following non-catastrophic drift episodes in streams with nearby upstream colonization sources have been on the order of two to six weeks (Wallace, 1990; Clearwater et al., 2011), it is likely that our collections were affected by the storm pulse 4-days earlier. A pulse of Cu from storm runoff would also be expected to be accompanied by an increase in DOC which could mitigate Cu toxicity. For instance, in Silver Bow Creek, Montana, Balistrieri et al. (2012) captured an increase in dissolved Cu from about 6 to 27 μg/L and an increase in DOC from about 4 to 13 mg/L during the first 1.5h of a rain storm. The extent to which an increase in DOC would mitigate Cu toxicity would likely be influenced by whether the pulses were in synchrony, their relative concentrations, and contact time between the DOC and Cu.

## Cobalt

While the risks of Co in aquatic ecosystems are less well known than those of Cu, Co in the Blackbird Creek drainage is elevated more than two orders of magnitude above background concentrations. In 2013 concentrations in Panther Creek at site PA-km37 exceeded 40 μg/L during September stable flow periods (Figure 3). In contrast with Cu, Co concentrations are lowest during the brief spring snowmelt period, and highest and fairly stable during the low flow periods.

Ecotoxicology data for Co are sparse relative to Cu, but what there are suggest very different aquatic toxicity profiles for Co and Cu. Whereas both fish and invertebrates may be sensitive to low Cu concentrations that were elevated only 3 to 5X above background concentrations (e.g., Besser et al., 2007; Mebane and Arthaud, 2010; Clements et al., 2013), we found no reports of Co having any direct adverse effects to fish at environmentally relevant concentrations. In a 60-day growth and survival test of Rainbow Trout fry using dilution water from upper Panther Creek, the lowest observed effect was a 5% reduction in growth at 242 μg/L (Pacific EcoRisk, 2005). Marr et al. (1998) compared the relative toxicity of Cu and Co to Rainbow Trout in 14-day toxicity tests (test pH 7.6, DOC 0.2 mg/L, hardness 25 mg/L). Copper was both a more potent and faster acting toxicant with incipient lethal levels >20X lower than those for Co (14 vs. 346 μg/L respectively). In mixture tests with Cu and Co with Rainbow Trout, the acute toxicity of Cu was not consistently increased by the presence of Co or vice versa (Marr et al. 1998).

In contrast to the apparent indifference of fish to Co, adverse effects to freshwater invertebrates from Co have been observed at concentrations almost three orders of magnitude lower than those adverse to fish. Norwood et al. (2007) obtained a 28-day LC25 of only 4 μg/L with the amphipod *Hyalella azteca* (test pH 8.2, DOC 1.1 mg/L, water hardness 122 mg/L). Similar low effects concentrations with Co have been obtained with daphnids and snails (Environment Canada, 2013). However, we are only aware of two long-term exposures of stream-resident aquatic insect species with Co. A 20-day exposure of *Chironomus dilutus* larvae to Co in Panther Creek water only produced a 20% reduction in survival at 216 μg/L (Pacific EcoRisk, 2005). These insensitive results were congruent with field observations in Panther Creek. Prior to restoration efforts, Chironomids were abundant in Panther Creek at PA-km37 with >6,000 individuals/m$^2$ and average Co concentrations of about 90 μg/L (Beltman et al. 1999; Table 2). In contrast to the insensitive *Chironomus* results, exposure of mayfly *Ephemerella ignita* nymphs for 28-days to 33 μg/L Co in the Ricklea

River, Sweden, resulted in only 26% survival compared to 77% survival in the controls, and a 48% reduction in growth (as wet weight) relative to controls. Low effects (7% reduction in weight; a 4-day delay in median emergence times) were observed in the lowest 5.2 µg/L Co treatment (Södergren, 1976). These results suggest that Co concentrations in the range of 10–50 µg/L in Panther Creek could contribute to the scarcity of *Ephemerella* and other taxa (Figure 3; Figure 5). Cobalt toxicity is moderated by Ca and DOC (Richards and Playle, 1998), and Co potency in Panther Creek is probably roughly comparable to that in the Rickleå River tests. Compared to the Rickleå River water, Panther Creek has higher Ca which would make Co relatively less toxic that in Södergren's (1976) tests, but lower DOC which would have the opposite effects (Ca about 12 mg/L vs. 3.5 mg/L, and DOC about 2 mg/L vs. about 10 mg/L during base flows for Panther Creek and the Rickleå River respectively, with DOC estimated from Hoppe et al. (2015)).

## Arsenic

Arsenic has been persistently elevated in sediment and periphyton in Panther Creek, and in some years has been elevated in invertebrate tissues. However, the high arsenic concentrations in sediment and periphyton did not appear to pass through to aquatic macroinvertebrate tissues (Figure 2), suggesting that the arsenic sorbed to sediments or periphyton might have low bioavailability. In toxicity testing of Panther Creek sediments with the benthic invertebrate *Hyalella azteca*, correlations between reduced biomass were much weaker with arsenic than with Cu (r = -0.27 and -0.84, respectively) suggesting arsenic had less influence on benthic communities than Cu or Co (Mebane, 1994). Arsenic and Fe have been shown to be strongly correlated in Panther Creek sediments (r >0.9), suggesting sequestration in Fe oxyferrihydroxides in stream sediment (Mok and Wai, 1989; Mebane, 1994; Gray and Eppinger, 2012). Erickson et al. (2010), noted that inorganic arsenic in the diet of trout at about 20 mg/kg dry weight (dw) or higher has been correlated with reduced growth, and in their feeding study with live invertebrate diets enriched with arsenic, 26 mg/kg dw or higher arsenic in the diet was directly demonstrated to impair growth in Rainbow Trout. Arsenic residues in Panther Creek invertebrate tissue were slightly above 20 mg/kg dw in some years, although the years with lower condition factors in fish did not match years with elevated arsenic in invertebrate tissues (Figures 2 and 10). Our 2012 targeting of specific taxa showed decreasing concentrations with increasing trophic level, suggesting bio-dilution through trophic transfer.

## Biological factors

The recoveries of benthic macroinvertebrate communities have been uneven as Cu and Co stress have lessened over time. The community at one site, BD-km5.3, has become mostly similar to that of the nearby upstream reference site. However, species richness at other sites remains lower than that at reference sites, even though major groups such as the mayflies and stoneflies have become well represented throughout Big Deer and Panther Creeks.

Distance to source populations and interspecific competition could influence aquatic insect recolonization. Dispersal abilities by air or drift vary greatly among freshwater invertebrates, but are usually reported as <1 km per generation, and dispersal is often from downstream to upstream (Mackay, 1992; Elliott, 2003; MacNeale et al., 2005). Perennial tributaries with clean-water refugia areas are present along Panther Creek (Figure 1), suggesting dispersal is unlikely a persistent limiting factor in recovery of stream insects in our study area, unlike some areas (Masters et al., 2007; Milner et al., 2008; Brederveld et al., 2011). Still, sites in lower Panther Creek (PA-km17 and km22) could be distant enough from large clean water tributaries that dispersal distances contribute to a lag in the recovery of benthic communities following water quality improvements.

Taxa richness is greater in mine-influenced sections of Big Deer Creek than those in Panther Creek (Figure 3, Figure 4). Colonization began on nearly bare substrates in Big Deer Creek, whereas in Panther Creek prior to water-quality restoration, benthic macroinvertebrates were about as abundant as at upstream reference sites, but were dominated by a few taxa, usually chironomids and *Brachycentrus* caddisflies (Mebane, 1994; Beltman et al., 1999). *Brachycentrus* is a strong competitor that once established can exclude other insects (Peterson et al., 1993). This suggests the possibility of biotic resistance from extant or early colonizing taxa that could compete with later arrivals. Competition and trophic changes may lead to indirect biological effects that can either mask or amplify direct effects through modifying competition for limited food or space resources (Fleeger et al., 2003; Johnston and Keough, 2003). After following the colonization and succession of macroinvertebrates for several years after re-watering of a river channel, Minshall et al. (1983) suggested that the final structure of the benthic community may be determined to a large extent by which species become established first. Conceptually, succession and resilience dynamics in macroinvertebrate communities could result in new stable states, and resist return to reference conditions (McAuliffe, 1984; Fisher, 1990).

## Benthic macroinvertebrates and fish populations

Fish populations in mining-influenced streams may be indirectly constrained by reduced macroinvertebrate prey availability (Hogsden and Harding, 2012), which could be reflected in reduced abundances or reduced condition factor (Munkittrick and Dixon, 1989). However, a few benthic macroinvertebrate taxa probably are of disproportionate importance to fish populations. Chinook Salmon, other stream-resident salmonids, and sculpin have been shown to feed primarily on *Baetis* mayflies, chironomids, simuliids, as well as other bite-sized, abundant taxa such as the stonefly *Zapada* (Bailey, 1952; Allan, 1983; Esteban and Marchetti, 2004). These taxa were early colonizers in Panther Creek and remain abundant. However, salmonids are opportunistic and will eat any invertebrates that are about the right size for their gape and are palatable. Conceptually, increased benthic diversity might decrease the reliance of any particular taxa and reduce the likelihood of seasonal shortages following hatches. While as of 2013 benthic macroinvertebrate diversities downstream of mine-influenced tributaries remain about 10 to 30% lower than those at the reference sites, juvenile fish growth, as inferred from body condition and length vs. weight regressions in salmonids, increased from the headwaters sites downstream (not shown), as would be expected with increasing temperatures in least-disturbed natural streams. Panther Creek in 2013 had the highest overall fish densities measured during this project, and a decline in invertebrate biomass relative to 2012. Whether salmonids can effectively depress benthos is uncertain (Allan, 1983), but sculpin certainly can, at least for some vulnerable taxa such as chironomids and *Baetis* mayflies (Brocksen et al., 1968; Flecker, 1984). In sum, fish populations do not appear to be any more limited by prey availability in mine-influenced downstream reaches than upstream reference reaches.

## 4.5 Natural ecological variability and recovery

Natural stream communities are predicted to change along longitudinal gradient as the physical features of streams change from headwaters to mid-order streams to large rivers (Vannote et al., 1980). This presents a challenge in stream pollution ecology studies because reference sites may need to be located upstream of anthropogenic disturbances in river basins, resulting in overlying longitudinal natural and anthropogenic gradients. For instance, fish species richness increases with increasing stream size, and steep gradient, head-water streams are commonly only inhabited by a single salmonid species. As the streams become less steep and larger and riffle habitats become more common, sculpin will appear and become numerically dominant in mid (3rd and 4th order streams). As streams transition to rivers, shallow riffles give way to deeper run and pool habitats, and the numerical dominance of sculpins and salmonids will decline as minnows and suckers become abundant (Platts, 1979; Mebane et al., 2003).

Several lines of evidence allow us to attribute biological changes to water quality changes rather than co-occurring streamflow or temperature differences. First, even between our most distant sites (PA-km4, ~40 km downstream from reference sites), summertime high temperatures are about 2°C warmer but habitats are otherwise fundamentally similar. Shorthead Sculpin and Rainbow Trout remain abundant, and temperature sensitive genera such as *Epeorus* and *Zapada* and metals sensitive taxa such as *Cinygmula*, and *Ephemerella* were common at both PA-km4 and at upstream reference sites (Mebane et al., 2015). Second, the tributary sites had particularly well matched reference and assessment sites. In Big Deer Creek upstream of the waterfall, the physical stream characteristics were very similar between the reference and mining-influenced sites.

Changing correlation patterns over time between physicochemical and biological metrics give clues to relative influences of mining and natural factors. Wiens and Parker (1995) suggest "impact" can be defined as a statistically significant correlation between injury and exposure, and "recovery" can then be defined as the disappearance of such a correlation through time. During the 1993–2002 period which bracketed the onset of restoration activities, there were strong, negative correlations between mayfly and stonefly densities with Cu (r values of -0.89 and -0.76 respectively), whereas during the latter 2003–2013 recovery period, the correlations were much weaker or reversed, with r values of -0.17 and 0.25, respectively. In contrast, taxa richness remained strongly correlated with Cu (r -0.80) but not with temperature (r -0.15). The weak correlation between taxa richness and temperature in Panther Creek is consistent with natural patterns from least-disturbed streams elsewhere in the Salmon River basin. Species richness in 33 streams selected for their minimal anthropogenic disturbances was only weakly correlated with annual maximum weekly maximum temperatures (r -0.38) despite a span of 12°C (Ott and Maret, 2003). In contrast, the difference between the matched furthest upstream and furthest downstream sites in Panther Creek was only about 2°C.

Interpretations of perceived changes may always be open to some debate in ecological field studies and inferences of effects and recovery rely on coherence of lines of evidence rather than strict causality (Parker and Wiens, 2005; Munkittrick, 2009). Prior to the restoration efforts that began in 1995, there were clear differences between the reference and mine-influenced sites in metals exposures and associated biological measures. Throughout the post-restoration recovery period, declines in metals concentrations were accompanied by or followed by declines in apparent biological effects attributable to metals, and correlations between exposure and effects have weakened or disappeared.

# 5. Conclusions

The changes that we observed in the Panther Creek watershed gave insights on ecosystem recovery patterns that would not necessarily have been predicted from chemical monitoring or small-scale toxicity testing alone. These include:

1.  Fish populations recovered rapidly once the limiting water quality constraint was relaxed, on the order of about three generations to reach reference densities;

2.  Speed of recovery differed greatly for species with different traits. The larger bodied, comparatively pelagic, salmonid species recovered much faster than the less motile smaller-bodied, benthic sculpin species;

3.  Recovery needed to be viewed on a landscape scale. Recovery of anadromous fish was limited by factors outside of the watershed, in addition to any within-watershed constraints such as water-quality, and the influences of disturbances other than those that were the focus of the study (e.g., wildfire) had to be considered;

4.  When investigating the effects of disturbances in streams such as mine pollution from discrete tributary sources, conditions upstream of the affected environment are an obvious point of reference. Yet in flowing waters, a study design relying on upstream-downstream, reference-comparison monitoring sites introduces the complication that potential effects of pollution could be masked by or mistaken for natural longitudinal changes in the stream ecology;

5.  Water quality suitability was not a binary, yes/no question hinging on criteria being met or not. Fish began to move into marginally suitable habitats while Cu concentrations were still suboptimal (~3X criterion), yet it is possible that even at <1X criterion, Cu continued to influence the insect community;

6.  The benthic macroinvertebrate community gained species as Cu and Co declined in patterns that were congruent with concepts of sensitivities of insects to ionoregulatory disturbance and critical body residues of metals;

7.  Cu and Co concentrations in water, sediment, periphyton, and macroinvertebrate tissues were all correlated, yet taxa richness was most strongly correlated with Co in water followed by Cu in water;

8.  Lower taxa richness and scarcity of some invertebrate taxa relative to nearby reference sites suggests that Co and/or Cu may be toxic to a minority of taxa at concentrations lower than numeric recovery targets. There was little evidence that this reduced taxa richness measurably affected recovery of fish populations, which in turn probably depend more on overall abundance of common taxa such as baetid mayflies, simuliids, and chironomids;

9.  Distance and direction to source areas of colonists both appeared to influence recovery times with at least Shorthead Sculpin appearing to extend their range mostly from downstream to upstream; and

10. For both fish and invertebrates, the presence and abundance of species are probably nonexclusively influenced by both internal factors (intrinsic physiological sensitivity to metals), and external factors (competition from early colonists inhibits later arrivals).

Reviews of recovery of freshwater ecosystems have emphasized times to recovery (Niemi et al., 1990; Detenbeck et al., 1992; Jones and Schmitz, 2009). In some pulse disturbances (such as accidental spills or deliberate fish poisonings, experimental insecticide treatments, or re-watering a de-watered channel), there is a discrete point in time in which the direct disturbance ended, and the time to recovery began. However, water quality improvements from longstanding impairments on the physical scale of the Blackbird Mine remediation are progressive and adaptive (Gustavson et al., 2007; USEPA, 2013). Without discrete starting or finish lines to measure time to recovery, recovery times are decidedly ambiguous.

The Panther Creek restoration largely meets Palmer et al.'s (2005) five criteria for ecologically successful river restoration. These criteria were 1) the project design should be based on a specified guiding image of a more dynamic, healthy river that could be attained [most germane to projects with physical or hydrological manipulations]; 2) the river's ecological condition must be measurably improved; 3), the river system must be more self-sustaining and resilient to external perturbations so that only minimal follow-up maintenance is needed; 4) during the construction phase, no lasting harm should be inflicted on the ecosystem, and 5) both pre- and post-assessments must be completed and data made publicly available (Palmer et al., 2005). We think criteria 1, 2, 4, and 5 have been met. For the criterion no. 3, while the river's ecology is self-sustaining so long as active pollution controls are maintained, the costs of pollution controls are nontrivial. The water treatment plant must continue to operate for the foreseeable future, and the various runoff controls require annual upkeep. Finally, the past environmental costs and subsequent financial costs of the Panther Creek watershed restoration were substantial.

# References

Adams SM, Hill WR, Peterson MJ, Ryon MG, Smith JG, et al. 2002. Assessing recovery in a stream ecosystem: applying multiple chemical and biological endpoints. *Ecol Appl* **12**(5): 1510–1527. doi:10.1890/1051-0761(2002)012[1510:-ariase]2.0.co;2.

Adams WJ, Blust R, Borgmann U, Brix KV, DeForest DK, et al. 2011. Utility of tissue residues for predicting effects of metals on aquatic organisms. *Integr Environ Assess Manage* **7**(1): 75–98. doi:10.1002/ieam.108.

Allan JD. 1983. Predator-prey relationships in streams, in Barnes JR, Minshall GW, eds., *Stream Ecology: Application and Testing of General Ecological Theory*. New York and London: Plenum Press: pp. 191–229. doi: 10.1007/978-1-4613-3775-1_9.

Bailey JE. 1952. Life history and ecology of the sculpin *Cottus bairdi punctulatus* in southwestern Montana. *Copeia* **1952**(4): 243–255. doi:10.2307/1439271.

Balistrieri LS, Mebane CA, Schmidt TS, Keller WB. 2015. Expanding metal mixture toxicity models to natural stream and lake invertebrate communities. *Environ Toxicol Chem.* doi:10.1002/etc.2824.

Balistrieri LS, Nimick DA, Mebane CA. 2012. Assessing time-integrated dissolved concentrations and predicting toxicity of metals during diel cycling in streams. *Sci Total Environ* **425**: 155–168. http://dx.doi.org/10.1016/j.scitotenv.2012.03.008.

Beltman DJ, Lipton J, Cacela D, Clements WH. 1999. Benthic invertebrate metals exposure, accumulation, and community-level effects downstream from a hard-rock mine site. *Environ Toxicol Chem* **18**(2): 299–307. doi:10.1002/etc.5620180229.

Bernhardt ES, Palmer MA, Allan JD, Alexander G, Barnas K, et al. 2005. Synthesizing U.S. river restoration efforts. *Science* **308**(5722): 636–637. doi:10.1126/science.1109769.

Besser JM, Mebane CA, Mount DR, Ivey CD, Kunz JL, et al. 2007. Relative sensitivity of mottled sculpins (*Cottus bairdi*) and rainbow trout (*Oncorhynchus mykiss*) to toxicity of metals associated with mining activities. *Environ Toxicol Chem* **26**(8): 1657–1665. doi:10.1897/06-571R.1.

Boag TD. 1987. Food habits of bull char, *Salvelinus confluentus*, and rainbow trout, *Salmo gairdneri*, coexisting in a foothills stream in northern Alberta. *Canadian Field Naturalist* **101**(1): 56–62.

Brederveld RJ, Jähnig SC, Lorenz AW, Brunzel S, Soons MB. 2011. Dispersal as a limiting factor in the colonization of restored mountain streams by plants and macroinvertebrates. *J Appl Ecol* **48**(5): 1241–1250. doi:10.1111/j.1365-2664.2011.02026.x.

Breen MJ, Ruetz CRI, Thompson KJ, Kohler SL. 2009. Movements of mottled sculpins (*Cottus bairdii*) in a Michigan stream: how restricted are they? *Can J Fish Aquat Sci* **66**(1): 31–41. doi:10.1139/F08-189.

Brinkman SF, Johnston WD. 2008. Acute toxicity of aqueous copper, cadmium, and zinc to the mayfly *Rhithrogena hageni*. *Arch Environ Con Tox* **54**(3): 466–72. doi:10.1007/s00244-007-9043-z.

Brix KV, DeForest DK, Adams WJ. 2011. The sensitivity of aquatic insects to divalent metals: A comparative analysis of laboratory and field data. *Sci Total Environ* **409**(20): 4187–4197. doi:10.1016/j.scitotenv.2011.06.061.

Brocksen RW, Davis GE, Warren CE. 1968. Competition, food consumption, and production of sculpins and trout in laboratory stream communities. *J Wildlife Manage* **32**(1): 51–75. doi:10.2307/3798237.

Brown L, Downhower JF. 1982. Summer movements of mottled sculpins, *Cottus bairdi* (Pisces: Cottidae). *Copeia* **1982**(2): 450–455.

Buchwalter DB, Cain DJ, Martin CA, Xie L, Luoma SN, et al. 2008. Aquatic insect ecophysiological traits reveal phylogenetically based differences in dissolved cadmium susceptibility. *Proc Natl Acad Sci* **105**(24): 8321–8326. doi:10.1073/pnas.0801686105.

Byrne P, Wood PJ, Reid I. 2012. The impairment of river systems by metal mine contamination: a review including remediation options. *Crit Rev Env Sci Tec* **42**(19): 2017–2077. doi:10.1080/10643389.2011.574103.

Cain DJ, Luoma SN, Wallace WG. 2004. Linking metal bioaccumulation of aquatic insects to their distribution patterns in a mining-impacted river. *Environ Toxicol Chem* **23**(6): 1463–1473. doi:10.1897/03-291.

Campbell PGC, Hontela A, Rasmussen JB, Giguère A, Gravel A, et al. 2003. Differentiating between direct (physiological) and food-chain mediated (bioenergetic) effects on fish in metal-impacted lakes. *Hum Ecol Risk Asses* **9**(4): 847–866. doi:10.1080/713610012.

Cedergreen N, Nørhave NJ, Nielsen K, Johansson HKL, Marcussen H, et al. 2013. Low temperatures enhance the toxicity of copper and cadmium to *Enchytraeus crypticus* through different mechanisms. *Environ Toxicol Chem* **32**(10): 2274–2283. doi:10.1002/etc.2274.

Chapman DJ, Julius BE. 2005. The use of preventative projects as compensatory restoration. *J Coastal Res* **40**: 120–131. doi:10.2307/25736620.

Chapman GA. 1978. Toxicities of cadmium, copper, and zinc to four juvenile stages of chinook salmon and steelhead. *Trans Am Fish Soc* **107**(6): 841–847. doi:10.1577/1548-8659(1978)107<841:TOCCAZ>2.0.CO;2.

Chapman PM. 2001. Utility and relevance of aquatic oligochaetes in ecological risk assessment. *Hydrobiol* **463**(1–3): 149–169. doi:10.1023/a:1013103708250.

Clark JL, Clements WH. 2006. The use of in situ and stream microcosm experiments to assess population- and community-level responses to metals. *Environ Toxicol Chem* **25**(9): 2306–2312. doi:10.1897/05-552.1.

Clearwater SJ, Jellyman PG, Biggs BJF, Hickey CW, Blair N, et al. 2011. Pulse-dose application of chelated copper to a river for *Didymosphenia geminata* control: Effects on macroinvertebrates and fish. *Environ Toxicol Chem* **30**(1): 181–195. doi:10.1002/etc.369.

Clements WH. 2004. Small-scale experiments support causal relationships between metal contamination and macroinvertebrate community composition. *Ecol Appl* **14**(3): 954–967.

Clements WH, Cadmus P, Brinkman SF. 2013. Responses of aquatic insects to Cu and Zn in stream microcosms: understanding differences between single species tests and field responses. *Environ Sci Technol* **47**(13): 7506–7513. doi:10.1021/es401255h.

Clements WH, Carlisle DM, Lazorchak JM, Johnson PC. 2000. Heavy metals structure benthic communities in Colorado mountain streams. *Ecol Appl* **10**(2): 626–638. doi:10.1890/1051-0761(2000)010[0626:HMSBCI]2.0.CO;2.

Clements WH, Vieira NKM, Church SE. 2010. Quantifying restoration success and recovery in a metal-polluted stream: a 17-year assessment of physicochemical and biological responses. *J Appl Ecol* **47**(4): 899–910. doi:10.1111/j.1365-2664.2010.01838.x.

Cleveland WS, Devlin SJ. 1988. Locally weighted regression: an approach to regression analysis by local fitting. *J Am Stat Assoc* **83**(403): 596–610. doi:10.1080/01621459.1988.10478639.

Connell JH, Slatyer RO. 1977. Mechanisms of succession in natural communities and their role in community stability and organization. *Am Nat* **111**(982): 1119–1144. doi:10.2307/2460259.

Copeland T, Bumgarner JD, Byrne A, Denny L, Hebdon JL, et al. 2013. Reconstruction of the 2010/2011 Steelhead Spawning Run into the Snake River Basin. *Report to Bonneville Power Administration, Portland, Oregon.* http://www.efw.bpa.gov.

Cormier SM, Suter GW, Zheng L. 2013. Derivation of a benchmark for freshwater ionic strength. *Environ Toxicol Chem* **32**(2): 263–271. doi:10.1002/etc.2064.

Costello DM, Burton GA. 2014. Response of stream ecosystem function and structure to sediment metal: Context-dependency and variation among endpoints. *Elem Sci Anth* **2**(1): 000030. doi:10.12952/journal.elementa.000030.

Courtney LA, Clements WH. 2002. Assessing the influence of water and substratum quality on benthic macroinvertebrate communities in a metal-polluted stream: an experimental approach. *Freshwater Biol* **47**(9): 1766–1778. doi:10.1046/j.1365-2427.2002.00896.x.

Cunjak RA, Prowse TD, Parrish DL. 1998. Atlantic salmon (*Salmo salar*) in winter: "the season of parr discontent"? *Can J Fish Aquat Sci* **55**(S1): 161–180. doi:10.1139/cjfas-55-S1-161.

Detenbeck NE, DeVore PW, Niemi GJ, Lima A. 1992. Recovery of temperate-stream fish communities from disturbance - a review of case studies and synthesis of theory. *Environ Manage* **16**(1): 33–53. doi:10.1007/BF02393907.

Di Stefano J, Fidler F, Cumming G. 2005. Effect size estimates and confidence intervals: An alternative focus for the presentation and interpretation of ecological data, in Burk AR, ed., *New trends in ecology research.*, New York: Nova Science Publishers: pp. 71–102.

Ebel JD, Marcarelli AM, Kohler AE. 2014. Biofilm nutrient limitation, metabolism, and standing crop responses to experimental application of salmon carcass analog in Idaho streams. *Can J Fish Aquat Sci* **71**(12): 1796–1804. doi:10.1139/cjfas-2014-0266.

Elliott JM. 2003. A comparative study of the dispersal of 10 species of stream invertebrates. *Freshwater Biol* **48**(9): 1652–1668. doi:10.1046/j.1365-2427.2003.01117.x.

Environment Canada. 2013. Federal Environmental Quality Guidelines: Cobalt. Environment Canada. 10 pp. http://www.ec.gc.ca/ese-ees/default.asp?lang=En&n=92F47C5D-1#a8.

Eppinger RG, Briggs PH, Rieffenberger B, Dorn CV, Brown ZA, et al. 2003. Geochemical data for stream sediment and surface water samples from Panther Creek, the Middle Fork of the Salmon River, and the Main Salmon River, collected before and after the Clear Creek, Little Pistol, and Shellrock wildfires of 2000 in central Idaho. *U.S. Geological Survey Open-File Report 2003-152.* http://pubs.usgs.gov/of/2003/152/ [Accessed July 2014].

Erickson RJ, Mount DR, Highland TL, Hockett JR, Leonard EN, et al. 2010. Effects of copper, cadmium, lead, and arsenic in a live diet on juvenile fish growth. *Can J Fish Aquat Sci* **67**(11): 1816–1826. doi:10.1139/F10-098.

Esteban EM, Marchetti MP. 2004. What's on the menu? Evaluating a food availability model with young-of-the-year Chinook Salmon in the Feather River, California. *Trans Am Fish Soc* **133**(3): 777–788. doi:10.1577/t03-115.1.

Farag AM, Stansbury MA, Bergman HL, Hogstrand C, MacConnell E. 1995. The physiological impairment of free-ranging brown trout exposed to metals in the Clark Fork River, Montana. *Can J Fish Aquat Sci* **52**(9): 2038–2050. doi:10.1139/f95-795.

Fisher S. 1990. Recovery processes in lotic ecosystems: Limits of successional theory. *Environ Manage* **14**(5): 725–736. doi:10.1007/bf02394721.

Flecker AS. 1984. The effects of predation and detritus on the structure of a stream insect community: a field test. *Oecologia* **64**(3): 300–305. doi:10.1007/bf00379125.

Fleeger JW, Carman KR, Nisbet RM. 2003. Indirect effects of contaminants in aquatic ecosystems. *Sci Total Environ* **317**(1–3): 207–233. doi:10.1016/S0048-9697(03)00141-4.

Gauthier C, Campbell PGC, Couture P. 2009. *Condition and pyloric caeca as indicators* of food web effects in fish living in metal-contaminated lakes. *Ecotox Environ Safe* **72**(8): 2066–2074. doi:10.1016/j.ecoenv.2009.08.005.

Gibbons WN, Munkittrick KR, Taylor WD. 1998. Monitoring aquatic environments receiving industrial effluent using small fish species 1: response of spoonhead sculpin (*Cottus ricei*) downstream of a bleached-kraft pulp mill. *Environ Toxicol Chem* **17**(11): 2227–2237. doi:10.1002/etc.5620171113.

Gray JE, Eppinger RG. 2012. Distribution of Cu, Co, As, Fe, and Mn in sediment, soil, and water in and around mineral deposits and mines of the Idaho Cobalt Belt, USA. *Appl Geochem* **27**(6): 1053–1062. doi:10.1016/j.apgeochem.2012.02.001.

Gustavson KE, Barnthouse LW, Brierley CL, Clark EH, II, and Ward CH. 2007. Superfund and mining megasites. *Environ Sci Technol* **41**(8): 2667–2672. doi:10.1021/es0725091.

Hamilton SK. 2012. Biogeochemical time lags may delay responses of streams to ecological restoration. *Freshwater Biol* **57**: 43–57. doi:10.1111/j.1365-2427.2011.02685.x.

Hansen JA, Lipton J, Holmes J, Bergman HL. 1995. Caged Fish Bioassay Studies, Panther Creek Idaho, Spring 1993. Report to the National Oceanic and Atmospheric Administration. Boulder, Colorado: RCG/Hagler Bailly. 100 pp.

Hansen JA, Marr JCA, Lipton J, Bergman HL. 1999. Differences in neurobehavioral responses of chinook salmon (*Oncorhynchus tshawytscha*) and rainbow trout (*Oncorhynchus mykiss*) exposed to copper and cobalt: behavioral avoidance. *Environ Toxicol Chem* **18**(9): 1972–1978. doi:10.1002/etc.5620180916.

Harris GP. 2012. Introduction to the special issue: 'Achieving ecological outcomes'. Why is translational ecology so difficult? *Freshwater Biol* **57**: 1–6. doi:10.1111/j.1365-2427.2012.02773.x.

Helsel DR. 2005. *Nondetects and data analysis: statistics for censored environmental data.* Hoboken, New Jersey: Wiley Interscience.

Hilderbrand RH, Watts AC, Randle AM. 2005. The myths of restoration ecology. *Ecology and Society* **10**(1): 19. http://www.ecologyandsociety.org/vol10/iss1/art19/.

Hogsden KL, Harding JS. 2012. Consequences of acid mine drainage for the structure and function of benthic stream communities: a review. *Freshwater Science* **31**(1): 108–120. doi:10.1899/11-091.1.

Hoppe S, Gustafsson JP, Borg H, Breitholtz M. 2015. Evaluation of current copper bioavailability tools for soft freshwaters in Sweden. *Ecotox Environ Safe* **114**(0): 143–149. doi:10.1016/j.ecoenv.2015.01.023.

Hornberger MI, Luoma SN, Johnson ML, Holyoak M. 2009. Influence of remediation in a mine-impacted river: metal trends over large spatial and temporal scales. *Ecol Appl* **19**(6): 1522–1535. http://dx.doi.org/10.1890/08-1529.1.

Jackson JK, Füreder L. 2006. Long-term studies of freshwater macroinvertebrates: A review of the frequency, duration and ecological significance. *Freshwater Biol* **51**(3): 591–603. doi:10.1111/j.1365-2427.2006.01503.x.

Jähnig SC, Lorenz AW, Hering D, Antons C, Sundermann A, et al. 2011. River restoration success: a question of perception. *Ecol Appl* **21**(6): 2007–2015. doi:10.1890/10-0618.1.

Johnson DW, Cannamela DA, Gasser KW. 1983. Food habits of the shorthead sculpin (*Cottus confusus*) in the Big Lost River, Idaho. *Northwest Sci* **57**(3): 229–239.

Johnston EL, Keough MJ. 2002. Direct and indirect effects of repeated pollution events on marine hard-substrate assemblages. *Ecol Appl* **12**(4): 1212–1228. doi:10.1890/1051-0761(2002)012[1212:daieor]2.0.co;2.

Johnston EL, Keough MJ. 2003. Competition modifies the response of organisms to toxic disturbance. *Mar Ecol Prog Ser* **251**: 15–26. doi:10.3354/meps251015.

Jones HP, Schmitz OJ. 2009. Rapid recovery of damaged ecosystems. *PLoS ONE* **4**(5): e5653. doi:10.1371/journal.pone.0005653.

Kennedy P, Apperson KA, Flinders J, Corsi M, Johnson J, et al. 2013. Monitoring relative abundance and age composition of spring-summer Chinook Salmon on the spawning grounds in Idaho. *Natural Production Monitoring and Evaluation, 2012 Annual Report. Idaho Department of Fish and Game, IDFG Report Number 13–12*, pp. 8–24. https://collaboration.idfg.idaho.gov/FisheriesTechnicalReports/Forms/AllItems.aspx.

Lake PS, Bond N, Reich P. 2007. Linking ecological theory with stream restoration. *Freshwater Biol* **52**(4): 597–615. doi:10.1111/j.1365-2427.2006.01709.x.

LeJeune K, Lipton J, Walsh WA, Cacela D, Jensen S, et al. 1995. Fish population survey, Panther Creek, Idaho. *Report by RCG/Hagler Bailly, Boulder, CO to the State of Idaho and National Oceanic and Atmospheric Administration*. 214 pp.

Mackay RJ. 1992. Colonization by lotic macroinvertebrates: A review of processes and patterns. *Can J Fish Aquat Sci* **49**(3): 617–628. doi:10.1139/f92-071.

Macneale KH, Peckarsky BL, Likens GE. 2005. Stable isotopes identify dispersal patterns of stonefly populations living along stream corridors. *Freshwater Biol* **50**(7): 1117–1130. doi:10.1111/j.1365-2427.2005.01387.x.

MacRae RK, Maest AS, Meyer JS. 1999. Selection of an organic acid analogue of dissolved organic matter for use in toxicity testing. *Can J Fish Aquat Sci* **56**(8): 1484–1493. doi:10.1139/f99-090.

Malaj E, Grote M, Schäfer RB, Brack W, von der Ohe PC. 2012. Physiological sensitivity of freshwater macroinvertebrates to heavy metals. *Environ Toxicol Chem* **31**(8): 1754–1764. doi:10.1002/etc.1868.

Maraldo K, Christensen B, Strandberg B, Holmstrup M. 2006. Effects of copper on enchytraeids in the field under differing soil moisture regimes. *Environ Toxicol Chem* **25**(2): 604–612. doi:10.1897/05-076r.1.

Marr JCA, Hansen JA, Meyer JS, Cacela D, Podrabsky TL, et al. 1998. Toxicity of cobalt and copper to rainbow trout: application of a mechanistic model for predicting survival. *Aquat Toxicol* **43**: 225–237. doi:10.1016/S0166-445X(98)00061-7.

Marr JCA, Lipton J, Cacela D, Hansen JA, Bergman HL, et al. 1996. Relationship between copper exposure duration, tissue copper concentration, and rainbow trout growth. *Aquat Toxicol* **36**(1): 17–30. doi:10.1016/S0166-445X(96)00801-6.

Marr JCA, Lipton J, Cacela D, Hansen JA, Meyer JS, et al. 1999. Bioavailability and acute toxicity of copper to rainbow trout (*Oncorhynchus mykiss*) in the presence of organic acids simulating natural dissolved organic carbon. *Can J Fish Aquat Sci* **56**(8): 1471–1483. doi:10.1139/f99-089.

Martínez-Abraín A. 2007. Are there any differences? A non-sensical question in ecology. *Acta Oecol* **32**(2): 203–206. doi:10.1016/j.actao.2007.04.003.

Masters Z, Peteresen I, Hildrew AG, Ormerod SJ. 2007. Insect dispersal does not limit the biological recovery of streams from acidification. *Aquat Conserv* **17**(4): 375–383. doi:10.1002/aqc.794.

McAuliffe JR. 1984. Competition for space, disturbance, and the structure of a benthic stream community. *Ecology* **65**(3): 894–908. doi:10.2307/1938063.

Mebane CA. 1994. Preliminary Natural Resource Survey - Blackbird Mine, Lemhi County, Idaho. National Oceanic and Atmospheric Administration, Hazardous Materials Assessment and Response Division, Seattle, WA. http://dx.doi.org/10.13140/2.1.4116.3840 [Accessed December 2014].130 pp

Mebane CA, Arthaud DL. 2010. Extrapolating growth reductions in fish to changes in population extinction risks: copper and Chinook salmon. *Hum Ecol Risk Asses* **16**(5): 1026–1065. doi:10.1080/10807039.2010.512243.

Mebane CA, Dillon FS, Hennessy DP. 2012. Acute toxicity of cadmium, lead, zinc, and their mixtures to stream-resident fish and invertebrates. *Environ Toxicol Chem* **31**(6): 1334–1348. doi:10.1002/etc.1820.

Mebane CA, Eakins RJ, Fraser BG, Adams WJ. 2015. Data from: Recovery of a mining-damaged stream ecosystem. *Dryad Digital Repository*. doi:10.5061/dryad.67n20.

Mebane CA, Hennessy DP, Dillon FS. 2008. Developing acute-to-chronic toxicity ratios for lead, cadmium, and zinc using rainbow trout, a mayfly, and a midge. *Water Air Soil Poll* **188**(1–4): 41–66. doi:10.1007/s11270-007-9524-8.

Mebane CA, Maret TR, Hughes RM. 2003. An index of biological integrity (IBI) for Pacific Northwest rivers. *Trans Am Fish Soc* **132**(2): 239–261. doi:10.1577/1548-8659(2003)132<0239:AIOBII>2.0.CO;2.

Meyer JS, Adams WJ. 2010. Relationship between biotic ligand model-based water quality criteria and avoidance and olfactory responses. *Environ Toxicol Chem* **29**(9): 2096–2103. doi:10.1002/etc.254.

Meyer KA, High B. 2011. Accuracy of removal electrofishing estimates of trout abundance in Rocky Mountain streams. *N Am J Fish Manage* **31**(5): 923–933. doi:10.1080/02755947.2011.633684.

Mills KH, Chalanchuk SM, Allan DJ. 2000. Recovery of fish populations in Lake 223 from experimental acidification. *Can J Fish Aquat Sci* **57**(1): 192–204. doi:10.1139/f99-186.

Milner AM, Robertson AL, Brown LE, Sønderland SH, McDermott M, et al. 2011. Evolution of a stream ecosystem in recently deglaciated terrain. *Ecology* **92**(10): 1924–1935. doi:10.1890/10-2007.1.

Milner AM, Robertson AL, Monaghan KA, Veal AJ, and Flory EA. 2008. Colonization and development of an Alaskan stream community over 28 years. *Front Ecol Environ* **6**(8): 413–419. doi:10.1890/060149.

Milner NJ, Elliott JM, Armstrong JD, Gardnier R, Welton JS, et al. 2003. The natural control of salmon and trout populations in streams. *Fish Res* **62**(2): 111–125. doi:10.1016/S0165-7836(02)00157-1.

Minshall GW, Andrews DA, Manuel-Faler CY. 1983. Application of island biogeographic theory to streams: macroinvertebrate recolonization of the Teton River, Idaho, in Barnes JR, Minshall GW, eds., *Stream Ecology: Application and Testing of General Ecological Theory*. New York and London: Plenum Press: pp. 279–297.

Mok WM, Wai CM. 1989. Distribution and mobilization of arsenic species in the creeks around the Blackbird mining district, Idaho. *Water Res* **23**(1): 7–13. doi:10.1016/0043-1354(89)90054-7.

Moore JN, Langner HW. 2012. Can a river heal itself? Natural attenuation of metal contamination in river sediment. *Environ Sci Technol* **46**(5): 2616–2623. doi:10.1021/es203810j.

Mullan JW, Williams KR, Rhodus G, Hillman TW, McIntyre JD. 1992. Production and habitat of salmonids in mid-Columbia River tributary streams. U.S. Fish and Wildlife Service, Monograph I. U.S. Government Printing Office. Washington, D.C. 489 pp.

Munkittrick KR. 2009. Ubiquitous criticisms of ecological field studies. *Hum Ecol Risk Assess* **15**(4): 647–650. doi:10.1080/10807030903050616.

Munkittrick KR, Dixon DG. 1989. A holistic approach to ecosystem health using fish population characteristics. *Hydrobiol* **188/189**(1): 123–135. doi:10.1007/BF00027777.

Murphy JF, Winterbottom JH, Orton S, Simpson GL, Shilland EM, et al. 2014. Evidence of recovery from acidification in the macroinvertebrate assemblages of UK fresh waters: A 20-year time series. *Ecol Indic* **37** Part B(0): 330–340. doi:10.1016/j.ecolind.2012.07.009.

Niemi GJ, Detenbeck NE, Perry JA. 1993. Comparative analysis of variables to measure recovery rates in streams. *Environ Toxicol Chem* **12**(9): 1541–1547. doi:10.1002/etc.5620120904.

Niemi GJ, DeVore P, Detenbeck N, Taylor D, Lima A, et al. 1990. Overview of case studies on recovery of aquatic systems from disturbance. *Environ Manage* **14**(5): 571–587. doi:10.1007/BF02394710.

Norwood WP, Borgmann U, Dixon DG. 2007. Chronic toxicity of arsenic, cobalt, chromium and manganese to *Hyalella azteca* in relation to exposure and bioaccumulation. *Environ Pollut* **147**(1): 262–272. doi:10.1016/j.envpol.2006.07.017.

Orcutt DR, Pulliam BR, Arp A. 1968. Characteristics of steelhead trout redds in Idaho streams. *Trans Am Fish Soc* **97**(1): 42–45. doi:10.1577/1548-8659(1968)97[42:costri]2.0.co;2.

Ormerod SJ. 2003. Restoration in applied ecology: editor's introduction. *J Appl Ecol* **40**(1): 44–50. doi:10.1046/j.1365-2664.2003.00799.x.

Ormerod SJ, Boole P, McCahon CP, Weatherley NS, Pascoe D, et al. 1987. Short-term experimental acidification of a Welsh stream: comparing the biological effects of hydrogen ions and aluminium. *Freshwater Biol* **17**(2): 341–356. doi:10.1111/j.1365-2427.1987.tb01054.x.

Ott DS, Maret TR. 2003. Aquatic assemblages and their relation to temperature variables of least-disturbed streams in the Salmon River basin, Idaho, 2001. U.S. Geological Survey, Water-Resources Investigative Report 03-4076. Boise, Idaho. http://pubs.er.usgs.gov/publication/wri034076 [Accessed July 2012]. 45 pp.

Pacific EcoRisk. 2005. An evaluation of the acute toxicity of cobalt in Panther Creek water to three resident invertebrate species *(Brachycentrus americanus, Centroptilum conturbatum,* and *Serratella tibialis)* and the acute and chronic toxicity of cobalt in Panther Creek water to *Chironomus tentans* and *Oncorhynchus mykiss*. Pacific EcoRisk, Martinez, California. 111 pp.

Palmer MA. 2009. Reforming watershed restoration: science in need of application and applications in need of science. *Estuaries and Coasts* **32**(1): 1–17. doi:10.1007/s12237-008-9129-5.

Palmer MA, Ambrose RF, Poff NL. 1997. Ecological theory and community restoration ecology. *Restoration Ecology* **5**(4): 291–300. doi:10.1046/j.1526-100X.1997.00543.x.

Palmer MA, Bernhardt ES, Allan JD, Lake PS, Alexander G, et al. 2005. Standards for ecologically successful river restoration. *J Appl Ecol* **42**(2): 208–217. doi:10.1111/j.1365-2664.2005.01004.x.

Parker KR, Wiens JA. 2005. Assessing recovery following environmental accidents: environmental variation, ecological assumptions, and strategies. *Ecol Appl* **15**(6): 2037–2051. doi:10.1890/04-1723.

Peterson BJ, Deegan L, Helfrich J, Hobbie JE, Hullar M, et al. 1993. Biological responses of a tundra river to fertilization. *Ecology* **74**(3): 653–672. doi:10.2307/1940794.

Platts WS. 1979. Relationships among stream order, fish populations, and aquatic geomorphology in an Idaho river drainage. *Fisheries* **4**(2): 5–9. doi:10.1577/1548-8446(1979)004<0005:RASOFP>2.0.CO;2.

Poteat MD, Buchwalter DB. 2014. Four reasons why traditional metal toxicity testing with aquatic insects is irrelevant. *Environ Sci Technol* **48**(2): 887–888. doi:10.1021/es405529n.

Power M. 1997. Assessing the effects of environmental stressors on fish populations. *Aquat Toxicol* **39**(2): 151–169. doi:10.1016/S0166-445X(97)00020-9.

Quinn TP. 2005. The Behavior and Ecology of Pacific Salmon and Trout. American Fisheries Society and University of Washington. Seattle, Washington, USA. 328 pp.

Raddum GG, Fjellheim A. 2003. Liming of River Audna, Southern Norway: A large-scale experiment of benthic invertebrate recovery. *Ambio* **32**(3): 230–234. doi:10.1579/0044-7447(2003)032[0230:lorasn]2.0.co;2.

Railsback SF, Rose KA. 1999. Bioenergetics modeling of stream trout growth: temperature and food consumption effects. *Trans Am Fish Soc* **128**(2): 241–256. doi:10.1577/1548-8659(1999)128<0241:bmostg>2.0.co;2.

Rainbow PS, Hildrew AG, Smith BD, Geatches T, Luoma SN. 2012. Caddisflies as biomonitors identifying thresholds of toxic metal bioavailability that affect the stream benthos. *Environ Pollut* **166**(0): 196–207. doi:10.1016/j.envpol.2012.03.017.

Reiser DW. 1986. Habitat Rehabilitation - Panther Creek, Idaho. Bonneville Power Administration, Division of Fish and Wildlife, BPA Pro. No. 84-29. Report No: DOE/BP/17449-1 NTIS No: DE86015222/HDM, Portland, Oregon. 446 pp.

Richards C, Cernera PJ. 1989. Dispersal and abundance of hatchery-reared and naturally spawned juvenile Chinook Salmon in an Idaho stream. *N Am J Fish Manage* 9(3): 345–351. doi:10.1577/1548-8675(1989)009<0345:daaohr>2.3.co;2.

Richards JG, Playle RC. 1998. Cobalt binding to gills of rainbow trout (*Oncorhynchus mykiss*): an equilibrium model. Comparative Biochemistry and Physiology Part C: Pharmacology. *Toxicology and Endocrinology* 119(2): 185–197. doi:10.1016/S0742-8413(97)00206-5.

Riehle MD, Griffith JS. 1993. Changes in habitat use and feeding chronology of juvenile rainbow trout *(Oncorhynchus mykiss)* in fall and the onset of winter in Silver Creek, Idaho. *Can J Fish Aquat Sci* 50(10): 2119–2128. doi:10.1139/f93-237.

Schindler DW, Mills KH, Malley DF, Findlay DL, Shearer JA, et al. 1985. Long-term ecosystem stress - the effects of years of experimental acidification on a small lake. *Science* 228(6): 1395–1401. doi:10.2307/1695685.

Schmetterling DA, Adams SB. 2004. Summer movements within the fish community of a small montane stream. *N Am J Fish Manage* 24(4): 1163–1172. doi:10.1577/M03-025.1.

Sloat M, Fraser D, Dunham J, Falke J, Jordan C, et al. 2014. Ecological and evolutionary patterns of freshwater maturation in Pacific and Atlantic salmonines. *Rev Fish Biol Fish* 24(3): 689–707. http://dx.doi.org/10.1007/s11160-014-9344-z.

Smith M, Von Bargen J, Denny L. 2012. Genetic analysis of the origin of Chinook salmon in Panther Creek, Idaho. U.S. Fish and Wildlife Service, Abernathy Fish Technology Center and Shoshone-Bannock Tribes, Fish and Wildlife Department. Longview, WA and Fort Hall, ID. 30 pp.

Södergren S. 1976. Ecological effects of heavy metal discharge in a salmon river. Institute of Freshwater Research (Drottingholm, Sweden) 55: 91–131.

Stewart-Oaten A, Bence JR, Osenberg CW. 1992. Assessing effects of unreplicated perturbations: no simple solutions. *Ecology* 73(4): 1396–1404. http://dx.doi.org/10.2307/1940685.

Suttle KB, Power ME, Levine JM, McNeely C. 2004. How fine sediment in riverbeds impairs growth and survival of juvenile salmonids. *Ecol Appl* 14(4): 969–974. http://dx.doi.org/10.1890/03-5190.

U.S. Geological Survey. 2012, The StreamStats program: U.S. Geological Survey database. Available from http://streamstats.usgs.gov [Accessed January 28, 2014].

USEPA. 2003. Record of Decision: Blackbird Mine Superfund Site, Lemhi County, Idaho. U.S. Environmental Protection Agency. Seattle, WA. http://yosemite.epa.gov/R10/cleanup.nsf/sites/Blackbird. 159 pp.

USEPA. 2007. Aquatic life ambient freshwater quality criteria - copper, 2007 revision. *U.S. Environmental Protection Agency, EPA-822-R-07-001 (March 2, 2007)*, Washington, DC. http://www.epa.gov/waterscience/criteria/copper/ [Accessed 30 March 2008]. 208 pp.

USEPA. 2013. Second Five-Year Review Report, Blackbird Mine Site, August 2013. U.S. Environmental Protection Agency. Seattle, WA. http://yosemite.epa.gov/R10/cleanup.nsf/sites/Blackbird [Accessed July 2014]. 281 pp.

Vannote RL, Minshall GW, Cummins KW, Sedell JR, Cushing CE. 1980. The river continuum concept. *Can J Fish Aquat Sci* 37(1): 130–137. http://dx.doi.org/10.1139/f80-017.

von der Ohe PC, Liess M. 2004. Relative sensitivity distribution of aquatic invertebrates to organic and metal compounds. *Environ Toxicol Chem* 23(1): 150–156. http://dx.doi.org/10.1897/02-577.

Wallace JB. 1990. Recovery of lotic macroinvertebrate communities from disturbance. *Environ Manage* 14(5): 605–620. http://dx.doi.org/10.1007/BF02394712.

Ward RC, Loftis JC, McBride GB. 1986. The "data-rich but information-poor" syndrome in water quality monitoring. *Environ Manage* 10(3): 291–297. http://dx.doi.org/10.1021/es062253j.

Wiens JA, Parker KR. 1995. Analyzing the effects of accidental environmental impacts: approaches and assumptions. *Ecol Appl* 5(4): 1069–1083. http://dx.doi.org/10.2307/2269355.

Woodward DF, Brumbaugh WG, Lelonay AJ, Little EE, Smith CE. 1994. Effects on rainbow trout fry of a metals-contaminated diet of benthic invertebrates from the Clark Fork River, Montana. *Trans Am Fish Soc* 123(1): 51–62. http://dx.doi.org/10.1577/1548-8659(1994)123<0051:EORTFO>2.3.CO;2.

Woody CA, Hughes RM, Wagner EJ, Quinn TP, Roulson LH, et al. 2010. The Mining Law of 1872: Change is Overdue. *Fisheries* 37(7): 321–331. http://dx.doi.org/10.1577/1548-8446-35.7.321.

## Contributions

- Contributed to conception and design: RJE, BGF, CAM
- Contributed to acquisition of data: RJE, BGF, CAM
- Contributed to analysis and interpretation of data: CAM, RJE
- Drafted and/or revised the article: CAM, RJE, BGF, WJA
- Approved the submitted version for publication: CAM, RJE, BGF, WJA


## Acknowledgments

The study of the biological recovery of the Panther Creek watershed was initially designed and led by Paul McKee, who died in 2007. We acknowledge Dan Myers, as well as others (too many to name) who have supported the data collection over the years, often under arduous conditions. Cathy Smith, Golder Associates, Redmond, WA assisted with many data requests. We thank Jason B. Dunham and two anonymous reviewers for their constructive criticisms of early versions. Unpublished references may be obtained by writing to the corresponding author. Use of firm, trade, or brand names in this paper is for identification purposes only and does not constitute endorsement by the U.S. Geological Survey.


## Funding information

Funding for the site restoration efforts, monitoring and assessment was from the Blackbird Mine Site Group, which in turn is funded by a group of mining companies and the U.S. government. Funding to write this article was provided by Rio Tinto. C.A. Mebane's involvement in investigations has been supported in part by the U.S. National Oceanic and

Atmospheric Administration (NOAA) and USEPA from 1991 to 1995, by the Idaho Department of Environmental Quality from 1995 to 2003, and by NOAA and the U.S. Geological Survey from 2003 to 2013.

## Competing interests

CAM, RJE, and BGF have no competing interests. WJA is employed by Rio Tinto, a metals and mining corporation with partial financial responsibility for the Panther Creek restoration efforts.

## Supplementary material

- Figure S1. Interactive study area detail map, viewable with Internet-based, two-dimensional map and three-dimensional Earth browsers such as the Google Earth application (http://www.google.com/intl/en/earth/index.html).
- Text S2. More information on: (1) Methods, including concepts for evaluating recovery, and (2) Supplemental findings relating to statistical analyses, Blackbird Creek, native fishes, and wildfire influences.
- Figure S3. Photographs of habitat features of the study sites, including substrates and organisms.

## Data accessibility statement

Major datasets generated for this study are provided via the Dryad Digital Repository http://doi.org/10.5061/dryad.67n20. These include stream chemistry, temperature, and streamflow data, metals concentrations in tissues and sediments, and fish and benthic macroinvertebrate community data.

# Permissions

All chapters in this book were first published in ESA, by BioOne; hereby published with permission under the Creative Commons Attribution License or equivalent. Every chapter published in this book has been scrutinized by our experts. Their significance has been extensively debated. The topics covered herein carry significant findings which will fuel the growth of the discipline. They may even be implemented as practical applications or may be referred to as a beginning point for another development.

The contributors of this book come from diverse backgrounds, making this book a truly international effort. This book will bring forth new frontiers with its revolutionizing research information and detailed analysis of the nascent developments around the world.

We would like to thank all the contributing authors for lending their expertise to make the book truly unique. They have played a crucial role in the development of this book. Without their invaluable contributions this book wouldn't have been possible. They have made vital efforts to compile up to date information on the varied aspects of this subject to make this book a valuable addition to the collection of many professionals and students.

This book was conceptualized with the vision of imparting up-to-date information and advanced data in this field. To ensure the same, a matchless editorial board was set up. Every individual on the board went through rigorous rounds of assessment to prove their worth. After which they invested a large part of their time researching and compiling the most relevant data for our readers.

The editorial board has been involved in producing this book since its inception. They have spent rigorous hours researching and exploring the diverse topics which have resulted in the successful publishing of this book. They have passed on their knowledge of decades through this book. To expedite this challenging task, the publisher supported the team at every step. A small team of assistant editors was also appointed to further simplify the editing procedure and attain best results for the readers.

Apart from the editorial board, the designing team has also invested a significant amount of their time in understanding the subject and creating the most relevant covers. They scrutinized every image to scout for the most suitable representation of the subject and create an appropriate cover for the book.

The publishing team has been an ardent support to the editorial, designing and production team. Their endless efforts to recruit the best for this project, has resulted in the accomplishment of this book. They are a veteran in the field of academics and their pool of knowledge is as vast as their experience in printing. Their expertise and guidance has proved useful at every step. Their uncompromising quality standards have made this book an exceptional effort. Their encouragement from time to time has been an inspiration for everyone.

The publisher and the editorial board hope that this book will prove to be a valuable piece of knowledge for researchers, students, practitioners and scholars across the globe.

# List of Contributors

**David M. Karl**
Center for Microbial Oceanography: Research and Education, University of Hawaii, Honolulu, Hawaii, United States

**Kevin R. Arrigo**
Stanford University, Stanford, California, United States

**Zachary W. Brown**
Stanford University, Stanford, California, United States

**Matthew M. Mills**
Stanford University, Stanford, California, United States

**Lisa A. Miller**
Institute of Ocean Sciences, Fisheries and Oceans Canada, Sidney, British Columbia, Canada

**Francois Fripiat**
Laboratoire de Glaciologie, Université Libre de Bruxelles, Brussels, Belgium Analytical, Environmental and Geo-Chemistry, Earth Sciences Research Group, Vrije Universiteit Brussel, Brussels, Belgium

**Brent G.T. Else**
Department of Geography, University of Calgary, Calgary, Alberta, Canada Centre for Earth Observation Science, University of Manitoba, Winnipeg, MB, Canada

**Jeff S. Bowman**
School of Oceanography, University of Washington, Seattle, Washington, United States

**Kristina A. Brown**
Department of Earth, Ocean and Atmospheric Sciences, University of British Columbia, Vancouver, British Columbia, Canada

**R. Eric Collins**
School of Fisheries and Ocean Sciences, University of Alaska Fairbanks, Fairbanks, Alaska, United States

**Marcela Ewert**
School of Oceanography, University of Washington, Seattle, Washington, United States

**Agneta Fransson**
Norwegian Polar Institute, Fram Centre, Tromsø, Norway

**Michel Gosselin**
Institut des sciences de la mer, Université du Québec à Rimouski, Rimouski, Quebec, Canada

**Delphine Lannuzel**
Institute for Marine and Antarctic Studies, University of Tasmania, IMAS–Sandy Bay, Hobart, Tasmania, Australia

**Klaus M. Meiners**
Institute for Marine and Antarctic Studies, University of Tasmania, IMAS–Sandy Bay, Hobart, Tasmania, Australia Australian Antarctic Division, Dept. of the Environment, Kingston, Tasmania, Australia

**Christine Michel**
Freshwater Institute, Fisheries and Oceans Canada, Winnipeg, Manitoba, Canada

**Jun Nishioka**
Institute of Low Temperature Science, Hokkaido University, Sapporo, Japan

**Daiki Nomura**
Institute of Low Temperature Science, Hokkaido University, Sapporo, Japan

**Stathys Papadimitriou**
School of Ocean Sciences, Bangor University, Menai Bridge, Anglesey, United Kingdom

**Lynn M. Russell**
Scripps Institution of Oceanography, La Jolla, California, United States

**Lise Lotte Sørensen**
Department of Environmental Science, Aarhus University, Roskilde, Denmark
Arctic Research Centre, Aarhus University, Aarhus, Denmark

**David N. Thomas**
School of Ocean Sciences, Bangor University, Menai Bridge, Anglesey, United Kingdom
Arctic Research Centre, Aarhus University, Aarhus, Denmark
Finnish Environment Institute (SYKE), Helsinki, Finland

**Jean-Louis Tison**
Laboratoire de Glaciologie, Université Libre de Bruxelles, Brussels, Belgium

**Maria A. van Leeuwe**
Laboratory of Plant Physiology, University of Groningen, Groningen, The Netherlands

**Martin Vancoppenolle**
Laboratoire d'Océanographie et du Climat (LOCEAN-IPSL), Sorbonne Universités (UPMC Paris 6, CNRS, IRD, MNHN), Paris, France

**Eric W. Wolff**
Department of Earth Sciences, University of Cambridge, Cambridge, United Kingdom

**Jiayun Zhou**
Unité d'océanographie chimique, Université de Liège, Liège, Belgium
Centre for Earth Observation Science, University of Manitoba, Winnipeg, MB, Canada

**D. G. Barber**
Centre for Earth Observation Science, University of Manitoba, Winnipeg, Manitoba, Canada

**G. McCullough**
Centre for Earth Observation Science, University of Manitoba, Winnipeg, Manitoba, Canada

**D. Babb**
Centre for Earth Observation Science, University of Manitoba, Winnipeg, Manitoba, Canada

**A. S. Komarov**
Centre for Earth Observation Science, University of Manitoba, Winnipeg, Manitoba, Canada

**SL. M. Candlish**
Centre for Earth Observation Science, University of Manitoba, Winnipeg, Manitoba, Canada

**J. V. Lukovich**
Centre for Earth Observation Science, University of Manitoba, Winnipeg, Manitoba, Canada

**M. Asplin**
Centre for Earth Observation Science, University of Manitoba, Winnipeg, Manitoba, Canada

**S. Prinsenberg**
Coastal Ocean Science Bedford Institute of Oceanography, Department of Fisheries and Oceans Canada, Dartmouth

**I. Dmitrenko**
Centre for Earth Observation Science, University of Manitoba, Winnipeg, Manitoba, Canada

**S. Rysgaard**
Centre for Earth Observation Science, University of Manitoba, Winnipeg, Manitoba, Canada
Greenland Climate Research Centre, Greenland Institute of Natural Resources, Nuuk, Greenland
Arctic Research Centre (ARC), Aarhus University, Aarhus, Denmark

**Anne-Carlijn Alderkamp**
Department of Environmental Earth System Science, Stanford University, Stanford, California, United States

**Gert L. van Dijken**
Department of Environmental Earth System Science, Stanford University, Stanford, California, United States

**Kate E. Lowry**
Department of Environmental Earth System Science, Stanford University, Stanford, California, United States

**Tara L. Connelly**
Department of Marine Sciences, University of Georgia, Athens, Georgia, United States
Marine Science Institute, The University of Texas at Austin, Port Aransas, Texas, United States

**Maria Lagerström**
Department of Marine and Coastal Sciences, Rutgers University, New Brunswick, New Jersey, United States
Department of Applied Environmental Science (ITM), Stockholm University, Stockholm, Sweden

**Robert M. Sherrell**
Department of Marine and Coastal Sciences, Rutgers University, New Brunswick, New Jersey, United States
Department of Earth and Planetary Sciences, Rutgers University, New Brunswick, New Jersey, United States

**Christina Haskins**
Department of Marine and Coastal Sciences, Rutgers University, New Brunswick, New Jersey, United States

**Emily Rogalsky**
Department of Marine and Coastal Sciences, Rutgers University, New Brunswick, New Jersey, United States

**Oscar Schofield**
Department of Marine and Coastal Sciences, Rutgers University, New Brunswick, New Jersey, United States

**Sharon E. Stammerjohn**
Institute of Arctic and Alpine Research, University of Colorado, Boulder, Colorado, United States

**Patricia L. Yager**
Department of Marine Sciences, University of Georgia, Athens, Georgia, United States

**Kevin R. Arrigo**
Department of Environmental Earth System Science, Stanford University, Stanford, California, United States

**Kira Homola**
School of Oceanography, University of Washington, Seattle, Washington, United States
Graduate School of Oceanography, University of Rhode Island, Narragansett, Rhode Island, United States

**H. Paul Johnson**
School of Oceanography, University of Washington, Seattle, Washington, United States

**Casey Hearn**
School of Oceanography, University of Washington, Seattle, Washington, United States
Graduate School of Oceanography, University of Rhode Island, Narragansett, Rhode Island, United States

**Stephanie E. Wilson**
Bangor University, School of Ocean Sciences, Menai Bridge, United Kingdom

**Rasmus Swalethorp**
Department of Biology and Environmental Sciences, University of Gothenburg, Göteborg, Sweden
DTU Aqua, Technical University of Denmark, Charlottenlund, Denmark

**Sanne Kjellerup**
Department of Biology and Environmental Sciences, University of Gothenburg, Göteborg, Sweden
DTU Aqua, Technical University of Denmark, Charlottenlund, Denmark
Greenland Climate and Research Center, Greenland Institute of Natural Resources, Nuuk, Greenland

**Megan A. Wolverton**
Arizona State University, Tempe, Arizona, United States

**Hugh W. Ducklow**
Lamont-Doherty Earth Observatory, Columbia University, Palisades, New York, United States

**Patricia L. Yager**
University of Georgia, Athens, Georgia, United States

**David T. Long**
Department of Geological Sciences, Michigan State University, East Lansing, Michigan, United States
Department of Civil and Environmental Engineering, Michigan State University, East Lansing, Michigan, United States

**Thomas C. Voice**
Department of Civil and Environmental Engineering, Michigan State University, East Lansing, Michigan, United States
Department of Geological Sciences, Michigan State University, East Lansing, Michigan, United States

**Ao Chen**
Department of Civil and Environmental Engineering, Michigan State University, East Lansing, Michigan, United States

**Fangli Xing**
Department of Civil and Environmental Engineering, Michigan State University, East Lansing, Michigan, United States

**Shu-Guang Li**
Department of Civil and Environmental Engineering, Michigan State University, East Lansing, Michigan, United States

**Inga Richert**
Department of Ecology and Genetics, Limnology and Science for Life Laboratory, Uppsala University, Uppsala, Sweden
Department of Environmental Microbiology, Helmholtz Centre for Environmental Research – UFZ, Microbial Ecosystem Services Group, Leipzig, Germany

**Julie Dinasquet**
Marine Biological Section, University of Copenhagen, Helsingør, Denmark
Marine Biology Research Division, Scripps Institution of Oceanography, UCSD, San Diego, California, United States

**Ramiro Logares**
Institute of Marine Sciences, CSIC, Barcelona, Spain

**Lasse Riemann**
Marine Biological Section, University of Copenhagen, Helsingør, Denmark

**Patricia L. Yager**
University of Georgia, Department of Marine Science, Athens, Georgia, United States

**Annelie Wendeberg**
Department of Ecology and Genetics, Limnology and Science for Life Laboratory, Uppsala University, Uppsala, Sweden
Department of Environmental Microbiology, Helmholtz Centre for Environmental Research – UFZ, Microbial Ecosystem Services Group, Leipzig, Germany

**Stefan Bertilsson**
Department of Ecology and Genetics, Limnology and Science for Life Laboratory, Uppsala University, Uppsala, Sweden

**Christopher A. Mebane**
Idaho Water Science Center, U.S. Geological Survey, Boise, Idaho, United States

**Robert J. Eakins**
EcoMetrix Incorporated, Mississauga, Ontario, Canada
Rio Tinto, Lake Point, Utah, United States

**Brian G. Fraser**
EcoMetrix Incorporated, Mississauga, Ontario, Canada
Rio Tinto, Lake Point, Utah, United States

**William J. Adams**
Rio Tinto, Lake Point, Utah, United States

www.ingramcontent.com/pod-product-compliance
Lightning Source LLC
Chambersburg PA
CBHW050442200326
41458CB00014B/5035